Proceedings of the Eighth International Colloquium on Differential Equations

Also available from VSP

Proceedings of the Seventh International Colloquium on Differential Equations
Plovdiv, Bulgaria, 18 - 23 August 1996

Proceedings of the Sixth International Colloquium on Differential Equations
Plovdiv, Bulgaria, 18 - 23 August 1995

Proceedings of the Fifth International Colloquium on Differential Equations
Plovdiv, Bulgaria, 18 - 23 August 1994

Proceedings of the Fourth International Colloquium on Differential Equations
Plovdiv, Bulgaria, 18 - 22 August 1993

Proceedings of the Third International Colloquium on Differential Equations
Plovdiv, Bulgaria, 18 - 23 August 1992

Proceedings of the Third International Colloquium on Numerical Analysis
Plovdiv, Bulgaria, 13 - 17 August 1994

Proceedings of the Second International Colloquium on Numerical Analysis
Plovdiv, Bulgaria, 13 - 17 August 1993

Proceedings of the First International Colloquium on Numerical Analysis
Plovdiv, Bulgaria, 13 - 17 August 1992

Proceedings of the Eighth International Colloquium on Differential Equations

Plovdiv, Bulgaria, 18 - 23 August, 1997

Editor: **D. Bainov**

///VSP///

Utrecht, The Netherlands, 1998

VSP BV
P.O. Box 346
3700 AH Zeist
The Netherlands

First published in 1998

ISBN 90-6764-279-7

Printed in The Netherlands by Ridderprint bv, Ridderkerk.

CONTENTS

Preface

The Eighth International Colloquium on Differential Equations was organized by the Institute for Basic Science of Inha University, the International Federation of Nonlinear Analysts, the Mathematical Society of Japan, Pharmaceutical Faculty of the Medical University of Sofia, University of Catania and UNESCO, with the cooperation of the Association Suisse d'Informatique, the Canadian Mathematical Society, Kyushu University, the London Mathematical Society, Technical University of Plovdiv, and it was partially sponsored by the Bulgarian Ministry of Education, Science and Technologies under Grant MM-702.

The Colloquim was held on August 18-23, 1997, in Plovdiv, Bulgaria. The subsequent colloquia will take place each year from 18 to 23 of August in Plovdiv, Bulgaria.

The publishing of this volume is fully supported by the Bulgarian Ministry of Education, Science and Technologies under Grant MM-702.

Organizing Committee:

D. Bainov (Chairman), Q. H. Choi, J.Diblik, S. Hristova, J. Kajiwara (Vice Chairman), N. Kitanov (Secretary), D. Kolev (Secretary), V. Lakshmikantham, M. Marino, A. Maugeri (Vice Chairman), E. Minchev (Secretary), N. Popivanov, P. Popivanov, M. Robnik, M. Sambandham, H.M. Srivastava, R.U. Verma.

Scientific Committee:

D. Bainov (Bulgaria), C. Dafermos (USA), G. Da Prato (Italy), H. Fujita (Japan), R. Ivanov (Bulgaria), J. Kajiwara (Japan), A. Kaneko (Japan), N. Kenmochi (Japan), V. Lakshmikantham (USA), A. Maugeri (Italy), E. Minchev (Bulgaria), N. Pavel (USA), N. Popivanov (Bulgaria), P. Popivanov (Bulgaria), M. Robnik (Slovenia), M. Sambandham (USA), H.M. Srivastava (Canada), Fr. Stenger (USA), E. Titi (USA), M. Tsutsumi (Japan), R.U. Verma (USA).

The address of the Organizing Committee is:
Drumi Bainov, P.O.Box 45, 1504 Sofia, Bulgaria.

The Editor

The Scientific and Organizing Committees of the

EIGHTH INTERNATIONAL COLLOQUIUM ON DIFFERENTIAL EQUATIONS,

PLOVDIV, BULGARIA, AUGUST 18-23, 1997

adjudged the prize of the

International Colloquia on Differential Equations

to Professor Antonino Maugeri

University of Catania, Italy.

The prize is awarded for the whole scientific activity and research of the recipient, and for his special merits in the organization and yearly carrying out of the Colloquia on Differential Equations.

50th Anniversary of Prof. P. Popivanov

DRUMI BAINOV

Higher Medical Institute, Sofia 1504
P.O. Box 45, Bulgaria

The member of the Organizing Committee of the International Colloquia on Differential Equations in Plovdiv — Bulgaria, Prof. Petar Popivanov, D.Sci., corr. member of the Bulgarian Academy of Sciences completed 50 years. About 25 years of his life he devoted to the Partial Differential Equations (PDE), working in the domains of general theory of linear PDE, linear and nonlinear microlocal analysis, degenerate boundary value problems for semilinear elliptic equations, Monge-Ampere type equations, Cauchy problem for nonlinear hyperbolic equations and systems, PDE on the torus.

Since 1990 he is regularly and actively participating in the activities of our International Colloquium on Differential Equations. Many of his new results were published for the first time in the Proceedings of the Colloquium - see [1,2,3,4] and more than 12 scientific communications were given by him. Prof. P. Popivanov was a chairman of the Opening Ceremonies of several conferences, chairman of many round tables on Impulsive Differential Equations, on different aspects of the PDE and on new trends in the theory of Differential Equations. With his talks, tact and authority he contributed too much in organizing of 8 conferences on Differential Equations on a high professional level. Despite the heavy crisis in our country each year more than 200 participants are proposing their new achievements in the domain of Differential Equations. As the biggest part of them are coming from abroad the Colloquia in Plovdiv give best possibilities for international scientific contacts and scientific cooperation.

It is a pleasure for me as a chairman of the Organizing Committee and on behalf of it to congratulate Prof. P. Popivanov with his anniversary and to give him my best wishes for new successes in the domain of PDE.

References

[1] P. R. Popivanov, Removable singularities of the solutions of fully nonlinear systems of partial differential equations, *Proc. of the 2 Coll. on Diff. Eq.*, World Scientific, (1992), pp. 185–190.

[2] P. R. Popivanov, Nonexistence of isolated singularities for semilinear systems of partial differential equations, *Proc. of the 5 Coll. on Diff. Eq.*, VSP, (1995), pp. 283–286.

[3] P. R. Popivanov, Microlocal hypoellipticity of some classes of overdetermined systems of pseudodifferential operators, *Proc. of the 4 Coll. on Diff. Eq.*, VSP, (1994), pp. 229–233.

[4] P. R. Popivanov, D. K. Palagachev, On a degenerate boundary value problem for second order quasilinear elliptic operators, *Proc. of the 6 Coll. on Diff. Eq.*, VSP, (1996), pp. 197–208.

[5] P. R. Popivanov, D. K. Palagachev, *The Degenerate Oblique Derivative Problem for Elliptic and Parabolic Equations,* Akademie-Verlag, Berlin, (1997).

8th Int. Coll. on Differential Equations, pp. 1-8
D. Bainov (Ed)

Study of the Ince Equation

MAKHLOUF AMAR AND BAHLOUL TOUNSI
Institute of Mathematics, University of Annaba, B.P. 12, El-Hadjar, Annaba, Algeria

Abstract. In this work, we study the Ince equation:

$$(*) \qquad \cos(\nu)y''(\nu) + p\sin(\nu)y'(\nu) + q\cos(\nu)y(\nu) = 0.$$

First, we give the fundamental solutions of $(*)$ in the intervals $\left(-\frac{\pi}{2} + k\pi, \frac{\pi}{2} + k\pi\right)$ using the Gauss-Equation. We study the conditions for which $(*)$ has the solutions: $(\overline{C_0(\nu)}, \overline{S_0(\nu)})$, $(\overline{C_1(\nu)}, \overline{S_1(\nu)})$, $(\overline{C_0(\nu)}, \overline{S_1(\nu)})$, where $\overline{C_0(\nu)} = \sum_{p=0}^{n} A_{2p}\cos(2p\nu)$, $\overline{S_0(\nu)} = \sum_{p=1}^{n} B_{2p}\sin(2p\nu)$, $\overline{C_1(\nu)} = \sum_{p=0}^{n} A_{2p+1}\cos((2p+1)\nu)$, $\overline{S_1(\nu)} = \sum_{p=1}^{n} B_{2p+1}\sin((2p+1)\nu)$. We give the necessary and sufficient conditions for the equation $(*)$ to have solutions $\left(\overline{S_1(\nu)}, \nu\overline{S_1(\nu)} + \overline{C_1(\nu)}\right)$, $\left(\overline{C_0(\nu)}, \nu\overline{C_0(\nu)} + \overline{S_0(\nu)}\right)$ and the closed, truncated solutions.

Keywords: Ince equation, Gauss hypergeometric equation.

Introduction

We consider the Kepler problem:

$$\ddot{r} = -\frac{1}{r^2}, \qquad r = r(t), \qquad r(0) = 1, \qquad \dot{r}(0) = 0.$$

The normal variational equation of this problem is the singular Hill equation:

$$r^3(t)\frac{d^2\xi}{dt^2} + \lambda\xi(t) = 0.$$

By putting: $r(t) = \cos^2(\nu)$, $\xi(t) = Z(\nu)$, this equation becomes:

$$\cos^2(\nu)Z''(\nu) + 2\sin(\nu)\cos(\nu)Z'(\nu) + 2\lambda Z(\nu) = 0,$$

this equation is a particular case of a general equation which, we resolve:

$$\cos^2(\nu)Z''(\nu) + P\sin(\nu)\cos(\nu)Z'(\nu) + \left[R + Q\cos^2(\nu)\right]Z(\nu) = 0.$$

We consider the following singular Ince Equation:

$$(1) \qquad \left[1 + \cos(2\nu)\right]Z''(\nu) + B\sin(2\nu)Z'(\nu) + \left[C + D\cos(2\nu)\right]Z(\nu) = 0,$$

where B, C, D are reals. It can be written in the form

$$\cos^2(\nu)Z''(\nu) + P\sin(\nu)\cos(\nu)Z'(\nu) + \left[R + Q\cos^2(\nu)\right]Z(\nu) = 0, \tag{2}$$

where $P = B$, $Q = D$, $R = \dfrac{C-D}{2}$.

Proposition 1. *If $Z(\nu)$ is a solution of (2) then the function $y(\nu)$ defined by $Z(\nu) = [\cos(\nu)]^\mu y(\nu)$ verifies the equation:*

$$\cos^2(\nu)y''(\nu) + P_1\sin(\nu)\cos(\nu)y'(\nu) + \left[R_1 + Q_1\cos^2(\nu)\right]y(\nu) = 0, \tag{3}$$

where

$$\begin{cases} P_1 = -2\mu + P, \\ Q_1 = -\mu^2 + P\mu + Q, \\ R_1 = \mu^2 - (P+1)\mu + R. \end{cases} \tag{4}$$

Proof. By substitution. □

From equations (4), we deduce the differential system:

$$\begin{cases} \dfrac{dP_1}{d\mu} = -2 & P_1(0) = P, \\ \dfrac{dQ_1}{d\mu} = P_1 & Q_1(0) = Q, \\ \dfrac{dR_1}{d\mu} = -(P_1+1) & R_1(0) = R. \end{cases} \tag{5}$$

This system has the next two first integrals:

$$\begin{aligned} \chi_1 &= (P_1+1)^2 - 4R_1 = (P+1)^2 - 4R, \\ \chi_2 &= P_1^2 + 4Q_1 = P^2 + 4Q. \end{aligned} \tag{6}$$

Proposition 2. *If $X_1 \geq 0$, then the function $Z(\nu) = (\cos(\nu))^\mu y(\nu)$ where μ is a solution of the equation $R_1 = 0$, transforms (2) to the equation:*

$$\cos(\nu)y''(\nu) + p\sin(\nu)y'(\nu) + q\cos(\nu)y(\nu) = 0, \tag{7}$$

where $p = P - 2\mu$, $q = Q + P\mu - \mu^2$. The equation (7) is called the reduced singular Ince equation.

Proof. $R_1 = 0$ implies $\mu^2 - (P+1)\mu + R = 0$. We have $\Delta = (P+1)^2 - 4R = \chi_1$. If $\chi_1 \geq 0$ then it exist μ such that $R_1 = 0$, so the equation (2) becomes (7). □

We consider a solution $y(\nu)$ of (7). It is easy to verify that the function $\tilde{y}(\nu)$ defined by:

$$y(\nu) = \left[\cos(\nu)\right]^{p+1}\tilde{y}(\nu), \qquad \nu \in \left(-\frac{\pi}{2}, \frac{\pi}{2}\right),$$

verifies the equation:

$$\cos(\nu)\tilde{y}''(\nu) + p_1\sin(\nu)\tilde{y}'(\nu) + q_1\cos(\nu)\tilde{y}(\nu) = 0, \tag{8}$$

where $p_1 = -p - 2$, $q_1 = q - (p+1)$.

Solution of (7) in $I_k = \left(-\frac{\pi}{2} + k\pi, \frac{\pi}{2} + k\pi\right)$.

We consider the Gauss equation:

$$z(1-z)y'' + \left[c - (a+b+1)z\right]y'(z) - aby(z) = 0, \tag{9}$$

where $c = a + b + \frac{1}{2}$, which possesses the solution:

$$F_0^1(z) = F(a,b,c,z) = 1 + \frac{ab}{c}z + \frac{a(a+1)b(b+1)}{c(c+1)}z^2 + \cdots,$$

$$F_0^2(z) = z^{1-c}F(a+1,c,b+1,z),$$

$$F_1^1(z) = F(a,b,a+b+1-c,1-z),$$

$$F_1^2(z) = (1-z)^{c-a-b}F(c-a,c-b,c+1-a,1-z).$$

If we put $z = \cos^2(\nu)$, $\nu \in \left(-\frac{\pi}{2}, \frac{\pi}{2}\right)$ and $Y(z) = y(\nu)$, then the equation (9) becomes [1]:

$$\cos(\nu)y'' + p\sin(\nu)y' + q\cos(\nu)y = 0 \tag{10}$$

with $p = -2(a+b)$, $q = -4ab$.

Solution in the interval I_0.

The equation (7) has two fundamental solutions $f(\nu)$ and $g(\nu)$ where $\nu \in \left(-\frac{\pi}{2}, \frac{\pi}{2}\right)$ such that $f(0) = 1$, $f'(0) = 0$, $g(0) = 0$, $g'(0) = 1$.

a) For $\nu \in \left(0, \frac{\pi}{2}\right)$ to $y(\nu)$ there corresponds $Y(z)$: $z \in (0,1)$ is defined by $Y(z) = y(\arccos\sqrt{z})$ and to $Y(z)$ there corresponds $y(\nu)$ defined by $y(\nu) = Y(cos^2(\nu))$.

b) For $\nu \in \left(-\frac{\pi}{2}, 0\right)$ and $z \in (0,1)$, to $y(\nu)$ there corresponds $Y(z) = y(-\arccos\sqrt{z})$ and to $Y(z)$ there corresponds $y(\nu) = Y(cos^2(\nu))$.

Proposition 3. *In the interval I_0, the fundamental solutions of (7) are:*

$$f(\nu) = F_1^1(z) = F\left(a, b, \frac{1}{2}, \sin^2(\nu)\right),$$

$$g(\nu) = F_1^2(z) = \sin(\nu)F\left(b + \frac{1}{2}, a + \frac{1}{2}, \frac{3}{2}, \sin^2(\nu)\right).$$

Proof. $f(\nu)$ is a solution of (7) for $\nu \in I_0$, so we can compute $f''(0)$, $f'''(0), \ldots$. In the neighborhood of 0, we have:

$$f(\nu) = 1 - \frac{q}{2!}\nu^2 + \frac{q(q+2p)}{4!}\nu^4 + O(\nu^5).$$

We have also:

$$F_1^1(z) = F\left(a, b, \frac{1}{2}, \sin^2(\nu)\right) = 1 + ab\nu^2 + \frac{4(a^2b^2 + a^2b + ab^2)\nu^4}{6} + O(\nu^5)$$

and $p = 2(a+b)$, $q = -4ab$. Therefore,

$$F\left(a, b, \frac{1}{2}, \sin^2(\nu)\right) = 1 - \frac{q}{2!}\nu^2 + \frac{q(q+2p)}{4!}\nu^4 + O(\nu^5)$$

and by using a) and b), we have $f(\nu) = F_1^1(z)$, $z = \cos^2(\nu)$.

Analogously, we demonstrate that $g(\nu) = F_1^2(z)$, $z = \cos^2(\nu)$. □

Solutions in the interval I_1.

We consider the fundamental solutions $f_1(\nu)$ and $g_1(\nu)$ of (7) in $I_1 = \left(\frac{\pi}{2}, 3\frac{\pi}{2}\right)$ such that $f_1(\pi) = 1$, $f_1'(\pi) = 0$, $g_1(\pi) = 0$, $g_1'(\pi) = 1$.

We consider a solution $y(\nu)$ of (7) in I_1 and put $\nu = \pi + \mu$, $\mu \in I_0$ and $y(\nu) = \tilde{y}(\mu)$. It is easy to verify that $\tilde{y}(\mu)$ satisfies (7).

We put $y(\nu) = f_1(\nu)$ such that $y(\pi) = 1$, $y'(\pi) = 0$ so $\tilde{y}(0) = 1$, $\tilde{y}'(0) = 0$. We deduce that $\tilde{y}(\mu) = f(\mu)$ and hence $f_1(\nu) = f(\nu - \pi)$. We also prove that $g_1(\nu) = g(\nu - \pi)$. Thus, we can affirm:

Proposition 4. *The fundamental solutions are*

$$f_1(\nu) = F\left(a, b, \frac{1}{2}, \sin^2(\nu)\right), \qquad g_1(\nu) = -\sin(\nu) F\left(b + \frac{1}{2}, a + \frac{1}{2}, \frac{3}{2}, \sin^2(\nu)\right).$$

We can generalize this result in the interval I_k. We find that $f_k(\nu)$ and $g_k(\nu)$ are such that:

$$f_k(k\pi) = 1, \quad f_k'(k\pi) = 0, \qquad g_k(k\pi) = 0, \quad g_k'(k\pi) = 1.$$

Therefore, $f_k(\nu) = f(\nu - k\pi)$, $g_k(\nu) = g(\nu - k\pi)$ and this implies that:

$$f_k(\nu) = F\left(a, b, \frac{1}{2}, \sin^2(\nu)\right), \qquad g_k(\nu) = (-1)^k \sin(\nu) F\left(b + \frac{1}{2}, a + \frac{1}{2}, \frac{3}{2}, \sin^2(\nu)\right).$$

Closed solutions of (7).

Definition. The series $C_0(\nu) = \sum_{p=0}^{\infty} A_{2p} \cos(2p\nu)$ is closed of order n if $A_{2n} \neq 0$ and $A_{2p} = 0$ for each $p > n$. It is denoted by $\overline{C_0(\nu)}$. We set further

$$\overline{S_0(\nu)} = \sum_{p=0}^{n} B_{2p} \sin(2p\nu), \qquad \overline{C_1(\nu)} = \sum_{p=0}^{n-1} A_{2p+1} \cos((2p+1)\nu)$$

and

$$\overline{S_1(\nu)} = \sum_{p=0}^{n-1} B_{2p+1} \sin((2p+1)\nu).$$

The closed solutions are important because they are regular in $\nu = \frac{\pi}{2} + k\pi$, hence there is not continuation problem.

Proposition 5. *The equation (7) has the two closed solutions: $\overline{C_0(\nu)}$ and $\overline{S_0(\nu)}$ if and only if $\chi_1 = N'^2$, $\chi_2 = N^2$ with $N' = 2K' + 1$, $N = 2K$ ($N' < N$), K and K' have the same parity.*

Proof. We substitute $C_0(\nu)$ in (7), and obtain the system:

$$\begin{cases} 2A_0\phi(0) + A_2\phi(-2) = 0, \\ A_2\phi(2) + A_4\phi(-4) = 0, \\ \dots\dots\dots\dots\dots\dots\dots\dots\dots\dots, \\ A_{2n}\phi(2n) + A_{2n+2}\phi(-2n-2) = 0, \\ \dots\dots\dots\dots\dots\dots\dots\dots\dots\dots \end{cases}$$

If we substitute $S_0(\nu)$ in (7), we obtain the system:

$$\begin{cases} B_2\phi(-2) = 0, \\ B_2\phi(2) + B_4\phi(-4) = 0, \\ \dots\dots\dots\dots\dots\dots\dots\dots\dots\dots, \\ B_{2n}\phi(2n) + B_{2n+2}\phi(-2n-2) = 0, \\ \dots\dots\dots\dots\dots\dots\dots\dots\dots\dots, \end{cases}$$

where $\phi(x) = x^2 - px - q$.

We suppose that $C_0(\nu)$ is closed with $A_{2n+2} = 0$. Then $\phi(2n) = 0$ and $B_{2n+2} = 0$ with $\phi(-2n') = 0$ for $n' < n$.

Calculation of χ_1 and χ_2 gives $\phi(2n) = \phi(-2n') = 0$ and thus $p = X_1 + X_2 = 2(n - n') > 0$ and $q = -X_1 X_2 = 4nn'$. Therefore,

$$\chi_1 = (p+1)^2 = \left[2(n-n')+1\right]^2 = N'^2, \qquad \chi_2 = \left[2(n+n')\right]^2 = N^2.$$

Reciprocally, we consider $N' = 2K' + 1$, $N = 2K$, $N' < N$. The roots of $\phi(x) = 0$ are:

$$X_1 = \frac{P + \sqrt{P^2 + 4q}}{2}, \qquad X_2 = \frac{P - \sqrt{P^2 + 4q}}{2}$$

but $\chi_1 = (P+1)^2 = (2K'+1)^2 \Longrightarrow p = 2k'$; $\chi_2 = p^2 + 4q = (2K)^2$.

Then $X_1 = \dfrac{2k' + 2k}{2} = k + k'$, $X_2 = k' - k$. If k and k' have the same parity then x_1 is positive and even and x_2 negative and even. □

We demonstrate the following two propositions in the same fashion:

Proposition 6. *The equation* (7) *has the solutions* $\overline{C_1(\nu)}$ *and* $\overline{S_1(\nu)}$ *if and only if* $\chi_1 = (2k'+1)^2$ *and* $\chi_2 = (2k)^2$ *such that* k *and* k' *have different parity.*

Proposition 7. *The equation* (7) *has the solutions* $\overline{C_0(\nu)}$ *and* $\overline{S_1(\nu)}$ *if and only if* $\chi_1 = (2k)^2$ *and* $\chi_2 = (2k'+1)^2$ *with* $k - k' - 1 \geq 0$.

Proposition 8. *If* $\chi_1 = (2k'+1)^2$ *and* $\chi_2 = (2k)^2$ *such that* $k < k'$ *and* k *and* k' *have different parity, then the equation* (7) *has the two linearly independent solutions:* $y_1(\nu) = \overline{S_1(\nu)}$, $y_2(\nu) = \nu\overline{S_1(\nu)} + \overline{C_1(\nu)}$.

Proof. We consider a solution $y(\nu)$ of (7) and put $\tilde{y}(\nu) = \nu y(\nu) + Z(\nu)$. It is easy to verify that $\tilde{y}(\nu)$ is a solution of (7) if $Z(\nu)$ verifies:

$$\cos(\nu)Z'' + p\sin(\nu)Z' + q\cos(\nu)Z = -2\cos(\nu)y' - p\sin(\nu)y. \tag{11}$$

If $\phi(2n-1) = 0$, then (7) admits the solution:

$$y(\nu) = \sum_{p=1}^{n} B_{2p-1} \sin\left[(2p-1)\nu\right].$$

If we put

$$\cos(\nu)Z'' + p\sin(\nu)Z' + q\cos(\nu)Z = M_0 + M_2\cos(2\nu) + \ldots + M_{2n}\cos(2n\nu) + \ldots,$$

by identification we have:

$$\begin{cases} M_0 = -\left(\frac{p}{2}+1\right) B_1, \\ M_2 = \left(\frac{p}{2}-1\right) B_1 - \left(\frac{p}{2}+3\right) B_3, \\ \ldots\ldots\ldots\ldots\ldots\ldots\ldots\ldots\ldots\ldots\ldots\ldots, \\ M_{2m} = \left(\frac{p}{2}-(2m-1)\right) B_{2m-1} - \left(\frac{p}{2}+(2m+1)\right) B_{2m+1}, \\ \ldots\ldots\ldots\ldots\ldots\ldots\ldots\ldots\ldots\ldots\ldots\ldots, \\ M_{2n} = \left(\frac{p}{2}-(2n-1)\right) B_{2n-1} - \left(\frac{p}{2}+(2n+1)\right) B_{2n+1}. \end{cases} \tag{12}$$

If $B_{2n+1} = 0$ then $M_{2n} = \left(\frac{p}{2}-(2n-1)\right) B_{2n-1}$ and for $m > n$ we have $M_{2m} = 0$ because $B_{2n+1} = B_{2n+3} = \cdots = 0$. We look for Z in the form:

$$Z = C_1(\nu) = \sum_{p=0}^{\infty} R_{2p+1}\cos((2p+1)\nu).$$

We substitute Z in (11) and we identify

$$\begin{cases} M_0 = -\frac{1}{2} R_1\phi(-1), \\ M_2 = -\frac{1}{2}\left(R_1\phi(1) + R_3\phi(-3)\right), \\ \ldots\ldots\ldots\ldots\ldots\ldots\ldots\ldots\ldots\ldots\ldots\ldots, \\ M_{2m} = -\frac{1}{2}\left(R_{2m-1}\phi(2m-1) + R_{2m+1}\phi(-2m-1)\right). \end{cases} \tag{13}$$

By comparing (12) and (13), we find:

$$-\left(\frac{p}{2}+1\right) B_1 = -\frac{1}{2} R_1\phi(-1).$$

R_1 is known if $\phi(-1) \neq 0$ and we have $R_1 = \dfrac{2\left(\frac{p}{2}+1\right)}{\phi(-1)} B_1$,

$$\left(\frac{p}{2}-1\right) B_1 - \left(\frac{p}{2}+3\right) B_3 = -\frac{1}{2}(R_1\phi(1) + R_3\phi(-3)).$$

R_1 is known and we can calculate R_3 if $\phi(-3) \neq 0$. Generally, we have for $m \leq n$:

$$-\frac{1}{2}\Big[R_{2m-1}\phi(2m-1)+R_{2m+1}\phi(-2m-1)\Big] = \Big(\frac{p}{2}-(2m-1)\Big)\, B_{2m-1}-\Big(\frac{p}{2}+(2m+1)\Big)\, B_{2m+1}.$$

That is, R_{2m-1} is calculated then R_{2m+1} is known. If $m > n$ then we have $R_{2m+1} = R_{2m+3} = \ldots = 0$. We conclude that if $\phi(2n-1) = \phi(2m-1) = 0$ for $m > n$, then we find

$$Z(\nu) = \overline{C_1(\nu)} = R_1\cos(\nu) + \ldots + R_{2m-1}\cos((2m-1)\nu).$$

Thus, it is easily find $\chi_1 = (2(m+n)-1)^2$, $\chi_2 = 4(m-n)^2$.

Oppositely: Suppose that $\chi_1 = (2k'+1)^2$, $\chi_2 = (2k)^2$, $k < k'$. Then $p = 2k'+1-1 = 2k'$. We find $x_1 = k'+k$ and $x_2 = k'-k$. We conclude that $\phi(x)$ vanishes for positive integers x_1 and x_2 such that $x_2 < x_1$. We remark that if x_1 and x_2 have a different parity then x_1 and x_2 are odd. □

We can demonstrate in the same fashion the proposition:

Proposition 9. *The equation* (7) *has the solutions* $y_1(\nu) = \overline{C_0(\nu)}$ *and* $y_2(\nu) = \nu\overline{C_0(\nu)} + \overline{S_0(\nu)}$ *if and only if* $\chi_1 = (2k'+1)^2$, $\chi_2 = (2k)^2$ *with* $k \leq k'$ *and* k *and* k' *have the same parity.*

Closed, truncated solutions.

We consider the solution of (7): $C_0(\nu) = \sum\limits_{n=0}^{\infty} A_{2n}\cos(2n\nu)$. Then we have the system

$$\begin{cases} 2A_0\phi(0) + A_2\phi(-2) = 0, \\ A_2\phi(2) + A_4\phi(-4) = 0, \\ \ldots\ldots\ldots\ldots\ldots\ldots\ldots\ldots\ldots, \\ A_{2n}\phi(2n) + A_{2n+1}\phi(-2n-2) = 0, \\ \ldots\ldots\ldots\ldots\ldots\ldots\ldots\ldots\ldots \end{cases}$$

Proposition 10. *Suppose that* $\phi(2n) = \phi(-2n') = 0$ *and* $n \leq n'$. *Then the equation has the closed, truncated solutions:*

$$\overline{C_0(\nu)} = A_{2n'}\cos(2n'\nu) + \ldots + A_{2n'+2k}\cos((2n'+2k)\nu) + \ldots + A_{2n}\cos(2n\nu),$$

$$\overline{S_0(\nu)} = B_{2n'}\sin(2n'\nu) + \ldots + B_{2n'+2k}\sin((2n'+2k)\nu) + \ldots + B_{2n}\sin(2n\nu),$$

where $A_{2n'+2k} = \dfrac{(2n'+k-1)!(n-n')!(n+n')!}{(2n'-1)!k!(n-n'-k)!(n+n'+k)!}A_{2n'}$, $k \geq 1$ *and* $2n'+2k \leq 2n$. *We have the same expression for* $B_{2n'+2k}$.

Proof. We shall prove that $\overline{C_0(\nu)}$ is a solution. The same proof is valid for $\overline{S_0(\nu)}$. Thus $A_{2n'-2} = A_{2n'-4} = \ldots = A_4 = A_2 = 0$ and we have $A_{2n'-2}\phi(2n'-2) + A_{2n'}\phi(-2n') = 0$. We have $\phi(-2n') = 0$ and then $A_{2n'-2} = 0$ because of $\phi(2n'-2) \neq 0$. The equation $A_{2n'-4}\phi(2n'-4) + A_{2n'-2}\phi(-2n'-2) = 0$ implies $A_{2n'-4}\phi(2n'-4) = 0$ (because of $A_{2n'-2} = 0$), or $\phi(2n'-4) \neq 0$ then $A_{2n'-4} = 0$. Proceeding in the same fashion, we find that: $A_{2n'-6} = A_{2n'-8} = \ldots = A_4 = A_2 = A_0 = 0$. To prove that $A_{2n+2} = A_{2n+4} = \ldots = 0$ we have $A_{2n}\phi(2n) + A_{2n+2}\phi(-2n-2) = 0$, or we have $\phi(2n) = 0$. Then $A_{2n+2} = 0$ because of $\phi(-2n-2) \neq 0$. From the equation $A_{2n+2}\phi(2n+2) + A_{2n+4}\phi(-2n-4) = 0$ we have $A_{2n+4} = 0$. In the same fashion we find $A_{2n+2} = A_{2n+4} = \ldots = 0$.

Calculation of the coefficient $A_{2n'+2k}$: From the equation $A_{2n'}\phi(2n') + A_{2n'+2}\phi(-2n'-2) = 0$ we have $A_{2n'+2} = -\dfrac{\phi(2n')}{\phi(-2n'-2)}A_{2n'}$. We have $\phi(2n') = (2n')^2 - 2n'p - q$, where $p = 2n - 2n'$, $q = 4nn'$. Therefore, we obtain $\phi(2n') = 8n'(n'-n)$ and also $\phi(-2n'-2) = 4n' + 4n + 4$. Thus

$$A_{2n'+2} = \frac{2n'(n-n')}{n+n'+1}A_{2n'}.$$

Generally,

$$A_{2k+2} = \frac{-\phi(2k)}{\phi(-2k-2)}A_{2k} = \frac{(n-k)(n'+k)}{(n+k+1)(k+1-n')}A_{2k}, \qquad k \geq n'.$$

We find that

$$\begin{aligned} A_{2n'+2k} &= \frac{2n'(2n'+1)(2n'+2)\ldots(2n'+k-1)(n-n')(n-n'-1)\ldots(n-n'-k+1)}{(n+n'+1)(n+n'+2)\ldots(n+n'+k).1.2.3\ldots k}A_{2n'} \\ &= \frac{(2n'+k-1)!(n-n')!(n+n')!}{(2n'-1)!k!(n-n'-k)!(n+n'+k)!}A_{2n'}, \end{aligned}$$

where $2n' + 2k \leq 2n$, $k \in N^*$. □

By using an analogous proof, we obtain the proposition.

Proposition 11. *Suppose that $\phi(2n+1) = \phi(-2n'-1) = 0$, $n \geq n'$, then the equation (7) has the closed truncated solutions:*

$$\overline{C_1(\nu)} = A_{2n'+1}\cos((2n'+1)\nu) + \ldots + A_{2n+1}\cos((2n+1)\nu),$$

$$\overline{S_1(\nu)} = B_{2n'+1}\sin((2n'+1)\nu) + \ldots + B_{2n+1}\sin((2n+1)\nu).$$

References

1. M. Amar, Equation de Hill Dans le Domaine Complexe et Application à L'étude de L'équation aux Variations Normale au Voisinage des Solutions Périodiques Dans un Potentiel Homogène, Thèse d'état soutenue à l'université de Annaba en Septembre (1991).

2. M. Amar, C.R. Acad. Sci., Paris, série 2, 313, 183–186 (1991).

3. H. Yoshida, Celestial Mechanics, 40, 55–61 (1987).

8th Int. Coll. on Differential Equations, pp. 9-16
D. Bainov (Ed)

Trace Formula for the Sturm-Liouville Operator with Singularity

R. KH. AMIROV AND Y. ÇAKMAK
Cumhuriyet University, Dept. of Mathematics, 58140, SIVAS-TURKEY

Abstract. Let $\mu_1, \mu_2, \ldots, \mu_n, \ldots$ be the Dirichlet spectrum of the operator $-d^2/dx^2 + q(x)$ acting on $L^2(0,\pi)$. In the special case where $q(x) \equiv 0, \mu_n = n^2$. In the [1] and others discovered the asymptotic formula

$$\mu_n = n^2 + \frac{1}{\pi}\int_0^\pi q(x)dx + O(n^{-2})$$

and the trace formula

$$\sum_n [\mu_n - n^2] = \frac{q(0)+q(\pi)}{4},$$

provided that $\int_0^\pi q(x)dx = 0$, where $q(x) \in C^2[0,\pi]$. These are beautiful formulas with many applications for example in solving inverse problems. In this work, the above mentioned problem has been studied for a Sturm-Liouville operator with A/x (A is real) singularity at $x = 0$.
Keywords: Trace formula, spectrum.
AMS subject classifications. 35J10, 35J40, 35K35

Introduction: Learning about the mation of electrons moving under Coulomb potential is of significance in quantum theory. Solving these type of problems provides us to find energy levels of not only Hidrogen atom but also single valance electron atoms such as Sodium.

For Hidrogen atom, Coulomb potential is given by $U = -\frac{e^2}{r}$, where r is the radius of the nucleus, e is electronic charge. Accordingly, we use time dependent Schrödinger equation

$$i\hbar\frac{\partial\Psi}{\partial t} = -\frac{\hbar^2}{2m}\frac{\partial^2\Psi}{\partial x^2} + U(x,y,z,t)\Psi, \quad \int_{R^3} |\Psi|^2 \, dxdydz = 1.$$

Here,Ψ is the wave function, $\hbar$ is Planck's constant and m is the mass of electron. Here, of we make the necessary transformations we can get a Sturm-Liouville equation with Coulomb potential. $-y'' + \left(\frac{A}{x} + q(x)\right)y = \lambda y$ where λ is a parameter which corresponds to the energy.

As for the class of function on which this type of singular expressions are defined $y'(0)$ is infinite, the class satisfying discrete boundary conditions $y'(0) = 0,\ y'(\pi) - Hy(\pi) = 0$

will be taken as the domain for the differential expressions under study. In consequence,

$$L = \begin{cases} -y'' + \left(\frac{A}{x} + q(x)\right) y & = \lambda y \\ y(0) & = 0 \\ y'(\pi) - Hy(\pi) & = 0. \end{cases}$$

The domain of operator L is taken as

$$D(L) = \{y|Ly \in L_2[0,\pi],\ q(x) \in L_2[0,\pi],\ A, H\ real\ const.,\ y(0) = 0,\ y'(\pi) - Hy(\pi) = 0\}.$$

Eigenfunctions and eigenvalues of the operator L are given in the [1].

Now,let $A : H \longrightarrow H$ be an operator.H is a Hilbert space and as it is separable we can use orthonormal systems. $\{e_n\} \in H$is an orthonormal system and $\|e_n\|_H = 1$.

If we take $\{e_n\}$ as the eigenfunctions of A , $(Ae_n = \lambda_n e_n)$, we find

$$\sum_{n=1}^{\infty} < Ae_n, e_n > = \sum_{n=1}^{\infty} < \lambda_n e_n, e_n > = \sum_{n=1}^{\infty} \lambda_n < e_n, e_n > = \sum_{n=1}^{\infty} \lambda_n.$$

If $\sum_{n=1}^{\infty} \lambda_n < +\infty$ then it is the trace of A, shortly $tr(A) = \sum_{n=1}^{\infty} \lambda_n$.

As $L : L_2[0,\pi] \longrightarrow L_2[0,\pi]$ and is a Hilbert space, $\{y_n\}$ can be thought of an orthonormal system. Taking $Ly_n = \lambda_n y_n$, one obtains

$$\sum_{n=1}^{\infty} < Ly_n, y_n > = \sum_{n=1}^{\infty} < \lambda_n y_n, y_n > = \sum_{n=1}^{\infty} \lambda_n < y_n, y_n > = \sum_{n=1}^{\infty} \lambda_n.$$

However,this series is not convergent. Thus, we search for the regularized trace.To this end, we take out some terms from λ_n . These traces are important in solving inverse problems with two spectrums.

Calculation of the regularized trace of Sturm-Liouville operator with Coulomb singularities.

Given the differential equation

$$-y'' + \left(\frac{A}{x} + q(x)\right) y = \lambda y \tag{1}$$

$$y(0) = 0 \tag{2}$$

$$y'(\pi) - H_1 y(\pi) = 0 \tag{3}$$

$$(y'(\pi) - H_2 y(\pi) = 0) \tag{3'}$$

where $q(x) \in C^2[0,\pi]$; A, H_1 and H_2 real numbers. Let $\{\lambda_n\}$ and $\{\mu_n\}$respectively represent the spectrum of (1)-(2)-(3) and (1)-(2)-(3').For the problem of (1)-(2)-(3), the following theorem has been proved.

Theorem: If $q(x) \in C^2[0,\pi]$ then

$$\sum_{n=0}^{\infty} (\lambda_n - \mu_n) = \frac{q(\pi) - q(0)}{4} + \frac{A}{4\pi} + AHM - \frac{H}{2}\int_0^{\pi} q(t)dt - \frac{AHN}{2} - \frac{Aln\pi}{2\pi} - S$$

$$-\frac{AHln\pi}{2} - \frac{AN}{2\pi} - \frac{1}{\pi}\left[\frac{1}{2}\int_0 \pi q(t)dt - AM - H\right]$$

$$+\left[\frac{AN}{2} + \left(\frac{1}{2}\int_0^{\pi} q(t)dt - AM - H\right) + \frac{Aln\pi}{2}\right]^2.$$

Here M, N and S are known constants,$H = H_1 - H_2$.

Proof: To find them we notice that the above summation is convergent and $\varphi(x, \lambda)$ is an entire function of $\frac{1}{2}$ order with respect to λ parameter. Weierstrass theorem gives

$$\varphi'(\pi, \lambda) - H\varphi(\pi, \lambda) = A\Phi(\lambda).$$

With $\lambda_1, \lambda_2, \dots \lambda_n, \dots$ are the zeros of $\Phi(\lambda)$ and $\Phi(\lambda)$ is an entire function, $\Phi(\lambda) = \prod_{n=0}^{\infty}\left(1 - \frac{\lambda}{\lambda_n}\right)$ holds.

If $\lambda = -\mu$ and $\mu > 0$ is satisfied then we can use the equation $\varphi'(\pi, -\mu) - H\varphi(\pi, -\mu) = A\Phi(-\mu)$. First, we find the asymptotic of $\Phi(-\mu)$.

$$\Phi(\lambda) = \prod_{n=0}^{\infty}\left(1 - \frac{\lambda}{\lambda_n}\right) \Rightarrow \Phi(-\mu) = \prod_{n=0}^{\infty}\left(1 + \frac{\mu}{\lambda_n}\right) = \frac{\prod_{n=0}^{\infty}\left(1 + \frac{\mu}{\lambda_n}\right)}{\prod_{n=0}^{\infty}\left(1 + \frac{\mu}{(n+1/2)}\right)} \cosh\sqrt{\mu}\pi$$

$$= \prod_{n=0}^{\infty}\left(\frac{(n+1/2)^2}{\lambda_n}\right)\prod_{n=0}^{\infty}\left[\frac{\lambda_n + \mu}{\mu + (n+1/2)^2}\right] \cosh\sqrt{\mu}\pi$$

$\Phi(-\mu) = c\Psi(\mu)\cosh\sqrt{\mu}\pi$, where $c = \prod_{n=0}^{\infty}\left(\frac{(n+1/2)^2}{\lambda_n}\right)$. $\Psi(\mu)$ is given by

$\Psi(\mu) = \prod_{n=0}^{\infty}\left[\frac{\lambda_n + \mu}{\mu + (n+1/2)^2}\right] = \prod_{n=0}^{\infty}\left[1 - \frac{(n+1/2)^2 - \lambda_n}{\mu + (n+1/2)^2}\right]$ and $\Psi(\mu)$ is convergent. Therefore,

$$\ln\Psi(\mu) = \sum_{n=0}^{\infty}\ln\left[1 - \frac{(n+1/2)^2 - \lambda_n}{\mu + (n+1/2)^2}\right]$$

$$= \ln\left(1 - \frac{1/4 - \lambda_0}{\mu + 1/4}\right) + \sum_{n=0}^{\infty}\ln\left[1 - \frac{(n+1/2)^2 - \lambda_n}{\mu + (n+1/2)^2}\right].$$

Combining these, we find

$$\ln\Psi(\mu) = \frac{\lambda_0 - 1/4}{\mu} + \sum_{n=0}^{\infty}\ln\left[1 - \frac{(n+1/2)^2 - \lambda_n}{\mu + (n+1/2)^2}\right] + O\left(\frac{1}{\mu^2}\right).$$

As $\sum_{n=1}^{\infty}\frac{\ln n}{n^2}$ is convergent, above series is convergent , too. For $n \longrightarrow \infty$, general sum goes to zero, enabling us to perform the following operators. We obtain

$$\ln\Psi(\mu) = \frac{\lambda_0 - 1/4}{\mu} - \sum_{n=1}^{\infty}\sum_{k=1}^{\infty}\frac{1}{k}\left(\frac{(n+1/2)^2 - \lambda_n}{\mu + (n+1/2)^2}\right)^k + O\left(\frac{1}{\mu^2}\right)$$

$$= \frac{\lambda_0 - 1/4}{\mu} - \sum_{k=1}^{\infty}\frac{1}{k}\sum_{n=1}^{\infty}\left(\frac{(n+1/2)^2 - \lambda_n}{\mu + (n+1/2)^2}\right)^k + O\left(\frac{1}{\mu^2}\right).$$

Let us take the expression $\left(\frac{(n+1/2)^2 - \lambda_n}{\mu + (n+1/2)^2}\right)^k$. Adding and substracting the term

$\frac{A}{\pi}\ln\left(n+\frac{1}{2}\right)$. We obtain

$$\left[\frac{(n+1/2)^2+\frac{A}{\pi}\ln(n+1/2)-\lambda_n-\frac{A}{\pi}\ln(n+1/2)}{\mu+(n+1/2)^2}\right]^k$$

$$=\left[\frac{(n+1/2)^2+\frac{A}{\pi}\ln(n+1/2)-\lambda_n}{\mu+(n+1/2)^2}-\frac{A}{\pi}\frac{\ln(n+1/2)}{\mu+(n+1/2)^2}\right]^k$$

$$=\sum_{j=0}^{k}\binom{k}{j}\left[\frac{(n+1/2)^2+\frac{A}{\pi}\ln(n+1/2)-\lambda_n}{\mu+(n+1/2)^2}\right]^{k-j}\left(\frac{A}{\pi}\right)^j\left(\frac{\ln(n+1/2)}{\mu+(n+1/2)^2}\right)^j.$$

Inserting its place in $\ln\Psi(\mu)$, we have

$$\ln\Psi(\mu)=-\sum_{k=1}^{\infty}\frac{1}{k}\sum_{n=1}^{\infty}\sum_{j=0}^{k}\binom{k}{j}\left[\frac{(n+1/2)^2+\frac{A}{\pi}\ln(n+1/2)-\lambda_n}{\mu+(n+1/2)^2}\right]^{k-j}\times$$

$$\times\left(\frac{A}{\pi}\right)^j\left(\frac{\ln(n+1/2)}{\mu+(n+1/2)^2}\right)^j+\frac{\lambda_0-1/4}{\mu}+O\left(\frac{1}{\mu^2}\right)$$

$$=\frac{\lambda_0-1/4}{\mu}-\sum_{k=1}^{\infty}\frac{1}{k}\sum_{n=1}^{\infty}\left[\frac{(n+1/2)^2+\frac{A}{\pi}\ln(n+1/2)-\lambda_n}{\mu+(n+1/2)^2}\right]^k$$

$$-\sum_{k=1}^{\infty}\sum_{n=1}^{\infty}\frac{1}{k}k\left[\frac{(n+1/2)^2+\frac{A}{\pi}\ln(n+1/2)-\lambda_n}{\mu+(n+1/2)^2}\right]^{k-1}\frac{A}{\pi}\frac{\ln(n+1/2)}{\mu+(n+1/2)^2}$$

$$-\sum_{n=1}^{\infty}\sum_{k=1}^{\infty}\frac{1}{k}\sum_{j=2}^{\infty}\binom{k}{j}\left[\frac{(n+1/2)^2+\frac{A}{\pi}\ln(n+1/2)-\lambda_n}{\mu+(n+1/2)^2}\right]^{k-j}\left(\frac{A}{\pi}\right)^j\left(\frac{\ln(n+1/2}{\mu+(n+1/2)^2}\right)^j$$

$$+O\left(\mu^{-2}\right).$$

Now take the expression $\frac{A}{\pi}\sum_{n=1}^{\infty}\left[\frac{(n+1/2)^2+\frac{A}{\pi}\ln(n+1/2)-\lambda_n}{\mu+(n+1/2)^2}\right]^{k-1}\frac{\ln(n+1/2}{\mu+(n+1/2)^2}$.

From the eigenvalue expression we know that $\left|\lambda_n-\left(n+\frac{1}{2}\right)^2-\frac{A}{\pi}\ln\left(n+\frac{1}{2}\right)\right|\le a$ (a constant).

$$\frac{A}{\pi}\sum_{n=1}^{\infty}\left[\frac{(n+1/2)^2+\frac{A}{\pi}\ln(n+1/2)-\lambda_n}{\mu+(n+1/2)^2}\right]^{k-1}\frac{\ln(n+1/2)}{\mu+(n+1/2)^2}$$

$$\leq \frac{A}{\pi}\sum_{n=1}^{\infty}\frac{a^{k-1}}{[\mu+(n+1/2)^2]^{k-1}}\frac{\ln(n+1/2)}{\mu+(n+1/2)^2}=\frac{A}{\pi}a^{k-1}\sum_{n=1}^{\infty}\frac{\ln(n+1/2)}{[\mu+(n+1/2)^2]^k}$$

$$=\frac{A}{\pi}\frac{a^{k-1}}{\mu^k}\sum_{n=1}^{\infty}\frac{\ln(n+1/2)}{\left[1+\left(\frac{n+1/2}{\sqrt{\mu}}\right)^2\right]^k}.$$

That is,

$$\ln\Psi(\mu) = -\sum_{n=1}^{\infty}\sum_{k=1}^{\infty}\frac{1}{k}\left(\frac{(n+1/2)^2+\frac{A}{\pi}\ln(n+1/2)-\lambda_n}{\mu+(n+1/2)^2}\right)^k$$

$$-\sum_{n=1}^{\infty}\sum_{k=1}^{\infty}\frac{1}{k}\sum_{j=2}^{k}\binom{k}{j}\left[\frac{(n+1/2)^2+\frac{A}{\pi}\ln(n+1/2)-\lambda_n}{\mu+(n+1/2)^2}\right]^{k-j}\left(\frac{A}{\pi}\right)^j\left(\frac{\ln(n+1/2)}{\mu+(n+1/2)^2}\right)^j$$

$$-\sum_{k=1}^{\infty}\frac{A}{\pi}\frac{a^{k-1}}{\mu^k}\sum_{n=1}^{\infty}\frac{\ln(n+1/2)}{\left[1+\left(\frac{n+1/2}{\sqrt{\mu}}\right)^2\right]^k}+\frac{\lambda_0-1/4}{\mu}+O\left(\frac{1}{\mu^2}\right)$$

$$=-\sum_{n=1}^{\infty}\left(\frac{(n+1/2)^2+\frac{A}{\pi}\ln(n+1/2)-\lambda_n}{\mu+(n+1/2)^2}\right)-\frac{A}{\pi}\frac{1}{\mu}\sum_{n=1}^{\infty}\frac{\ln(n+1/2)}{1+\left(\frac{n+1/2}{\sqrt{\mu}}\right)^2}$$

$$-\sum_{k=2}^{\infty}\frac{1}{k}\sum_{n=1}^{\infty}\left(\frac{(n+1/2)^2+\frac{A}{\pi}\ln(n+1/2)-\lambda_n}{\mu+(n+1/2)^2}\right)^k-\sum_{k=2}^{\infty}\frac{A}{\pi}\frac{a^{k-1}}{\mu^k}\sum_{n=1}^{\infty}\frac{\ln(n+1/2)}{1+\left(\frac{n+1/2}{\sqrt{\mu}}\right)^2}$$

$$-\sum_{n=1}^{\infty}\sum_{k=1}^{\infty}\frac{1}{k}\sum_{j=2}^{k}\binom{k}{j}\left[\frac{(n+1/2)^2+\frac{A}{\pi}\ln(n+1/2)-\lambda_n}{\mu+(n+1/2)^2}\right]^{k-j}\left(\frac{A}{\pi}\right)^j\left(\frac{\ln(n+1/2)}{\mu+(n+1/2)^2}\right)^j$$

$$+\frac{\lambda_0-1/4}{\mu}+O\left(\frac{1}{\mu^2}\right).$$

Therefore,

$$\ln\Psi(\mu) = -\sum_{n=1}^{\infty}\left(\frac{(n+1/2)^2+\frac{A}{\pi}\ln(n+1/2)-\lambda_n}{\mu+(n+1/2)^2}\right)+\frac{A}{2}\frac{\ln\sqrt{\mu}}{\sqrt{\mu}}+\frac{3A}{2\mu\pi}(1+\ln(3/2))$$

$$-\sum_{k=2}^{\infty}\frac{1}{k}\sum_{n=1}^{\infty}\left(\frac{(n+1/2)^2+\frac{A}{\pi}\ln(n+1/2)-\lambda_n}{\mu+(n+1/2)^2}\right)^k-\sum_{k=2}^{\infty}\frac{A}{\pi}\frac{a^{k-1}}{\mu^k}\sum_{n=1}^{\infty}\frac{\ln(n+1/2)}{1+\left(\frac{n+1/2}{\sqrt{\mu}}\right)^2}$$

$$-\sum_{n=1}^{\infty}\sum_{k=1}^{\infty}\frac{1}{k}\sum_{j=2}^{k}\binom{k}{j}\left[\frac{(n+1/2)^2+\frac{A}{\pi}\ln(n+1/2)-\lambda_n}{\mu+(n+1/2)^2}\right]^{k-j}\left(\frac{A}{\pi}\right)^j\left(\frac{\ln(n+1/2)}{\mu+(n+1/2)^2}\right)^j$$

$$+\frac{\lambda_0-1/4}{\mu}+O\left(\frac{\ln\sqrt{\mu}}{\mu^2}\right).$$

Making suitable addings and substractions to this expression

$$\ln\Psi(\mu)=\sum_{n=1}^{\infty}\left(\lambda_n-(n+1/2)^2-\frac{A}{\pi}\ln(n+1/2)-2c_0-2c_1\frac{\ln(n+1/2)}{(n+1/2)}-\frac{2c_2}{(n+1/2)}\right.$$

$$\left.-\gamma_1\frac{\ln^2(n+1/2)}{(n+1/2)^2}-\gamma_2\frac{\ln(n+1/2)}{(n+1/2)^2}\right)\frac{1}{\mu+(n+1/2)^2}$$

$$+2c_0\sum_{n=1}^{\infty}\frac{1}{\mu+(n+1/2)^2}+2c_1\sum_{n=1}^{\infty}\frac{\ln(n+1/2)}{(n+1/2)(\mu+(n+1/2)^2)}$$

$$+2c_2\sum_{n=1}^{\infty}\frac{1}{(n+1/2)(\mu+(n+1/2)^2)}+\gamma_1\sum_{n=1}^{\infty}\frac{\ln^2(n+1/2)}{(n+1/2)^2(\mu+(n+1/2)^2)}$$

$$+\gamma_2\sum_{n=1}^{\infty}\frac{\ln(n+1/2)}{(n+1/2)^2(\mu+(n+1/2)^2)}+\frac{\lambda_0-1/4}{\mu}+\frac{A}{2}\frac{\ln\sqrt{\mu}}{\sqrt{\mu}}$$

$$+\frac{3A}{2\mu\pi}(1+\ln(3/2))+O\left(\frac{\ln\sqrt{\mu}}{\mu^2}\right).$$

If necessary processes are done,

$$\ln\Psi(\mu)=\frac{1}{\mu}\sum_{n=0}^{\infty}(\lambda_n-(n+1/2)^2-\frac{A}{\pi}\ln(n+1/2)-2c_0-2c_1\frac{\ln(n+1/2)}{(n+1/2)}$$

$$-\frac{2c_2}{(n+1/2)}-\gamma_1\frac{\ln^2(n+1/2)}{(n+1/2)^2}-\gamma_2\frac{\ln(n+1/2)}{(n+1/2)^2}\Bigg)$$

$$-\frac{1}{\mu}(\lambda_0-\frac{1}{4}+\frac{A}{2\pi}\ln 2-2c_0+4c_1\ln 2-4c_2-2\gamma_1\ln^2 2+4\gamma_2\ln 2)$$

$$-\frac{1}{\mu}\sum_{n=1}^{\infty}(\lambda_n-(n+1/2)^2-\frac{A}{\pi}\ln(n+1/2)-2c_0-2c_1\frac{\ln(n+1/2)}{(n+1/2}-\frac{2c_2}{(n+1/2)}$$

$$-\gamma_1\frac{\ln^2(n+1/2)}{(n+1/2)^2}-\gamma_2\frac{\ln(n+1/2)}{(n+1/2)^2}\Bigg)\frac{(n+1/2)^2}{\mu+(n+1/2)^2}$$

$$+2c_0\sum_{n=1}^{\infty}\frac{1}{\mu+(n+1/2)^2}+2c_1\sum_{n=1}^{\infty}\frac{\ln(n+1/2)}{(n+1/2)(\mu+(n+1/2)^2)}$$

$$+2c_2\sum_{n=1}^{\infty}\frac{1}{(n+1/2)(\mu+(n+1/2)^2)}+\gamma_1\sum_{n=1}^{\infty}\frac{\ln^2(n+1/2)}{(n+1/2)^2(\mu+(n+1/2)^2)}$$

$$+\gamma_2\sum_{n=1}^{\infty}\frac{\ln(n+1/2)}{(n+1/2)^2(\mu+(n+1/2)^2)}+\frac{\lambda_0-1/4}{\mu}+\frac{A}{2}\frac{\ln\sqrt{\mu}}{\sqrt{\mu}}$$

$$+\frac{3A}{2\mu\pi}(1+\ln(3/2))+O\left(\frac{\ln\sqrt{\mu}}{\mu^2}\right).$$

Now, we find the asymptotic values of these sums.

$$\sum_{n=1}^{\infty}\frac{1}{\mu+(n+1/2)^2}=\frac{\pi}{2\sqrt{\mu}}-\frac{1}{2\mu}+O(e^{-2\sqrt{\mu}\pi})$$

$$\sum_{n=1}^{\infty}\frac{\ln(n+1/2)}{(n+1/2)(\mu+(n+1/2)^2)}=\frac{\ln^2(3/2)-2\ln(3/2)\ln\sqrt{\mu}+2\ln^2\sqrt{\mu}+\pi^2/24}{\mu}$$
$$+O\left(\frac{\ln\sqrt{\mu}}{\mu^2}\right)$$

$$\sum_{n=1}^{\infty}\frac{1}{(n+1/2)(\mu+(n+1/2)^2)}=\frac{\ln\sqrt{\mu}-\ln(3/2)}{\mu}+\frac{9}{8\mu^2}+O\left(\frac{1}{\mu^2}\right)$$

$$\sum_{n=1}^{\infty}\frac{\ln^2(n+1/2)}{(n+1/2)^2(\mu+(n+1/2)^2)}=\frac{4}{3}\left(\frac{\ln(3/2)\ln\sqrt{\mu}-\ln^2\sqrt{\mu}+\ln\sqrt{\mu}}{\mu}\right)+\frac{3\ln\sqrt{\mu}}{\mu^2}$$

$$+\frac{2\ln\sqrt{\mu}\ln(3/2)\arctan(3/2\sqrt{\mu})-2\ln^2\sqrt{\mu}\arctan(3/2\sqrt{\mu})}{\mu^{3/2}}+O\left(\frac{\ln\sqrt{\mu}}{\mu^3}\right)$$

$$\sum_{n=1}^{\infty}\frac{\ln(n+1/2)}{(n+1/2)^2(\mu+(n+1/2)^2)}=-\frac{\pi}{2}\frac{\ln\sqrt{\mu}}{\mu^{3/2}}+\frac{2\ln\sqrt{\mu}}{3\mu}+\frac{3}{2\mu^2}+O\left(\frac{\ln\sqrt{\mu}}{\mu^3}\right)$$

$$\sum_{n=1}^{\infty}\Big(\lambda_n-(n+1/2)^2-\frac{A}{\pi}\ln(n+1/2)-2c_0-2c_1\frac{\ln(n+1/2)}{(n+1/2)}-\frac{2c_2}{(n+1/2)}$$

$$\left.-\gamma_1\frac{\ln^2(n+1/2)}{(n+1/2)^2}-\gamma_2\frac{\ln(n+1/2)}{(n+1/2)^2}\right)\frac{(n+1/2)^2}{\mu+(n+1/2)^2}=-\frac{\pi}{2}\frac{\gamma_3}{\mu^{3/2}}+\frac{3\gamma_3}{2\mu^2}+O\left(\frac{\ln^2\sqrt{\mu}}{\mu^3}\right).$$

We also obtain the asymptotic expression of $\varphi'(\pi,-\mu)-H\varphi(\pi,-\mu)$.

$$\varphi'(\pi,-\mu)-H\varphi(\pi,-\mu)=\frac{e^{\sqrt{\mu}\pi}}{2}\left[1+\frac{A}{2}\frac{\ln\sqrt{\mu}}{\sqrt{\mu}}+\frac{1}{\sqrt{\mu}}\left(\frac{A\alpha_1}{2}+(\beta_2-H)+\frac{A\ln\pi}{2}\right)\right.$$

$$\left.-\frac{AH}{2}\frac{\ln\sqrt{\mu}}{\mu}+\frac{1}{\mu}\left(\beta_1+AH\alpha_2-\frac{H}{2}\int_0^{\pi}q(t)dt-\frac{AH\alpha_1}{2}-\frac{AH\ln\pi}{2}\right)\right]+O\left(\frac{\ln\sqrt{\mu}}{\mu^3}\right)$$

The trace is obtained from the equation $\varphi'(\pi,\lambda)-H\varphi(\pi,\lambda)=A\Phi(\lambda)$.

References

1. I. M. Gelfand, B. M. Levitan, Dokl.Akad.Nauk SSSR. 88, 593-599 (1953).

2. M. G. Gasimov, R. Kh. Amirov, Dokl.Akad.Nauk Az.SSR, 41, no. 8 (1985).

8th Int. Coll. on Differential Equations, pp. 17–24
D. Bainov (Ed)

Inverse Spectral Problem for the Differential Equation of the Second Order with Singularity

R. KH. AMIROV AND S. GÜLYAZ
Cumhuriyet University, Dept. of Mathematics, 58140, SIVAS-TURKEY

Abstract. In this article, for given values of $\{\lambda_n\}$ and $\{\mu_n\}$ constructing a differential operator, which has a singularity, has exploined. In order to acomplish this, λ_n and μ_n have to have some certain asymptotic. This kind of asymptotic formulae depend on the regularity order of the function $q(x)$. This connection can be seen in the solution of the inverse problem which depend on the series of $\{\lambda_n\}$ and $\{\mu_n\}$.

Keywords: Inverse problem, regulary, singulary.
AMS subject classifications. Primary 34A55, Secondary 34B24, 34L05.

Introduction : The main subject of the present paper is the solution of the following problem; given two sequences of real numbers $\lambda_0, \lambda_1, \dots, \lambda_n, \dots$ and $\mu_0, \mu_1, \dots, \mu_n, \dots$ what are necessary and sufficient conditions for these sequences to be two spectra of one and the same Strum-Liouville operator of the from

$$y'' + \left\{\lambda - \frac{\ell(\ell+1)}{x^2} - \frac{A}{x} - q(x)\right\} y = 0, \ (0 \le x \le \pi)$$

under different boundary conditions? Here $q(x)$ is a real function integrable over every interval .$(0, b')$, $b' < \pi$. Before formulating the principal results of the paper we shall briefly describe the main steps in the development of the so-called inverse Strum-Liouville problem. V. A. Marchenko [1] has proven that two different spectrum of Sturm-Liouville equation with a singularity determines this equation uniquely. Subsequently, M. G. Gasimov and B. M. Levitan [2] have produced an effective method to determine this equation. However, all these methods proposed have some restrictions such that $\lambda_0 < \lambda_1 < \dots$ and $\mu_0 < \mu_1 < \dots$ must be the eigenvalues of given equation. To this end, the sequences $\{\lambda_n\}$ and $\{\mu_n\}$ should have such conditions that they will be the different spectrum of a given equation. The problems mentioned above have all been completely solved for regular equations, but for those equations with the type of singularity $\ell(\ell+1)/(\pi - x)^2$, which is singular at π in the interval $[0, \pi]$, M. G. Gasimov [3], given two spectrum, has solved the inverse problem. In $[0, \pi]$, the singularity of A/x (A is real) at $x = 0$ has been studied by M. G. Gasimov and R. Kh. Amirov [4] and the inverse problem with two spectrum has been solved. In this paper, again in $[0, \pi]$, at $x = 0$, for the equations with the singularity $\left(\frac{A}{x} + \frac{\ell(\ell+1)}{x^2}\right)$, a theorem giving the inverse problem solution with two spectrum has been proved.

1. Expression for the normalizing constants in terms of the spectra
We consider the differential equation

$$-y'' + \left\{\frac{\ell(\ell+1)}{x^2} + \frac{A}{x} + q(x)\right\} y = \lambda y, \tag{1.1}$$

and the boundary conditions

$$y(0) = 0\ ,\ y'(\pi) + H_1 y(\pi) = 0, \tag{1.2}$$

$$y(0) = 0\ ,\ y'(\pi) + H_2 y(\pi) = 0. \tag{1.3}$$

We assume that $q(x) \in L_2[0,\pi]$;ℓ is a positive integer, A, H_1 and H_2 are real numbers, with $H_1 \neq H_2$. We denote the eigenvalues of the boundary value problems (1.1)-(1.2) and (1.1)-(1.3) by $\lambda_0 < \lambda_1 < \dots$ and $\mu_0 < \mu_1 < \dots$ respectively. Let $\varphi(x,\lambda)$ and $\psi(x,\lambda)$ be the solutions of (1.1) satisfying the initial conditions,

$$\varphi(0,\lambda) = 0\ ,\ \varphi'(0,\lambda) = 1 \tag{1.4}$$

and

$$\psi(0,\lambda) = 0\ ,\ \psi'(0,\lambda) = 1. \tag{1.5}$$

Then it is clear that the eigenvalues $\lambda_0, \lambda_1, \dots$ and $\mu_0, \mu_1, \dots$ of the boundary value problems (1.1)-(1.2) and (1.1)-(1.3) coincide with the zeros of the functions

$$\Phi_1(\lambda) = \varphi'(\pi,\lambda) - H_1\varphi(\pi,\lambda), \tag{1.6}$$

$$\Phi_2(\lambda) = \psi'(\pi,\lambda) - H_2\psi(\pi,\lambda). \tag{1.7}$$

Clearly $\varphi(x,\lambda_n) = \varphi_n(x)$ is an eigenfunction of the boundary value problem (1.1)-(1.2) and its norm is equal to

$$\alpha_n = \int_0^\pi \varphi_n^2(x)dx. \tag{1.8}$$

The numbers $\alpha_0, \alpha_1, \dots$ are called the normalizing constants of the boundary value problem (1.1)-(1.2). In this section we shall derive a formula for calculating the numbers $\alpha_0, \alpha_1, \dots$ in terms of the two spectra $\lambda_0, \lambda_1, \dots$ and $\mu_0, \mu_1, \dots$.

Because $\varphi(x,\lambda)$ is the solution of the equation (1.1) that satisfying then (1.4) condition

$$\int_0^\pi \varphi^2(x,\lambda)dx = \varphi'(\pi,\lambda)\frac{\partial}{\partial\lambda}\varphi(\pi,\lambda) - \varphi(\pi,\lambda)\frac{\partial}{\partial\lambda}\varphi'(\pi,\lambda). \tag{1.9}$$

The initial conditions (1.4) and (1.5) also $\Phi_1(\lambda)$ and $\Phi_2(\lambda)$ are entire in λ for fixed x. It is clear from this that $\Phi_1(\lambda)$ and $\Phi_2(\lambda)$ are entire functions of order one half and therefore are determined by their zeros, to within a constant multiplying factor. Therefore,

$$\varphi'(\pi,\lambda) - H_1\varphi(\pi,\lambda) = C_1\prod_{k=0}^{\infty}\left(1 - \frac{\lambda}{\lambda_k}\right) = \Phi_1(\lambda), \tag{1.10}$$

and

$$\varphi'(\pi,\lambda) - H_2\varphi(\pi,\lambda) = C_2\prod_{k=0}^{\infty}\left(1 - \frac{\lambda}{\mu_k}\right) = \Phi_2(\lambda), \tag{1.11}$$

where C_1 and C_2 are constants. Combining (1.10) and (1.11) we obtain

$$\int_0^\pi \varphi^2(x,\lambda_n)dx = \frac{1}{H_2 - H_1}\acute{\Phi}_1(\lambda_n)\Phi_2(\lambda_n).$$

Let $\varphi_0(x,\lambda)$ be solution of equation (1.1) when $q(x)=0$ satisfying the condition (1.4). Respectively $\lambda_0^0, \lambda_1^0, \dots$ and $\mu_0^0, \mu_1^0, \dots$ eigenvalues problems (1.1)-(1.4) and (1.1)-(1.5) when $q(x)=0$. Then it is clear that the eigenvalues $\lambda_0^0, \lambda_1^0, \dots$ and $\mu_0^0, \mu_1^0, \dots$ of the boundary value problems (1.1)-(1.2) and (1.1)-(1.3) coincide with the zeros of the functions

$$\varphi_0'(\pi,\lambda) - H_1\varphi_0(\pi,\lambda) = C_1^0\prod_{k=0}^{\infty}\left(1-\frac{\lambda}{\lambda_k^0}\right) = \Phi_{01}(\lambda), \tag{1.12}$$

and

$$\varphi_0'(\pi,\lambda) - H_2\varphi_0(\pi,\lambda) = C_2^0\prod_{k=0}^{\infty}\left(1-\frac{\lambda}{\mu_k^0}\right) = \Phi_{02}(\lambda). \tag{1.13}$$

Clearly $\varphi_0(x,\lambda_n^0) = \varphi_{0n}(x)$ is an eigenfunction of the boundary value problem (1.1)-(1.2) and its norm is equal to

$$\alpha_n^0 = \int_0^\pi \varphi_{0n}^2(x)dx. \tag{1.14}$$

Because $\varphi_0(x,\lambda)$ is the solution of the equation (1.1) when $q(x)=0$ satisfying the condition (1.4).

$$\int_0^\pi \varphi_0^2(x,\lambda_n^0)dx = \frac{1}{H_2-H_1}\dot{\Phi}_{01}(\lambda_n^0)\Phi_{02}(\lambda_n^0).$$

Therefore

$$\alpha_n = \frac{1}{H_2-H_1}\dot{\Phi}_1(\lambda_n)\Phi_2(\lambda_n), \tag{1.15}$$

and

$$\alpha_n^0 = \frac{1}{H_2-H_1}\dot{\Phi}_{01}(\lambda_n^0)\Phi_{02}(\lambda_n^0). \tag{1.16}$$

Then we obtain $\dot{\Phi}_1(\lambda_n) = -\frac{C_1}{\lambda_n}\prod_{k=0}^{\infty}{}'\left(1-\frac{\lambda_n}{\lambda_k}\right)$, $\dot{\Phi}_{01}(\lambda_n^0) = -\frac{C_1^0}{\lambda_n^0}\prod_{k=0}^{\infty}{}'\left(1-\frac{\lambda_n^0}{\lambda_k^0}\right)$ where (here and henseforth) the symbol $\prod'$ means that the factor $k=n$ is absent from the infinite product. These formula and (1.15)-(1.16) imply that

$$\alpha_n = -\frac{C_1C_2}{H_2-H_1}\frac{1}{\lambda_n}\prod_{k=0}^{\infty}{}'\left(1-\frac{\lambda_n}{\lambda_k}\right)\prod_{k=0}^{\infty}\left(1-\frac{\lambda_n}{\mu_k}\right), \tag{1.17}$$

and

$$\alpha_n^0 = -\frac{C_1^0C_2^0}{H_2-H_1}\frac{1}{\lambda_n^0}\prod_{k=0}^{\infty}{}'\left(1-\frac{\lambda_n^0}{\lambda_k^0}\right)\prod_{k=0}^{\infty}\left(1-\frac{\lambda_n^0}{\mu_k^0}\right), \tag{1.18}$$

Since

$$\Phi_2(\lambda_n) = C_2\prod_{k=0}^{\infty}\left(1-\frac{\lambda_n}{\mu_k}\right) = C_2\prod_{k=0}^{\infty}\left(1-\frac{\lambda_n}{\mu_k^0}\right)\frac{\prod_{k=0}^{\infty}\left(1-\frac{\lambda_n}{\mu_k}\right)}{\prod_{k=0}^{\infty}\left(1-\frac{\lambda_n}{\mu_k^0}\right)}$$

$$= C_2\prod_{k=0}^{\infty}\left(1-\frac{\lambda_n}{\mu_k^0}\right)\prod_{k=0}^{\infty}\frac{\mu_k^0}{\mu_k}\prod_{k=0}^{\infty}\frac{\lambda_n-\mu_k}{\lambda_n-\mu_k^0}$$

$$= B_2C_2\prod_{k=0}^{\infty}\left(1-\frac{\lambda_n}{\mu_k^0}\right)\prod_{k=0}^{\infty}\frac{\lambda_n-\mu_k}{\lambda_n-\mu_k^0} = B_2C_2\prod_{k=0}^{\infty}\left(1-\frac{\lambda_n^0}{\mu_k^0}\right)\frac{\prod_{k=0}^{\infty}\left(1-\frac{\lambda_n}{\mu_k^0}\right)}{\prod_{k=0}^{\infty}\left(1-\frac{\lambda_n^0}{\mu_k^0}\right)}\prod_{k=0}^{\infty}\frac{\lambda_n-\mu_k}{\lambda_n-\mu_k^0}$$

$$= B_2C_2\prod_{k=0}^{\infty}\left(1-\frac{\lambda_n^0}{\mu_k^0}\right)\prod_{k=0}^{\infty}\frac{\lambda_n-\mu_k^0}{\lambda_n^0-\mu_k^0}\prod_{k=0}^{\infty}\frac{\lambda_n-\mu_k}{\lambda_n-\mu_k^0}.$$

In the same way we can prove that

$$\dot{\Phi}_1(\lambda_n) = -\frac{B_1C_1}{\lambda_n^0}\prod_{k=0}^{\infty}{}'\left(1-\frac{\lambda_n^0}{\lambda_k^0}\right)\prod_{k=0}^{\infty}{}'\frac{\lambda_n-\lambda_k}{\lambda_n^0-\lambda_k^0}\prod_{k=0}^{\infty}{}'\frac{\lambda_n-\lambda_k}{\lambda_n-\lambda_k^0}.$$

These formula and (1.15) imply that

$$\alpha_n = \frac{1}{H_2-H_1}\dot{\Phi}_1(\lambda_n)\Phi_2(\lambda_n) = -\frac{1}{H_2-H_1}\frac{B_1C_1}{\lambda_n^0}\prod_{k=0}^{\infty}{}'\left(1-\frac{\lambda_n^0}{\lambda_k^0}\right)\prod_{k=0}^{\infty}{}'\frac{\lambda_n-\lambda_k^0}{\lambda_n^0-\lambda_k^0}\prod_{k=0}^{\infty}{}'\frac{\lambda_n-\lambda_k}{\lambda_n-\lambda_k^0}$$

$$B_2C_2\prod_{k=0}^{\infty}\left(1-\frac{\lambda_n^0}{\mu_k^0}\right)\prod_{k=0}^{\infty}\frac{\lambda_n-\mu_k^0}{\lambda_n^0-\mu_k^0}\prod_{k=0}^{\infty}\frac{\lambda_n-\mu_k}{\lambda_n-\mu_k^0}$$

$$= -\frac{B_1B_2C_1C_2}{H_2-H_1}\frac{1}{\lambda_n^0}\prod_{k=0}^{\infty}{}'\left(1-\frac{\lambda_n^0}{\lambda_k^0}\right)\prod_{k=0}^{\infty}\left(1-\frac{\lambda_n^0}{\mu_k^0}\right)\prod_{k=0}^{\infty}{}'\frac{\lambda_n-\lambda_k^0}{\lambda_n^0-\lambda_k^0}\prod_{k=0}^{\infty}{}'\frac{\lambda_n-\lambda_k}{\lambda_n-\lambda_k^0}$$

$$\prod_{k=0}^{\infty}\frac{\lambda_n-\mu_k^0}{\lambda_n^0-\mu_k^0}\prod_{k=0}^{\infty}\frac{\lambda_n-\mu_k}{\lambda_n-\mu_k^0}$$

$$= \frac{C_1C_2}{C_1^0C_2^0}B_1B_2\alpha_n^0\frac{\lambda_n-\mu_n}{\lambda_n^0-\mu_n^0}\prod_{k=0}^{\infty}{}'\frac{\lambda_n-\lambda_k^0}{\lambda_n^0-\lambda_k^0}\prod_{k=0}^{\infty}{}'\frac{\lambda_n-\lambda_k}{\lambda_n-\lambda_k^0}\prod_{k=0}^{\infty}\frac{\lambda_n-\mu_k^0}{\lambda_n^0-\mu_k^0}\prod_{k=0}^{\infty}\frac{\lambda_n-\mu_k}{\lambda_n-\mu_k^0}.$$

We now show that

$$\frac{C_1C_2}{C_1^0C_2^0}B_1B_2 = 1. \tag{1.19}$$

To do this we observe that $\lim\limits_{\lambda\to-\infty}\dfrac{\Phi_1(\lambda)}{\Phi_{01}(\lambda)} = 1$. Therefore

$$\lim_{\lambda\to-\infty}\frac{C_1}{C_1^0}\prod_{k=0}^{\infty}\left(1-\frac{\lambda}{\lambda_k}\right)\left(1-\frac{\lambda}{\lambda_k^0}\right)^{-1} = \frac{C_1}{C_1^0}\prod_{k=0}^{\infty}\frac{\lambda_k^0}{\lambda_k}\lim_{\lambda\to-\infty}\prod_{k=0}^{\infty}\frac{\lambda_k-\lambda}{\lambda_k^0-\lambda} = 1. \tag{1.20}$$

We now show that

$$\lim_{\lambda\to-\infty}\prod_{k=0}^{\infty}\frac{\lambda_k-\lambda}{\lambda_k^0-\lambda} = \lim_{\lambda\to-\infty}\prod_{k=0}^{\infty}\left(1+\frac{\lambda_k-\lambda_k^0}{\lambda_k^0-\lambda}\right) = 1. \tag{1.21}$$

It follows from the asymptotic formulae that $\lambda_k = (k+\alpha_0(\ell))^2 + \frac{A}{\pi}\ln\left(k+\frac{1}{2}\right) + O(1)$ and
$\lambda_k^0 = (k+\alpha_0(\ell))^2 + \frac{A}{\pi}\ln\left(k+\frac{1}{2}\right) + O(1)$.

Therefore $\lambda_k - \lambda_k^0 = O(1)$ and the series $\sum\limits_{k=0}^{\infty}\left|\dfrac{\lambda_k-\lambda_k^0}{\lambda_k^0-\lambda}\right|$ converges uniformly in λ for $-\infty < \lambda \leq -1$. Therefore the infinite product (1.21) converges uniformly in λ for $-\infty < \lambda \leq -1$, and so we can pass to the limit $\lambda \to -\infty$ in this product. If we do this, we find that $\lim\limits_{\lambda\to-\infty}\prod\limits_{k=0}^{\infty}\left(1+\dfrac{\lambda_k-\lambda_k^0}{\lambda_k^0-\lambda}\right) = 1$. Therefore $\dfrac{C_1}{C_1^0}\prod\limits_{k=0}^{\infty}\dfrac{\lambda_k^0}{\lambda_k} = \dfrac{C_1}{C_1^0}B_1 = 1$. This means that $\dfrac{C_1C_2}{C_1^0C_2^0}B_1B_2 = 1$. This formula and the formula (1.20) imply (1.19), which proves

$$\alpha_n = \alpha_n^0\frac{\lambda_n-\mu_n}{\lambda_n^0-\mu_n^0}\prod_{k=0}^{\infty}{}'\frac{\lambda_n-\lambda_k^0}{\lambda_n^0-\lambda_k^0}\prod_{k=0}^{\infty}{}'\frac{\lambda_n-\lambda_k}{\lambda_n-\lambda_k^0}\prod_{k=0}^{\infty}\frac{\lambda_n-\mu_k^0}{\lambda_n^0-\mu_k^0}\prod_{k=0}^{\infty}\frac{\lambda_n-\mu_k}{\lambda_n-\mu_k^0}. \tag{1.22}$$

2. Asymptotic formulae for the numbers α_n.

Suppose that we are given two sequences of numbers $\lambda_0, \lambda_1, \dots$, $\mu_0, \mu_1, \dots$, and that we know them to be the eigenvalues of the boundary value problems (1.1)-(1.2) and (1.1)-(1.3) with an unknown function $q(x)$ and with the real numbers H_1, H_2. Suppose that the following asymptotic formulae hold:

$$\lambda_n = (n+\alpha_0(\ell))^2 + \frac{A}{2\pi}\ln(n+1/2) + 2a_0 + 2a_1\frac{\ln(n+1/2)}{(n+1/2)} + \frac{2a_2}{(n+1/2)} + \left(2a_3 + \frac{A^2}{4\pi^2}\right)\frac{\ln^2(n+1/2)}{(n+1/2)^2} + \left(2a_4 + \frac{Aa_0}{\pi}\right)\frac{\ln(n+1/2)}{(n+1/2)^2} + \frac{a_0^2+2a_5}{(n+1/2)^3} + O\left(\frac{\ln^2 n}{n^3}\right) \tag{2.1}$$

$$\mu_n = (n+\alpha_0(\ell))^2 + \frac{A}{2\pi}\ln(n+1/2) + 2a_0' + 2a_1'\frac{\ln(n+1/2)}{(n+1/2)} + \frac{2a_2'}{(n+1/2)} + \left(2a_3' + \frac{A^2}{4\pi^2}\right)\frac{\ln^2(n+1/2)}{(n+1/2)^2} + \left(2a_4' + \frac{Aa_0'}{\pi}\right)\frac{\ln(n+1/2)}{(n+1/2)^2} + \frac{a_0'^2+2a_5'}{(n+1/2)^3} + O\left(\frac{\ln^2 n}{n^3}\right) \tag{2.2}$$

Then, by (2.1) and (2.2) $a_0 = \frac{1}{\pi}\left[-H_1 + \frac{1}{2}\int_0^\pi q(t)dt\right]$, $a_0' = \frac{1}{\pi}\left[-H_2 + \frac{1}{2}\int_0^\pi q(t)dt\right]$. Therefore

$$H_1 - H_2 = \pi(a_0 - a_0'). \tag{2.3}$$

By hypothesis the numbers $\{\lambda_n\}$ and $\{\mu_n\}$ are distinct spectra of one and the same equation. Hence it follows from the results of the previous section that the normalizing consants of the boundary value problem (2.1)-(2.2) are given by the formula (1.22) .

In the rest of this section we shall find the asymptotic formula for the numbers α_n in terms of the known asymptotic formulae for the numbers λ_n and μ_n. We shall carry out this derivation in several stages. We first consider the infinite product $\Psi(\lambda_n) = \prod_{k=0}^{\infty}{}' \left(1 + \frac{\lambda_k - \lambda_k^0}{\lambda_k^0 - \lambda_n}\right)$. Clearly $\ln\Psi(\lambda_n) = \sum_{k=1}^{\infty}{}' \ln\left(1 + \frac{\lambda_k - \lambda_k^0}{\lambda_k^0 - \lambda_n}\right)$. For sufficiently large n and $k \neq n$ $\left|\frac{\lambda_k - \lambda_k^0}{\lambda_k^0 - \lambda_n}\right| < \frac{C}{n+1/2}$, (in what follows C will denote a constant, not necessarily the same one). Therefore

$$\ln\Psi(\lambda_n) = -\sum_{k=1}^{\infty}{}' \left[\sum_{p=1}^{\infty}\frac{(-1)^p}{p}\left(\frac{\lambda_k - \lambda_k^0}{\lambda_k^0 - \lambda_n}\right)^p\right]. \tag{2.4}$$

We have

LEMMA: Let $|\lambda_k - \lambda_k^0| \leq a$ $(k = 1, 2, \dots)$. Then

$$\sum_{k=1}^{\infty}{}' \left|\frac{\lambda_k - \lambda_k^0}{\lambda_k^0 - \lambda_n}\right|^p \leq \begin{cases} C\dfrac{\ln(n+1/2)}{n+1/2}, & p = 1 \\ C\dfrac{a^p}{(n+1/2)^p}, & p \geq 2. \end{cases} \tag{2.5}$$

Proof of Lemma is same as the proof of [2].

Let us study $\ln\Psi(\lambda_n)$ a little further. It follows from the estimate (2.5) that

$$\begin{aligned} &|\sum_{k=1}^{\infty}{}' \sum_{p=3}^{\infty} \frac{(-1)^p}{p}\left(\frac{\lambda_k-\lambda_k^0}{\lambda_k^0-\lambda_n}\right)^p| \le \sum_{p=3}^{\infty}\frac{1}{p}\sum_{k=1}^{\infty}{}' |\frac{\lambda_k-\lambda_k^0}{\lambda_k^0-\lambda_n}|^p \\ &\le C\sum_{p=3}^{\infty}\frac{a^p}{(n+1/2)^p} = \frac{Ca^3}{(n+1/2)^3}\sum_{p=0}^{\infty}\frac{a^p}{(n+1/2)^p} = O\left(\frac{1}{n^3}\right). \end{aligned} \tag{2.6}$$

This estimate and the formula (2.4) show that

$$\ln\Psi(\lambda_n) = \sum_{k=1}^{\infty}{}' \frac{\lambda_k-\lambda_k^0}{\lambda_k^0-\lambda_n} - \frac{1}{2}\sum_{k=1}^{\infty}{}' \left(\frac{\lambda_k-\lambda_k^0}{\lambda_k^0-\lambda_n}\right)^2 + O\left(\frac{1}{n^3}\right). \tag{2.7}$$

We now consider the behaviour of the sums appearing in this formula.

Using the asymptotic formulae (2.1) and (2.2) we see that the first sum satisfies. If we use the expression λ_k and λ_k^0 asymptotic, we get the expression

$$\begin{aligned} \sum_{k=1}^{\infty}{}' \frac{\lambda_k-\lambda_k^0}{\lambda_k^0-\lambda_n} &= A(C_0-C_0^0)\left(\frac{\pi}{4}-\frac{1}{3\pi}\right)\frac{\ln(n+1/2)}{(n+1/2)^2} + \{M_\lambda + (C_0-C_0^0) \\ &\left[\frac{3}{2}+\frac{A\ln(3/2)}{3\pi}+\frac{\pi^2(2C_0-C_0^0)}{3\pi}\right]\Big\}\frac{1}{(n+1/2)^2} + O\left(\frac{\ln n}{n^3}\right), \end{aligned} \tag{2.8}$$

$$-\frac{1}{2}\sum_{k=1}^{\infty}{}' \left(\frac{\lambda_k-\lambda_k^0}{\lambda_k^0-\lambda_n}\right)^2 = -(C_0-C_0^0)^2\frac{\pi^2}{4}\frac{1}{(n+1/2)^2} O\left(\frac{1}{n^3}\right). \tag{2.9}$$

Therefore, the formulae (2.7), (2.8) and (2.9) imply that

$$\begin{aligned} \ln\Psi(\lambda_n) &= A(C_0-C_0^0)\left(\frac{\pi}{4}-\frac{1}{3\pi}\right)\frac{\ln(n+1/2)}{(n+1/2)^2} + \Big\{M_\lambda + \frac{1}{2}\Big[3+\pi^2(C_0-C_0^0) \\ &+\frac{2A\ln(3/2)}{3\pi}\Big](C_0-C_0^0) + \frac{\pi^2(C_0-C_0^0)^2)}{4}\Big\}\frac{1}{(n+1/2)^2} + O\left(\frac{\ln^2 n}{n^3}\right). \end{aligned}$$

Hence

$$\begin{aligned} \Psi(\lambda_n) &= 1 + A(C_0-C_0^0)\left(\frac{\pi}{4}-\frac{1}{3\pi}\right)\frac{\ln(n+1/2)}{(n+1/2)^2} + \Big\{M_\lambda + \frac{1}{2}\Big[3+\pi^2(C_0-C_0^0) \\ &+\frac{2A\ln(3/2)}{3\pi}\Big](C_0-C_0^0) + \frac{\pi^2(C_0-C_0^0)^2)}{4}\Big\}\frac{1}{(n+1/2)^2} + O\left(\frac{\ln^2 n}{n^3}\right). \end{aligned} \tag{2.10}$$

Therefore

$$\begin{aligned} \prod_{k=0}^{\infty}{}' \frac{\lambda_n-\lambda_k}{\lambda_n-\lambda_k^0} &= 1 + A(C_0-C_0^0)\left(\frac{\pi}{4}-\frac{1}{3\pi}\right)\frac{\ln(n+1/2)}{(n+1/2)^2} + \Big\{M_\lambda + \frac{1}{2}\Big[3+\pi^2(C_0-C_0^0) \\ &+\frac{2A\ln(3/2)}{3\pi}\Big](C_0-C_0^0) + \frac{\pi^2(C_0-C_0^0)^2)}{4}\Big\}\frac{1}{(n+1/2)^2} + O\left(\frac{\ln^2 n}{n^3}\right). \end{aligned}$$

We now consider the behaviour of the infinity products

$$\varphi(\lambda_n) = \prod_{k=0}^{\infty}{}' \frac{\lambda_n - \mu_k}{\lambda_n - \mu_k^0} = \prod_{k=0}^{\infty}{}' \left(1 + \frac{\mu_k - \mu_k^0}{\mu_k^0 - \lambda_n}\right)$$

$$\prod_{k=0}^{\infty}{}' \frac{\lambda_n - \mu_k}{\lambda_n - \mu_k^0} = 1 + A(C_0' - C_0'^0)\left(\frac{\pi}{4} - \frac{1}{3\pi}\right)\frac{\ln(n+1/2)}{(n+1/2)^2} + \left\{M_\mu + \frac{1}{2}\left[3 + \pi^2(C_0' - C_0'^0)\right.\right.$$

$$\left.\left. + \frac{2A\ln(3/2)}{3\pi}\right](C_0' - C_0'^0) + \frac{\pi^2(C_0' - C_0'^0)^2)}{4}\right\}\frac{1}{(n+1/2)^2} + O\left(\frac{\ln^2 n}{n^3}\right). \quad (2.11)$$

Here $M_\mu = \sum_{k=1}^{\infty}\{\mu_k - \mu_k^0 - 2(C_0' - C_0'^0)\}$, $\Psi(\lambda_n^0) = \prod_{k=0}^{\infty}{}' \frac{\lambda_n - \lambda_k^0}{\lambda_n^0 - \lambda_k^0} = \prod_{k=0}^{\infty}{}' \left(1 + \frac{\lambda_n - \lambda_n^0}{\lambda_n^0 - \lambda_k^0}\right)$.

In the same way we can prove that

$$\begin{aligned}\Psi(\lambda_n^0) &= 1 + A(C_0 - C_0^0)\left(\frac{\pi}{4} - \frac{1}{3\pi}\right)\frac{\ln(n+1/2)}{(n+1/2)^2} + \left\{\frac{(C_0 - C_0^0)}{2}\right.\\ &\left.\left[3 + \frac{\pi^2 C_0^0}{2} - \frac{A\ln(3/2)}{3\pi}\right] - \frac{\pi^2(C_0 - C_0^0)^2)}{4}\right\}\frac{1}{(n+1/2)^2} + O\left(\frac{\ln^2 n}{n^3}\right).\end{aligned} \quad (2.12)$$

$$\varphi(\lambda_n^0) = \prod_{k=0}^{\infty}{}' \frac{\lambda_n - \mu_k^0}{\lambda_n^0 - \mu_k^0} = \prod_{k=0}^{\infty}{}' \left(1 + \frac{\lambda_n - \lambda_n^0}{\lambda_n^0 - \mu_k^0}\right)$$

$$\begin{aligned}\varphi(\lambda_n^0) &= 1 + A(C_0 - C_0^0)\left(\frac{\pi}{4} - \frac{1}{3\pi}\right)\frac{\ln(n+1/2)}{(n+1/2)^2} + \left\{\frac{(C_0 - C_0^0)}{2}\right.\\ &\left.\left[3 + \frac{\pi^2 C_0^0}{2} - \frac{A\ln(3/2)}{3\pi}\right] - \frac{\pi^2(C_0 - C_0^0)^2)}{4}\right\}\frac{1}{(n+1/2)^2} + O\left(\frac{\ln^2 n}{n^3}\right).\end{aligned} \quad (2.13)$$

It is clear from the asymptotic formulae (2.4) and (2.5) that

$$\frac{\lambda_n - \mu_n}{\lambda_n^0 - \mu_n^0} = 1 + \left\{\frac{1}{2}\left[(C_0 - C_0^0) + (C_0' - C_0'^0)\right] - \frac{C_2 - C_2'}{C_0 - C_0'}\frac{C_2^0 - C_2'^0}{C_0^0 - C_0'^0}\right\}\frac{1}{(n+1/2)^2} + O\left(\frac{\ln^2 n}{n^3}\right), \quad (2.14)$$

$$\frac{\lambda_n - \lambda_0}{\lambda_n^0 - \lambda_0^0} = 1 + \frac{(C_0^0 - C_0) + (\lambda_0^0 - \lambda_0)}{(n+1/2)^2)} + O\left(\frac{\ln n}{n^3}\right), \quad (2.15)$$

$$\frac{\lambda_n - \mu_0}{\lambda_n^0 - \mu_0^0} = 1 + \frac{(C_0^0 - C_0) + (\mu_0^0 - \mu_0)}{(n+1/2)^2} + O\left(\frac{\ln n}{n^3}\right). \quad (2.16)$$

It is well known that

$$\alpha_n^0 = \frac{\pi}{2} + \frac{A\pi^2}{4}\frac{1}{n+1/2} + \frac{A}{4}\frac{\ln(n+1/2)}{(n+1/2)^2} + \frac{\gamma'}{(n+1/2)^2} + O\left(\frac{\ln^2 n}{n^3}\right). \quad (2.17)$$

Therefore we obtain the formulae (2.10), (2.11), (2.12), (2.13), (2.14), (2.15), (2.16) and (2.17) the following asymptotic formula for the numbers α_n:

$$\begin{aligned}\alpha_n &= \frac{\pi}{2} + \frac{A\pi^2}{4}\frac{1}{n+1/2} + \frac{A(3\pi^2 - 4)}{24}\left\{\left[3(C_0 - C_0^0) + (\gamma_1 - \gamma_1^0) + (C_0' - C_0'^0) + (\gamma_1' - \gamma_1'^0)\right]\right.\\ &\left. + \frac{6}{3\pi^2 - 4}\right\}\frac{\ln(n+1/2)}{(n+1/2)^2} + \frac{\pi}{2}\left\{M_\lambda + M_\mu - (\lambda_0 + \mu_0) + (\lambda_0^0 + \mu_0^0) + \frac{3}{2}(C_0 - C_0^0)\right.\end{aligned}$$

$$+\frac{1}{2}(C_0' - C_0'^0) - \frac{C_2 - C_2'}{C_0 - C_0'}\frac{C_2^0 - C_2'^0}{C_0^0 - C_0'^0} + \frac{C_0 - C_0^0}{2}\left(3 + \frac{\pi^2(C_0^0 - C_0'^0)}{2} - \frac{A\ln(3/2)}{3\pi}\right)$$
$$-\frac{\pi^2(C_0 - C_0^0)}{4} + \frac{1}{2}\left[3 + \pi^2(C_0 - C_0^0) + \frac{2A\ln(3/2)}{3\pi}\right][(C_0 - C_0^0) + (\gamma_1 - \gamma_1^0)]$$
$$+\frac{1}{2}\left[3 + \pi^2(C_0 - C_0'^0) + \frac{A\ln(3/2)}{3\pi}\right]\left[(C_0' - C_0'^0) + (\gamma_1' - \gamma_1'^0)\right] + \frac{2\gamma'}{\pi}$$
$$+\frac{\pi^2}{4}\left[(C_0 - C_0^0)^2 + (C_0' - C_0'^0)^2\right]\Bigg\}\frac{1}{(n+1/2)^2} + O\left(\frac{\ln^2 n}{n^3}\right). \tag{2.18}$$

REMARK 1. The asymptotic formula (2.18) for the α_n holds from some sufficiently large n onwards. For small n the numbers α_n must be calculated directly from the formula (1.22), where the difference $H_2 - H_1$ is given by equation (2.3).

REMARK 2. If we specify further terms in the asymptotic formulae (2.1) and (2.2) for the λ_n and μ_n, then this method can be used to calculate further terms in the asymptotic formula for the α_n, but we shall not give these here because of the heavy computations.

3.The Inverse Sturm-Liouville Problem.

In this section we consider the inverse Sturm-Liouville problem.

THEOREM: Suppose that we are given two sequence of numbers $\{\lambda_n\}$ and $\{\mu_n\}$, $(n = 0,\ 1, \dots)$. Satisfying the following conditions:

1.The numbers λ_n and μ_n interlace;

2. $\lambda_n = (n+\alpha_0(\ell))^2 + a\ln(n+\alpha_0(\ell)) + c_0 + c_n$, $\mu_n = (n+\alpha_0(\ell))^2 + a'\ln(n+\alpha_0(\ell)) + c_0' + c_n'$ asymptotic formulae valid. Here $c_0 \neq c_0'$ and $\sum c_n^2$, $\sum c_n'^2$ are both convergent series.

Then there exist $q(x) \in C[a,\ b]$ and H_1, H_2 real numbers so that $\{\lambda_n\}$ is a spectrum of (1.1)-(1.2) and $\{\mu_n\}$ is a spectrum of (1.1)-(1.3) and $H_1 - H_2 = \pi(c_0' - c_0)$.

PROOF: Suppose that we are given sequence of numbers $\{\lambda_n\}$ and $\{\mu_n\}$with the above properties. We construct the sequence of α_n by means of the formula (1.22). It satisfies the asymptotic estimate (2.18). We shall show that all the $\alpha_n > 0$. The first condition of the theorem shows that λ_n and μ_n are interlaced. Therefore, for $n = 0,\ 1, \dots$ $\mu_n > \lambda_n$ or $\mu_n < \lambda_n$. α_n's are given by formula (1.22) or

$$\alpha_n = \alpha_n^0 \frac{\lambda_n - \mu_n}{\lambda_n^0 - \mu_n^0}\frac{\lambda_n - \lambda_0}{\lambda_n^0 - \lambda_k^0}\frac{\lambda_n - \mu_0}{\lambda_n^0 - \mu_0^0}\prod_{k=1}^{\infty}{}' \frac{\lambda_n - \lambda_k}{\lambda_n^0 - \lambda_k^0}\prod_{k=1}^{\infty}{}' \frac{\lambda_n - \mu_k}{\lambda_n^0 - \mu_k^0}.$$

Here, we can write $\prod_{k=1}^{\infty}{}' \frac{\lambda_n - \lambda_k}{\lambda_n^0 - \lambda_k^0} > 0$, $\prod_{k=1}^{\infty}{}' \frac{\lambda_n - \mu_k}{\lambda_n^0 - \mu_k^0} > 0$.

Since $\alpha_n^0 > 0$ and λ_n, μ_n and λ_n^0, μ_n^0 are interlaced. It is obvious from α_n formula that $n = 0,\ 1, \dots,\ \alpha_n > 0$. Since Sturm-Liouville operators can be built according to $\{\lambda_n\}$ and $\{\mu_n\}$ sequences. As a result, we have proven that for given $\{\lambda_n\}$ and $\{\mu_n\}$ sequences Sturm-Liouville operators can be built [4].

References

1. V. A. Marchenko, Tr. Mosk. Mat. Obsh., 1, 327 (1952).
2. M. G. Gasimov, B. M. Levitan, Matem. sborn., 63 (105), no. 3, 445 (1964).
3. M. G. Gasimov, Dokl. Akad., Nauk SSSR, vol. 161. no. 2 (1965).
4. M. G. Gasimov, R. Kh. Amirov, Dokl. Akad., Nauk Az.SSR, vol. 41., no.8 (1985).

8th Int. Coll. on Differential Equations, pp. 25-34
D. Bainov (Ed)

Application of the Laplace, Mellin and Stieltjes Transformations in the Evaluation of Integrals

ALEXANDER APELBLAT
Department of Chemical Engineering, Ben Gurion University of the Negev, Beer Sheva, Israel

Abstract. The evaluation of definite and improper integrals using the Laplace, Mellin and Stieltjes transformations is discussed. The proposed procedures are illustrated by considering a large number of logarithmic, trigonometric, Bessel functions and other integrals.

Keywords: *Integral transformations, integrals of elementary and special functions*

1. INTRODUCTION

Integral transforms serve as a starting point in the evaluation of definite and improper integrals of elementary and special functions by applying various mathematical operations under the integral sign, frequently in conjunction with rules and theorems of operational calculus (see for example [1]). For some classes of integrals [2–4], special techniques were developed, mainly applying the Laplace transformation while the Mellin and Stieltjes transformations are only scarcely considered in the literature. In this work these integral transforms are used to obtain integrals which are not tabulated in standard compilations of integrals and integral transforms [5–11].

2. EVALUATION OF INTEGRALS USING THE LAPLACE TRANSFORMATION

If the function $f(t)$ possesses the Laplace transform $F(s)$

$$L\{f(t)\} = \int_0^\infty e^{-st} f(t)dt = F(s) \tag{1}$$

and is of the form $f(t) = u(t-a)h(t)$, where $u(t-a)$ is the shifted step unit function, then by changing the integration variable in (1)

$$\int_0^\alpha h\left(-\frac{1}{s}\ln x\right) dx = sF(s), \qquad a > 0, \qquad \alpha = e^{-sa}. \tag{2}$$

The case for which $a = 0$ and $\alpha = 1$ was already considered in [4].

Similarly, if $f(t) = [u(t) - u(t-a)]h(t)$, equation (1) gives

$$\int_\alpha^1 h\left(-\frac{1}{s}\ln x\right) dx = sF(s), \qquad a > 0, \qquad \alpha = e^{-sa}. \tag{3}$$

These relations are illustrated by a number of examples. If $f(t) = u(t-a)t^\nu$ and $f(t) = [u(t) - u(t-a)]t^\nu$, then from [9, p. 137, (4.3.2) and (4.3.3)] it follows that

$$\int_0^\alpha \left[\ln\left(\frac{1}{x}\right)\right]^\nu dx = \Gamma(\nu+1, \lambda), \qquad \lambda = \ln\left(\frac{1}{\alpha}\right), \tag{4a}$$

$$\int_\alpha^1 \left[\ln\left(\frac{1}{x}\right)\right]^\nu dx = \gamma(\nu+1, \lambda), \qquad 0 < \alpha < 1, \qquad \operatorname{Re}\nu > -1 \tag{4b}$$

where $\Gamma(a, z)$ and $\gamma(a, z)$ are the incomplete gamma functions [12].

If the same functions are shifted, $f(t) = u(t-a)(t-a)^\nu$ and $f(t) = [u(t) - u(t-a)](t-a)^\nu$, then from [9, p. 137, (4.3.5) and (4.3.6)] one has

$$\int_0^\alpha \left[\ln\left(\frac{\alpha}{x}\right)\right]^\nu dx = \alpha\Gamma(\nu+1), \qquad 0 < \alpha < 1, \qquad \operatorname{Re}\nu > -1, \tag{5a}$$

$$\int_\alpha^1 \left[\ln\left(\frac{x}{\alpha}\right)\right]^\nu dx = \alpha\gamma(\nu+1, -\lambda). \tag{5b}$$

Similarly derived, without going into details about functions $f(t)$ and $F(s)$, the following group of logarithmic integrals is expressed in terms of the incomplete gamma function $\Gamma(a, z)$ and the exponential integral $Ei(z)$ [12].

$$\int_0^\alpha \frac{\left[\ln\left(\frac{\alpha}{x}\right)\right]^\nu}{\ln\left(\frac{1}{x}\right)} dx = \Gamma(\nu+1)\lambda^\nu\Gamma(-\nu, \lambda), \quad 0 < \alpha < 1, \quad \operatorname{Re}\nu > -1, \tag{6a}$$

$$\int_0^\alpha \frac{dx}{\ln\left(\frac{\beta}{x}\right)} = -\beta Ei\left[\ln\left(\frac{\alpha}{\beta}\right)\right], \quad 0 < \alpha < 1, \quad \beta > 0, \tag{6b}$$

$$\int_0^\alpha \frac{dx}{\ln(\beta x)\ln\left(\frac{x}{\beta}\right)} = \frac{1}{2\ln\beta}\left\{\beta Ei\left[\ln\left(\frac{\alpha}{\beta}\right)\right] - \frac{1}{\beta}Ei[\ln(\alpha\beta)]\right\}, \quad \alpha\beta \neq 1, \tag{6c}$$

$$\int_0^\alpha \frac{x^{\pm\beta}}{\ln x} dx = Ei[(1\pm\beta)\ln\alpha], \tag{6d}$$

$$\int_\alpha^1 \frac{x^\beta - x^{-\beta}}{\ln x} dx = \left\{\ln\left(\frac{1+\beta}{1-\beta}\right) + Ei[(1-\beta)\ln\alpha] - Ei[(1+\beta)\ln\alpha]\right\}, \tag{6e}$$

$$0 < \alpha < 1, \quad 0 < \beta < 1.$$

The second group of logarithmic integrals is expressed with the aid of the complementary error function, $erfc(z)$ [12]

$$\int_0^\alpha \frac{dx}{\left[\ln\left(\frac{1}{x}\right)\right]^{3/2}} = \frac{2\alpha}{\sqrt{\lambda}} - 2\sqrt{\pi}\,\mathrm{erfc}\,(\sqrt{\lambda}), \quad 0 < \alpha < 1, \tag{7a}$$

$$\int_0^\alpha \frac{dx}{\ln\left(\frac{1}{x}\right)\sqrt{\ln\left(\frac{\alpha}{x}\right)}} = \frac{\pi}{\sqrt{\lambda}}\mathrm{erfc}\,(\sqrt{\lambda}), \tag{7b}$$

$$\int_0^\alpha \frac{\sqrt{\ln\left(\frac{\alpha}{x}\right)}}{\ln\left(\frac{1}{x}\right)}dx = \pi\left[\alpha - \sqrt{\lambda}\,\mathrm{erfc}\,(\sqrt{\lambda})\right] \tag{7c}$$

where λ is defined in (4a).

The last group of logarithmic integrals, related to the Laplace transform, is associated with the modified Bessel functions $I_\nu(z)$ and $K_\nu(z)$ and the Struve function $L_\nu(z)$ [12]

$$\int_0^\alpha \frac{\ln x}{\sqrt{\ln(\alpha x)\ln\left(\frac{x}{\alpha}\right)}}dx = -\lambda K_1(\lambda), \quad 0 < \alpha < 1, \tag{8a}$$

$$\int_0^\alpha \ln\left(\frac{1}{\alpha x}\right)\sqrt{\ln\left(\frac{\alpha}{x}\right)}dx = -\frac{\alpha\sqrt{\pi}}{4} + \sqrt{2}\lambda^{3/2}K_{3/2}(\lambda). \tag{8b}$$

$$\int_0^\alpha \left[\ln(\alpha x)\ln\left(\frac{x}{\alpha}\right)\right]^{\nu-1/2}dx = \frac{(2\lambda)^\nu\Gamma\left(\nu+\frac{1}{2}\right)}{\sqrt{\pi}}K_\nu(\lambda), \quad \mathrm{Re}\,\nu > -1/2, \tag{8c}$$

$$\int_\alpha^1 \left[\ln\left(\frac{x}{\alpha}\right)\ln\left(\frac{1}{\alpha x}\right)\right]^{\nu-1/2}dx = 2^{\nu-1}\sqrt{\pi}\Gamma\left(\nu+\frac{1}{2}\right)\lambda^\nu[I_\nu(\lambda) - L_\nu(\lambda)]. \tag{8d}$$

3. EVALUATION OF INTEGRALS USING THE MELLIN TRANSFORMATION

If in the Mellin transform of the function $f(t)$

$$M\{f(t)\} = \int_0^\infty t^{s-1}f(t)dt = M(s), \tag{9}$$

the integration variable is changed to $x = e^{-t}$, then equation (9) becomes

$$\int_0^1 \frac{1}{x}\left[\ln\left(\frac{1}{x}\right)\right]^{s-1} f(-\ln x)dx = M(s). \tag{10}$$

The equivalent formulas to (2) and (3), for $f(t) = u(t-a)h(t)$ and $f(t) = [u(t) - u(t-a)]h(t)$, are

$$\int_0^\alpha \frac{1}{x}\left[\ln\left(\frac{1}{x}\right)\right]^{s-1} h(-\ln x)dx = M(s,\lambda), \quad \lambda = \ln\left(\frac{1}{\alpha}\right) > 0, \tag{11a}$$

$$\int_\alpha^1 \frac{1}{x}\left[\ln\left(\frac{1}{x}\right)\right]^{s-1} h(-\ln x)dx = M(s,\lambda). \tag{11b}$$

These relations are illustrated by considering the function-Mellin transform pairs, e.g., if $f(t) = \ln(1+e^{-at})$ [11, p. 38, (4.33) and (4.34)] then

$$\int_0^1 \frac{1}{x}\left[\ln\left(\frac{1}{x}\right)\right]^{\beta-1} \ln(1+x^\alpha)dx = \frac{\Gamma(\beta)}{\alpha^\beta}(1-2^{-\beta})\zeta(\beta+1), \quad \alpha,\ \beta > 0, \tag{12a}$$

$$\int_0^1 \frac{1}{x}\left[\ln\left(\frac{1}{x}\right)\right]^{\beta-1} \ln(1-x^\alpha)dx = -\frac{\Gamma(\beta)}{\alpha^\beta}\zeta(\beta+1). \tag{12b}$$

In the second example, $f(t) = [u(t) - u(t-a)]e^{-bt}$ [11, p. 25, (3.2)] and using (11b)

$$\int_\alpha^1 x^{\mu-1}\left[\ln\left(\frac{1}{x}\right)\right]^{\nu-1} dx = \frac{1}{\mu^\nu}\gamma(\nu,\mu\lambda), \quad \lambda = \ln\left(\frac{1}{\alpha}\right), \quad \operatorname{Re}\mu,\ \nu > 0. \tag{13}$$

If $\nu = 1/2$, the integrals (13) can be expressed in terms of the error functions

$$\int_0^\alpha \frac{x^{\mu-1}}{\sqrt{\ln\left(\frac{1}{x}\right)}}dx = \sqrt{\frac{\pi}{\mu}}\operatorname{erfc}(\sqrt{\lambda\mu}), \tag{14a}$$

$$\int_\alpha^1 \frac{x^{\mu-1}}{\sqrt{\ln\left(\frac{1}{x}\right)}}dx = \sqrt{\frac{\pi}{\mu}}\operatorname{erfc}(\sqrt{\lambda\mu}). \tag{14b}$$

In the case of $f(t) = 1/(1+e^{\alpha t})$ [11, p. 28, (3.26)], using (10) leads to

$$\int_0^1 \frac{x^{\alpha-1}}{1+x^\alpha}\left[\ln\left(\frac{1}{x}\right)\right]^{\beta-1} dx = \frac{\Gamma(\beta)}{\alpha^\beta}(1-2^{1-\beta})\zeta(\beta), \quad \alpha > 0, \quad \beta > 1. \tag{15}$$

However, applying the operational rule of the Mellin transformation

$$M\{f^{(n)}(t)\} = (-1)^n\frac{\Gamma(s)}{\Gamma(s-n)}M(s-n), \qquad n = 0,1,2,3,\ldots \tag{16}$$

to the first and second derivatives of the function $f(t)$

$$f'(t) = -\frac{\alpha e^{\alpha t}}{(1+e^{\alpha t})^2}, \qquad f''(t) = \alpha f'(t) + \frac{2\alpha^2 e^{2\alpha t}}{(1+e^{\alpha t})^3} \tag{17}$$

it follows from (16) that

$$M\left\{-\frac{\alpha e^{\alpha t}}{(1+e^{\alpha t})^2}\right\} = -(s-1)M(s-1) \tag{18}$$

and

$$(s-1)(s-2)M(s-2) = -\alpha(s-1)M(s-1) + M\left\{\frac{2\alpha^2 e^{2\alpha t}}{(1+e^{\alpha t})^3}\right\} \tag{19}$$

where $M(s) = M(\beta)$ is given in (15). Thus, from (10), (15), (18) and (19) one has

$$\int_0^1 \frac{x^{\alpha-1}}{(1+x^\alpha)^2}\left[\ln\left(\frac{1}{x}\right)\right]^{\beta-1} dx = \frac{\Gamma(\beta)}{\alpha^\beta}(1-2^{2-\beta})\varsigma(\beta-1), \quad \alpha > 0,\ \beta > 2, \tag{20a}$$

$$\int_0^1 \frac{x^{\alpha-1}}{(1+x^\alpha)^3}\left[\ln\left(\frac{1}{x}\right)\right]^{\beta-1} dx = \frac{\Gamma(\beta)}{2\alpha^\beta}\Big[(1-2^{2-\beta})\varsigma(\beta-1) + (1-2^{3-\beta})\varsigma(\beta-2)\Big], \tag{20b}$$

$$\alpha > 0, \quad \beta > 3.$$

Similarly, introducing $f(t) = 1/(e^{\alpha t}-1)$ [11, p. 28, (3.19)] it is possible to obtain

$$\int_0^1 \frac{x^{\alpha-1}}{1-x^\alpha}\left[\ln\left(\frac{1}{x}\right)\right]^{\beta-1} dx = \frac{\Gamma(\beta)}{\alpha^\beta}\varsigma(\beta), \quad \alpha > 0,\ \beta > 1, \tag{21a}$$

$$\int_0^1 \frac{x^{\alpha-1}}{(1-x^\alpha)^2}\left[\ln\left(\frac{1}{x}\right)\right]^{\beta-1} dx = \frac{\Gamma(\beta)}{\alpha^\beta}\varsigma(\beta-1), \quad \alpha > 0,\ \beta > 2, \tag{21b}$$

$$\int_0^1 \frac{x^{2\alpha-1}}{(1-x^\alpha)^3}\left[\ln\left(\frac{1}{x}\right)\right]^{\beta-1} dx = \frac{\Gamma(\beta)}{2\alpha^\beta}[\varsigma(\beta-1)+\varsigma(\beta-2)], \quad \alpha > 0,\ \beta > 3. \tag{21c}$$

The results in (20) and (21) can be generalized

$$\int_0^1 \frac{x^{a+b-1}}{1+x^a}\left[\ln\left(\frac{1}{x}\right)\right]^{c-1} dx = \frac{\Gamma(c)}{(2a)^c}\left\{\varsigma\left(c, \frac{a+b}{2a}\right) - \varsigma\left(c, \frac{2a+b}{2a}\right)\right\} \tag{22a}$$

$$\int_0^1 \frac{x^{a+b-1}}{1-x^a}\left[\ln\left(\frac{1}{x}\right)\right]^{c-1} dx = \frac{\Gamma(c)}{a^c}\varsigma\left(c, \frac{a+b}{a}\right) \tag{22b}$$

$$\int_0^1 \frac{x^{ab-1}}{d-x^b}\left[\ln\left(\frac{1}{x}\right)\right]^{c-1} dx = \frac{\Gamma(c)}{b^c d}\Phi\left(\frac{1}{d}, c, a\right), \quad a, b > 0,\ c, d > 1 \tag{22c}$$

and

$$\int_0^1 \frac{x^{a+b-1}}{(1+x^a)^2}\left[\ln\left(\frac{1}{x}\right)\right]^{c-1} dx = \tag{23a}$$

$$= \frac{\Gamma(c)}{(2a)^c}\left\{2\left[\varsigma\left(c-1, \frac{a+b}{2a}\right) - \varsigma\left(c-1, \frac{2a+b}{2a}\right)\right] + \frac{b}{a}\left[\varsigma\left(c, \frac{a+b}{2a}\right) - \varsigma\left(c, \frac{2a+b}{2a}\right)\right]\right\}$$

$$\int_0^1 \frac{x^{a+b-1}}{(1-x^a)^2}\left[\ln\left(\frac{1}{x}\right)\right]^{c-1} dx = \frac{\Gamma(c)}{a^c}\left\{\zeta\left(c-1, \frac{a+b}{a}\right) - \frac{b}{a}\zeta\left(c, \frac{a+b}{a}\right)\right\} \tag{23b}$$

$$\int_0^1 \frac{x^{ab-1}}{(d-x^b)^2}\left[\ln\left(\frac{1}{x}\right)\right]^{c-1} dx = \frac{\Gamma(c)}{b^c d^2}\left[\Phi\left(\frac{1}{d}, c, a\right) - \Phi\left(\frac{1}{d}, c-1, a\right)\right] \tag{23c}$$

$$a,\ b > 0, \quad c > 2, \quad d > 1$$

where $\zeta(z, a)$ and $\Phi(z, \alpha, \beta)$ are the generalized zeta and the Lerch zeta functions [12].

In the next example, the operational rule of the Mellin transformation

$$M\{f(t) + tf'(t)\} = -(s-1)M(s) \tag{24}$$

is applied to the function $f(t) = e^{-t}\ln t$ [9, p. 315, (6.4.13)]. After a number of steps,

$$\int_0^1 \left[\ln\left(\frac{1}{x}\right)\right]^{\alpha} \ln\left[\ln\left(\frac{1}{x}\right)\right] dx = \Gamma(\alpha)[1 + \alpha\psi(\alpha)], \qquad \alpha > 0 \tag{25}$$

where $\psi(z)$ is the psi (digamma) function [12].

If the Mellin transforms can be separated into the real and imaginary parts, $M(\sigma + i\omega) = A(\sigma, \omega) + iB(\sigma, \omega)$, then it follows from (9) that

$$\int_0^\infty t^{\sigma-1}\cos(\omega \ln t) f(t) dt = A(\sigma, \omega), \tag{26a}$$

$$\int_0^\infty t^{\sigma-1}\sin(\omega \ln t) f(t) dt = B(\sigma, \omega). \tag{26b}$$

By introducing $t = e^{-x}$ into (26) one has

$$\int_{-\infty}^\infty e^{-\sigma x}\cos(\omega x) f(e^{-x}) dx = A(\sigma, \omega), \tag{27a}$$

$$\int_{-\infty}^\infty e^{-\sigma x}\sin(\omega x) f(e^{-x}) dx = -B(\sigma, \omega) \tag{27b}$$

which has the form of a two-sided Laplace transformation [13] (in the case of the finite Mellin transforms, equations (27) become the ordinary Laplace transforms).

These sets of equations will be illustrated by three examples. For $f(t) = 1/(a+t)$ [11, p. 13, (2.4)] after long but rather elementary operations it is possible to obtain

$$\int_{-\infty}^\infty \frac{e^{-\sigma x}\cos(\omega x)}{a + e^{-x}} dx = \pi \frac{e^{\pi\omega}\sin(\pi\sigma + \omega \ln a) + e^{-\pi\omega}\sin(\pi\sigma - \omega\ln a)}{a^{1-\sigma}[\cosh(2\pi\omega) - \cos(2\pi\sigma)]}, \tag{28a}$$

$$\int_{-\infty}^\infty \frac{e^{-\sigma x}\sin(\omega x)}{a + e^{-x}} dx = \pi \frac{e^{\pi\omega}\cos(\pi\sigma + \omega \ln a) - e^{-\pi\omega}\cos(\pi\sigma - \omega\ln a)}{a^{1-\sigma}[\cosh(2\pi\omega) - \cos(2\pi\sigma)]}, \tag{28b}$$

$$a > 0, \quad 0 < \sigma < 1.$$

For the cases with the Bessel functions, $f(t) = [u(t) - u(t-1)]a^{-\nu}J_\nu(-a\ln t)$ [11, p. 171, (2.44)] and $f(t) = [u(t) - u(t - e^{-b})]J_0\{a(\ln t)^2 - b^2]^{1/2}\}$ [11, p. 175, (3.15)] the following equations were obtained

(29a) $$\int_0^\infty e^{-\sigma x}\cos(\omega x)J_\nu(ax)dx = \frac{1}{b}\left(\frac{a}{R}\right)^\nu \cos(\psi + \nu\phi),$$

(29b) $$\int_0^\infty e^{-\sigma x}\sin(\omega x)J_\nu(ax)dx = \frac{1}{b}\left(\frac{a}{R}\right)^\nu \sin(\psi + \nu\phi),$$

where

(30a) $$b = \left[(a^2+\sigma^2-\omega^2)^2 + (2\sigma\omega)^2\right]^{1/4},$$

(30b) $$R = \left[(\sigma + b\cos\psi)^2 + (\omega + b\sin\psi)^2\right]^{1/2},$$

(30c) $$\psi = \frac{1}{2}\tan^{-1}\left[\frac{2\sigma\omega}{a^2+\sigma^2-\omega^2}\right], \qquad \phi = \tan^{-1}\left[\frac{\omega + b\sin\psi}{\sigma + b\cos\psi}\right]$$

and

(31a) $$\int_b^\infty e^{-\sigma x}\cos(\omega x)J_0\left(a\sqrt{x^2-b^2}\right)dx = \frac{1}{R}e^{-bR\cos\phi}\cos(\phi + bR\sin\phi),$$

(31b) $$\int_b^\infty e^{-\sigma x}\sin(\omega x)J_0\left(a\sqrt{x^2-b^2}\right)dx = -\frac{1}{R}e^{-bR\cos\phi}\sin(\phi + bR\sin\phi).$$

4. EVALUATION OF INTEGRALS USING THE STIELTJES TRANSFORMATIONS

The Stieltjes transformation (iterated Laplace transformation) and the generalized Stieltjes transformations are

(32a) $$S\{f(t)\} = \int_0^\infty \frac{f(t)}{s+t}dt = G(s),$$

(32b) $$S_\rho\{f(t)\} = \int_0^\infty \frac{f(t)}{(s+t)^\rho}dt = G(s;\rho).$$

By introducing $x = e^{-t/s}$ or $x = e^{-t}$, these transformations take the form

(33a) $$\int_0^1 \frac{1}{x(1-\ln x)} f(-s\ln x)dx = G(s),$$

(33b) $$\int_0^1 \frac{1}{x\ln\left(\frac{\beta}{x}\right)} f(-\ln x)dx = G(\ln\beta), \quad s = \ln\beta$$

(33c) $$\int_0^1 \frac{1}{x(1-\ln x)^{\rho}} f(-s\ln x)dx = s^{\rho-1}G(s;\rho),$$

(33d) $$\int_0^1 \frac{1}{x\left[\ln\left(\frac{\beta}{x}\right)\right]^{\rho}} f(-\ln x)dx = G(\ln\beta;\rho).$$

In the first example, $f(t) = t^{-\nu}e^{-at}$ [9, p. 217, (14.2.14), p. 233, (14.4.11)] and therefore

(34a) $$\int_0^1 \frac{x^{\alpha-1}dx}{(1-\ln x)\left[\ln\left(\frac{1}{x}\right)\right]^{\nu}} = \Gamma(1-\nu)e^{\alpha}\Gamma(\nu,\alpha),$$

(34b) $$\int_0^1 \frac{x^{\alpha-1}dx}{\ln\left(\frac{\beta}{x}\right)\left[\ln\left(\frac{1}{x}\right)\right]^{\nu}} = \frac{\Gamma(1-\nu)\beta^{\alpha}\Gamma(\nu,\alpha\ln\beta)}{(\ln\beta)^{\nu}},$$

(34c) $$\int_0^1 \frac{x^{\alpha-1}dx}{\left[\ln\left(\frac{\beta}{x}\right)\ln\left(\frac{1}{x}\right)\right]^{\nu}} = \frac{\Gamma(1-\nu)\beta^{\alpha/2}}{\sqrt{\pi}}\left(\frac{\alpha}{\ln\beta}\right)^{\nu} K_{\nu-1/2}\left(\frac{\alpha}{2}\ln\beta\right),$$

$$\alpha > 0, \quad \beta > 1, \qquad \mathrm{Re}\,\nu < 1.$$

For $f(t) = 1/(a+t)$ [9, p. 216, (14.2.2)] and using repeatedly, the operational rule,

(35) $$S\{f'(t)\} = -\frac{dG(s)}{ds} - \frac{1}{s}f(0)$$

it is possible to obtain a general result

(36a) $$\int_0^1 \frac{dx}{x(1-\ln x)^{n+1}} = \frac{1}{n}, \qquad n = 1,2,3,\dots,$$

(36b) $$\int_0^1 \frac{dx}{x(1-\ln x)(\alpha-\ln x)^{n+1}} = \frac{(-1)^n}{n!}\frac{d^n}{d\alpha^n}\left(\frac{\ln\alpha}{\alpha-1}\right), \qquad \alpha > 0.$$

Similarly, applying $f(t) = e^{-at}$ [9, p. 217, (14.2.11)] and

(37a) $$S\{tf(t)\} = \int_0^{\infty} f(t)dt - sG(s),$$

(37b) $$S_{\rho}\{tf(t)\} = G(s;\rho-1) - sG(s;\rho)$$

one has

(38a) $$\int_0^1 \frac{x^{\alpha-1}\ln x}{1-\ln x}dx = -e^{\alpha}Ei(-\alpha) - \frac{1}{\alpha}, \qquad \alpha > 0,$$

(38b) $$\int_0^1 \frac{x^{\alpha-1}\ln x}{(1-\ln x)^{\rho}}dx = \alpha^{\rho-2}e^{\alpha}\Big[\alpha\Gamma(1-\rho,\alpha) - \Gamma(2-\rho,\alpha)\Big].$$

If both Stieltjes transforms of $f(t)$ and $tf(t)$ are known and the integral in (37a) can be independently determined, then new functional special functions can be derived.

In the case of the Bessel function $J_\nu(at)$ [9, p. 224, (14.3.2)], it follows that

(39a) $$S\{J_\nu(at)\} = \frac{\pi}{\sin(\pi\nu)}\Big[\mathbf{J}_\nu(as) - J_\nu(as)\Big], \quad a > 0, \quad \operatorname{Re}\nu > -1,$$

(39b) $$S\{tJ_\nu(at)\} = \frac{\pi s}{\sin(\pi\nu)}J_\nu(as) + \frac{1}{a}{}_1F_2\left(1; \frac{1-\nu}{2}, \frac{1+\nu}{2}; -\frac{a^2s^2}{4}\right) - $$
$$- \frac{s}{\nu}{}_1F_2\left(1; \frac{2-\nu}{2}, \frac{2+\nu}{2}; -\frac{a^2s^2}{4}\right),$$

(39c) $$\int_0^\infty J_\nu(at)dt = \frac{1}{a},$$

where $\mathbf{J}_n(z)$ is the Anger function [12]. From (37a) and (39) the Anger function can be expressed in terms of the generalized hypergeometric series

(40)
$$\mathbf{J}_\nu(z) = \frac{\sin(\pi\nu)}{\pi z}F(\nu, z),$$
$$F(\nu, z) = 1 + \frac{z}{\nu}{}_1F_2\left(1; \frac{2-\nu}{2}, \frac{2+\nu}{2}; -\frac{z^2}{4}\right) - {}_1F_2\left(1; \frac{1-\nu}{2}, \frac{1+\nu}{2}; -\frac{z^2}{4}\right).$$

Comparison of the integral representations of the Anger function [12] with (40) yields

(41a) $$\int_0^\pi \cos(\nu\theta - z\sin\theta)d\theta = \frac{\sin(\pi\nu)}{z}F(\nu, z),$$

(41b) $$\int_0^\infty e^{-z\sinh\theta - \nu\theta}d\theta = \frac{1}{z}F(\nu, z) - \pi\operatorname{cosec}(\pi\nu)\mathbf{J}_\nu(z).$$

For $\nu = 0$, the integral (42b) is reduced to the well-known result [11, p. 17, (2.38)]

(42) $$\int_0^\infty e^{-z\sinh\theta}d\theta = \int_0^\infty \frac{e^{-zy}}{\sqrt{1+y^2}}dy = \frac{\pi}{2}\Big[H_0(z) - Y_0(z)\Big], \quad \operatorname{Re} z > 0,$$

where $H_0(z)$ and $Y_0(z)$ are the Struve and Bessel functions of the second kind and of the zero order [12]. For $\nu = 1$, a new Laplace transform is derived

(43) $$\int_0^\infty \frac{e^{-zy}}{\sqrt{1+y^2}\Big[y + \sqrt{1+y^2}\Big]}dy = \frac{z+1}{z} - \frac{\pi}{2}\Big[H_1(z) - Y_1(z)\Big], \quad \operatorname{Re} z > 0.$$

References

1. M. Parodi, Introduction à l'Etude de l'Analyse Symbolique, Gauthier-Villars, Paris (1957).
2. A. Apelblat, IMA J. Appl. Math., **27**, 481–496 (1981).
3. A. Apelblat, IMA J. Appl. Math., **34**, 173–186 (1985).
4. A. Apelblat, J. Math. Anal. Appl., **186**, 237–286 (1994).
5. I. S. Gradshtein and I. Ryzhik, Tables of Series, Products and Integrals, Verlag Harri Deutsch, Frankfurt (1964).
6. A. Apelblat, Table of Definite and Infinite Integrals, Elsevier Sci. Publ. Co., Amsterdam (1983).
7. A. P. Prudnikov, Yu. A. Brychkov and O. I. Marichev, Integrals and Series, Gordon and Breach Sci. Publ., New York (1988).
8. A. Apelblat, Tables of Integrals and Series, Verlag Harri Deutsch, Frankfurt (1996).
9. A. Erdélyi, W. Magnus, F. Oberhettinger and F. G. Tricomi, Tables of Integral Transforms, McGraw-Hill Book Co., Inc., New York (1954).
10. F. Oberhettinger and L. Badii, Table of Laplace Transforms, Springer-Verlag, Berlin (1973).
11. F. Oberhettinger, Table of Mellin Transforms, Springer-Verlag, Berlin (1974).
12. A. Erdélyi, W. Magnus, F. Oberhettinger and F. G. Tricomi, Higher Transcendential Functions, McGraw-Hill Book Co., Inc., New York (1955).
13. B. van der Pol and H. Bremmer, Operational Calculus Based on the Two-Sided Laplace Integral, Cambridge University Press, Cambridge (1964).

8th Int. Coll. on Differential Equations, pp. 35–41
D. Bainov (Ed)

Asynchronous Multisplitting Algorithms for Differential-Algebraic Systems Discretized by Runge-Kutta Methods

J. BAHI AND J.-C. MIELLOU

Scientific Calculus Laboratory. U.M.R. - C.N.R.S. 6623, I.U.T Belfort-Montbéliard, BP 527, 90016 Belfort cedex, France
jacques.bahi@iut-bm.univ-fcomte.fr, miellou@comte.univ-fcomte.fr

Abstract. We present iterative parallel asynchronous multisplitting algorithms in order to treat index one differential-algebraic systems discretized by suitable Runge-Kutta methods. The theoretical framework is based on El tarazi's theorem which concerns general fixed point iterative methods in the case of a product space and here include nonlinear waveform relaxation methods (Jacobi, Gauss-Seidel), and also various parallel extensions mixing point and block relaxation implemented in synchronous or asynchronous mode.

Keywords: Asynchronous algorithms, multisplitting methods, differential algebraic systems, Runge-Kutta methods.

1 Introduction

We present iterative parallel asynchronous multisplitting algorithms in order to treat index one differential-algebraic systems discretized by suitable Runge-Kutta methods. Differential-algebraic systems arise in many scientific problems, particularly from electrical network modeling, where the computer simulation of the time behavior requires the numerical integration of very large differential-algebraic equations, see [1], [2].

In [3] we defined a continuous fixed point mapping which the associated asynchronous algorithms converge to the solution of the considered differential-algebraic problem. In this paper we are interested with the discretization by Runge Kutta methods of the above mentioned continuous fixed point mapping. We prove, not only the convergence of the associated asynchronous algorithms but also the convergence of associated asynchronous multisplitting algorithms.
This aim is achieved by using a general formulation of asynchronous multisplitting methods which we introduced in [4], we obtain in this way the convergence of the particular cases corresponding to O'Leary and White kind multisplittings and Schwarz alternating methods. The theoretical framework is based on El tarazi's theorem which concerns general fixed point iterative methods in the case of a product space and here include nonlinear waveform relaxation methods (Jacobi, Gauss-Seidel), and also various parallel extensions mixing point and block relaxation implemented in synchronous or asynchronous mode.

2 The problems and their discretizations

We consider differential-algebraic systems in the form:

$$(1)\qquad \begin{cases} f(x'(t), x(t), y(t), t) = 0 \\ g(x(t), y(t), t) = 0 \\ x(t_0) = x_0 \end{cases}$$

$$t \in [t_0, t_0 + T], \quad x = (x_1, \ldots, x_n), \quad y = (y_1, \ldots, y_m)$$
$$f = (f_1, \ldots, f_n), \quad g = (g_1, \ldots, g_m)$$

Consider the following continuous fixed point mapping the fixed point of which coincides with the solution of the continuous problem (1), see [3]

$$T(w, u, v) = (z, x, y) \quad such\ that$$

$$(2)\qquad \begin{cases} f_i(w_1(t), \ldots, w_{i-1}(t), z_i(t), w_{i+1}(t), \ldots, w_n(t), u(t), v(t), t) = 0 \\ g_i(u_1(t), \ldots, u_n(t), v_1(t), \ldots, v_{i-1}(t), y_i, v_{i+1}(t), \ldots, v_m(t), t) = 0 \\ x_i'(t) = z_i(t) \\ x_i(t_0) = x_{0i} \end{cases}$$

$[t_0, t_0 + T]$ is discretized as follows: $t_0 < t_1 < \ldots < t_{N-1} = t_0 + T$, with the constant time window $h = t_i - t_{i-1}$ for $i \in \{1, \ldots, N-1\}$. Denote:

$$(3)\qquad A = (a_{ij})_{1 \le i,j \le M}, b = (b_1, \ldots, b_M), c = (c_1, \ldots, c_M)$$

consider the Runge-Kutta method defined by the scheme $\begin{pmatrix} c & A \\ & b^t \end{pmatrix}$. We suppose that $c_j \le 1\ \forall j \in \{1, \ldots, M\}$ and that A is invertible. We suppose also that the Runge-Kutta method is at least of order 1.

In the sequel we denote: $t_{nl} = t_n + c_l h$, and $u_{i,t}$ the approximate value of $u_i(t)$.

The discretization by the Runge-Kutta method (3) of the fixed point mapping (2) gives:

$$T_h(w, u, v) = (z, x, y)$$

such that to compute $z_{i,t_{n+1}}, x_{i,t_{n+1}}, y_{i,t_{n+1}}$, we solve for $l = 1, \ldots, M$:

$$(4)\qquad \begin{cases} F_i(w_1, .., w_{i-1}, z_{i,t_n} + h \sum_{j=1}^{M} a_{lj} z'_{i,t_{nj}}, w_{i+1}, .., w_n, u, v, t_{nl}) = 0 \\ z_{i,t_n} + h \sum_{j=1}^{M} a_{lj} z'_{i,t_{nj}} = x'_{i,t_{nl}} \\ G_i(u_1, \ldots, u_n, v_1, \ldots, v_{i-1}, y_{i,t_n} + h \sum_{j=1}^{M} a_{lj} y'_{i,t_{nj}}, v_{i+1}, \ldots, v_m, t_{nl}) = 0 \end{cases}$$

$$(5)\qquad \begin{cases} z_{i,t_{n+1}} = z_{i,t_n} + h \sum_{j=1}^{M} b_j z'_{i,t_{nj}} \\ x_{i,t_{n+1}} = x_{i,t_n} + h \sum_{j=1}^{M} b_j x'_{i,t_{nj}} \\ y_{i,t_{n+1}} = y_{i,t_n} + h \sum_{j=1}^{M} b_j y'_{i,t_{nj}} \end{cases}$$

We consider L sets I_k such that $\bigcup_{k=1}^{L} I_k = \{1, ..., 2n+m\}$, let I_l^C be their complementaries in $\{1, ..., 2n+m\}$.
For $k \in \{1, ..., L\}$, consider the implicit fixed point mapping associated to T

(6) $$\widetilde{T}^k_{h,i}(w,u,v) = T_{h,i}\left(\sigma_i^k\left(\widetilde{T}^k_h(w,u,v),(w,u,v)\right)\right)$$

where

(7) $$\begin{cases} \sigma_i^k(x,y) = (z_1^k, ..., z_n^k) \text{ such that} \\ z_j^k = x_j \text{ if } (i,j \in I_k) \text{ or } \left(i,j \in I_k^C\right) \\ z_j^k = y_j \text{ elsewhere} \end{cases}$$

The fixed point of $\widetilde{T}^k_h$ (if it exists) coincides with the fixed point of T_h, we define in this way L splittings of the fixed point mapping T_h, see [4].

3 Convergence

Let l denote either n or m, equip R^l with the norm:

(8) $$|\varphi| = \max_{1\le i\le l} |\varphi_i|$$

and $E = C\left([t_0, t_0+T], R^l\right)$ with the norm:

$$\|\varphi_i\| = \max_{t_0\le t_d\le t_N} |\varphi_i(t_d)|$$

where either:

$$t_d = t_n, \text{ or } t_d = t_{nl} = t_n + c_l h, (n,l) \in \{0,..,N-1\} \times \{1,..,M\}$$

We suppose that f and g satisfy the following hypotheses:

(9) $$\left|f_i(z,x,y,t) - f_i(z,\widetilde{x},y,t)\right| \le L_1 \left|x - \widetilde{x}\right|$$

(10) $$\left|f_i(z,x,y,t) - f_i(z,x,\widetilde{y},t)\right| \le L_2 \left|y - \widetilde{y}\right|$$

(11) $$\left|g_i(x,y,t) - g_i(\widetilde{x},y,t)\right| \le L \left|x - \widetilde{x}\right|$$

(12) $$\begin{cases} \text{If } N^*_{i,t} = \left|f_i(z,x,y,t) - f_i(\widetilde{z},x,y,t)\right| \text{ then there exist} \\ \gamma_1^*(t), ..., \gamma_n^*(t), \ \nu^* \text{ and } c^* \text{ such that :} \\ \frac{|\widetilde{z}_i(t)-z_i(t)|}{\gamma_i^*(t)} \le \nu^* \max_{j\ne i} \frac{|\widetilde{z}_j(t)-z_j(t)|}{\gamma_j^*(t)} + c^* N^*_{i,t} \\ 0 < \nu^* < 1 \end{cases}$$

(13) $$\begin{cases} \text{If } N_{i,t} = \left|g_i(x,y,t) - f_i(x,\widetilde{y},t)\right| \text{ then there exist} \\ \gamma_1(t), ..., \gamma_m(t), \ \nu \text{ and } c \text{ such that :} \\ \frac{|\widetilde{y}_i(t)-y_i(t)|}{\gamma_i(t)} \le \nu \max_{j\ne i} \frac{|\widetilde{y}_j(t)-y_j(t)|}{\gamma_j(t)} + c N_{i,t} \\ 0 < \nu < 1 \end{cases}$$

Consider the norms:

$$\|\varphi\|_K = \max_{1\leq i\leq n} \max_{0\leq n\leq N-2} \max_{t_n\leq t_d<t_{n+1}} e^{-Kt_n} |\varphi_{i,t_d}|_i$$

$$\|\varphi\|_{\gamma^*,K} = \max_{1\leq i\leq n} \max_{0\leq n\leq N-2} \max_{t_n\leq t_d<t_{n+1}} e^{-Kt_n} \frac{|\varphi_{i,t_d}|_i}{\gamma^*_{i,d}}$$

$$\|\varphi\|_{\gamma,K} = \max_{1\leq i\leq m} \max_{0\leq n\leq N-2} \max_{t_n\leq t_d<t_{n+1}} e^{-Kt_n} \frac{|\varphi_{i,t_d}|_i}{\gamma_{i,d}}$$

Equip $E \times E \times E$ by the vectorial norm:

$$\left(\|\varphi\|_{\gamma^*,K} \ , \ \|\varphi\|_K \ , \ \|\varphi\|_{\gamma,K} \right)$$

denote:

(14)
$$\begin{cases} \gamma = \max_{1\leq i\leq n} \max_{t_0\leq t_d\leq t_0+T} \gamma_{i,t_d} \\ \gamma^* = \max_{1\leq i\leq m} \max_{t_0\leq t_d\leq t_0+T} \gamma^*_{i,t_d} \end{cases}$$

Proposition 1 *If the Runge-Kutta method and K satisfy:*

(15)
$$\begin{cases} \exists p \in \{1,\dots,M\} \ \ such\ that: \\ b_j = a_{pj}, \forall j \in \{1,\dots,M\} \end{cases}$$

and

(16)
$$h\left(\frac{1-e^{-(N-1)Kh}}{1-e^{-Kh}}\right) < \frac{(1-\nu)(1-\nu^*)}{\gamma^* c^* \left((1-\nu)L_1 + c\gamma L L_2\right)}$$

then $\forall l \in \{1,\dots,L\}$, $\widetilde{T}^l_h$ *(defined in (6)) is contractive.*

Proof. Let:

$$(z,x,y) = T_h(w,u,v), (\widetilde{z},\widetilde{x},\widetilde{y}) = T_h(\widetilde{w},\widetilde{u},\widetilde{v})$$

thus:

$$\begin{cases} |f_i(w_1,..,w_{i-1},z_{i,t_{nl}},w_{i+1},..,w_n,u,v,t_{nl}) - \\ f_i(\widetilde{w}_1,..,\widetilde{w}_{i-1},\widetilde{z}_{i,t_{nl}},\widetilde{w}_{i+1},..,\widetilde{w}_n,u,v,t_{nl})| \\ = \\ |f_i(\widetilde{w}_1,..,\widetilde{w}_{i-1},\widetilde{z}_{i,t_{nl}},\widetilde{w}_{i+1},..,\widetilde{w}_n,\widetilde{u},\widetilde{v},t_{nl}) - \\ f_i(\widetilde{w}_1,..,\widetilde{w}_{i-1},\widetilde{z}_{i,t_{nl}},\widetilde{w}_{i+1},..,\widetilde{w}_n,u,v,t_{nl})| \end{cases}$$

therefore from (9) and (10):

$$N^*_{i,t_{nl}} \leq L_1 |\widetilde{u}_{t_{nl}} - u_{t_{nl}}| + L_2 |\widetilde{v}_{t_{nl}} - v_{t_{nl}}|$$

the hypothesis (12) implies that $\forall l \in \{1,..,M\}$:

$$\frac{\left|\widetilde{z}_{i,t_{nl}} - z_{i,t_{nl}}\right|}{\gamma^*_{i,t_{nl}}} \leq \Phi_{i,t_{nl}}$$

where Φ_{i,t_d} denotes the quantities

$$\nu^* \max_{k\neq i} \frac{|w_{k,t_d} - \widetilde{w}_{k,t_d}|}{\gamma^*_{k,t_d}} + c^* \left(L_1 |\widetilde{u}_{t_d} - u_{t_d}| + L_2 \gamma \frac{|\widetilde{v}_{i,t_d} - v_{i,t_d}|}{\gamma_{i,t_d}} \right)$$

where γ is defined in (14). Hence:

$$e^{-Kt_n}\frac{\left|\tilde{z}_{i,t_{nl}}-z_{i,t_{nl}}\right|}{\gamma^*_{i,t_{nl}}}\leq e^{-Kt_n}\Phi_{i,t_{nl}} \tag{17}$$

since $a_{pj}=b_j$ we deduce:

$$z_{i,t_p} = z_{i,t_{n+1}}$$

we have:

$$e^{-Kt_{n+1}}\frac{\left|\tilde{z}_{i,t_{n+1}}-z_{i,t_{n+1}}\right|}{\gamma^*_{i,t_{n+1}}}\leq e^{-Kt_n}\Phi_{i,t_{nl}} \tag{18}$$

we have from (17) and (18): $\forall 0\leq n\leq N-2, \forall t_n\leq t_d\leq t_{n+1}$:

$$e^{-Kt_n}\frac{\left|\tilde{z}_{i,t_d}-z_{i,t_d}\right|}{\gamma^*_{i,t_d}}\leq \max_{0\leq k\leq N-2}\ \max_{t_k\leq t_l<t_{k+1}} e^{-Kt_k}\Phi_{i,t_l}$$

so:

$$\left\|\tilde{z}-z\right\|_{\gamma^*,K}\leq \nu^*\left\|\tilde{w}-w\right\|_{\gamma^*,K}+c^*L_1\left\|\tilde{u}-u\right\|_K+c^*L_2\gamma\left\|\tilde{v}-v\right\|_{\gamma,K} \tag{19}$$

On the other hand, from (5), we obtain:

$$\left|\tilde{x}_{i,t_{n+1}}-x_{i,t_{n+1}}\right| \leq \left|\tilde{x}_{i,t_n}-x_{i,t_n}\right|+\left|h\sum_{j=1}^{M}b_j\left(\tilde{z}_{i,t_{nj}}-z_{i,t_{nj}}\right)\right|$$

denote:

$$\Delta_i^n=\left|\tilde{x}_{i,t_n}-x_{i,t_n}\right|$$

then from the order one condition we have successively

$$\begin{aligned}
e^{-Kt_{n+1}}\Delta_i^{n+1} &\leq e^{-Kh}e^{-Kt_n}\Delta_i^n+he^{-Kh}e^{-Kt_n}\left|\tilde{z}_{i,t_{nj}}-z_{i,t_{nj}}\right|\\
e^{-Kt_{n+1}}\Delta_i^{n+1} &\leq e^{-Kh}\left(e^{-Kt_n}\Delta_i^n+h\gamma^*\left\|\tilde{z}-z\right\|_{\gamma^*,K}\right)
\end{aligned}$$

$\tilde{x}_{i,0}=x_{i,0}=x_{0,i}$ so the above recurrence formula gives: $\forall 0\leq n\leq N-2$:

$$e^{-Kt_{n+1}}\left|\tilde{x}_{i,n+1}-x_{i,n+1}\right|\leq he^{-Kh}\gamma^*\frac{1-e^{-(n+1)Kh}}{1-e^{-Kh}}\left\|\tilde{z}-z\right\|_{\gamma^*,K} \tag{20}$$

$\forall 0\leq n\leq N-2, \forall n<n_j<n+1$ we have:

$$\left|\tilde{x}_{i,t_{nj}}-x_{i,t_{nj}}\right| \leq \left|\tilde{x}_{i,t_n}-x_{i,t_n}\right|+\left|h\sum_{l=1}^{M}a_{jl}\left(\tilde{z}_{i,t_{nl}}-z_{i,t_{nl}}\right)\right|$$

order zero condition gives:

$$e^{-Kt_n}\left|\tilde{x}_{i,t_{nj}}-x_{i,t_{nj}}\right| \leq e^{-Kt_n}\Delta_i^n x_i+h\gamma^*\max_{t_n\leq t_d<t_{n+1}}e^{-Kt_n}\frac{\left|\tilde{z}_{i,t_d}-z_{i,t_d}\right|}{\gamma^*_{i,d}}$$

$$e^{-Kt_n}\left|\tilde{x}_{i,t_{nj}}-x_{i,t_{nj}}\right|\leq h\gamma^*\left(\frac{1-e^{-(n+1)Kh}}{1-e^{-Kh}}\right)\left\|\tilde{z}-z\right\|_{\gamma^*,K} \tag{21}$$

finally (20) and (21) give:

$$\left\|\widetilde{x}-x\right\|_K \leq h\gamma^*\left(\frac{1-e^{-(N-1)Kh}}{1-e^{-Kh}}\right)\left\|\widetilde{z}-z\right\|_{\gamma^*,K} \tag{22}$$

In the same way as (19), we can prove that:

$$\left\|\widetilde{y}-y\right\|_{\gamma,K} \leq cL\left\|\widetilde{u}-u\right\|_K+\nu\left\|\widetilde{v}-v\right\|_{\gamma,K} \tag{23}$$

put $\omega = h\gamma^*\left(\frac{1-e^{-(N-1)Kh}}{1-e^{-Kh}}\right)$ then:

$$\begin{pmatrix}\left\|\widetilde{z}-z\right\|_{\gamma^*,K}\\ \left\|\widetilde{x}-x\right\|_K\\ \left\|\widetilde{y}-y\right\|_{\gamma,K}\end{pmatrix} \leq \overbrace{\begin{pmatrix}\nu^* & c^*L_1 & c^*L_2\gamma\\ \omega\nu^* & \omega c^*L_1 & \omega c^*L_2\gamma\\ 0 & cL & \nu\end{pmatrix}}^{A}\begin{pmatrix}\left\|\widetilde{w}-w\right\|_{\gamma^*,K}\\ \left\|\widetilde{u}-u\right\|_K\\ \left\|\widetilde{v}-v\right\|_{\gamma,K}\end{pmatrix}$$

we can easily prove that if condition (16) is satisfied then the spectral radius of A is less than one, hence T_h is contractive. Since $\lim\limits_{K\to+\infty} h\left(\frac{1-e^{-(N-1)Kh}}{1-e^{-Kh}}\right)=h$ then in order to satisfy (16) it is sufficient to choose K sufficiently large, and h sufficiently small. Now by the application of corollary 4.3 of [4] we conclude that $\widetilde{T}_h^l$ is contractive.

Proof.

Denote

$$\begin{cases} W=\left(w^1,\ldots,w^L\right)\in\left(R^n\right)^L\\ U=\left(u^1,\ldots,u^L\right)\in\left(R^n\right)^L\\ V=\left(v^1,\ldots,v^L\right)\in\left(R^m\right)^L\end{cases}$$

Define the extended fixed point mapping

$$\begin{cases} \hat{T}_h:\left(R^n\right)^L\times\left(R^n\right)^L\times\left(R^m\right)^L & \longrightarrow & \left(R^n\right)^L\times\left(R^n\right)^L\times\left(R^m\right)^L\\ (W,U,V) & \longrightarrow & (X,Y,Z)\end{cases}$$

such that for $l\in\{1,..,L\}$

$$\begin{cases}\left(z^l,x^l,y^l\right)=\widetilde{T}_h^l z^l\\ z^l=\sum\limits_{k=1}^{L}E_{lk}(w^k,u^k,v^k)^t\end{cases} \tag{24}$$

where E_{lk} are diagonal, positive matrices satisfying

$$\sum_{k=1}^{L}E_{lk}=I\ \ (\textit{identity matrix in } R^n\times R^n\times R^m),\ \ \forall l\in\{1,\ldots,L\} \tag{25}$$

T describes the multisplitting methods associated to T_h^l,and defined by (24), (25).

Remark 1 *We usually take*

$$\left(E_{lk}\right)_{ij}=0\ \ for\ \ j\in I_k^C$$

Proposition 2 *Any asynchronous algorithm corresponding to the multisplitting method defined by ($\hat{T}_h$) converges to the discrete solution of (1).*

Proof. Since $\widetilde{T}_h$ is contractive Proposition 3.2 of [4] ensure that every asynchronous algorithm corresponding to the multisplitting method defined by (6) and (24) converges to the fixed point of T_h.

4 Some particular algorithms

i) O'Leary and White algorithms

Take the diagonal positive matrices $E_{lk}(X)$ depending only on k

$$E_{lk}(X) = E_k$$

and satisfying

$$(E_k)_{ii} = 0, \quad \forall i \notin I_k \tag{26}$$

the asynchronous iterations corresponding to O'Leary and White multisplitting are defined by the fixed point mapping (24) which we denote $\hat{T}^{OW}$ and where the weighting matrices satisfy (26).

ii) Discrete analogue of the Schwarz alternating method

Suppose $I_1 \bigcap I_2 \neq \emptyset$, so we have an overlap between the 1st and the 2nd subdomains. Consider the matrices E_{lk} such that

$$\begin{aligned} (E_{11})_{ii} &= \begin{cases} 1 \ \forall i \in I_1 \\ 0 \ \forall i \notin I_1 \end{cases}, \quad (E_{12})_{ii} = \begin{cases} 0 \ \forall i \in I_1 \\ 1 \ \forall i \notin I_1 \end{cases} \\ (E_{21})_{ii} &= \begin{cases} 1 \ \forall i \notin I_2 \\ 0 \ \forall i \in I_2 \end{cases}, \quad (E_{22})_{ii} = \begin{cases} 0 \ \forall i \notin I_2 \\ 1 \ \forall i \in I_2 \end{cases} \end{aligned} \tag{27}$$

Then the additive discrete analogue of the Schwarz alternating method corresponds to the successive approximation method applied to the fixed point mapping (24) which we denote here $\hat{T}^S$ and where the weighting matrices satisfy (27), and the multiplicative discrete analogue of the Schwarz alternating method corresponds to the block nonlinear Gauss-Seidel method applied to $\hat{T}^S$.

References

1. L.Petersen and S.Mattisson, IEEE Trans. onCAD, 12, (1993).

2. P.Debefve, F.Odeh and A.Ruehli, Circuit Analysis, Simulation and Design, Ruehli. Ed., North Holland, Elsevier part 2 (1987).

3. J.Bahi, E.Griepentrog and J.C.Miellou, SIAM Journal on Numerical Analysis, 23, 1969-1980 (1996).

4. J. Bahi, J.-C. Miellou and K. Rhofir, Numerical Algorithms, (in press).

5. K.E.Brenan, S.L.Campbell and L.R.Petzold, Numerical Solution of Initial-Value Problems in Differential-Algebraic Equations, North Holland (1989).

6. M.N.El Tarazi, Contraction et ordre partiel pour l'étude d'algorithmes synchrones et asynchrones en analyse numérique, Thesis. Univ. Franche-comt. (1981).

8th Int. Coll. on Differential Equations, pp. 43–49
D. Bainov (Ed)

Forced Oscillations of Solutions of Parabolic Equations of Neutral Type with Maxima

DRUMI BAINOV AND EMIL MINCHEV
Medical University of Sofia, P.O. Box 45, 1504 Sofia, Bulgaria

Abstract. Oscillation properties of solutions of parabolic equations of neutral type with maxima are studied and sufficient conditions for forced oscillations of the solutions of a boundary value problem are obtained.
Keywords: *Parabolic equations with maxima, forced oscillations, neutral type.*
AMS Subject Classification: 35B05, 35R10.

1 Introduction

The oscillation theory of partial differential equations with deviating arguments is an object of traditional interest [1]–[5]. In the present paper we investigate the oscillation properties of solutions of neutral parabolic equations which depend on the maximum of the absolute value of the unknown function. Let us note that these equations can be successfully used in the mathematical simulation of many processes in population dynamics, control theory and chemical technologies.

2 Preliminary notes

In this paper we consider forced oscillations of solutions of parabolic equations with maxima

$$\frac{\partial}{\partial t}\left[u+\sum_{i=1}^{m}\lambda_i(t)u(x,t-\tau_i)\right] = a(t)\Delta u + \sum_{j=1}^{k} a_j(t)\Delta u(x,\rho_j(t)) -$$

$$(1) \qquad - c(x,t,u(x,t), \max_{s\in M(t)}|u(x,s)|) + f(x,t), \quad (x,t)\in\Omega\times(0,\infty)\equiv G,$$

where Ω is a bounded domain in $\mathbb{R}^n$ with a piecewise smooth boundary $\partial\Omega$, Δ is the Laplacian in the Euclidean n-space $\mathbb{R}^n$ and $M(t)\subset[0,\infty)$ for $t\geq 0$.

We consider two kinds of boundary conditions:

$$(2) \qquad u(x,t)=\varphi(x,t), \qquad (x,t)\in\partial\Omega\times\mathbb{R}_+,$$

$$(3) \qquad \frac{\partial u}{\partial n}(x,t)+\gamma(x,t)u(x,t)=g(x,t), \qquad (x,t)\in\partial\Omega\times\mathbb{R}_+,$$

where $\mathbb{R}_+=[0,\infty)$, n is the unit exterior normal vector to $\partial\Omega$.

Introduce the following assumptions:

H1. $a(t),\ a_j(t) \in C(\mathbb{R}_+, \mathbb{R}_+),\ \lambda_i(t) \in C^1(\mathbb{R}_+, \mathbb{R}_+),\ j = 1, \ldots, k;\ i = 1, \ldots, m.$

H2. $\rho_j(t) \in C(\mathbb{R}_+, \mathbb{R}),\ \lim_{t\to\infty} \rho_j(t) = \infty,\ j = 1, \ldots, k;\ \tau_i = \text{const} > 0,\ i = 1, \ldots, m.$

H3. $c(x, t, \eta, \xi) \in C(G \times \mathbb{R} \times \mathbb{R}_+, \mathbb{R})$,

$$c(x,t,\eta,\xi) \geq 0 \quad \text{for} \quad (x,t) \in G,\ \eta \geq 0,\ \xi \geq 0,$$

$$c(x,t,\eta,\xi) \leq 0 \quad \text{for} \quad (x,t) \in G,\ \eta \leq 0,\ \xi \geq 0.$$

H4. $f(x,t) \in C(G, \mathbb{R})$.

H5. $M(t)$ is a closed and bounded set for each $t \in [0, \infty)$ and $\lim\limits_{t\to\infty} \min\limits_{s\in M(t)} s = \infty$.

H6. $\varphi(x,t) \in C(\partial\Omega \times \mathbb{R}_+, \mathbb{R})$.

H7. $g(x,t) \in C(\partial\Omega \times \mathbb{R}_+, \mathbb{R})$.

H8. $\gamma(x,t) \in C(\partial\Omega \times \mathbb{R}_+, \mathbb{R}_+)$.

Definition 1 The solution $u \in C^2(G) \cap C^1(\overline{G})$ of problem (1), (2) ((1), (3)) is said to oscillate in the domain $G = \Omega \times (0, \infty)$ if for any positive number μ there exists a point $(x_0, t_0) \in \Omega \times [\mu, \infty)$ such that the equality $u(x_0, t_0) = 0$ holds.

It is known that the first eigenvalue α_0 of the problem

$$\Delta U + \alpha U = 0 \quad \text{in } \Omega,$$

$$U = 0 \quad \text{on } \partial\Omega$$

is positive and the corresponding eigenfunction Φ is positive in Ω. Without loss of generality we may assume that Φ is normalized, i.e., $\int\limits_\Omega \Phi(x)dx = 1$.

3 Main results

3.1 Oscillation of solutions of problem (1), (2)

Theorem 1 *Let the following conditions hold:*

1. *Assumptions* H1–H6 *are fulfilled.*
2. *For every sufficiently large number* $t_0 \geq 0$ *we have*

$$\liminf_{t\to\infty} \int\limits_{t_0}^{t} \left\{ \int\limits_\Omega f(x,s)\Phi(x)dx - \right.$$

$$\left. - \int\limits_{\partial\Omega} \left[a(s)\varphi(x,s) + \sum_{j=1}^{k} a_j(s)\varphi(x,\rho_j(s)) \right] \frac{\partial\Phi}{\partial n} dS \right\} ds = -\infty \tag{4}$$

and

$$\limsup_{t\to\infty} \int\limits_{t_0}^{t} \left\{ \int\limits_\Omega f(x,s)\Phi(x)dx - \right.$$

$$\left. - \int\limits_{\partial\Omega} \left[a(s)\varphi(x,s) + \sum_{j=1}^{k} a_j(s)\varphi(x,\rho_j(s)) \right] \frac{\partial\Phi}{\partial n} dS \right\} ds = \infty. \tag{5}$$

Then every solution $u \in C^2(G) \cap C^1(\overline{G})$ *of problem* (1), (2) *oscillates in* G.

Proof. Suppose that the conclusion of the theorem is not true and let $u \in C^2(G) \cap C^1(\overline{G})$ be a solution of problem (1), (2) which has no zero in $\Omega \times [t_0, \infty)$ for some $t_0 > 0$.

Let $u(x,t) > 0$ in $\Omega \times [t_0, \infty)$. It follows from H2 that there exists $t_1 \geq t_0$ such that

$$u(x, t - \tau_i) > 0, \quad u(x, \rho_j(t)) > 0$$

for $(x,t) \in \Omega \times [t_1, \infty)$, $i = 1, \dots, m$; $j = 1, \dots, k$. Assumption H3 implies that

$$c(x, t, u(x,t), \max_{s \in M(t)} |u(x,s)|) \geq 0 \tag{6}$$

for $(x,t) \in \Omega \times [t_1, \infty)$. Thus, we have from (6) that

$$\begin{aligned} \frac{\partial}{\partial t}\left[u + \sum_{i=1}^{m} \lambda_i(t) u(x, t - \tau_i)\right] \leq a(t)\Delta u + \\ + \sum_{j=1}^{k} a_j(t) \Delta u(x, \rho_j(t)) + f(x,t), \quad (x,t) \in \Omega \times [t_1, \infty). \end{aligned} \tag{7}$$

Multiplying (7) by Φ and integrating over Ω, we obtain

$$\begin{aligned} \frac{d}{dt}\left[\int_\Omega u\Phi dx + \sum_{i=1}^{m} \lambda_i(t) \int_\Omega u(x, t - \tau_i)\Phi(x) dx\right] \leq \\ \leq a(t) \int_\Omega \Delta u \Phi dx + \sum_{j=1}^{k} a_j(t) \int_\Omega \Delta u(x, \rho_j(t)) \Phi(x) dx + \\ + \int_\Omega f(x,t)\Phi(x) dx, \qquad t \geq t_1. \end{aligned} \tag{8}$$

From (2) and Green's formula it follows that

$$\begin{aligned} \int_\Omega \Delta u \Phi dx = \int_{\partial\Omega} \left(\frac{\partial u}{\partial n}\Phi - u\frac{\partial \Phi}{\partial n}\right) dS + \int_\Omega u \Delta \Phi dx = \\ = - \int_{\partial\Omega} \varphi \frac{\partial \Phi}{\partial n} dS - \alpha_0 \int_\Omega u \Phi dx, \quad t \geq t_1, \end{aligned} \tag{9}$$

$$\begin{aligned} \int_\Omega \Delta u(x, \rho_j(t)) \Phi(x) dx = \int_{\partial\Omega} \left[\frac{\partial u}{\partial n}(x, \rho_j(t))\Phi(x) - u(x, \rho_j(t)) \frac{\partial \Phi}{\partial n}(x)\right] dS + \\ + \int_\Omega u(x, \rho_j(t)) \Delta \Phi(x) dx = \\ = - \int_{\partial\Omega} \varphi(x, \rho_j(t)) \frac{\partial \Phi}{\partial n}(x) dS - \alpha_0 \int_\Omega u(x, \rho_j(t)) \Phi(x) dx, \\ t \geq t_1, \ j = 1, \dots, k. \end{aligned} \tag{10}$$

Thus, by (8)–(10) we obtain

$$\frac{d}{dt}\left[V(t)+\sum_{i=1}^{m}\lambda_i(t)V(t-\tau_i)\right]+\alpha_0 a(t)V(t)+$$

$$+\alpha_0\sum_{j=1}^{k}a_j(t)V(\rho_j(t))\le$$

$$\le -a(t)\int_{\partial\Omega}\varphi\frac{\partial\Phi}{\partial n}dS- \tag{11}$$

$$-\sum_{j=1}^{k}a_j(t)\int_{\partial\Omega}\varphi(x,\rho_j(t))\frac{\partial\Phi}{\partial n}dS+\int_{\Omega}f(x,t)\Phi(x)dx,\quad t\ge t_1,$$

where $V(t)=\int_{\Omega}u(x,t)\Phi(x)dx,\ t\ge t_0$.

Setting

$$F(t)=-a(t)\int_{\partial\Omega}\varphi\frac{\partial\Phi}{\partial n}dS-\sum_{j=1}^{k}a_j(t)\int_{\partial\Omega}\varphi(x,\rho_j(t))\frac{\partial\Phi}{\partial n}dS+\int_{\Omega}f(x,t)\Phi(x)dx,$$

we obtain from (11) that

$$\frac{d}{dt}\left[V(t)+\sum_{i=1}^{m}\lambda_i(t)V(t-\tau_i)\right]\le F(t),\quad t\ge t_1. \tag{12}$$

Integrating inequality (12) over the interval $[t_1,t]$, we get

$$V(t)+\sum_{i=1}^{m}\lambda_i(t)V(t-\tau_i)\le C_0+\int_{t_1}^{t}F(s)ds,$$

where C_0 is a constant. Moreover,

$$V(t)\le V(t)+\sum_{i=1}^{m}\lambda_i(t)V(t-\tau_i)\le C_0+\int_{t_1}^{t}F(s)ds,\quad t\ge t_1. \tag{13}$$

Conditions (4) and (13) imply that

$$\liminf_{t\to\infty}V(t)=-\infty,$$

which contradicts the fact that $V(t)$ is positive for $t\ge t_1$.

If $u(x,t)<0$ for $(x,t)\in\Omega\times[t_0,\infty)$, then by H3 the function $-u(x,t)$ satisfies the relations

$$\frac{\partial}{\partial t}\left[u+\sum_{i=1}^{m}\lambda_i(t)u(x,t-\tau_i)\right]\le$$

$$\le a(t)\Delta u+\sum_{j=1}^{k}a_j(t)\Delta u(x,\rho_j(t))-f(x,t),\quad (x,t)\in\Omega\times[t_0,\infty),$$

$$u(x,t)=-\varphi(x,t),\quad (x,t)\in\partial\Omega\times\mathbb{R}_+.$$

Proceeding as in the case $u>0$ and using condition (5) we are led to a contradiction. ■

Corollary 1 *Let the following conditions hold:*

1. *Assumptions* H1–H5 *are fulfilled.*
2. *For every sufficiently large number* $t_0 \geq 0$ *we have*

$$\liminf_{t\to\infty} \int_{t_0}^{t} \left(\int_{\Omega} f(x,s)\Phi(x)dx \right) ds = -\infty$$

and

$$\limsup_{t\to\infty} \int_{t_0}^{t} \left(\int_{\Omega} f(x,s)\Phi(x)dx \right) ds = \infty.$$

Then every solution $u \in C^2(G) \cap C^1(\overline{G})$ *of the problem*

$$\frac{\partial}{\partial t}\left[u + \sum_{i=1}^{m} \lambda_i(t)u(x,t-\tau_i)\right] = a(t)\Delta u + \sum_{j=1}^{k} a_j(t)\Delta u(x,\rho_j(t)) -$$

$$- c(x,t,u(x,t), \max_{s\in M(t)} |u(x,s)|) + f(x,t), \quad (x,t) \in \Omega \times (0,\infty),$$

$$u = 0 \quad on\ \partial\Omega \times \mathbb{R}_+$$

oscillates in G.

3.2 Oscillation of solutions of problem (1), (3)

Theorem 2 *Let the following conditions hold:*

1. *Assumptions* H1–H5, H7, H8 *are fulfilled.*
2. *For every sufficiently large number* $t_0 \geq 0$ *we have*

$$\liminf_{t\to\infty} \int_{t_0}^{t} \left\{ \int_{\partial\Omega} \left[a(s)g(x,s) + \sum_{j=1}^{k} a_j(s)g(x,\rho_j(s)) \right] dS + \right.$$

$$\left. + \int_{\Omega} f(x,s)dx \right\} ds = -\infty \tag{14}$$

and

$$\limsup_{t\to\infty} \int_{t_0}^{t} \left\{ \int_{\partial\Omega} \left[a(s)g(x,s) + \sum_{j=1}^{k} a_j(s)g(x,\rho_j(s)) \right] dS + \right.$$

$$\left. + \int_{\Omega} f(x,s)dx \right\} ds = \infty. \tag{15}$$

Then every solution $u \in C^2(G) \cap C^1(\overline{G})$ *of problem* (1), (3) *oscillates in* G.

Proof. Suppose that the problem (1), (3) has a solution $u \in C^2(G) \cap C^1(\overline{G})$ which has no zero in $\Omega \times [t_0, \infty)$ for some $t_0 > 0$.

Assume that $u(x,t) > 0$ in $\Omega \times [t_0, \infty)$. Then there exists $t_1 \geq t_0$ such that $u(x, t - \tau_i) > 0$, $u(x, \rho_j(t)) > 0$ for $(x,t) \in \Omega \times [t_1, \infty)$, where $i = 1, \dots, m$; $j = 1, \dots, k$. Therefore, from (1) and (3) we obtain

$$\frac{d}{dt}\left[\int_\Omega u dx + \sum_{i=1}^m \lambda_i(t) \int_\Omega u(x, t - \tau_i) dx\right] \leq a(t) \int_\Omega \Delta u dx +$$

$$+ \sum_{j=1}^k a_j(t) \int_\Omega \Delta u(x, \rho_j(t)) dx + \int_\Omega f(x,t) dx =$$

$$= a(t) \int_{\partial\Omega} (-\gamma u + g) dS + \sum_{j=1}^k a_j(t) \int_{\partial\Omega} \Big(-\gamma(x, \rho_j(t)) u(x, \rho_j(t)) + g(x, \rho_j(t)) \Big) dS +$$

(16)

$$+ \int_\Omega f(x,t) dx \leq$$

$$\leq a(t) \int_{\partial\Omega} g dS + \sum_{j=1}^k a_j(t) \int_{\partial\Omega} g(x, \rho_j(t)) dS + \int_\Omega f(x,t) dx, \quad t \geq t_1.$$

It follows from (16) that the function $W(t) = \frac{1}{|\Omega|} \int_\Omega u(x,t) dx$, $t \geq t_0$, $|\Omega| = \int_\Omega dx$, satisfies

$$\frac{d}{dt}\left[W(t) + \sum_{i=1}^m \lambda_i(t) W(t - \tau_i)\right] \leq$$

$$\leq \frac{1}{|\Omega|}\left[a(t) \int_{\partial\Omega} g dS + \sum_{j=1}^k a_j(t) \int_{\partial\Omega} g(x, \rho_j(t)) dS + \int_\Omega f(x,t) dx\right] \stackrel{\text{def.}}{=}$$

$$\stackrel{\text{def.}}{=} H(t), \qquad t \geq t_1,$$

that is,

$$\frac{d}{dt}\left[W(t) + \sum_{i=1}^m \lambda_i(t) W(t - \tau_i)\right] \leq H(t), \quad t \geq t_1. \tag{17}$$

Integrating inequality (17) over the interval $[t_1, t]$, we get

$$W(t) + \sum_{i=1}^m \lambda_i(t) W(t - \tau_i) \leq C_0 + \int_{t_1}^t H(s) ds,$$

where C_0 is a constant. Therefore, by (14) we have

$$\liminf_{t \to \infty} W(t) = -\infty,$$

which contradicts to the fact that $W(t) > 0$ for $t \geq t_1$.

If $u(x,t) < 0$ for $(x,t) \in \Omega \times [t_0, \infty)$, then by H3 the function $-u(x,t)$ satisfies the relations

$$\frac{\partial}{\partial t}\left[u + \sum_{i=1}^{m} \lambda_i(t) u(x, t - \tau_i)\right] \leq a(t)\Delta u + \sum_{j=1}^{k} a_j(t) \Delta u(x, \rho_j(t)) - f(x,t),$$

$$(x,t) \in \Omega \times [t_0, \infty),$$

$$\frac{\partial u}{\partial n}(x,t) + \gamma(x,t) u(x,t) = -g(x,t), \quad (x,t) \in \partial\Omega \times \mathbb{R}_+.$$

Similarly to the case $u > 0$ we can get a contradiction using condition (15). ∎

Corollary 2 *Let the following conditions hold:*

1. *Assumptions* H1–H5 *are fulfilled.*
2. *For every sufficiently large number* $t_0 \geq 0$ *we have*

$$\liminf_{t \to \infty} \int_{t_0}^{t} \left(\int_{\Omega} f(x,s) dx \right) ds = -\infty$$

and

$$\limsup_{t \to \infty} \int_{t_0}^{t} \left(\int_{\Omega} f(x,s) dx \right) ds = \infty.$$

Then every solution $u \in C^2(G) \cap C^1(\overline{G})$ *of the problem*

$$\frac{\partial}{\partial t}\left[u + \sum_{i=1}^{m} \lambda_i(t) u(x, t - \tau_i)\right] = a(t)\Delta u + \sum_{j=1}^{k} a_j(t) \Delta u(x, \rho_j(t)) -$$

$$- c(x, t, u(x,t), \max_{s \in M(t)} |u(x,s)|) + f(x,t), \quad (x,t) \in \Omega \times (0, \infty),$$

$$\frac{\partial u}{\partial n} = 0 \quad \text{on } \partial\Omega \times \mathbb{R}_+$$

oscillates in G.

Acknowledgements

The present investigation was partially supported by the Bulgarian Ministry of Education, Science and Technologies under grant MM–512.

References

1. K. Kreith, Oscillation Theory, Lecture Notes in Mathematics, 324, Springer Verlag, Berlin (1973).
2. N. Yoshida, Math. Journal of Toyama University, 15, 131–142 (1992).
3. N. Yoshida, Bull. Austral. Math. Soc., 36, 289–294 (1987).
4. N. Yoshida, Hiroshima Math. Journal, 16, 305–314 (1986).
5. N. Yoshida, Tamkang Journal of Math., 26, 131–139 (1995).

8th Int. Coll. on Differential Equations, pp. 51-56
D. Bainov (Ed)

On the Stability of Solutions of Impulsive Parabolic Differential-Functional Equations

DRUMI BAINOV AND EMIL MINCHEV
Medical University of Sofia, P.O. Box 45, 1504 Sofia, Bulgaria

Abstract. Stability and asymptotic stability of the solutions of impulsive parabolic differential-functional equations are studied via the method of differential-functional inequalities.
Key words and phrases: *stability of solutions, impulsive parabolic differential-functional equations.*
AMS Subject Classification: 35R10.

1 Introduction

The theory of the impulsive ordinary differential equations (ODE) started in 1960 with the pioneer work of V. Mil'man and A. Myshkis [1]. First this theory developed slowly due to difficulties of technical and theoretical character, related to the presence of phenomena as: "beating", "merging", "dying" of solutions, loss of property of autonomy, etc. However, in the last decade the theory of the impulsive ODE marked rapid development [2], [3], [4]-[7].

The theory of the impulsive partial differential equations (PDE) is new but it attracted the attention of many specialists. The bigger part of the results obtained in the last five years can be found in the monograph [8] and the survey papers [9]-[11], [12]. The big interest is due to the fact that the impulsive PDE can be successfully used for mathematical simulation of many processes in nature which are characterized by the fact that the system parameters are subject to short-term perturbations in time.

In the present paper we investigate the stability and the asymptotic stability of the solutions of impulsive parabolic differential-functional equations via the method of differential-functional inequalities.

2 Preliminary notes

Denote by $C(X, Y)$ the class of all continuous functions from X into Y, where X and Y are metric spaces. Let $Q = (0, +\infty) \times (-b, b)$, where $b = (b_1, \dots, b_n) \in \mathbb{R}^n_+$, $\mathbb{R}_+ = [0, +\infty)$. Suppose that $h \in \mathbb{R}_+$ and

$$\Gamma = ([-h, +\infty) \times [-b, b]) \setminus Q,$$

$$\Gamma_0 = [-h, 0] \times [-b, b],$$

$$\partial_0\Gamma = ((0, +\infty) \times [-b, b]) \setminus Q.$$

Suppose that $0 = x_0 < x_1 < x_2 < \ldots < x_k < \ldots$ are given numbers such that $\lim_{k\to\infty} x_k = +\infty$. Define

$$Q_k = (x_k, x_{k+1}) \times (-b, b), \quad k = 0, 1, \ldots, \qquad Q_{imp} = Q \setminus \bigcup_{k=0}^{\infty} Q_k.$$

Let $C_{imp}[\partial_0\Gamma, \mathbb{R}]$ be the class of all functions $z\colon \partial_0\Gamma \to \mathbb{R}$ such that:

(i) $z\big|_{P_k}$ are continuous functions, where $P_k = ((x_k, x_{k+1}) \times \mathbb{R}^n) \cap \partial_0\Gamma$, $k = 0, 1, \ldots$.

(ii) For each k, $k = 1, 2, \ldots$, $(x_k, y) \in \partial_0\Gamma$, there exists the limit

$$\lim_{\substack{(t,s)\to(x_k,y)\\ t<x_k}} z(t, s) = z(x_k^-, y). \tag{1}$$

(iii) For each k, $k = 0, 1, \ldots$, $(x_k, y) \in \partial_0\Gamma$, there exists the limit

$$\lim_{\substack{(t,s)\to(x_k,y)\\ t>x_k}} z(t, s) = z(x_k^+, y) \tag{2}$$

and $z(x_k, y) = z(x_k^+, y)$.

Denote by $C_{imp}[\Gamma \cup Q, \mathbb{R}]$ the class of all functions $z\colon \Gamma \cup Q \to \mathbb{R}$ such that:

(i) $z\big|_{\Gamma_0}$, $z\big|_{Q_k}$, $k = 0, 1, \ldots$, are continuous functions and $z\big|_{\partial_0\Gamma} \in C_{imp}[\partial_0\Gamma, \mathbb{R}]$.

(ii) For each k, $k = 1, 2, \ldots$, $y \in (-b, b)$, there exists the limit (1).

(iii) For each k, $k = 0, 1, \ldots$, $y \in (-b, b)$, there exists the limit (2) and $z(x_k, y) = z(x_k^+, y)$.

Let $M[n]$ be the class of all matrices $A = [a_{ij}]_{i,j=1,\ldots,n}$ with $a_{ij} \in \mathbb{R}$ and $a_{ij} = a_{ji}$, $i, j = 1, \ldots, n$. Define

$$\Omega = (Q \setminus Q_{imp}) \times \mathbb{R} \times C_{imp}[\Gamma \cup Q, \mathbb{R}] \times \mathbb{R}^n \times M[n],$$

$$\tilde{\Omega} = Q_{imp} \times C_{imp}[\Gamma \cup Q, \mathbb{R}].$$

Suppose that $f : \Omega \to \mathbb{R}$, $g : \tilde{\Omega} \to \mathbb{R}$, $\varphi_0 : \Gamma_0 \to \mathbb{R}$, $\varphi : \partial_0\Gamma \to \mathbb{R}$ are given functions with $\varphi_0 \in C(\Gamma_0, \mathbb{R})$, $\varphi \in C_{imp}[\partial_0\Gamma, \mathbb{R}]$.

Consider the initial-boundary value problem (IBVP)

$$z_x(x, y) = f(x, y, z(x, y), z(\cdot), z_y(x, y), z_{yy}(x, y)), \quad (x, y) \in Q \setminus Q_{imp}, \tag{3}$$

$$z(x, y) = \varphi_0(x, y), \qquad (x, y) \in \Gamma_0, \tag{4}$$

$$z(x, y) = \varphi(x, y), \qquad (x, y) \in \partial_0\Gamma, \tag{5}$$

$$\Delta z(x, y) = g(x, y, z(\cdot)), \qquad (x, y) \in Q_{imp}, \tag{6}$$

where $\Delta z(x, y) = z(x, y) - z(x^-, y)$, $z_y(x, y) = (z_{y_1}(x, y), \ldots, z_{y_n}(x, y))$, $z_{yy}(x, y) = [z_{y_i y_j}(x, y)]_{i,j=1,\ldots,n}$.

Denote by $C_{imp}^{1,2}[\Gamma \cup Q, \mathbb{R}]$ the class of all functions $z : \Gamma \cup Q \to \mathbb{R}$ such that $z \in C_{imp}[\Gamma \cup Q, \mathbb{R}]$, there exist the derivatives $z_x(x, y)$, $z_y(x, y)$, $z_{yy}(x, y)$ for $(x, y) \in Q \setminus Q_{imp}$ and the functions $z_x\big|_{Q_k}$, $z_y\big|_{Q_k}$, $z_{yy}\big|_{Q_k}$, $k = 0, 1, \ldots$, are continuous.

Definition 1 A function $z : \Gamma \cup Q \to \mathbb{R}$ is said to be a solution of the IBVP (3)–(6) if $z \in C_{imp}^{1,2}[\Gamma \cup Q, \mathbb{R}]$, z satisfies (3) for $(x, y) \in Q \setminus Q_{imp}$ and conditions (4)–(6) hold true.

For $r,\ \tilde{r} \in M[n]$ we write $r \leq \tilde{r}$ if for each $\lambda = (\lambda_1, \dots, \lambda_n) \in \mathbb{R}^n$ we have

$$\sum_{i,j=1}^{n} (r_{ij} - \tilde{r}_{ij})\lambda_i\lambda_j \leq 0.$$

A function $z : \Gamma \cup Q \to \mathbb{R}$ is called parabolic on Q with respect to f if $z \in C^{1,2}_{imp}[\Gamma \cup Q, \mathbb{R}]$ and for $r,\ \tilde{r} \in M[n]$ such that $r \leq \tilde{r}$ we have

$$f(x, y, z(x, y), z(\cdot), z_y(x, y), r) \leq f(x, y, z(x, y), z(\cdot), z_y(x, y), \tilde{r}),$$

where $(x, y) \in Q \setminus Q_{imp}$.

In the sequel we will consider parabolic solutions of IBVP (3)–(6).

We adopt the following definitions of stability.

Definition 2 The trivial solution $z(x, y) \equiv 0$, $(x, y) \in \Gamma \cup Q$ of the IBVP (3)–(6) is said to be stable if for every $\varepsilon > 0$ there exists $\delta = \delta(\varepsilon) > 0$ such that $|\varphi_0(x, y)| < \delta$ on Γ_0 and $|\varphi(x, y)| < \delta$ on $\partial_0\Gamma$ imply $|z(x, y)| < \varepsilon$ on $Q \cup \partial_0\Gamma$.

Definition 3 The trivial solution $z(x, y) \equiv 0$, $(x, y) \in \Gamma \cup Q$ of the IBVP (3)–(6) is said to be asymptotically stable if:

(i) It is stable.

(ii) There exists a positive number δ_0 such that to every $\varepsilon > 0$ there corresponds $X(\varepsilon)$ such that $|\varphi_0(x, y)| < \delta_0$ on Γ_0 and $|\varphi(x, y)| < \delta_0$ on $\partial_0\Gamma$ imply $|z(x, y)| < \varepsilon$ for $x \in X(\varepsilon)$, $y \in [-b, b]$.

Let $I_0 = [-h, 0]$, $I = (0, +\infty)$, $I_{imp} = \{x_k\}_{k=1}^{\infty}$, $K_k = (x_k, x_{k+1})$, $k = 0, 1, \dots$.

Denote by $C_{imp}[I_0 \cup I, \mathbb{R}_+]$ the class of all functions $\alpha\colon I_0 \cup I \to \mathbb{R}_+$ such that:

(i) $\alpha\big|_{I_0}$, $\alpha\big|_{K_k}$ are continuous for each $k = 0, 1, \dots$.

(ii) For each k, $k = 1, 2, \dots$, there exists the limit

$$\lim_{\substack{t \to x_k \\ t < x_k}} \alpha(t) = \alpha(x_k^-).$$

(iii) For each k, $k = 0, 1, \dots$, there exists the limit

$$\lim_{\substack{t \to x_k \\ t > x_k}} \alpha(t) = \alpha(x_k^+)$$

and $\alpha(x_k) = \alpha(x_k^+)$.

3 Main results

Introduce the following assumptions:

H1. The function $\sigma\colon (I \setminus I_{imp}) \times \mathbb{R}_+ \times C_{imp}[I_0 \cup I, \mathbb{R}_+] \to \mathbb{R}_+$ is continuous, nondecreasing with respect to the functional argument and it satisfies the following Volterra condition: if $(x, p) \in (I \setminus I_{imp}) \times \mathbb{R}_+$, w, $\overline{w} \in C_{imp}[I_0 \cup I, \mathbb{R}_+]$ and $w(\xi) = \overline{w}(\xi)$, $\xi \in [-h, x]$, we have $\sigma(x, p, w) = \sigma(x, p, \overline{w})$.

H2. The function $\tilde{\sigma}\colon I_{imp} \times C_{imp}[I_0 \cup I, \mathbb{R}_+] \to \mathbb{R}_+$ is continuous, nondecreasing with respect to the functional argument and it satisfies the following condition $V^{(-)}$: if $(x, p) \in I_{imp} \times \mathbb{R}_+$, w, $\overline{w} \in C_{imp}[I_0 \cup I, \mathbb{R}_+]$ and $w(\xi) = \overline{w}(\xi)$, $\xi \in [-h, x)$, we have $\tilde{\sigma}(x, p, w) = \tilde{\sigma}(x, p, \overline{w})$.

H3. The function $\eta : I_0 \to \mathbb{R}_+$ is continuous and the maximal solution $\omega(\cdot;\eta)$ of the problem

$$\alpha'(x) = \sigma(x, \alpha(x), \alpha(\cdot)), \qquad x \in I \setminus I_{imp}, \tag{7}$$

$$\alpha(x) = \eta(x), \qquad x \in I_0, \tag{8}$$

$$\Delta\alpha(x) = \tilde{\sigma}(x, \alpha(\cdot)), \qquad x \in I_{imp} \tag{9}$$

is defined on $I_0 \cup I$.

Lemma 1 *Let the following conditions hold:*

1. *Assumptions* H1–H3 *are fulfilled.*
2. $\gamma \in C_{imp}[I_0 \cup I, \mathbb{R}_+]$ *and it satisfies the inequality*

$$D_-\gamma(x) \le \sigma(x, \gamma(x), \gamma(\cdot)) \tag{10}$$

for $x \in E = \left\{x \in I \setminus I_{imp} : \ \gamma(x) > \omega(x;\eta)\right\}$, *where* D_- *is the left-hand lower Dini derivative.*

3. *The estimate for the impulses*

$$\Delta\gamma(x) \le \tilde{\sigma}(x, \gamma(\cdot)), \qquad x \in I_{imp} \tag{11}$$

and the initial estimate

$$\gamma(x) \le \eta(x), \qquad x \in I_0 \tag{12}$$

are satisfied.

Then we have

$$\gamma(x) \le \omega(x;\eta) \qquad on \quad I_0 \cup I.$$

We omit the proof of Lemma 1.

For $w \in C_{imp}[\Gamma \cup Q, \mathbb{R}]$ we define a function $Tw : I_0 \cup I \to \mathbb{R}_+$ in the following way:

$$(Tw)(x) = \max\left\{|w(x,y)| : \quad y \in [-b, b]\right\}.$$

It is easy to see that if $w \in C_{imp}[\Gamma \cup Q, \mathbb{R}]$ then $Tw \in C_{imp}[I_0 \cup I, \mathbb{R}_+]$.

Theorem 1 *Let the following conditions hold:*

1. *Assumptions* H1–H3 *are fulfilled.*
2. *For each* $(x, y, p, w) \in (Q \setminus Q_{imp}) \times \mathbb{R} \times C_{imp}[\Gamma \cup Q, \mathbb{R}]$ *we have*

$$\left|f(x, y, p, w, 0, 0)\right| \le \sigma(x, |p|, Tw).$$

3. *The inequality*

$$\left|g(x, y, w)\right| \le \tilde{\sigma}(x, Tw)$$

is valid for $(x, y, w) \in \tilde{\Omega}$.

4. *The function* $u \in C^{1,2}_{imp}[\Gamma \cup Q, \mathbb{R}]$ *is a parabolic solution of the IBVP* (3)–(6) *and*

$$|u(x,y)| \le \eta(x) \qquad for \quad (x,y) \in \Gamma_0, \tag{13}$$

where $\eta \in C(I_0, \mathbb{R}_+)$.

5. $\omega(\cdot;\eta) : I_0 \cup I \to \mathbb{R}_+$ *is the maximal solution of the problem* (7)–(9) *and*

$$|u(x,y)| \le \omega(x;\eta) \qquad for \quad (x,y) \in \partial_0\Gamma. \tag{14}$$

Then

$$|u(x,y)| \le \omega(x;\eta) \qquad for \quad (x,y) \in \Gamma \cup Q.$$

Proof. Define

$$\gamma(x) = \max\left\{|u(x,y)| : \quad y \in [-b,b]\right\}, \qquad x \in I_0 \cup I.$$

Then $\gamma \in C_{imp}[I_0 \cup I, \mathbb{R}_+]$. We will prove that γ satisfies all conditions of Lemma 1.

It follows from (13) and from condition 2 of the theorem that the estimate for the impulses (11) and the initial inequality (12) are fulfilled. Suppose that $\tilde{x} \in E$. Then there exists $\tilde{y} \in [-b,b]$ such that $\gamma(\tilde{x}) = |u(\tilde{x},\tilde{y})|$. Assume that $\gamma(\tilde{x}) = u(\tilde{x},\tilde{y})$. It follows from (14) that $y \in (-b,b)$ and consequently

$$u_y(\tilde{x},\tilde{y}) = 0, \qquad u_{yy}(\tilde{x},\tilde{y}) \le 0.$$

Then we have

$$\begin{aligned} D_-\gamma(\tilde{x}) &\le u_x(\tilde{x},\tilde{y}) = f(\tilde{x},\tilde{y},u(\tilde{x},\tilde{y}),u(\cdot),u_y(\tilde{x},\tilde{y}),u_{yy}(\tilde{x},\tilde{y})) \le \\ &\le f(\tilde{x},\tilde{y},u(\tilde{x},\tilde{y}),u(\cdot),0,0) \le \sigma(\tilde{x},|u(\tilde{x},\tilde{y})|,Tu) = \\ &= \sigma(\tilde{x},\gamma(\tilde{x}),\gamma(\cdot)). \end{aligned}$$

Thus we see that

$$D_-\gamma(x) \le \sigma(x,\gamma(x),\gamma(\cdot)), \qquad x \in E.$$

In a similar way we can prove that the above inequality holds true in the case when $\gamma(\tilde{x}) = -u(\tilde{x},\tilde{y})$. From Lemma 1 we get $\gamma(x) \le \omega(x;\eta)$ on $I_0 \cup I$ and we proved the conclusion of the theorem. ■

Theorem 2 *Let the following conditions hold:*

1. *Conditions* 1, 2, 3 *of Theorem 1 are fulfilled.*
2. *The maximal solution* $\omega(\cdot;\eta)$ *of the problem* (7) – (9) *is defined on* $\mathbb{R}_+$.
3. $f(x,y,0,0,0,0) = 0$ *for* $(x,y) \in Q \setminus Q_{imp}$, $g(x,y,0) = 0$ *for* $(x,y) \in Q_{imp}$, $\sigma(x,0,0) = 0$ *for* $x \in I \setminus I_{imp}$, $\tilde{\sigma}(x,0) = 0$ *for* $x \in I_{imp}$.

Then the stability or asymptotic stability of the trivial solution of the problem (7)–(9) *implies stability or asymptotic stability of the trivial solution of IBVP* (3) – (6).

Proof. Suppose that the trivial solution of the problem (7)-(9) is stable. Then for every $\varepsilon > 0$ there exists $\delta = \delta(\varepsilon) > 0$ such that $\eta(x) < \delta$, $x \in I_0$ implies that $\omega(x;\eta) < \varepsilon$ on I. Then by Theorem 1 it follows that

$$|u(x,y)| \le \omega(x;\eta) \qquad \text{on} \quad \Gamma \cup Q$$

provided that $|\varphi_0(x,y)| \le \eta(x)$ for $(x,y) \in \Gamma_0$, $|\varphi(x,y)| \le \omega(x;\eta)$ for $(x,y) \in \partial_0\Gamma$ and $u \in C^{1,2}_{imp}[\Gamma \cup Q, \mathbb{R}]$ is a parabolic solution of the IBVP (3)-(6). This proves the stability of the trivial solution of the IBVP (3)-(6).

Analogously it can be proved that the asymptotic stability of the trivial solution of (7)-(9) implies asymptotic stability of the trivial solution of (3)-(6). ■

Acknowledgements

The present investigation was partially supported by the Bulgarian Ministry of Education, Science and Technologies under grant MM-512.

References

1. V. Mil'man and A. Myshkis, Sib. Math. J., 1, 233-237 (1960), (in Russian).
2. D. Bainov and V. Covachev, Impulsive Differential Equations with a Small Parameter, World Scientific Publishers, Singapore (1994).
3. D. Bainov and S. Kostadinov, Abstract Impulsive Differential Equations, Descartes Press, Koriyama (1996).
4. D. Bainov and P. Simeonov, Systems with Impulse Effect, Stability, Theory and Applications, Ellis Horwood, Chichester (1989).
5. D. Bainov and P. Simeonov, Theory of Impulsive Differential Equations: Asymptotic Properties of the Solutions, World Scientific Publishers, Singapore (1995).
6. D. Bainov and P. Simeonov, Theory of Impulsive Differential Equations: Periodic Solutions and Applications, Longman, Harlow (1993).
7. V. Lakshmikantham, D. Bainov and P. Simeonov, Theory of Impulsive Differential Equations, World Scientific Publishers, Singapore (1989).
8. D. Bainov, Z. Kamont and E. Minchev, Approximate Solutions of Impulsive Hyperbolic Equations, Academic Publishers, Calcutta (1996).
9. D. Bainov and E. Minchev, Journal of Henan University (Nat. Sci.), 25, 9–18 (1995).
10. D. Bainov and E. Minchev, Journal of Henan University (Nat. Sci.), 26, 1–13 (1996).
11. D. Bainov and E. Minchev, Nonlinear World, 3, 357–384 (1996).
12. E. Minchev and I. E. Okoroafor, Comm. Appl. Anal., 1, 351–369 (1997).

8th Int. Coll. on Differential Equations, pp. 57-66
D. Bainov (Ed)

Approximate Analysis of Non-Cyclic Multi-Queue Systems with Batch Markovian Arrival Streams

D. BAUM
University of Trier, Dept. IV, 54286 Trier, Germany

Abstract. We consider infinite capacity multi-queue systems with distinct batch Markovian arrival processes, which are characterized by the requirement, that only one job per arrival stream is admitted at a time into a central queue-server facility. Jobs which are not allowed into the central facility have to wait in stream-specific queues until the service of the respective predecessor of this stream has been completed. Service time is assumed to be stream-independent and constant. Such models typically are non-cyclic in their dynamic behaviour, and polling analysis, in general, fails. Although playing an important role in performance analysis of distributed systems, hardly any studies are available. In this paper we present an improved version of a recently proposed approach for the case of Poisson arrival streams [1], applying it to a model with non-Markovian inputs.
Keywords: *Meulti-Queue Systems, BMAPs, Matrix-Analysis*
AMS classification: 60K25, 90B22, 60J25, 68M10.

1. Introduction

Performance studies on constituents of integrated services data networks (communication hard-ware devices, channel assignment strategies, etc.) commonly lead to queueing models which consist of several input queues and one or more central service facility. Multi-queue systems with one common server have been considered mainly within the context of polling analysis, usually under the assumption of Poisson arrival characteristics. Polling analysis is based on the fact, that the system under consideration exhibits some sort of a cyclic behaviour. Systems within the above mentioned framework, but, do not behave that way. Examples are the DQDB network access protocol, or ATM switching units, etc., where one central processing unit admits only a limited amount of data from each input buffer and accepts those data in any order. A representative model of this kind is a queueing system with infinite total capacity being fed by several distinct arrival streams (defining job classes), and containing a central queue-server facility which gives admittance to enter to only one job per stream at a time. Jobs being not allowed into the central station have to wait in class-specific local queues of infinite capacity, until an associated job service has been completed. For the case of Poisson arrival streams this type of queueing model has been investigated in [1]. As opposed to that in this paper we consider nonMarkovian inputs, i.e. the model under consideration is a multi-queue/1-limited service system with batch Markovian arrival processes (BMAPs).

At a first glance it may be tempting — in spite of the before mentioned constraints — to apply polling analysis methods for approximate solutions. We mention some of the relevant results. P. Kuehn [2] (1979) considered asymmetric multi-queue systems with nonexhaustive cyclic service for a variety of service and interarrival time distributions. Comparable work has been performed by Boxma [3,4], Konheim [5], and Takagi [6]. Takine et al [7] provide exact results for asymmetric polling systems with single buffers. Lee and Sengupta [8] have studied polling models with limited service, where reservation features are incorporated. This may be of particular interest for bus access protocols, but is not generally applicable, since information exchange is required between the central station and the local queues. Tsai and Rubin [9] have devoted a note to a two level priority polling scheme with limited service for one priority level. The applicability of this work is confined to symmetric systems only. A survey on cyclic service strategies has been provided by Watson in 1984 [10], and more recent results may be found in Takagi's book [11]. To sum up, it turns out that results on polling systems with more than two queues and limited service are available only for symmetric models with Poisson input, and are restricted even in this case to expressions for first moments only (of queue lenghts and sojourn times). This is why *polling analysis fails* in case of the problem under discussion. If turning away from cyclic service systems there remain but few studies on multi-queue models. One, which had been motivated through the DQDB analysis problem, is a paper written by Potter and Zukerman [12]. There an analysis of a discrete, Poisson input multi-priority system is presented, in which a central service facility serves a number of stations according to the Processor Sharing (PS) discipline. Although well suited for DQDB performance description, the method seems not to allow to depart from the PS discipline or to drop the requirement of inherent level-specific symmetry.

In this paper class specific performance measures are approximated by those of $BMAP/G/1$ queues, interpreting, for each class r job, the central station sojourn time $T_0^{(r)}$ as the service time. The calculation of $T_0^{(r)}$ itself requires approximations, and we return partly to the approach presented in [1]. The paper is organized as follows. In section 2 we specify the queueing model and recapitulate the calculation of the distribution of sojourn time at the central station S_0 as applied in [1]. Section 3 contains an improved procedure to approximate the key parameters, i.e. steady state idle probabilities $P_j^{(r)}(0)$ corresponding to the absence of class j jobs in the central queue S_0 at those epochs, when a class r job enters S_0 ($1 \leq r \leq R$). In section 4 we summarize the approximate solution procedure for the general model based on class specific $BMAP/G/1$ analyses. Section 5 is devoted to a numerical example, including the comparison of analytical results with available exact measures. Section 6 contains a short resumé.

2. The multi-queue / 1-limited service model

The generic model consists of R infinite capacity input queues $Q_1, \ldots, Q_R$ corresponding to R job classes, and one central queue-server facility S_0 (see Figure 1). The central server station may contain no more than one job per input class; additional jobs have to wait in their respective input queue until the corresponding job in S_0 has been served. Service time is stream-independent deterministic of duration d (typically equaling one slot length in most applications), and service is offerred to the jobs in S_0 in FCFS order.

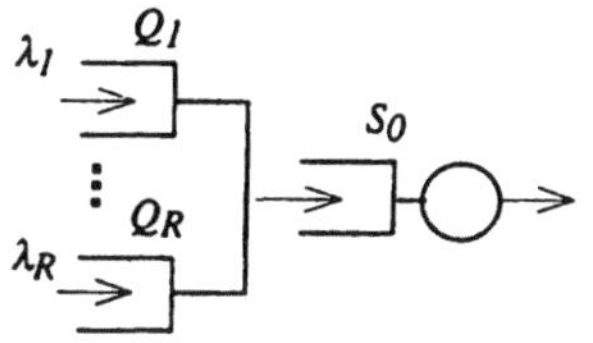

Figure 1: Multi-queue model with R input streams

This model can be interpreted as a resource sharing model in which a job has to capture a resource in order to be admitted into the central station. Exchanging the roles of jobs and resources the model can be considered as a closed multi-class network, the so called equivalent closed network (see Dallery [13], and Perros, Dallery, and Pujolle [14]).

The basic idea to approximately analyze the model consists in calculating the central queue sojourn time distribution $H_r(t) = prob\{T_0^{(r)} \leq t\}$ for each class r and solving for the class specific $BMAP/G/1$-model with service time distribution $H_r(t)$. With $1_{[X \in M]}$ denoting the indicator function for a r.v. X (i.e. $1_{[X \in M]} = 1$ for $X \in M$, $1_{[X \in M]} = 0$ otherwise), the r. v. $T_0^{(r)}$ of sojourn time in S_0 for a class r job is given is $T_0^{(r)} = d + 1_{[K^{(r)}=k>0]} \cdot \cdot \{C^{(r)} + (k-1) \cdot d\}$, and, consequently, its Laplace-Stieltjes-transform (LST) reads

$$T_0^{(r)*}(s) = e^{-s \cdot d} \cdot \Pi^{(r)}(0) + C^{(r)*}(s) \cdot \sum_{k=1}^{R-1} e^{-s \cdot k \cdot d} \cdot \Pi^{(r)}(k), \tag{2.1}$$

where $C^{(r)}$ is the r.v. of residual service time for the job in service immediately before a class r job enters S_0, and $\Pi^{(r)}(k)$ denotes the stationary probability for $K^{(r)} = k$, $K^{(r)}$ the number of jobs in S_0 seen by a class r job upon entering S_0. $C^{(r)}$ in case of $K^{(r)} > 0$ equals one whole service slot d, if the class r customer under consideration enters S_0 immediately after a class r service completion, i.e. if it belongs to those class r jobs, which find another class r job in the system upon arrival. The probability for this event, which we call *event* $\boldsymbol{A}$, is $1 - P_r^A(0) =: \nu_r$, with $P_r^A(k) = prob$ {*k class r jobs in system immediately before a class r arrival*} $(1 \leq r \leq R)$. On the other hand, if an arriving class r job does not find another class r job in the system (*event* $\boldsymbol{B}$, occurring with probability $1 - \nu_r$), the residual service time $C^{(r)}$ may be approximated by a random variable being uniformly distributed over the open interval $[0, d)$ (exact for Poisson inputs). With this approximation in mind the LST of $C^{(r)}$ reads

$$C^{(r)*}(s) = \nu_r \cdot e^{-s \cdot d} + (1 - \nu_r) \cdot \frac{1 - e^{-s \cdot d}}{s \cdot d}. \tag{2.2}$$

For Poisson inputs, PASTA property and symmetry property[1] guarantee $\nu_r = \rho_r$ (ρ_r the steady state property for the event that there is at least one class r job in the system, $1 \le r \le R$). Therefore, for the model with Poisson input streams — as was considered in [1]) -, equation (2.2) reads $C^{(r)*}(s) = \rho_r \cdot e^{-s \cdot d} + (1 - \rho_r) \cdot (1 - e^{-s \cdot d})/s \cdot d$.

In order to compute $T_0^{(r)*}(s)$, the probabilities $\Pi^{(r)}(k)$ have to be determined. In [1], these probabilities have been expressed as functions of the "idle probabilities" $P_j^{(r)}(0)$, $1 \le j \le R$, $j \ne r$, where $P_j^{(r)}(0)$ is the stationary probability for the event that there are no class j jobs in S_0 immediately before a class r job enters S_0. The resulting expression reads

$$\Pi^{(r)}(k) = \sum_{n=0}^{k} (-1)^n \binom{R - 1 - k + n}{n} \eta_{R-1-k+n}(z_1, \ldots, z_{R-1}), \tag{2.3}$$

with $z_j = P_j^{(r)}(0)$ for $1 \le j \ne r \le R$, where $\eta_i(z_1, \ldots, z_m)$ denotes the ith elementary symmetric function of m non-negative numbers z_n (defined by $\eta_0(z_1, \ldots, z_m) = 1$, and $\eta_i(z_1, \ldots, z_m) = \sum_{l=i}^{m} z_l \cdot \eta_{i-1}(z_1, \ldots, z_{l-1})$ for $1 \le i \le m$). As is easily seen from the equivalent closed network model, the class r arrival rate λ_r satisfies the equation $\lambda_r \cdot (E[T_r] + E[T_0^{(r)}]) = 1$ where T_r is the time a token in the closed model has to wait in its synchronization station ($r \in \{1, \ldots, R\}$). T_r equals zero, if an external arrival had occurred before, and equals the residual time X_r until the next class r arrival, if this was not the case. Thus, with $P_r^D(0)$ denoting the stationary probability for the event, that there are no class r jobs left in the system upon a class r departure,

$$\lambda_r \cdot E[T_0^{(r)}] = 1 - \lambda_r \cdot \frac{E[X_r^2]}{2 \cdot E[X_r]} \cdot P_r^D(0), \qquad 1 \le r \le R. \tag{2.4}$$

For Poisson inputs this results in $\lambda_r \cdot E[T_0^{(r)}] = \rho_r$, and the same is true for a class r specific $BMAP/G/1$ queue with service time distribution $H_r(t) = prob\{T_0^{(r)} \le t\}$. In view of the approximation technique to be applied here it is but consequent to assume, that $\lambda_r \cdot E[T_0^{(r)}] = \rho_r$ holds generally for $1 \le r \le R$. Using this approximation, equations (2.1) to (2.3) offer the possibility to compute the R unknown utilization factors $\rho_1, \ldots, \rho_R$. From (2.1) the expression

$$E[T_0^{(r)}] = d \cdot \Pi^{(r)}(0) + E[C^{(r)}] \cdot (1 - \Pi^{(r)}(0)) + d \cdot \sum_{k=1}^{R-1} k \cdot \Pi^{(r)}(k) \tag{2.5}$$

is obtained, and (2.2) yields $E[C^{(r)}] = \nu_r \cdot d + (1 - \nu_r) \cdot (d/2)$. Therefore, for each $r \in \{1, \ldots, R\}$, $\lambda_r \cdot E[T_0^{(r)}]$ is a function of ν_r and the $P_j^{(r)}(0)$ ($1 \le j \ne r \le R$). In [1], the

[1]This is the property that the steady state probability to see a system state by an arriving customer is equal to the steady state probability to leave behind exactly this state upon departure.

probabilities $P_j^{(r)}(0)$ have been roughly approximated by $P_j^{(r)}(0) \approx P_j(0) = 1 - \rho_j$ for all admissible indices j, and since each ν_r equals ρ_r for Poisson inputs, the following system of R nonlinear equations for the unknowns $\rho_1, \dots, \rho_R$ was deduced for the Poisson version: $\rho_r = \lambda_r \cdot d \cdot (R - 1 - \eta_1) + ((\lambda_r \cdot d)/2) \cdot (1 + \rho_r) \cdot (1 - \eta_{R-1}) + \lambda_r \cdot d \cdot \eta_{R-1}$, $1 \le r \le R$ (here η_i is a short cut for $\eta_i(P_1^{(r)}(0), \dots, P_{r-1}^{(r)}(0), P_{r+1}^{(r)}(0), \dots, P_R^{(r)}(0))$.

3. Calculation of the idle probabilities $P_j^{(r)}(0)$

Conditioning upon the above mentioned events $\boldsymbol{A}$ and $\boldsymbol{B}$ the idle probabilities $P_j^{(r)}(0)$ are expressible as

$$P_j^{(r)}(0) = \nu_r \cdot P_j^{(r)}(0|\boldsymbol{A}) + (1 - \nu_r) \cdot P_j^{(r)}(0|\boldsymbol{B}), \tag{3.1}$$

where $\nu_r = 1 - P_r^A(0)$. We approximate ν_r by ρ_r, and $P_j^{(r)}(0|\boldsymbol{B})$ by $1 - \rho_j$ $(1 \le j \ne r \le R)$, although no ASTA property holds for BMAPs. Thus it remains to determine the $P_j^{(r)}(0|\boldsymbol{A})$.

Let α_n $(n = 1, 2, \dots)$ denote the successive time epochs of class r arrivals at station S_0 caused by *such* class r jobs which had been waiting already in their external queue Q_r. We consider the embedded Markov chain $\{N_j^{(r)}(\alpha_n), J(\alpha_n) : n \ge 0\}$, where $N_j^{(r)}(\alpha_n)$ is the number of class j jobs in system, and $J(\alpha_n)$ is the phase, immediately before time epoch α_n. If *all* epochs α_n could be associated with arrivals of class r jobs in S_0, which had been waiting in Q_r for service completion of *such* a class r job, *which in turn had previously been waiting in* Q_r for some time interval of positive length, then the method to calculate the probabilities $P_j^{(r)}(0|\boldsymbol{A})$ would resemble the common approach to determine steady state probabilities at departure instants for a $BMAP/G/1$ system (since during each sojourn time $T_0^{(r)}$ only one class j job could leave the system). Unfortunately, this association is not possible in general, and assuming it we still involve an approximation (which is the more accurate, the higher the class r arrival rate is). For Poisson inputs the probabilities $P_j^{(r)}(0|\boldsymbol{A})$ are obtained as $P_j^{(r)}(0|\boldsymbol{A}) = 1 - \lambda_j \cdot E[T_0^{(r)}]$, resulting in $P_j^{(r)}(0) = 1 - \{((\lambda_j \cdot \rho_r^2)/\lambda_r) + (1 - \rho_r) \cdot \rho_j\}$. In the general case, for any input class j, $1 \le j \ne r \le R$, the steady state probabilities at class r departures can be computed by determining the "arrival matrices" $A_n^{(j)}$ $(n = 0, 1, 2, \dots)$ in the form

$$A_n^{(j)} = \int_0^\infty \tilde{P}_n^{(j)}(t) \cdot dH_r(t), \tag{3.2}$$

where the hk-entry of the matrix $\tilde{P}_n^{(j)}(t)$ is the probability, that up to time t there have counted n j arrivals, and the phase is k, provided that the class j BMAP started at time zero in phase h, and the counter was zero at that time. The knowledge of the matrices $A_n^{(j)}$ $(n = 0, 1, 2, \dots)$ allow to compute exactly all quantities which are necessary to determine the steady state probability vector $\boldsymbol{x}_0^{(j,r)}$, whose k^{th} component $x_{0;k}^{(j,r)}$, denotes the steady state probability for the event, that there are no class j jobs in the system, and the phase of the class j BMAP is k, at the time epoch of a class r departure. The probabilities $P_j^{(r)}(0|\boldsymbol{A})$ are now obtained as

$$P_j^{(r)}(0|\boldsymbol{A}) \approx \boldsymbol{x}_0^{(j,r)} \cdot \boldsymbol{e} = \sum_{k=1}^{m_j} x_{0;k}^{(j,r)} =: \left\langle \boldsymbol{x}_0^{(j,r)}, \boldsymbol{e} \right\rangle, \tag{3.3}$$

resulting in

$$P_j^{(r)}(0) = \rho_r \cdot \left\langle \boldsymbol{x}_0^{(j,r)}, \boldsymbol{e} \right\rangle + (1-\rho_r)\cdot(1-\rho_j), \qquad 1 \le j \ne r \le R.^2 \tag{3.4}$$

The vector $\boldsymbol{x}_0^{(j,r)}$ in (3.4) in turn depends on $H_r(t)$ and — according to equations (2.1) and (2.3) — on all parameters $z_j = P_j^{(r)}(0)$ for $1 \le j \ne r \le R$; thus it is necessary to solve a system of $R \cdot (R-1)$ nonlinear equations for the $R \cdot (R-1)$ unknowns $P_1^{(r)}(0), \dots, P_{r-1}^{(r)}(0), P_{r+1}^{(r)}(0), \dots, P_R^{(r)}(0)$, $1 \le r \le R$.

4. The $BMAP/H_r(t)/1$-analysis

Let $\boldsymbol{x}$ be the stationary probability vector of the embedded Markov chain at departure epochs which, assuming a total traffic intensity less than 1, is given as $\boldsymbol{x}_n = \boldsymbol{x}_0 \cdot B_n + \sum_{\nu=1}^{n+1} \boldsymbol{x}_\nu \cdot A_{n+1-\nu}$ $(n \ge 0)$. As is well known (cf. [15]), the main step in determining steady state probabilities requires the evaluation of the *fundamental period matrix* G of the $BMAP/G/1$-queue, which is stochastic, if the total traffic intensity ρ is less than or equal to one. G can be computed as the minimal non-negative solution of the non-linear matrix equation $G = \sum_{n=0}^{\infty} A_n \cdot G^n$, or from matrix convolution calculus (cf. [16-18]) as $G = S \cdot A_0$, where S is the upper left block of the sum matrix $\sum_{n=0}^{\infty} \mathcal{A}^n$, with $\mathcal{A}$ the block Hessenberg, block Toeplitz matrix whose upper row is $A_1, A_2, A_3, \dots$, and whose other i,j-matrix entries equal A_{j-i+1} for $j \ge i-1 \ge 0$ (and are the zero matrix otherwise).

We distinguish two parts in analyzing the model: The first part consists in computing the distribution functions $H_r(t) = prob\{T_0^{(r)} \le t\}$ by iteration, and the second part concerns the actual $BMAP_r/H_r(t)/1$ analysis, both for $1 \le r \le R$.

Part 1: As far as only $\boldsymbol{x}_0$ is required, as is the case in equation (3.3), the computation of G could be performed without computing explicitely all the matrices A_n (up to an adequate index n_0), as was shown by Lucantoni [16]; his algorithm for the G computation still is one of the fastest, and easy to implement. Although Bini and Meini [19] proposed algorithms, which exploit FFT techniques leading to even better performance, we do not rely on them here. As is obvious, the iteration to determine $H_r(t)$ for the rth input stream requires the knowledge of all other functions $H_j(t)$ $(1 \le j \ne r \le R)$. Starting with $\Pi^{(r)}(0) = \tilde{\lambda}_r$, and $\Pi^{(r)}(k) = 1 - \tilde{\lambda}_r/(R-1)$ for $1 \le k \le R-1$, where $\tilde{\lambda}_r = \lambda_r/(\sum_{i=1}^{R} \lambda_i)$, and $\rho_r = \lambda_r \cdot d$ for $1 \le r \le R$, and setting $\nu_r \approx \rho_r$, the probability density functions $h_r(t)$

[2] e denotes the column vector with each component equal to one, and m_j represents the number of phases for the class j BMAP.

associated with $H_r(t)$ are obtained from (2.1) and (2.2) as[3]

$$h_r(t) = \delta(t-d)\cdot\Pi^{(r)}(0) + \rho_r \cdot \sum_{k=1}^{R-1} \delta(t-[k+1]\cdot d)\cdot \Pi^{(r)}(k) + \tag{4.1}$$

$$+ (1-\rho_r)\cdot \sum_{k=1}^{R-1} c_k(t)\cdot \Pi^{(r)}(k),$$

wherein $c_k(t)$ is the density function of residual service time C_r in the interval $[k\cdot d,(k+1)\cdot d)$, taken here as uniformly distributed r.v. The values $E[T_0^{(r)}]$ from (4.1) (equal for all r in the initial iteration step) can be used to adjust the values $\rho_r = \lambda_r \cdot E[T_0^{(r)}]$, and this is done in form of a subiteration, until the fixed points for ρ_r are reached. Next, for any fixed $r \in \{1,\dots,R\}$ and all j, $1 \le j \ne r \le R$ the probabilities $P_j^{(r)}(0)$ are computed according to (3.4), and are used in (2.3) to determine iterated values for the probabilities $\Pi^{(r)}(k)$, $0 \le k \le R-1$, leading to the next iteration step for the computation of $h_r(t)$. After performing these steps for each $r \in \{1,\dots,R\}$, the iteration cycle is closed, and the stopping condition $\left|E[T_0^{(r)}]_\nu - E[T_0^{(r)}]_{\nu+1}\right| < \varepsilon$ is tested (index ν indicating the number of performed cycles, and ε the chosen degree of accuracy).

Part 2: Now that the distribution functions $H_r(t) = prob\{T_0^{(r)} \le t\}$ are known, the analysis of each $BMAP_r/H_r(t)/1$-queue is done as usual. The calculation of the matrices A_n can be performed according to

$$A_n = \int_0^\infty \sum_{i=0}^\infty \frac{t^i}{i!}\cdot \Delta_n^{*i}\cdot dH_r(t) \qquad \text{for} \quad n = 0,1,2,\dots, \tag{4.2}$$

where Δ_n^{*i} represents n^{th} component of the i-fold convolution of the sequence $\Delta = \{D_\nu\}_{\nu\in\mathbb{N}_0}$, defined through $\Delta_n^{*i} = \sum_{l=0}^n \Delta_l^{*i-1} * D_{n-l}$ for $i \ge 1$, $\Delta_n^{*0} = \begin{cases} I & \text{for } n=0,\\ O & \text{else}\end{cases}$ (see [17,18]). By scaling the sequence Δ, such that $\Delta = \sigma\cdot(\hat{\Delta} - \mathbf{1})$, $\mathbf{1} = \{I,O,\dots,O\}$, with $\sigma = \max\{-D_{0;hh} : h \in \{1,\dots,m_r\}\}$, this results in exactly the same procedure for computing the A_n as proposed by Lucantoni [15] on the basis of his uniformization arguments.

5. Numerical example

The proposed method has been tested for $R = 3$ input streams with different total loads. Results for a small number of queues — if satisfactorily validated — provide sufficient certainty with respect to the approximations involved, since a small number of competing streams should lead to significant mutual influences. The superposition of all R BMAPs constitutes a $BMAP/D/1$- queue which can be analyzed exactly, opening the possibility to examine the accuracy of the method. All quantities obtainable from single $BMAP_r/H_r(t)/1$-behaviour serve as comparative measures. Here we used the mean

[3] $\delta(t-c)$ denotes the Dirac delta function (unit impulse function) with an impulse at poit c.

number $\bar{L}$ of jobs waiting or being served. The following parameter values have been selected (the $D_n^{(i)}$ denote the term matrices of the phase process generators, the superscript indicating the stream number, the subscript n for the right hand matrices running from 1 to infinity):

$$D_0^{(1)} = \begin{pmatrix} -1.0 \cdot \gamma_1^{(1)} & 0.3 \cdot \gamma_1^{(1)} \\ 0.08 \cdot \gamma_2^{(1)} & -1.0 \cdot \gamma_2^{(1)} \end{pmatrix}, \qquad D_n^{(1)} = \begin{pmatrix} 2.4 \cdot 0.2^n \cdot \gamma_1^{(1)} & 0.9 \cdot 0.1^n \cdot \gamma_1^{(1)} \\ 0.03 \cdot 0.8^n \cdot \gamma_2^{(1)} & 1.2 \cdot 0.4^n \cdot \gamma_2^{(1)} \end{pmatrix},$$

$$D_0^{(2)} = \begin{pmatrix} -1.0 \cdot \gamma_1^{(2)} & 0.7 \cdot \gamma_1^{(2)} \\ 0.3 \cdot \gamma_2^{(2)} & -1.0 \cdot \gamma_2^{(2)} \end{pmatrix}, \qquad D_n^{(2)} = \begin{pmatrix} 0.8 \cdot 0.2^n \cdot \gamma_1^{(2)} & 0.7 \cdot 0.125^n \cdot \gamma_1^{(2)} \\ 0.3 \cdot 0.4^n \cdot \gamma_2^{(2)} & 0.5 \cdot 0.5^n \cdot \gamma_2^{(2)} \end{pmatrix},$$

$$D_0^{(3)} = \begin{pmatrix} -1.0 \cdot \gamma_1^{(3)} & 0.1 \cdot \gamma_1^{(3)} & 0.4 \cdot \gamma_1^{(3)} \\ 0.15 \cdot \gamma_2^{(3)} & -1.0 \cdot \gamma_2^{(3)} & 0.28 \cdot \gamma_2^{(3)} \\ 0.3 \cdot \gamma_3^{(3)} & 0.15 \cdot \gamma_3^{(3)} & -1.0 \cdot \gamma_3^{(3)} \end{pmatrix},$$

$$D_n^{(3)} = \begin{pmatrix} 0.2 \cdot 0.6^n \cdot \gamma_1^{(3)} & 0.1 \cdot 0.5^n \cdot \gamma_1^{(3)} & 0.4 \cdot 0.2^n \cdot \gamma_1^{(3)} \\ 0.15 \cdot 0.25^n \cdot \gamma_2^{(3)} & 0.1 \cdot 0.8^n \cdot \gamma_2^{(3)} & 0.28 \cdot 0.3^n \cdot \gamma_2^{(3)} \\ 0.3 \cdot 0.5^n \cdot \gamma_3^{(3)} & 0.1 \cdot 0.5^n \cdot \gamma_3^{(3)} & 0.9 \cdot 0.1^n \cdot \gamma_3^{(3)} \end{pmatrix}, \qquad n \geq 1.$$

Table 1: Class specific and total mean number of customers in system

	Mean no. $\bar{L}^{(r)}$ per stream			Total mean no. $\bar{L}$		relative
Load ρ	$r=1$	$r=2$	$r=3$	computed	BMAP/D/1	error
$\rho = 0.11$	0.153	0.085	0.042	0.279	0.269	3.53 %
$\rho = 0.16$	0.228	0.126	0.061	0.415	0.399	4.03 %
$\rho = 0.22$	0.349	0.192	0.091	0.632	0.600	5.34 %
$\rho = 0.32$	0.564	0.304	0.140	1.009	0.954	5.75 %
$\rho = 0.45$	1.012	0.523	0.227	1.762	1.662	5.97 %
$\rho = 0.64$	2.378	1.064	0.406	3.848	3.675	4.69 %
$\rho = 0.91$	10.772	4.274	0.961	16.007	15.421	3.80 %

The state depending rates $\gamma_i^{(r)}$ ($i = 1, \ldots, \dim(D^{(r)})$, $r = 1, 2, 3$) are used as variable parameters producing different load values. The total load is obtained from the superposition of the three BMAPs, multiplying the resulting mean arrival rate λ by d. Starting with the initial values $\gamma_1^{(1)} = 2130$, $\gamma_2^{(1)} = 3500$, $\gamma_1^{(2)} = 5000$, $\gamma_2^{(2)} = 3000$, $\gamma_1^{(3)} = 1200$, $\gamma_2^{(3)} = 570$, $\gamma_3^{(3)} = 490$, 7 configurations have been produced by multiplying stepwise by 1.42. In DQDB networks the line speed usually varies between 34 and 140 Mbps, and a slot corresponds to a cell of 53 octets, such that the slot duration d varies roughly between $12 \cdot 10^{-6}$ and $3 \cdot 10^{-6}$ seconds. Here d has been chosen to $d = 12.47058 \cdot 10^{-6}$.

Table 1 shows load depending results for the mean number $E[L] = \bar{L}$ of customers (cells) in the system (waiting or being served). The summed average numbers of customers are compared with exact results, computed for the $BMAP/D/1$-system with superponed arrival streams. Figure 2 depicts the function $\bar{L}(\rho)$ in logarithmic scale, where $\bar{L}(\rho)$ is approximately determined as the sum over the stream specific numbers $\bar{L}_r(\rho)$, and also exactly computed for the superposed $BMAP/D/1$-system. Note, that the computed results are pessimistic approximations to exact ones, providing confidence in the method applied. The error continues in acceptable bounds.

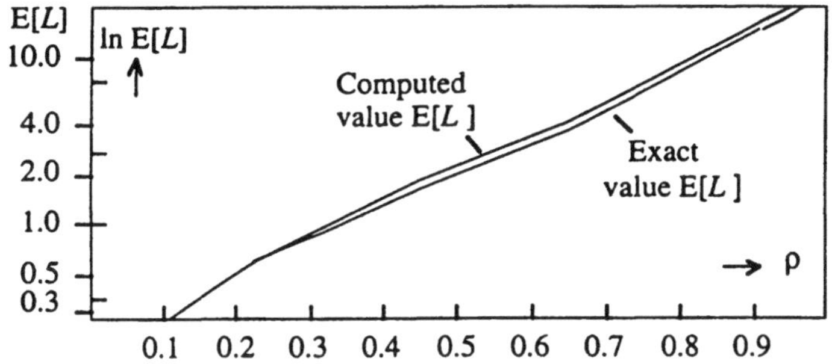

Figure 2: Computed and exact total mean number of customers in the system for different total loads.

6. Resume

This note presents a method to analyze the stream specific behaviour of infinite total / limited central capacity systems with batch Markovian arrival streams. The method is based on the determination of class specific central queue sojourn times, which allow to approximately apply $BMAP/G/1$-analyses separately for each input stream. Due to the non-cyclic (non-regenerative) dynamics as well as due to the inherent asymmetry of the model polling analysis methods are not applicable. Our method, in principle, is suitable also for systems with arbitrary capacity constraints. Serious problems, in case, may arise from numerical instabilities when performing the proposed recursive iterations for the computation of idle probabilities $P_j^{(r)}(0)$, requiring multiple runs for $BMAP/G/1$-analyses. Once these problems can be overcome, the method seems to provide fairly accurate results, as the examples show.

Acknowledgement

This work has been partly supported by a NATO grant under the project no. CRG 961111. The author wishes to thank Miss Huei-Er Yen and Mr. Lothar Breuer for performing the arduous task of numerically evaluating the above mentioned BMAP-multi-queue example.

References

1. D. Baum, in: Proc. IEEE Intern. Conf. on Communications, Vol. II, Seattle, USA (1995), pp. 788–792.
2. P. J. Kuehn, The Bell Systems Techn. Journ., 58 (3), 671–698 (1979).
3. O. J. Boxma and B. Meister, IBM Res. Rep. RZ 1348 (1984).
4. O. J. Boxma and B. Meister, Perf. Eval. Review (14), 254–262 (1986).
5. A. G. Konheim, IEEE Transactions on Communications, COM-24 (2), 203–210 (1976).
6. H. Takagi, Perf. Evaluation 5 (4), 271–277 (1985).
7. T. Takine, Y. Takahashi and T. Hasegawa, IEEE Transactions on Communications, COM-36 (10), 1119–1127 (1988).
8. D.-S. Lee and B. Sengupta, Queueing Systems 11 (1-2), 153–178 (1992).
9. Z. Tsai and I. Rubin, Queueing Systems 11 (6), 223–240 (1992).
10. K. S. Watson, in: Proc. Intern. Conf. Performance '84, E. Gelenbe (Ed.), Elsevier Science Publishers B. V., North-Holland (1984), pp. 521–533.
11. H. Takagi, Analysis of Polling Systems, MIT Press Series, Cambridge, Massachusetts (1986).
12. P. G. Potter and M. Zukerman, IEEE Journ. on Sel. Areas in Communications, 9 (2), 194–202 (1991).
13. Y. Dallery, Perform. Evaluation 11, 209–222 (1990).
14. H. G. Perros, Y. Dallery and G. Pujolle, in: Proc. IEEE INFOCOM '92 Conf., Vol. II, Florence (1992), pp. 0968–0977.
15. D. M. Lucantoni, Comm. Statist.-Stochast. Models, 7 (1), 1–46 (1991).
16. D. Baum, University of Trier Res. Rep. 96-10 (1996).
17. D. Baum, University of Trier Res. Rep. 96-22 (1996).
18. D. Baum, in: Proceedings XVIIth Intern. Seminar on Stability Problems for Stochastic Models, Debrecen, Hungary (1997), in press.
19. D. Bini and B. Meini, SIAM Journ. on Matrix Analysis and Applications, 17 (4), 906–926 (1996).

8th Int. Coll. on Differential Equations, pp. 67-71
D. Bainov (Ed)

Nonlocal Problems Modelling the Formation of Shear Bands

JERROLD BEBERNES
Department of Applied Mathematics, University of Colorado, Boulder, CO 80309-0526, USA

Abstract. The formation of shear bands in materials has important implications to a variety of physical processes. These bands are observed in very thin zones and are generally regarded as a precursor to material failure. Shear band formation is caused by the heat generated in regions with highest strain rate. With insufficient time for diffusion of this heat, a localized thermal softening of the material occurs which enhances plastic flow in a thin zone. This adiabatic strain localization can be modelled as nonlinear thermally-activated reaction-diffusion equations. This leads to a class of nonlocal parabolic problems and their associated time-independent steady-state counterparts. A discussion of the models and an analysis of the problems will be presented.

Keywords: Shear banding, localization, blowup, nonlocal parabolic partial differential equations.

1. INTRODUCTION

This lecture describes and explains some of the results found in Bebernes-Talaga [1], Bebernes-Lacey [2], and Bebernes-Li-Talaga [3]. The modelling discussed in the following is based on the work of Burns ([4],[5]).

Consider loading a thin-walled tube of metal of length d in torsion with ends held at constant temperature T_0 and the tube having initial temperature T_0. One end is fixed and the other end is twisted at a constant rate $v = v_0$. If z denotes the axial coordinate, t time, $w(z,t)$ the linear displacement, $v = w_t$ the velocity, $\gamma(z,t) = w_z(z,t)$ the shear strain, and $\tau(z,t)$ the shear stress, then the thermovisco-plastic shear model is given by the following system of conservation laws:

$$\begin{aligned} &\phi v_t = \tau_z && \text{(Momentum)} \\ &b\tau_t = v_z - \gamma_t && \text{(Elasticity)} \\ &T_z = \lambda T_{zz} + \mu^{p-1} \cdot \tau \cdot \gamma_t && \text{(Energy)} \\ &\gamma_t = \Phi(\tau, \gamma, T) && \text{(Constitutive)} \end{aligned} \tag{1.1}$$

where ϕ, b, λ, μ, and p are constants.

If $\phi << 1$ and $b << 1$, then the model simplifies to the quasi-static model:

$$\begin{aligned} &\tau_z = 0 \\ &v_z = \gamma_t \\ &T_t = \lambda T_{zz} + \mu^{p-1} \cdot T \cdot \gamma_t \\ &\gamma_t = \Phi(\tau, \gamma, T) \end{aligned} \tag{1.2}$$

from which we observe that the stress is only time-dependent, $\tau = \tau(t)$. When the stress-strain law is in the plastic regime (Marchand-Duffy, [6]), $\tau = \tau(t) = \tau_0$ is approximately constant. If the (plastic) strain rate is given by the Arrhenius law:

$$\gamma_t = v_z = \mu \exp\left(\tfrac{-\Delta H(\tau)}{KT}\right) \tag{1.3}$$

where ΔH is the activation enthalpy and K is Boltzman's constant, then, from (1.2),(1.3) the mathematical model for the shearing process reduces to a reaction-diffusion equation which describes the energy balance coupled with a compatibility equation

$$\begin{gathered} T_t - \lambda T_{zz} = \tau\mu^p \exp\left(\tfrac{-\Delta H}{KT}\right) \\ v_z = \mu \exp\left(\tfrac{-\Delta H}{KT}\right) \\ \begin{array}{ll} T(0,t) = T(d,t) = T_0 & v(0,t) = 0 \\ T(z,0) = T_0 & v(d,t) = v_0 \end{array} \end{gathered} \tag{1.4}$$

By integrating the compatibility equation, (1.4) reduces to the nonlocal problem

$$\begin{gathered} T_z = \lambda T_{zz} + \tau\mu^p \frac{\exp\left(\frac{-\Delta H}{KT}\right)}{\left(\int_0^d \exp\left(\frac{-\Delta H}{KT}\right) dz\right)^p} \\ T(0,t) = T(d,t) = T_0, \\ T(z,0) = T_0. \end{gathered} \tag{1.5}$$

This reduces to the nondimensional model

$$\begin{gathered} \theta_t - \theta_{xx} = \frac{\delta}{\left(\int_{-1}^1 \exp\left(\frac{-\beta}{1+\theta}\right) dx\right)^p} \cdot \exp\left(\frac{-\beta}{1+\theta}\right) \\ \theta(-1,t) = 0 = \theta(1,t) \\ \theta(x,0) = \theta_0(x) \geq 0 \end{gathered} \tag{1.6}$$

where $\beta = \dfrac{\Delta H(\tau_0)}{KT_0}$ and $\delta = \tau_0\mu^p \geq 0$.

It has been experimentally verified ([6]) that a typical value of β is 40 so $\epsilon = \beta^{-1} << 1$ and the reciprocal of the strain rate sensitivity can play a role in shear banding similar to the activation energy in combustion theory. Setting $\theta(x,t) = \epsilon u(x,t) + O(\epsilon^2)$, then to first-order, (1.6) becomes

$$\begin{gathered} u_t - u_{xx} = \frac{\delta}{\left(\int_{-1}^1 e^u dx\right)^p} \cdot e^u \\ u(-1,t) = 0 = u(1,t) \\ u(x,0) = u_0(x) \geq 0 \end{gathered} \tag{1.7}$$

where $\delta > 0$ and $p \geq 0$. Thus, the problem of shear band localization can be modelled by a nonlocal parabolic problem.

In the experimental study of shear band formation by Marchand and Duffy [6], a thin-walled tubular specimen of steel was loaded at a strain rate large enough to produce shear banding. During the shear band formation, temperature measurements were made and photographs taken of the specimens to provide strain measurements at several locations along the tube at different times. A narrow shear band is seen to form. This narrow band of high strain which often precedes failure in materials was seen to form near the axial midpoint on the surface of the tube as the temperature there increased dramatically.

The question of this lecture is does the nonlocal model (1.7) predict these experimental observations? Mathematically this would be answered affirmatively if solution to (1.7) blows up in finite time and if information can be given about the blow-up set.

2. NONLOCAL PROBLEMS

We are led to consider nonlocal problems of the form

$$\begin{aligned} u_t - \Delta u &= \frac{\delta f(u)}{(\int_\Omega f(u)dx)^p} \quad , \qquad x \in \Omega,\ t > 0 \\ u(x,0) &= u_0(x) \geq 0 \quad , \qquad x \in \Omega \\ u(x,t) &= 0 \quad , \qquad x \in \partial\Omega,\ t > 0 \end{aligned} \tag{2.1}$$

and the associated steady-state problem

$$\begin{aligned} -\Delta u &= \frac{\delta f(u)}{(\int_\Omega f(u)dx)^p} \quad , \qquad x \in \Omega \\ u(x) &= 0 \quad , \qquad x \in \partial\Omega \end{aligned} \tag{2.2}$$

where $p \geq 0$, $\delta > 0$, $\Omega \subset I\!R^n$ is a bounded domain with smooth boundary $\partial\Omega$, f is positive and Lipschitz continuous and $u_0(x) \geq 0$ is in $L^2(\Omega)$ with $u_0(x) = 0$ on $\partial\Omega$. The following standard results for classical nonlinear partial differential equations carry over to the nonlocal problems (2.1) and (2.2):

1) Any solution of IBVP(2.1) or BVP(2.2) is positive for $x \in \Omega$ with outer normal derivative $\frac{\partial u}{\partial N} \leq 0$, $x \in \partial\Omega$.

2) For $u_0 \in L^2(\Omega)$, $\sup u_0(x) < \infty$, IBVP(2.1) has a unique, nonextendable classical solution $u(x,t)$ on $\Omega \times [0,T)$ where either $T = +\infty$ or $T < +\infty$ and $\lim_{t\to T} \sup_{x\in\Omega} u(x,t) = +\infty$.

3) For $\Omega = B_1(0)$, (a) any solution of BVP(2.2) is radially symmetric and radially decreasing; (b) if $u_0(x)$ is radially symmetric and radially decreasing, then the solution of $u(x,t)$ of IBVP(2.1) is also for each $t \in [0,T)$.

3. EXISTENCE-NONEXISTENCE FOR BVP (2.2).

Let $f(u) = e^u$. The following theorems are proven in [2]. Nonexistence for BVP(2.2) should give information about nonexistence of global solutions for IBVP(2.1).

Theorem 1. For $\Omega = B_1(0) \subset I\!R^1$:

a) if $p \geq 1$, BVP (2.2) has a unique solution for all $\delta > 0$;

b) if $0 \leq p < 1$, then there exists $\delta^* > 0$ such that BVP (2.2) has: i) two solutions for $\delta < \delta^*$; ii) one solution for $\delta = \delta^*$; and iii) no solution for $\delta > \delta^*$.

Theorem 2. For $\Omega = B_1(0) \subset I\!R^2$,

a) if $p > 1$, BVP(2.2) has a unique solution for all $\delta > 0$.

b) if $p = 1$, BVP(2.2) has a unique solution for $\delta < \delta^* = 8\pi$ and no solution for $\delta \geq \delta^*$.

c) if $0 \leq p < 1$, then there exists $\delta^* > 0$ such that BVP(2.2) has: i) two solutions for $\delta < \delta^*$, ii) a unique solution for $\delta = \delta^*$, and iii) no solution for $\delta > \delta^*$.

By using Pohozaev's identity, nonexistence results can be extended to strictly star-shaped domains. A domain $\Omega \subset I\!R^n$, $n \geq 2$, is *strictly star-shaped* (containing 0) if there exists $a > 0$ such that

$$x \cdot N \geq a \int_{\partial\Omega} ds, \quad \text{for } x \in \partial\Omega,\ N \text{ unit outer normal.}$$

Theorem 3. For $p \leq 1$, spatial dimension $n \geq 2$, BVP(2.2) has no solution for $\delta > \frac{2n}{a}|\Omega|^{p-1}$.

4. FINITE TIME BLOWUP

Consider IBVP(2.1) and the associated steady-state problem BVP(2.2) when $f(u) = e^u$, $p < 1$, $n = 1$ or 2, and $\Omega \subset I\!R^n$ such that Theorem 1,2, or 3 is valid. Then there exists a critical $\delta^* > 0$ such that for $\delta > \delta^*$, no solution of BVP(2.2) exists.

For spatial dimensions $n = 1$ or 2, IBVP(2.1) defines a local semiflow in $H_0^1(\Omega)$. For $0 < p < 1$, this local semiflow has a Lyapanov functional given by

$$V[u](t) = \frac{1}{2}\int_\Omega |\nabla u|^2 dx + \frac{\delta}{p-1}\left(\int_\Omega e^u dx\right)^{1-p} \tag{4.1}$$

and the semiflow is gradient-like in the sense that for any $t \in [0, T)$

$$\int_0^t ||u_t||_2^2 + V[u](t) = V[u_0]. \tag{4.2}$$

Theorem 4. For $\delta > \delta^*$, the solution $u(t, u_0)$ of IBVP(2.1) blows up in finite time $T < \infty$.

The proof is given in [2] and is based on ideas of Marek [7]. An outline of the proof can be given by stating the following five lemmas.

Lemma 5. If u is a global solution of IBVP(2.1), then there exists $\kappa = \kappa(u_0)$ such that

$$||u(t, u_0)||_2 = \left(\int_\Omega |u|^2 dx\right)^{\frac{1}{2}} \leq \kappa \quad \text{for all } t \geq 0.$$

Lemma 6. If $||u(t, u_0)|| \equiv \left(\int_\Omega |\nabla u|^2 dx\right)^{\frac{1}{2}} \to \infty$ as $t \to t_m$, then $t_m < \infty$.

Lemma 7. If $u(t, u_0)$ is a global solution with the ω-limit set $\omega(u_0) \neq \phi$, and if $w \in \omega(u_0)$ is any equilibrium solution, then $||w|| \leq K(u_0)$.

Lemma 8. If $u(t, u_0)$ is a global solution with

$$\liminf_{t\to\infty} ||u(t, u_0)|| < \infty, \quad \limsup_{t\to\infty} ||u(t, u_0)|| = \infty,$$

then for any B sufficiently large, there exists an equilibrium solution w with $||w|| = B$.

Lemma 9. If $u(t, u_0)$ is global, then $\sup_{t\geq 0} ||u(t, u_0)|| < \infty$ and $\sup_{t\geq \tau} |u(t, u_0)| < \infty$ for any $\tau > 0$.

Lemma 9 follows from the previous lemmas. For if $u(t, u_0)$ is global, then, by Lemma 5, $||u(t, u_0)|| \not\to \infty$. By Lemmas 7 and 8, $\limsup_{t\to\infty} ||u|| \neq +\infty$ so $\sup_{t\geq 0} ||u(t, u_0)|| < \infty$. But then Lemma 9 implies solution of BVP(2.2) exists. But, by assumption $\delta > \delta^*$; therefore no such solution exists. Thus $u(t, u_0)$ must blow up in finite time.

5. SINGLE POINT BLOWUP

Theorem 4 for $\Omega = (-1, 1) \subset I\!R^1$, $\delta > \delta^*$, $p < 1$ where $u(x, t)$ represents the temperature perturbation for the shear banding model (1.7) tells us that thermal runaway or blowup occurs in finite time T. The question of where blow up occurs is answered by the following theorem.

Theorem 10. For $\Omega = B_1(0)$, initial data $u_0(x)$ radially symmetric and decreasing, $f(u) = e^u$, $n = 1$ or 2, $p < 1$, and $\delta > \delta^*$, then the blow up set for IBVP(2.1) consists of the single point $x = 0$.

The proof of this Theorem is given in [3].

References

1. J. Bebernes and P. Talaga, Comm. Appl. Nonlinear Anal., 3, 79-103 (1996).

2. J. Bebernes and A. Lacey, Advances in Diff. Equations, in press.

3. J. Bebernes, C. Li and P. Talaga, Single Point Blowup, submitted.

4. T.J. Burns, in: Shock Induced Transitions and Phase Structures in General Media, R. Fosdick et al., editors, Springer-Verlag, New York, (1992) Chapter 2.

5. T.J. Burns, Mech. Mat., 17, 261-271 (1994).

6. A. Machand and J. Duffy, J. Mech. Phys. Solids, 36, 251-283 (1988).

7. M. Fila, J. Differential Equations, 98, 226-240 (1992).

8th Int. Coll. on Differential Equations, pp. 73–79
D. Bainov (Ed)

Solution of the Dirichlet Problem for Laplace Equation on Domains with Analytic Boundaries

SERGEI BELOV AND NOBUO FUJII
Osaka Sangyo University, 3-1-1 Nakagaito, Daito-shi, Osaka 574, Japan
belov@h-ps011.ise.osaka-sandai.ac.jp, fujii@ise.osaka-sandai.ac.jp

Abstract. The boundary-value problems of Dirichlet type for Laplace equation on the star-shaped domains with analytic boundaries have been studied in this paper. The solution has been obtained as the sum of the infinite series of harmonic functions connected by the recurrence. The theorem about convergence of the series to the solution of the boundary-value problem has been established. The examples which illustrate applicability of the new method in practice are given.

Key words: Boundary-value problem, recurrence, variation of the solution, exact solution.

Introduction

In this paper we will consider the following problem: Let Ω_f be simply connected domain in R^2 with the analytic boundary Γ_f the equation of which in the polar coordinates is

$$r = f(\phi), \tag{1}$$

where f is an analytic function of the variable ϕ and the inequality $f(\phi) > 0$ holds for $\forall\phi$; we consider the boundary-value problem:

$$\Delta u_f = 0 \quad (x \in \Omega_f), \tag{2}$$
$$u_f = g \quad (x \in \Gamma_f). \tag{3}$$

Boundary condition (3) means that in the polar coordinates (r, ϕ),

$$u_f(f(\phi), \phi) = g(\phi), \tag{4}$$

where the boundary function g is also assumed to be analytic .

The problem (2), (3) is well-known and various results has been obtained. However, if someone tries to find the solution of this problem with some fixed function f he encounters with a lot of difficulties, for example, how to find Green's function or the conformal mapping of the domain Ω_f onto a circle .

As the main result of this paper we will obtain the formula for the solution of the problem (2), (3) in terms of f and g. For this purpose we propose a new approach to solving boundary-value problems. The main idea is to obtain the analogue of Taylor series with a circle as an initial point and the derivatives (variations of the solution) with respect to the boundary. Note here that Hadamard [1] was the first who obtained the formula for the first derivative of the Green's function with respect to the domain, using outward normal to describe a boundary variation. Present authors have obtained the formulas for first and second derivatives of the solution with respect to the domain (in various boundary-value problems) for the purposes of shape optimization [2], [3], [4], [5]. It should be noted that the first derivative of the solution was frequently used in shape optimization for quite a long time.

In this paper we will give for the first time the formula for the nth derivative of the solution of the boundary-value problem (2), (3) which will allow us to write out the formal "Taylor" series with a circle as an initial point. Then we will prove the theorem about convergence of this series to the solution u_f. Also, we will show how to apply the general formula to practical boundary-value problems.

Main result

First of all, let us reformulate the problem (2), (3). We can easily see that there exist a constant M such that $0 < f(\phi) < M, 0 \leq \phi \leq 2\pi$, since the function f is analytic. Without loss of generality we can put $M = 1$. Therefore we can rewrite the equation of the boundary (1) in the form

$$r = f(\phi) \equiv 1 + h(\phi), \qquad -1 < h(\phi) < 0. \tag{5}$$

Let Ω be the unit circle with its center at the origin, which contains the domain Ω_f. We denote the boundary of this circle by Γ. We can consider the unit circle Ω as the initial domain and the domain Ω_f as perturbed domain. Now we consider the boundary-value problem

$$\Delta F_0 = 0 \quad (x \in \Omega), \tag{6}$$

$$F_0 = g \quad (x \in \Gamma). \tag{7}$$

The formulas for the solution of this problem are well-known. Having solved the problem (6), (7) we can consider next boundary-value problem, defined on the unit circle Ω.

$$\Delta F_1 = 0 \quad (x \in \Omega), \tag{8}$$

$$F_1 = -\frac{\partial F_0}{\partial r} h \quad (x \in \Gamma). \tag{9}$$

The function F_1 is known in shape optimization theory as the first derivative (variation) of the solution with respect to domain [4]. It is obvious, that functions F_0, F_1 are also analytic. Now we define the function $F_n, n = 1, 2, \dots$ using the recurrence

$$\Delta F_n = 0 \quad (x \in \Omega), \tag{10}$$

$$F_n = -\sum_{j=1}^{n} \frac{1}{j!} \frac{\partial^j F_{n-j}}{\partial r^j} h^j \quad (x \in \Gamma). \tag{11}$$

It is not difficult to show that the F_n corresponds to the $n - th$ variation of the solution with respect to the boundary (the case $n = 2$ has been considered in [5]). But we will skip the proof of this fact, because it is out of focus of this paper.

Theorem 1. Let the series

$$\sum_{n=1}^{\infty} F_n \tag{12}$$

converges uniformly and absolutely on the boundary Γ of the unit circle Ω. Then $\sum_{i=0}^{\infty} F_i$ converges uniformly on $\overline{\Omega_f}$ to the solution of the boundary-value problem (2), (3):

$$u_f = \sum_{i=0}^{\infty} F_i, \qquad (x \in \overline{\Omega_f}). \tag{13}$$

Proof. Since all the functions F_n are continuous in $\overline{\Omega}$ and harmonic in Ω, the uniform convergence of the series $\sum_{i=0}^{\infty} F_i$ on the boundary Γ implies uniform convergence to harmonic function in $\overline{\Omega}$ (and consequently in $\overline{\Omega_f}$, $\Omega_f \subseteq \Omega$) according to well-known theorem. (see [6], pp. 192-193)

Let us now consider an arbitrary point x_f on the boundary Γ_f. We denote x the correspondent point on Γ which lies on the same polar ray as x_f does; the distance between x and x_f is equal to $|h(\phi)|$. Let consider the table (a_{ik}) which is infinite in both directions and has the elements $a_{ik} = \dfrac{1}{k!}\dfrac{\partial^k}{\partial r^k}(F_i(x))h^k(x)$. The sum of the elements of the i-th line is

$$F_i(x_f) - F_i(x) = \sum_{k=1}^{\infty} \frac{1}{k!}\frac{\partial^k}{\partial r^k}(F_i(x))h^k(\phi)) \tag{14}$$

and Taylor series in the right hand converges absolutely [6]. Moreover, since we have proved that the series $\sum_{i=0}^{\infty} F_i$ converges uniformly on $\overline{\Omega}$ we have that the series

$$\sum_{i=0}^{\infty}(F_i(x_f) - F_i(x)), \tag{15}$$

which corresponds the sum of all the lines of the table, also converges. Now we can see that the sum of the elements on the n-th diagonal of the table a_{ik} is equal to $-F_n(x)$ in (11). Changing the order of summation we will have

$$\begin{aligned} \sum_{n=0}^{\infty} F_n(x) &= F_0 - \sum_{n=1}^{\infty}\sum_{j=1}^{n} \frac{1}{j!}\frac{\partial^j F_{n-j}}{\partial r^j}(x)h^j(x) = F_0 - \sum_{i=0}^{\infty}\sum_{k=1}^{\infty} \frac{1}{k!}\frac{\partial^k}{\partial r^k}(F_i(x))h^k(x) \\ &= F_0(x) - \sum_{i=0}^{\infty}(F_i(x_f) - F_i(x)) = F_0(x) + \sum_{i=0}^{\infty} F_i(x) - \sum_{i=1}^{\infty} F_i(x_f). \end{aligned} \tag{16}$$

Interested reader can verify that the rearrangement of the terms leaves the sum of the series unchanged. Comparing the left hand and the right hand of (16), eliminating the equal terms we obtain

$$\sum_{i=0}^{\infty} F_i(x_f) = F_0(x) = g(x_f). \tag{17}$$

Applying the uniqueness theorem we have $u_f = \sum_{i=0}^{\infty} F_i$. Theorem 1 is thereby proved.

Examples.

Example 1. Let us consider an illustrative example for which we know the solution a priori:

$$\Delta u_f(x) = 0 \qquad (x \in \Omega_f), \tag{18}$$

$$u_f(x) = \cos\phi \qquad (x \in \Gamma_f), \tag{19}$$

$$\Gamma_f : r = 1 + a \qquad (-1 < a < 1). \tag{20}$$

In accordance with our method we will construct the sequence $F_n, n = 0, \cdots \infty$. For the function F_0 we have

$$\Delta F_0(x) = 0 \quad (x \in \Omega), \tag{21}$$
$$F_0(x) = \cos\phi \quad (x \in \Gamma), \tag{22}$$

where Ω is the unit circle. Hence, $F_0 = r\cos\phi$. Now we can obtain the function F_1 as the solution of the boundary-value problem

$$\Delta F_1(x) = 0 \quad (x \in \Omega), \tag{23}$$
$$F_1(x) = -a\cos\phi \quad (x \in \Gamma). \tag{24}$$

Hence, we have $F_1 = -ar\cos\phi$ and we can consequently obtain all the functions F_n. It is obvious that

$$F_n = (-a)^n r\cos\phi \tag{25}$$

and the series $\sum_{n=0}^{\infty} F_n$ converges uniformly and absolutely on Γ. Hence, in accordance with Theorem 1, it converges to the solution of the boundary-value problem (18), (19):

$$u_f = \sum_{n=0}^{\infty}(-a)^n r\cos\phi = \frac{r\cos\phi}{1+a}. \tag{26}$$

Remark 1. Note, the method works even in the case of positive a. Indeed, since in this case all the function F_n is defined not only on Ω but also on Ω_f, and we can prove the result similar to Theorem 1 using the convergence of the series $\sum_{i=0}^{\infty} F_i(x_f)$.

Example 2. Now, using Theorem 1, we will solve more complicated problem:

$$\Delta u_f(x) = 0 \quad (x \in \Omega_f), \tag{27}$$
$$u_f(x) = \cos\phi \quad (x \in \Gamma_f), \tag{28}$$
$$\Gamma_h : r = 1 + a\cos\phi \quad (-1 < a < 1). \tag{29}$$

For $h = a\cos\phi$ we have from (6),(7)

$$F_0 = r\cos\phi. \tag{30}$$

Now we can use F_0 for obtaining F_1. We have

$$\Delta F_1 = 0 \quad (x \in \Omega), \tag{31}$$
$$F_1 = -\frac{\partial F_0}{\partial r} a\cos\phi = -a(\cos\phi)^2 \quad (x \in \Gamma). \tag{32}$$

Hence,

$$F_1 = -\frac{a}{2} - \frac{a}{2} r^2 a\cos 2\phi \tag{33}$$

and now we can obtain F_2, etc. After some calculations we can obtain the general formula for $F_n, n \geq 2$:

$$F_n = (-1)^n \frac{a^n}{2^n} \sum_{i=0}^{n+1} r^i b_n^i \cos i\phi, \tag{34}$$

where $b_n^i = 0$, if $n-i$ is even and, otherwise,

$$b_n^i = \frac{(i+1)(i+2)...(i+n-1)}{n!} C_n^{\frac{n+1-i}{2}}, \tag{35}$$

where C_k^m are binomial coefficients.

We will prove the formula (34) by induction on n. Let the formula (34) be true for $k < n$ with coefficients given by (35) $0 \le i \le k+1$. We have to prove this formula for $k = n$, with coefficients $b_n^i, 0 \le i \le n+1$. We will consider the coefficient b_n^{n+1} (the other cases can be treated by the similar way).

First of all we note that $b_k^{k+1} = \frac{C_{2k}^k}{k+1}$ are exactly Catalan numbers. Substituting (34) in the right hand of (11) we can see that only coefficients $b_{n-j}^{n+1-j}, 1 \le j \le n$ are used to obtain b_n^{n+1}. Let us consider correspondent expression in $\frac{1}{j!}\frac{\partial^j F_{n-j}}{\partial r^j}h^j$. We have

$$\begin{aligned} &\frac{1}{j!}\frac{\partial^j}{\partial r^j}\left((-1)^{n-j}\frac{a^{n-j}}{2^{n-j}}r^{n+1-j}b_{n-j}^{n+1-j}\cos(n+1-j)\phi\right)\Bigg|_{r=1} a^j(\cos\phi)^j \\ &= (-1)^{n-j}\left(\frac{a^n}{2^{n-j}}C_{n+1-j}^j b_{n-j}^{n+1-j}\cos(n+1-j)\phi\right)(\cos\phi)^j. \end{aligned} \tag{36}$$

If we represent $(\cos\phi)^j$ as

$$(\cos\phi)^j = \sum_{i=0}^{j} d_i \cos i\phi \tag{37}$$

we can easily see that $d_j = \frac{1}{2^{j-1}}$. Note here, that we can write out now the expression which stands with $\cos(n+1)\phi$ in $\frac{1}{j!}\frac{\partial^j F_{n-j}}{\partial r^j}h^j$. Comparing all such terms for different j in (11) we will come to equality

$$B_n^{n+1} = -\sum_{j=1}^{n}(-1)^{n-j}\frac{a^n}{2^n}C_{n+1-j}^j b_{n-j}^{n+1-j} \tag{38}$$

where B_n^{n+1} is the coefficient for $\cos(n+1)\phi$ in F_n of (11). Substituting b_{n-j}^{n-j+1} from (35) and eliminating $(\frac{a}{2})^n$ we will have following equality to be proved:

$$\sum_{j=0}^{n}(-1)^{n-j}C_{n+1-j}^j\frac{1}{n+1-j}C_{2(n-j)}^{n-j} = 0, \tag{39}$$

where $j = 0$ corresponds b_n^{n+1}. As far as the authors know the equality (39) which connects Catalan numbers has not been obtained before.

Now we give the proof of this formula. Let us consider the sum in (39). First of all we consider the finite sum and infinite series ([7] pp. 520, 611)

$$\sum_{j=0}^{n}(-1)^j C_{n+1-j}^j x^j = \frac{(1+\sqrt{1-4x})^{n+2}-(1-\sqrt{1-4x})^{n+2}}{2^{n+2}\sqrt{1-4x}} \tag{40}$$

$$\sum_{j=0}^{\infty} C_{2j}^j x^j = \frac{1}{\sqrt{1-4x}}. \tag{41}$$

After integrating the series we obtain

$$\sum_{j=0}^{\infty}\frac{C_{2j}^j}{j+1}x^{j+1} = -\frac{\sqrt{1-4x}}{2}+\frac{1}{2}. \tag{42}$$

If we multiply the sum (40) and the series (42) we can see that the coefficient of x^{n+1} of the new series coincides with the left hand of (39), where $(-1)^j$ stands instead of $(-1)^{n-j}$. Therefore we have to prove that this coefficient is equal to 0. We have

$$\frac{(1+\sqrt{1-4x})^{n+2}-(1-\sqrt{1-4x})^{n+2}}{2^{n+2}\sqrt{1-4x}}\left(-\frac{\sqrt{1-4x}}{2}+\frac{1}{2}\right)=\frac{1}{2}\sum_{j=0}^{n}(-1)^j C_{n+1-j}^{j}x^j$$
$$-\frac{(1+\sqrt{1-4x})^{n+2}-(1-\sqrt{1-4x})^{n+2}}{2^{n+3}}. \tag{43}$$

In order to show that the coefficient of x^{n+1} is equal to 0 we have to consider only the second term in the right hand of (43) and we can change the sign in this expression, because the coefficient of x^{n+1} in $(1-\sqrt{1-4x})^{n+2}$ is equal to 0. Now if we take the first derivative we will obtain

$$\frac{\partial}{\partial x}\left((1+\sqrt{1-4x})^{n+2}+(1-\sqrt{1-4x})^{n+2}\right)$$
$$=-2(n+2)\frac{(1+\sqrt{1-4x})^{n+1}-(1-\sqrt{1-4x})^{n+1}}{\sqrt{1-4x}}$$
$$=-2^{n+2}(n+2)\sum_{j=0}^{n-1}(-1)^j C_{n-j}^{j}x^j. \tag{44}$$

Now it becomes obvious that the coefficient of x^{n+1} in

$$\left((1+\sqrt{1-4x})^{n+2}-(1-\sqrt{1-4x})^{n+2}\right) \tag{45}$$

is equal to 0 and the equality (39) is thereby proved.

Now, applying Weierstrass M-test and Stirling formula to F_n we immediately obtain uniform and absolute convergence of the series $\sum_{i=0}^{\infty}F_i$ on Γ, if $|a|<\frac{1}{4}$. Applying Theorem 1 we will get the formula for the solution of the boundary-value problem (27), (28):

$$u_f=\sum_{i=0}^{\infty}F_i, \tag{46}$$

where F_i are given by (30), (33), (34) (see also Remark 1).

It should be noted that it is easy to realize this method on the computer and it is also easy to get the estimates on the whole domain, using maximum principle, since we approximate the solution by harmonic functions.

References

1. M. J. Hadamard, in: Qeuvres de Jacques Hadamard, CNRS, Paris, France (1968) Vol. 2, pp. 281-368.

2. S. A. Belov and N. Fujii, Control and Cybernetics, 26, No. 1, 45-56 (1997).

3. S. A. Belov and S. N. Sluguin, Journal of Optimization Theory and Applications, 90, No. 2, 243-255 (1996).

4. N. Fujii, SIAM Journal on Control and Optimization, 24, No. 3, 346-360 (1986).

5. N. Fujii, Journal of Optimization Theory and Applications, 65, No. 2, 223-244 (1990).

6. M.A. Lavrentiev and B. V. Shabat, Methods of Theory of Complex Variable, Nauka, Moscow (1987).

7. K. D. Joshi, Foundations of Discrete Mathematics, Jonh Wiley and Sons, New York (1989)

8th Int. Coll. on Differential Equations, pp. 81-88
D. Bainov (Ed)

Ill-Conditioning in Optimal Control Computation

F. BENYAH AND L. S. JENNINGS
Centre for Applied Dynamics and Optimization, Department of Mathematics, University of Western Australia, WA 6907, Australia

Abstract. Optimal control computations are well known to be ill-conditioned, but there is little quantifiable knowledge on the topic. The mathematical machinery for the quantification of ill-conditioning in optimal control computation is presented. Implementation in the optimal control software MISER3 and computation of test examples are discussed. Some idea of how the condition numbers vary with the number of discretization parameters for controls of typical problems is given. Various constraint types are considered and a comparison of two methods for computing smooth controls is presented.
Keywords: optimal control, ill-conditioning, control parametrization, perturbations.

1 INTRODUCTION

The computation of solutions to optimal control problems is known to be ill-conditioned. That is, relatively large variations of parameter values are allowed for small variations of constraints and/or objective values. This is in fact a good engineering design feature, in that if the parameters represent in some way the operating parameters of some plant, then as the plant wears out and the optimal parameters need to change, the constraints and/or objective may only change a little. Putting a number on the extent of ill-conditioning for optimal control problems is more technical than putting a condition number on a set of linear equations. The reason is that constrained optimization algorithms are involved. Very few of these algorithms or their codes produce a condition number, be it a good estimate, a lower bound, an upper bound or a bad estimate. There are good computational reasons for this.

In this paper the mathematical background to the optimal control code MISER3 [1] which computes solutions using the control parametrization method is presented in sections 2 and 3. In section 4 the perturbation analysis for constrained optimization is presented in outline form, including its implementation in MISER3. To demonstrate its use some test problems are computed. In section 5 the condition number as a function of number of control parameters per control is investigated. Two methods for computing smooth controls are also investigated to see which has the smaller condition number. The folklore relied upon here is that smaller condition numbers imply better performance for constrained optimization codes.

2 OPTIMAL CONTROL PROBLEM TEMPLATE FOR *MISER3*

A form of the combined optimal control and optimal parameter selection problem is:

$$\text{(2.1)} \qquad \underset{\boldsymbol{u},\boldsymbol{z}}{\text{minimize}} \left\{ G_0(\boldsymbol{u},\boldsymbol{z}) = \phi_0(\boldsymbol{x}(t_f),\boldsymbol{z}) + \int_{t_s}^{t_f} g_0\left(t,\boldsymbol{x}(t),\boldsymbol{u}(t),\boldsymbol{z}\right) dt \right\}$$

subject to the state differential equation and initial conditions

$$\text{(2.2)} \qquad \dot{\boldsymbol{x}}(t) = \boldsymbol{f}\left(t,\boldsymbol{x}(t),\boldsymbol{u}(t),\boldsymbol{z}\right), \quad \boldsymbol{x}(0) = \boldsymbol{x}^0(\boldsymbol{z})$$

where $\boldsymbol{x}(t) = [x_1(t),\ldots,x_{n_s}(t)]^T \in R^{n_s}$, $\boldsymbol{u}(t) = [u_1(t),\ldots,u_{n_c}(t)]^T \in R^{n_c}$, and $\boldsymbol{z} = [z_1,\ldots,z_{n_z}]^T \in R^{n_z}$ are, respectively, the state function vector, the control function vector and system parameter vector. MISER3 is capable of dealing with a wide range of constraints. In the classical optimal control literature, constraints of different types are often treated differently. In the present unified framework, three standard forms of constraints are used. Any constraints involving the state variables or nonlinear in control functions are transformed into the standard canonical form:

$$\text{(2.3)} \quad G_k(\boldsymbol{u},\boldsymbol{z}) = \phi_k(\boldsymbol{x}(\tau_k),\boldsymbol{z}) + \int_{t_s}^{\tau_k} g_k\left(t,\boldsymbol{x}(t),\boldsymbol{u}(t),\boldsymbol{z}\right) dt \ \begin{smallmatrix} = \\ \geq \end{smallmatrix} \ 0, \quad k = 1,2,\ldots,n_{g_c}$$

where n_{g_c} is the total number of canonical constraints, and $\tau_k \in (0,t_f]$ is a known constant and is referred to as the 'characteristic time' associated with the constraint G_k. Note that $\tau_0 = t_f$ for the cost functional. The other two forms of constraints are firstly those which only involve the system parameters, and secondly those all-time linear constraints involving control functions only. Both of these forms are special in that they do not need costate variables to compute their gradients.

$$\text{(2.4)} \qquad G(\boldsymbol{z}) \ \begin{smallmatrix} \geq \\ = \end{smallmatrix} \ 0, \quad \text{and} \quad g(t) = \sum_{i=1}^{n_c} \alpha_i u_i(t) + \beta \ \begin{smallmatrix} \geq \\ = \end{smallmatrix} \ 0.$$

There are likely to be finite upper and lower bounds on all control functions and system parameters. More complicated upper and lower bounds on control which depend on time t usually have to be transformed into a constraint of type (2.3).

$$\text{(2.5)} \qquad u_i^L \leq u_i(t) \leq u_i^U, \quad \forall t \in [t_s,t_f], \quad i = 1,\ldots,n_c$$

$$\text{(2.6)} \qquad z_j^L \leq z_j \leq z_j^U, \quad j = 1,\ldots,n_z.$$

3 CONTROL PARAMETRIZATION

The central idea behind MISER3 is the concept of control parametrization, see section 2.6 of [1]. Essentially, each control component $u_i(t)$ is approximated by a zero-th order or first order spline function (that is, a piecewise constant function or a piecewise linear continuous function) defined on a set of knots $\left\{t_s = t_0^i, t_1^i, t_2^i, \ldots, t_{k_i}^i = t_f\right\}$, which could be unequally spaced and different for each control. Let $i_0 = 1$ for the case of piecewise constant basis functions and $i_0 = 0$ for piecewise linear continuous basis functions. Formally we write the i-th control as a sum of basis functions with coefficients or parameters $\{\sigma_{ij}, j = i_0,\ldots,k_i\}$

$$\text{(3.1)} \qquad u_i(t) = \sum_{j=i_0}^{k_i} \sigma_{ij} B_{ij}(t)$$

where $B_{ij}(t)$ is, for piecewise constant control, the indicator function for the j-th interval of the i-th set of knots, or for piecewise linear continuous control, the witches' hat functions. These basis functions have finite support and have the added property that the $\sigma_{ij} = u_i(t_j)$. This latter property is important for constraints of the form (2.5) and hence the direct refinement to (3.4), as well as for the all-time linear constraints of control only. Higher degree splines present further problems as we cannot have both small finite support and $\sigma_{ij} = u_i(t_j)$. We suspect also that the ill-conditioning inherent in optimal control problems could get worse if higher degree splines are used to approximate the controls. Note that the system parameters are constant for all t. After the control parametrization, the cost functional (2.1) as well as all the constraint functionals (2.3) - (2.6) can be regarded as (somewhat implicit) functions of parameter vector $\boldsymbol{\theta}$. More precisely, the corresponding approximate problem may be written as:

$$\text{minimize}_{\boldsymbol{\theta}} \tilde{G}_0(\boldsymbol{\theta}) \tag{3.2}$$

subject to the constraints:

$$\tilde{G}_i(\boldsymbol{\theta}) = 0, \quad i = 1, \ldots, n_e \qquad \tilde{G}_i(\boldsymbol{\theta}) \geq 0, \quad i = n_e + 1, \ldots, n_g \tag{3.3}$$

where $\boldsymbol{\theta} = [\sigma_{11_0}, \ldots, \sigma_{1k_1}, \sigma_{22_0}, \ldots, \sigma_{2k_2}, \ldots, \sigma_{n_c n_{c0}}, \ldots, \sigma_{n_c k_{n_c}}, z_1, z_2, \ldots, z_{n_z}^T \in R^{n_p}$. The control parameters and system parameters are bounded by a more general set of upper and lower bounds than (2.5) and (2.6):

$$u_{ij}^L < \sigma_{ij} < u_{ij}^U, \quad j = i_0, \ldots, k_i, \ i = 1, \ldots, n_c, \tag{3.4}$$

$$z_k^L \leq z_k \leq z_k^U, \quad k = 1, \ldots, n_z. \tag{3.5}$$

Clearly, this approximate problem can be regarded as a standard nonlinearly constrained mathematical programming problem. Note that bounds on σ_{ij} automatically put bounds on the control functions for all time as appropriate. This does not happen for higher order spline basis functions, another reason for not considering such basis functions. The problem posed by (3.2)-(3.5) is now a standard constrained nonlinear mathematical programming problem which can be solved by means of a sequential quadratic programming technique. See, for example, [2], [3]. Like many nonlinear programming techniques, these require the analytical gradients of the cost function $\tilde{G}_0$ as well as the constraint functions $\tilde{G}_k$, $k = 1, \ldots, n_{gc}$. However, due to the non-explicit dependency of $\tilde{G}_k$, $k = 0, \ldots, n_{gc}$ on $\boldsymbol{\theta}$ when state functions are present, the computation of gradients of canonical functionals has to be done in a somewhat roundabout way. See [1] and [4].

4 CONDITION NUMBERS FOR CONSTRAINED OPTIMIZATION

Sensitivity to perturbations for optimization problems can be reduced to considering a linear system linearized from the gradient of the Lagrange function. Assume the tangent planes of the active constraints are linearly independent and that there is a unique minimizer $\boldsymbol{x}^*$. Write the objective $f(\boldsymbol{x}^*)$ and constraints $\boldsymbol{h}(\boldsymbol{x}^*)$ in Taylor series about a computed solution point $\boldsymbol{x}$, where $\boldsymbol{x}^* = \boldsymbol{x} + \boldsymbol{z}$,

$$f(\boldsymbol{x}^*) \approx f(\boldsymbol{x}) + \nabla f(\boldsymbol{x})\boldsymbol{z} + \frac{1}{2}\boldsymbol{z}^T \boldsymbol{F}(\boldsymbol{x})\boldsymbol{z}, \quad \text{and} \quad 0 = \boldsymbol{h}(\boldsymbol{x}^*) \approx \boldsymbol{h}(\boldsymbol{x}) + \nabla \boldsymbol{h}(\boldsymbol{x})\boldsymbol{z},$$

where $F(x)$ is the Hessian of f at x. The perturbations to be considered are in those variables used to determine convergence of any optimization algorithm, namely, the constraint values and the Lagrange condition for optimality, the latter being the projected gradient values and the constraint values. For a constrained problem the sensitivity of Lagrange multipliers (and hence the Kuhn-Tucker conditions) to perturbations is also of interest. Let $H = \nabla h(x)$, $g(x) = \nabla f(x)^T$ and the non-compliance of the Kuhn-Tucker conditions be ζ, where at the computed solution point x,

(4.1) $$g(x) + H^T(x)\lambda = \zeta.$$

The true Lagrange multipliers are denoted by λ^*, and are assumed close to the computed value of the Lagrange multipliers λ. Let $\xi = \lambda^* - \lambda$, and consider the Taylor expansion of $g(x^*)$ about the computed solution point x. Then from the Kuhn-Tucker condition, $g(x^*) + H^T(x^*)\lambda^* = 0$, we get

$$\begin{aligned} g(x) + F(x)z + H^T(x^*)\lambda^* &= -H^T(x)\lambda + \zeta + F(x)z + H^T(x^*)\lambda^* \\ &= F(x)z + \left(H^T(x^*) - H^T(x)\right)\lambda + H^T(x^*)(\lambda^* - \lambda) + \zeta \\ &= \left(F(x) + \textstyle\sum_i \lambda_i \frac{\partial H_i}{\partial x}(x)\right) z + H^T(x^*)\xi + \zeta \\ &= \mathcal{O}(\|z\|^2). \end{aligned}$$

Ignoring second order small quantities,

(4.2) $$\begin{bmatrix} F(x) + \sum_i \lambda_i \frac{\partial H_i}{\partial x}(x) & H(x^*)^T \\ H(x) & 0 \end{bmatrix} \begin{bmatrix} z \\ \xi \end{bmatrix} = \begin{bmatrix} -\zeta \\ -h \end{bmatrix}.$$

This equation now gives the correction to the computed solution vector and Lagrange multipliers in terms of the measurable errors in the Kuhn-Tucker conditions and the constraint compliance. To this order of small quantities the (unknown) term $H(x^*)$ can be replaced by $H(x) = H$. Let $F = F(x) + \sum_i \lambda_i \frac{\partial H_i}{\partial x}(x)$. Factorize the matrix H to $[L \quad 0]Q$, using the QU factorization, where Q is orthogonal and L is lower triangular of full rank as the active constraints are assumed regular. Let $w = Qz$, $y = Q\zeta$ and partition w, y and QFQ^T to the partitioning of $[L \quad 0]$. This gives a transformation of (4.2), where the different components of the solution can be distinguished,

(4.3) $$\begin{bmatrix} F_{11} & F_{12} & L^T \\ F_{12}^T & F_{22} & 0 \\ L & 0 & 0 \end{bmatrix} \begin{bmatrix} w_1 \\ w_2 \\ \lambda \end{bmatrix} = \begin{bmatrix} -y_1 \\ -y_2 \\ -h \end{bmatrix}.$$

As Q is orthogonal, arbitrary perturbations in F, H and ζ are transformed to the same sized arbitrary perturbations in F_{11}, F_{12}, F_{22}, L, y_1 and y_2.

The component w_1, regarded as the component of z fixed by the constraints is given by

(4.4) $$Lw_1 = -h$$

and so has a condition number of $\chi(L)$. The component w_2, of z, gives the step to the minimizer of f in the subspace free of constraints (in the tangent space), and is given by

(4.5) $$F_{22}w_2 = -y_2 - F_{12}^T w_1.$$

Perturbations in this equation, through w_1, are proportional to $\chi(L)$, so an approximate condition number for w_2 is $\chi(L)\chi(F_{22})$, based on arbitrary perturbations in h, g, F and H. But notice that the sub-matrix F_{12} does effect the true condition number.

The Lagrange multipliers are given by

$$L^T\lambda = -y_1 - F_{11}w_1 - F_{12}w_2, \tag{4.6}$$

but as w_2 can have perturbations proportional to $\chi(L)\chi(F_{22})$ this gives an approximate condition number for λ of $\chi(L)^2\chi(F_{22})$. Of course for any particular problem the effect of F_{11} and F_{12} could lessen or worsen this condition number. This shows that the Lagrange multipliers are more likely to be affected by perturbations than the solution point. The extent of coupling between the three components of the solution can be seen in the inverse matrix description of (4.3).

$$\begin{bmatrix} w_1 \\ w_2 \\ \lambda \end{bmatrix} = \begin{bmatrix} 0 & 0 & L^{-1} \\ 0 & F_{22}^{-1} & -F_{22}^{-1}F_{12}^T L^{-1} \\ L^{-T} & -L^{-T}F_{12}F_{22}^{-1} & L^{-T}(F_{12}F_{22}^{-1}F_{12}^T - F_{11})L^{-1} \end{bmatrix} \begin{bmatrix} -y_1 \\ -y_2 \\ -h \end{bmatrix}. \tag{4.7}$$

This coupling is determined by the particular problem at hand. So if $F_{12}F_{22}^{-1}F_{12}^T - F_{11}$ is small the condition number for λ will be smaller than $\chi(L)^2\chi(F_{22})$. This also shows which perturbations are most important to reduce. Generally the active constraint equations should be the most accurate, $h = 0$ being ideal. Any perturbation from $h = 0$ will adversely affect the accuracy of λ and to a lessor extent w_2. In practice, practitioners may care little about exactly satisfying the constraints. The above analysis determines the inaccuracies that could be expected resulting from arbitrary perturbations in h, g, F and H.

The three scenarios, in the strongly coupled case of (i) L well-conditioned and F_{22} ill-conditioned, (ii) L ill-conditioned and F_{22} well-conditioned, and, (iii) both L and F_{22} ill-conditioned, are of interest as they give an overall ill-conditioned constrained optimization problem, in worsening order. In (i), the solution point and Lagrange multipliers are equally ill-determined through the unconstrained component of the solution. In cases (ii) and (iii) the Lagrange multipliers can be almost impossible to determine accurately as the solution component fixed by the constraints is ill-determined. This will most likely show as non-convergence of Kuhn-Tucker conditions, although this may depend on the algorithm. In cases (i) and (iii) large changes in iterates x are expected for little change in function values or gradient values. In cases (ii) and (iii) large changes in iterates x are expected for small changes in constraint compliance. In all cases large changes in Lagrange multiplier estimates and hence Kuhn-Tucker conditions are expected.

4.1 Testing accuracies for optimization

Assume that a user has asked for an accuracy of ϵ for the Kuhn-Tucker conditions and δ for the constraints, that is, the algorithm is considered converged if $\|\zeta\| \le \epsilon$ and $\|h\| \le \delta$. Then, close to a solution, it is to be expected that the iterates $x^{(k)}$ can still be changing by as much as

$$\|x^{(k+1)} - x^{(k)}\| \le \chi(L)\,(\delta + \chi(F_{22})\epsilon)\,.$$

The Kuhn-Tucker gradient $\nabla f + \lambda^T\nabla h = \zeta$ can have a norm as large as,

$$\|\zeta\| \le \delta\left(\chi^2(L)\chi(F_{22})\right)\|\nabla h\|,$$

but if λ is likely to have most of its error associated with the small singular values of ∇h, then $\|\lambda^T\nabla h\|_2 = \|L^T\lambda\|_2$ will vary like $\|-y_1 - F_{11}w_1 - F_{12}w_2\|_2$ varies, see (4.6), that is, proportional to

$$\|\zeta\| \le \delta\chi(L)\chi(F_{22}).$$

In this case any test for convergence on the Kuhn-Tucker gradient should be

$$\|\zeta\| \leq \max\{\epsilon,\ \delta\chi(\boldsymbol{L})\chi(\boldsymbol{F}_{22})\}.$$

4.2 Incorporation into Software

For ill-conditioned optimization problems (or even any optimization problem) most SQP algorithms do not attempt to find an accurate representation of $\boldsymbol{F}$. Hence an accurate representation of $\boldsymbol{F}_{22}$ from an SQP algorithm cannot be guaranteed. Instead $\boldsymbol{F}$ can be computed approximately from gradient perturbations using central difference formula. For example, suppose ∇f can be computed to an accuracy of at least 10^{-9}. Then if $\mu_i = 10^{-3}|x_i|$ ($|x_i| \neq 0$), let the $i-th$ row of $\hat{\boldsymbol{F}}$ be defined by

$$\rho_i(\hat{\boldsymbol{F}}) = \frac{1}{2\mu_i}\left(\nabla f(\boldsymbol{x} + \mu_i \boldsymbol{e}_i) - \nabla f(\boldsymbol{x} - \mu_i \boldsymbol{e}_i)\right), \quad i = 1, \ldots, n.$$

Now let the Hessian of f be approximated by $\boldsymbol{F} = \frac{1}{2}(\hat{\boldsymbol{F}} + \hat{\boldsymbol{F}}^T)$. This costs $2n$ gradient evaluations. An estimate of the error in this Hessian is $\max\left\{10^{-6},\ \|\hat{\boldsymbol{F}} - \hat{\boldsymbol{F}}^T\|\right\}$. Similar computations are done for the Hessians of the constraints. On the other hand $\boldsymbol{L}$ can be computed accurately. Perturbations to $\boldsymbol{h}$ and $\boldsymbol{\zeta}$ are known, as these should both be zero at a solution. Hence it should be possible to modify stopping criteria in the light of condition numbers of $\boldsymbol{L}$ and $\boldsymbol{F}_{22}$, and report the condition numbers at a possible solution.

5 EXAMPLES

***Example 1*: State internal and terminal constraints.**

$$\underset{u}{\text{minimize}}\left\{J(u) = \int_0^1 \left(x^2(t) + u^2(t)\right) dt\right\},$$

subject to the dynamics, $\dot{x}(t) = u(t)$, $x(0) = 1$, with terminal constraint $x(1) = 0.75$, and an interior point constraint $x(0.75) = 0.9$. This is a good test problem as it reduces to a quadratic programming problem, and is solved after only a few iterations.

***Example 2*: All-time state constraints**

$$\underset{u}{\text{minimize}}\left\{J(u) = \int_0^1 (x_1^2(t) + x_2^2(t) + 0.005u^2(t))\, dt\right\}$$

subject to the dynamics $\dot{x}_1(t) = x_2(t)$, $x_1(0) = 0$, $\dot{x}_2(t) = -x_2(t) + u(t)$, $x_2(0) = -1$, subject to all-time state inequality constraint $x_2(t) - 8(t-0.5)^2 + 0.5 \leq 0$, $\forall t \in [0, 1]$. This inequality is replaced by many internal state constraints namely, $x_2(t_i) - 8(t_i - 0.5)^2 + 0.5 \leq 0$, where t_i is chosen at all the knot points of the control function. At a solution slightly less than half of these constraints are active.

The graphs in Figures 5.1 and 5.2 show the plot of *condition number* versus *number of parameters*: an 'o' represents the control as a piecewise constant; a '·' represents control as piecewise linear with function values as parameters; a '+' represents control as piecewise linear with derivative values as parameters; a '×' represents control as a piecewise quadratic spline with derivative values as parameters; and a '*' represents control as a piecewise quadratic spline with second derivative values as parameters.

The number of control parameters are varied from 4, through 8, 16, 32 and to 64 to give an indication of how the condition number varies with the *number* of control

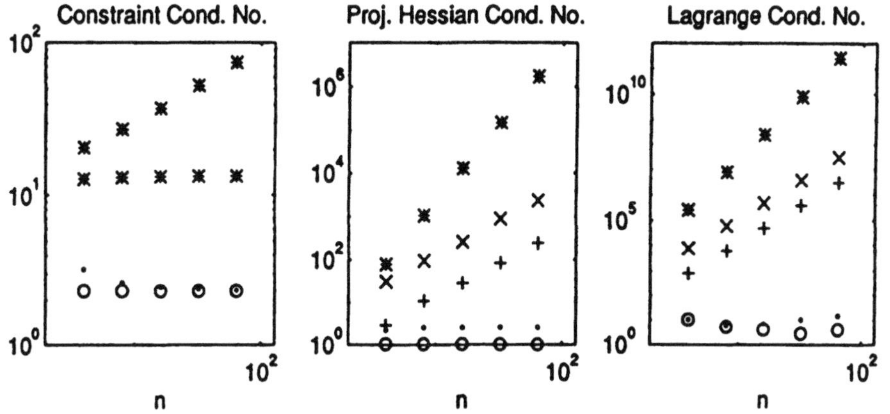

Figure 5.1: *In all three cases, where the condition number depends on the number n of parameters, the relationship appears to be* αn^β. *Note: In the first diagram,* (constraint condition no. *versus* number of parameters), the '+' and '×' *coincide, indicating that their respective condition numbers are approximately equal.*

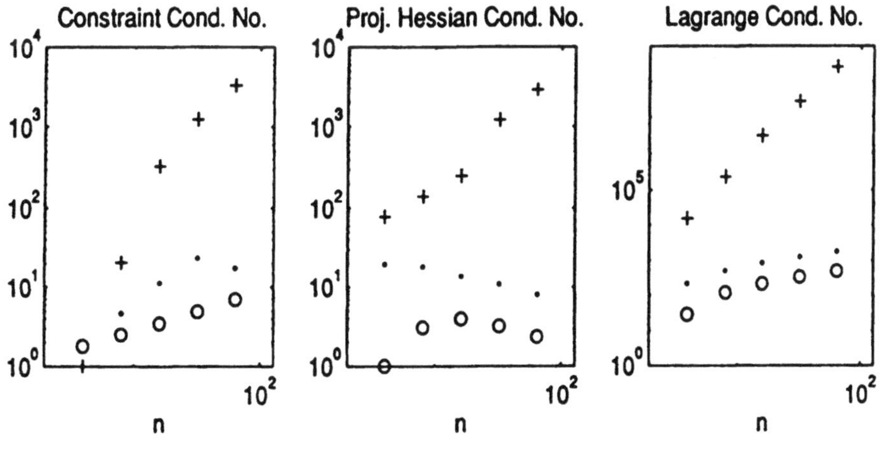

Figure 5.2: *As in Figure 5.1, the relationship appears to be* αn^β. *Note: We have not included the graphs of the condition numbers for '×' and '*', as they are extremely large; the condition numbers increase with smoothing.*

parameters. The other dimension used to model the control is its *smoothness*, that is, whether the control parameters represent functional values, derivative values, or second derivative values, at the knots. The condition number of active constraints $\chi(\boldsymbol{L})$, the condition number of the projected Hessian $\chi(\boldsymbol{F}_{22})$, and the overall condition number of the Lagrange condition, (4.2), are shown. These three condition numbers, as expected, increase in size as the degree of smoothing increases. In particular, if piecewise linear controls are required, it is better to use the *hat function basis* $(\cdot)$ than to integrate a piecewise constant control $(+)$.

6 CONCLUSIONS

The computation of solutions to optimal control problems is known to be ill-conditioned. We have examined the extent to which this ill-conditioning is influenced by the *number* of control parameters, as well as the *smoothness* of the control. This, we hope, provides some useful hints to be considered when discretizing the control function of an optimal control problem. In a forthcoming paper(s) we examine the effects of regularization on the above procedures.

Acknowledgments: This project was partially supported by a research grant from the Australian Research Council.

References

1. L. S. Jennings, M. E. Fisher, K. L. Teo and C. J. Goh, Advances in Engineering Software, 13, 190–196 (1991).

2. K. Schittkowski, NLPQL Version 2.3: An Implementation of a Sequential Quadratic Programming (SQP) Method for Solving Nonlinear Optimization Problems, Universitaet Bayreuth, 8580 Bayreuth, Germany, F.R. (1987).

3. J. L. Zhou, A. Tits and C. T. Lawrence, FFSQP Version 3.6: A Fortran Code for Solving Optimization Programs, University of Maryland, Technical Report SRC-TR-92-107r5, College Park, MD 20742 (1996).

4. K. L. Teo, C. J. Goh and K. H. Wong, A Unified Computational Approach to Optimal Control Problems, Longman Scientific and Technical, United Kingdom (1991).

5. F.Benyah L. S. Jennings, Regularization of Optimal Control Computation, ISIAC, Alaska (1998), (in press).

8th Int. Coll. on Differential Equations, pp. 89-96
D. Bainov (Ed)

Differentiability of Solutions of an Electro-Elasticity Model w.r.t. Parameters

NIKOLAI D. BOTKIN
Institute of Applied Mathematics of the Technical University of Munich, Dachauer str. 9a, D-80335 Munich, Germany

Abstract. A model describing a linear thin plate exited by a piezoelectric patch is considered. The peculiarity of this model consists in the presence of an unbounded input operator on the right-hand-side of the model's equation. The main purpose of the paper is to prove the existence of derivatives of solutions with respect to parameters. Such a differentiability is proved for an appropriate space, which enables us to handle pointwise measurements of data when applying a gradient descent method to the estimation of unknown parameters. It is shown that the derivatives can be computed using well-known variational equations. The paper is illustrated by computer simulations.

Keywords: *Linear thin plates, Piezoelectric actuators, Derivatives w.r.t. parameters, Estimation of parameters.*

1. Problem setting

We consider a linear system describing oscillations of a thin plate excited by a patch made of a piezoelectric ceramic. Assume that the plate occupies a domain S, where S is a bounded simple connected open set in R^2 with the piecewise infinitely differentiable boundary ∂S. The patch occupies a domain $S_P \subset S$ with the same properties as S. Denote $S_B = S \setminus S_P$. The equation describing the model reads:

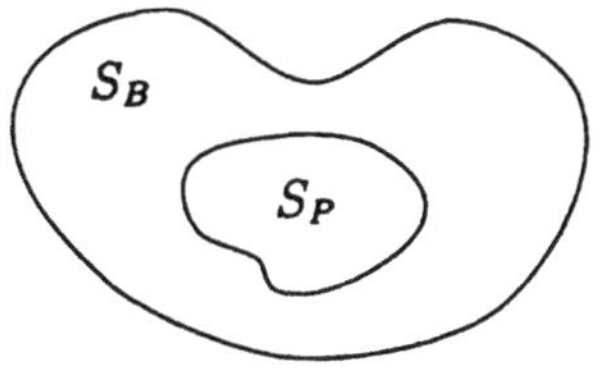

Figure 1: The plate S with the patch S_P, $S_B := S \setminus S_P$.

$$\rho\xi_{tt} + \Delta\Big(\gamma\Delta\xi\Big) - \mu\Delta\xi_t = v(t)K\Delta\Big(I_{S_P}\Big), \quad x \in S. \tag{1}$$

Here ξ is the vertical displacement of the plate, Δ is the Laplace operator, $v(t)$ is the voltage applied to the piezoelectric patch S_P (see Figure 1), I_{S_P} is the indicator function of the patch. The coefficients μ and K are constant; ρ and γ are piecewise constant functions defined as follows:

$$\rho = \begin{cases} \rho_P, & (x_1, x_2) \in S_P, \\ \rho_B, & (x_1, x_2) \in S_B, \end{cases} \qquad \gamma = \begin{cases} \gamma_P, & (x_1, x_2) \in S_P, \\ \gamma_B, & (x_1, x_2) \in S_B. \end{cases}$$

Indices P and B point out to the piezoelectric–base–mixed and base materials, respectively. It is assumed that $(\gamma_P, \gamma_B, \mu, K) \in \mathcal{P}$, where

$$\mathcal{P} := \{(\gamma_P, \gamma_B, \mu, K) \in R^4 : \gamma_P > 0, \gamma_B > 0, \mu \geq 0\}. \tag{2}$$

Let

$$v(\cdot) \in \begin{cases} L_2(0,T), & \text{if } \mu > 0, \\ H^1(0,T), & \text{if } \mu = 0. \end{cases} \tag{3}$$

Boundary and initial conditions are:

$$\xi|_{\partial S} = 0, \quad \partial\xi/\partial\vec{n}|_{\partial S} = 0, \quad \xi|_{t=0} = \xi_0, \quad \xi_t|_{t=0} = \xi_0'. \tag{4}$$

The model (1) is well known and was investigated in details in [1]. Note that (1) can be derived using the technique described in [2] and the theory of electro-elasticity developed in [3, 4, 5]. The equation (1) should be supplied with some interface conditions that hold on the boundary between S_P and S_B because of the integration by parts when deriving (1) from a weak formulation. We do not write down these conditions because the original weak formulation will be handled.

2. Properties of solutions

The weak formulation of (1) is given by the equality:

$$\iint\limits_{S_P} (\rho_P \xi_{tt}\varphi + \gamma_P \Delta\xi\Delta\varphi) + \iint\limits_{S_B} (\rho_B \xi_{tt}\varphi + \gamma_B \Delta\xi\Delta\varphi) + \tag{5}$$

$$+ \iint\limits_{S} \mu\nabla\xi_t\nabla\varphi = \iint\limits_{S} Kv(t)I_{S_P}(x)\Delta\varphi$$

for all $\varphi \in H_0^2(S)$. Consider the following generalization of the last equation:

$$\iint\limits_{S_P} (\rho_P \xi_{tt}\varphi + \gamma_P \Delta\xi\Delta\varphi) + \iint\limits_{S_B} (\rho_B \xi_{tt}\varphi + \gamma_B \Delta\xi\Delta\varphi) + \tag{6}$$

$$+ \iint\limits_{S} \mu\nabla\xi_t\nabla\varphi = \iint\limits_{S} g\Delta\varphi$$

which is obtained by the replacement of $Kv(t)I_{S_P}(x)$ by $g(t,x)$. Here

$$g \in \begin{cases} L_2(0,T; L_2(S)), & \text{if } \mu > 0, \\ H^1(0,T; L_2(S)), & \text{if } = 0. \end{cases}$$

Proposition 1. If $\xi_0 \in H_0^2(S)$ and $\xi_0' \in L_2(S)$, then the equation (6) has a unique solution ξ such that $(\xi, \xi_t) \in C([0,T]; H_0^2(S)) \times C([0,T]; L_2(S))$. The estimates hold:

$$\|\xi\|^2_{C([0,T];H_0^2(S))} + \|\xi_t\|^2_{C([0,T];L_2(S))} \leq \tag{7}$$

$$C_1\left(\|\xi_0\|^2_{H_0^2(S)}+\|\xi_0'\|^2_{L_2(S)}+\|g\|^2_{L_2(0,T;L_2(S))}\right),$$

if $\mu>0$ and

$$\|\xi\|^2_{C([0,T];H_0^2(S))}+\|\xi_t\|^2_{C([0,T];L_2(S))}\le \tag{8}$$

$$C_2\left(\|\xi_0\|^2_{H_0^2(S)}+\|\xi_0'\|^2_{L_2(S)}+\|g\|^2_{C([0,T];L_2(S))}+\|g_t\|^2_{L_2(0,T;L_2(S))}\right),$$

if $\mu\ge 0$.

Besides, C_1 is bounded if the parameters vector $(\gamma_P,\gamma_B,\mu,K)$ lies in a compact set $\Sigma\subset\text{int}(\mathcal{P})$; C_2 is bounded if $\Sigma\subset\mathcal{P}$. ■

Proof of Proposition 1. We restrict us to the case where $\mu\ge 0$. The proof is similar to that given in Chapter 8 of [6]. Let ω_i, $i=1,2...$, be a basis of $H_0^2(S)$. We define approximations $\xi^m(t)$ as follows: $\xi^m(t)=\sum_{i=1}^m r_i^m(t)\omega_i$. For each m, the unknown functions $r_i^m(t)$, $i=1,...,m$, satisfy the corresponding linear system of ordinary differential equations with initial conditions chosen such that $\sum_{i=1}^m r_i^m(0)\omega_i=a^m$ and $\sum_{i=1}^m d/dt\, r_i^m(0)\omega_i=b^m$, where $a^m\to\xi_0$ and $b^m\to\xi_0'$ in $H_0^2(S)$ and $L_2(S)$, respectively. One can easily obtain the following energy estimate:

$$\|\xi^m\|^2_{L_\infty(0,T;H_0^2(S))}+\|\xi_t^m\|^2_{L_\infty(0,T;L_2(S))}+\mu\|\nabla\xi_t^m\|^2_{L_2(0,T;L_2(S))}\le$$

$$C(\Omega)\left(\|a^m\|^2_{H_0^2(S)}+\|b^m\|^2_{L_2(S)}+\|g\|^2_{C([0,T];L_2(S))}+\|g_t\|^2_{L_2(0,T;L_2(S))}\right)$$

with $C(\Omega)$ bounded for any compact set $\Omega\subset\mathcal{P}$. Note that $g\in C([0,T];L_2(S))$ (see e.g. [7]). There exist a subsequence ξ^{m_k} and a function ξ such that $\xi^{m_k}\to\xi$ and $\xi_t^{m_k}\to\xi_t$, weak-$*$ in $L_\infty(0,T;H_0^2(S))$ and $L_\infty(0,T;L_2(S))$, respectively. Besides, $\nabla\xi_t^{m_k}\to\nabla\xi_t$ weakly in $L_2(0,T;L_2(S))$ if $\mu\ne 0$.

One proves that ξ is a unique solution of (6) which satisfies:

$$\|\xi\|^2_{L_\infty(0,T;H_0^2(S))}+\|\xi_t\|^2_{L_\infty(0,T;L_2(S))}+\mu\|\nabla\xi_t\|^2_{L_2(0,T;L_2(S))}\le$$

$$C(\Omega)\left(\|\xi(0)\|^2_{H_0^2(S)}+\|\xi_t(0)\|^2_{L_2(S)}+\|g\|^2_{C([0,T];L_2(S))}+\|g_t\|^2_{L_2(0,T;L_2(S))}\right).$$

Then one establishes the additional regularity of ξ with respect to t. To this end, one uses the energy equality:

$$\iint_{S_P}\rho_P\xi_t^2+\iint_{S_B}\rho_B\xi_t^2+\iint_{S_P}\gamma_P(\Delta\xi)^2+\iint_{S_B}\gamma_B(\Delta\xi)^2=$$

$$\iint_{S_P}\rho_P\xi_0'^2+\iint_{S_B}\rho_B\xi_0'^2+\iint_{S_P}\gamma_P(\Delta\xi_0)^2+\iint_{S_B}\gamma_B(\Delta\xi_0)^2-2\iint_{S}g(0)\Delta\xi_0+$$

$$+2\iint_{S}g\Delta\xi-2\int_0^t\iint_{S}g_t\Delta\xi-2\int_0^t\iint_{S}\mu(\nabla\xi_t)^2.$$

Note that the function $t\to\iint_S g(t)\Delta\xi(t)$ is continuous due to Lemma 8.2 of [6]. Since $\xi_0\in H_0^2(S)$ and $\xi_0'\in L_2(S)$, all terms on the right-hand-side are well defined and the function

$$t\to\iint_{S_P}\rho_P\xi_t^2+\iint_{S_B}\rho_B\xi_t^2+\iint_{S_P}\gamma_P(\Delta\xi)^2+\iint_{S_B}\gamma_B(\Delta\xi)^2$$

is continuous on $[0,T]$. Then, using the same arguments as in the proof of Theorem 8.2 of [6], one shows that $\xi\in C([0,T];H_0^2(S))$ and $\xi_t\in C([0,T];L_2(S))$. ■

3. Differentiability with respect to parameters

Let the initial data for (5) be fixed. Due to Proposition 1, the map

$$\mathcal{F} : (\gamma_P, \gamma_B, \mu, K) \to (\xi, \xi_t) : \quad \mathcal{P} \to C([0,T]; H_0^2(S)) \times C([0,T]; L_2(S)),$$

is defined. Here ξ is a solution of (5) corresponding to the vector $(\gamma_P, \gamma_B, \mu, K)$ of parameters. Now we prove differentiability of $\mathcal{F}$ and show that the partial derivatives can be computed using variational equations. The proof is based on the possibility to differentiate (6) w.r.t. t under some compatibility conditions (see e.g. [8]).

Proposition 2.

i) Let $\xi_0 \in H_0^2(S)$, $\xi_0' \in L_2(S)$ and $v \in L_2(0,T)$. Then the mapping $\mathcal{F}$ is infinitely differentiable at each point of $\mathcal{P}$ such that $\mu > 0$.

ii) If $\xi_0 = \xi_0' = 0$, $v \in H^{k+2}(0,T)$, and $v(0) = v^{(1)}(0) = \ldots = v^{(k)}(0) = 0$, then $\mathcal{F}$ is k times differentiable at points of $\mathcal{P}$ with $\mu = 0$.

The partial derivatives of $\mathcal{F}$ satisfy variational equations which are obtained using formal differentiation of (5).■

In particular, the partial derivatives $\delta^1 = \frac{\partial\xi}{\partial\gamma_P}$, $\delta^2 = \frac{\partial\xi}{\partial\gamma_B}$, $\delta^3 = \frac{\partial\xi}{\partial\mu}$, $\delta^3 = \frac{\partial\xi}{\partial K}$ at a point $(\gamma_P^0, \gamma_B^0, \mu^0, K^0)$ satisfy the following variational equations:

$$\iint\limits_{S_P} \left(\rho_P\delta^1_{tt}\varphi + \gamma_P^0\Delta\delta^1\Delta\varphi + \Delta\xi^0\Delta\varphi\right) + \iint\limits_{S_B} \left(\rho_B\delta^1_{tt}\varphi + \gamma_B^0\Delta\delta^1\Delta\varphi\right) + \iint\limits_{S} \mu^0\nabla\delta^1_t\nabla\varphi = 0,$$

$$\iint\limits_{S_P} \left(\rho_P\delta^2_{tt}\varphi + \gamma_P^0\Delta\delta^2\Delta\varphi\right) + \iint\limits_{S_B} \left(\rho_B\delta^2_{tt}\varphi + \gamma_B^0\Delta\delta^2\Delta\varphi + \Delta\xi^0\Delta\varphi\right) + \iint\limits_{S} \mu^0\nabla\delta^2_t\nabla\varphi = 0,$$

$$\iint\limits_{S_P} \left(\rho_P\delta^3_{tt}\varphi + \gamma_P^0\Delta\delta^3\Delta\varphi\right) + \iint\limits_{S_B} \left(\rho_B\delta^3_{tt}\varphi + \gamma_B^0\Delta\delta^3\Delta\varphi\right) + \iint\limits_{S} \left(\mu^0\nabla\delta^3_t\nabla\varphi + \nabla\xi^0_t\nabla\varphi\right) = 0,$$

$$\iint\limits_{S_P} \left(\rho_P\delta^4_{tt}\varphi + \gamma_P^0\Delta\delta^4\Delta\varphi\right) + \iint\limits_{S_B} \left(\rho_B\delta^4_{tt}\varphi + \gamma_B^0\Delta\delta^4\Delta\varphi\right) + \iint\limits_{S} \mu^0\nabla\delta^4_t\nabla\varphi = \iint\limits_{S_P} v(t)\Delta\varphi.$$

Remark. The differentiability of $\mathcal{F}$ implies the differentiability of the following mappings: $(\gamma_P, \gamma_B, \mu, K) \to \xi(t^*, x_1^*, x_2^*)$, where $(t^*, x_1^*, x_2^*) \in [0,T] \times S$ is a fixed point. This result follows immediately from the continuous embedding $C([0,T]; H_0^2(S)) \subset C([0,T] \times \bar{S})$.■

Proof of Proposition 2. Denote $\pi = (\gamma_P, \gamma_B, \mu, K)$ and $\pi^0 = (\gamma_P^0, \gamma_B^0, \mu^0, K^0)$. Define

$$\begin{aligned} w^0 &= \xi(\pi), \\ w^1 &= w^0 - \xi(\pi_0), \\ w^2 &= w^1 - \Phi^{(1)}(\pi_0)(\pi - \pi_0), \\ &\ldots \\ w^{k+1} &= w^k - \tfrac{1}{k!}\Phi^{(k)}(\pi_0)(\pi - \pi_0), \end{aligned} \tag{9}$$

where $\Phi^{(k)}(\pi_0)(\pi - \pi_0)$ is the k th formal derivative of the map $\pi \to \xi(\pi)$ at π^0 (k-linear form of $\pi - \pi_0$) which is obtained using formal differentiation of (5) k times with respect to π.

One can verify that w^{s+1}, $0 \le s \le k$, satisfy the following equations:

$$\iint\limits_{S_P} \left(\rho_P w^{s+1}_{tt}\varphi + \gamma_P^0\Delta w^{s+1}\Delta\varphi\right) + \iint\limits_{S_B} \left(\rho_B w^{s+1}_{tt}\varphi + \gamma_B^0\Delta w^{s+1}\Delta\varphi\right) + \iint\limits_{S} \mu^0\nabla w^{s+1}_t\nabla\varphi =$$

$$-\iint\limits_{S_P}(\gamma_P-\gamma_P^0)\Delta w^s\Delta\varphi-\iint\limits_{S_B}(\gamma_B-\gamma_B^0)\Delta w^s\Delta\varphi-\iint\limits_{S}(\mu-\mu^0)w_t^s\Delta\varphi+ \tag{10}$$

$$+\begin{cases}0, & \text{if } s>0,\\ \iint\limits_{S_P}(K-K_0)v(t)\Delta\varphi & \text{if } s=0\end{cases}$$

with the homogeneous initial conditions. Note that $w^0=\xi$ by the definition. Denote for brevity: $\|\cdot\|:=\|\cdot\|_{C([0,T];H_0^2(S))}$, $|\cdot|:=\|\cdot\|_{C([0,T];L_2(S))}$.

First, consider the case: $\mu^0>0$. Application of (7) to (10) yields:

$$\|w^{s+1}\|^2+|w_t^{s+1}|^2\le M[(\gamma_P-\gamma_P^0)^2+(\gamma_B-\gamma_B^0)^2+(\mu-\mu^0)^2]\times\left(\|w^s\|^2+|w_t^s|^2\right).$$

$$\|w^1\|^2+|w_t^1|^2\le M[(\gamma_P-\gamma_P^0)^2+(\gamma_B-\gamma_B^0)^2+(\mu-\mu^0)^2+(K-K_0)^2]\times\left(\|\xi\|^2+|\xi_t|^2+\|v\|_{L_2(0,T)}\right).$$

Therefore,

$$\|w^{s+1}\|^2+|w_t^{s+1}|^2\le L[(\gamma_P-\gamma_P^0)^2+(\gamma_B-\gamma_B^0)^2+(\mu-\mu^0)^2+(K-K_0)^2]^{s+1}$$

for $s\le k$. Since k is arbitrary, this proves the first part of Proposition 2.

Now consider the case $\mu^0=0$. Let auxiliary functions $\eta^{j,s+1}$ be defined for $j\le k-s$, $0\le s\le k$ as follows:

$$\iint\limits_{S_P}\left(\rho_P\eta_{tt}^{j,s+1}\varphi+\gamma_P^0\Delta\eta^{j,s+1}\Delta\varphi\right)+\iint\limits_{S_B}\left(\rho_B\eta_{tt}^{j,s+1}\varphi+\gamma_B^0\Delta\eta^{j,s+1}\Delta\varphi\right)= \tag{11}$$

$$-\iint\limits_{S_P}(\gamma_P-\gamma_P^0)\Delta\eta^{j,s}\Delta\varphi-\iint\limits_{S_B}(\gamma_B-\gamma_B^0)\Delta\eta^{j,s}\Delta\varphi-\iint\limits_{S}\mu\eta_t^{j,s}\Delta\varphi+$$

$$+\begin{cases}0, & \text{if } s>0,\\ \iint\limits_{S_P}(K-K_0)\dfrac{d^jv(t)}{dt^j}\Delta\varphi & \text{if } s=0,\end{cases}$$

where the initial conditions are homogeneous. We also define $\eta^{j,0}$ for $0\le j\le k+1$ as solutions of the equations:

$$\iint\limits_{S_P}\left(\rho_P\eta_{tt}^{j,0}\varphi+\gamma_P\Delta\eta^{j,0}\Delta\varphi\right)+\iint\limits_{S_B}\left(\rho_B\eta_{tt}^{j,0}\varphi+\gamma_B\Delta\eta^{j,0}\Delta\varphi\right)+ \tag{12}$$

$$+\iint\limits_{S}\mu\nabla\eta_t^{j,0}\nabla\varphi=\iint\limits_{S_P}K\frac{d^jv(t)}{dt^j}\Delta\varphi$$

with the homogeneous initial conditions (remember that the initial conditions for ξ are homogeneous). The following assertion is true.

<u>Assertion.</u> Let conditions of ii) in Proposition 2 be fulfilled. Then, for any $0\le s\le k$, the function w^{s+1} is $k-s$ times differentiable w.r.t. t, and $\dfrac{\partial^j w^{s+1}}{\partial t^j}=\eta^{j,s+1}$ for $j\le k-s$, $0\le s\le k$. The function ξ is $k+1$ times differentiable w.r.t. t, and $\dfrac{\partial^j\xi}{\partial t^j}=\eta^{j,0}$ for $0\le j\le k+1$.■

We outline the proof of the Assertion quite briefly. Let us prove the second part of the Assertion. Let

$$\hat{\xi}(t,\cdot) = \xi_0 + t\xi_0' + \int_0^t d\tau_j \ldots \int_0^{\tau_2} d\tau_1 \eta^{j,0}(\tau_1,\cdot), \tag{13}$$

where $\eta^{j,0}$ is a solution of (12). Integrate (12) j times to obtain:

$$\iint_{S_P} \left(\rho_P \hat{\xi}_{tt}\varphi + \gamma_P \Delta[\hat{\xi} - \xi_0 - t\xi_0']\Delta\varphi\right) + \iint_{S_B} \left(\rho_B \hat{\xi}_{tt}\varphi + \gamma_B \Delta[\hat{\xi} - \xi_0 - t\xi_0']\Delta\varphi\right) +$$

$$+ \iint_S \mu \nabla[\hat{\xi}_t - \xi_0']\nabla\varphi = \iint_{S_P} \left[\sum_{l=0}^{j-1} t^l/l! \frac{\partial^l v(0)}{dt^l} + v(t)\right] \Delta\varphi.$$

The homogeneous initial conditions for $\eta^{j,0}$ were used. If $\xi_0 = \xi_0' = 0$ and all of $\frac{\partial^l v(0)}{dt^l} = 0$, then $\hat{\xi}$ is a solution of (5) and, hence, $\hat{\xi} = \xi$. Taking into account (13) completes the proof.

Let us go back to the proof of Proposition 2. Applying (8) to (11) and (12) yields for $s \le k$, $j \le k - s$:

$$\left\|\frac{\partial^j w^{s+1}}{\partial t^j}\right\|^2 + \left|\frac{\partial^j w_t^{s+1}}{\partial t^j}\right|^2 \le M[(\gamma_P - \gamma_P^0)^2 + (\gamma_B - \gamma_B^0)^2 + \mu^2] \times$$

$$\times \left(\left\|\frac{\partial^j w^s}{\partial t^j}\right\|^2 + \left|\frac{\partial^j w_t^s}{\partial t^j}\right|^2 + \left\|\frac{\partial^{j+1} w^s}{\partial t^{j+1}}\right\|^2 + \left|\frac{\partial^{j+1} w_t^s}{\partial t^{j+1}}\right|^2\right).$$

For $s = 0$, $j \le k$:

$$\left\|\frac{\partial^j w^1}{\partial t^j}\right\|^2 + \left|\frac{\partial^j w_t^1}{\partial t^j}\right|^2 \le M[(\gamma_P - \gamma_P^0)^2 + (\gamma_B - \gamma_B^0)^2 + \mu^2 + (K - K_0)^2] \times$$

$$\times \left(\left\|\frac{\partial^j \xi}{\partial t^j}\right\|^2 + \left|\frac{\partial^j \xi_t}{\partial t^j}\right|^2 + \left\|\frac{d^j v(t)}{dt^j}\right\|^2_{C[0,T]} + \left\|\frac{\partial^{j+1} \xi}{\partial t^{j+1}}\right\|^2 + \left|\frac{\partial^{j+1} \xi_t}{\partial t^{j+1}}\right|^2 + \left\|\frac{d^{j+1} v(t)}{dt^{j+1}}\right\|^2_{L_2(0,T)}\right).$$

For $j \le k+1$: $\left\|\frac{\partial^j \xi}{\partial t^j}\right\|^2 + \left|\frac{\partial^j \xi_t}{\partial t^j}\right|^2 \le M \left(\left\|\frac{d^j v(t)}{dt^j}\right\|^2_{C[0,T]} + \left\|\frac{d^{j+1} v(t)}{dt^{j+1}}\right\|^2_{L_2(0,T)}\right).$

Induction arguments yield for $s \le k$:

$$\left\|w^{s+1}\right\|^2 + \left|w_t^{s+1}\right|^2 \le L[(\gamma_P - \gamma_P^0)^2 + (\gamma_B - \gamma_B^0)^2 + \mu^2 + (K - K_0)^2]^{s+1} \times$$

$$\times \left[\sum_{j=0}^{s+1} \binom{j}{s+1} \left\|\frac{d^j v(t)}{dt^j}\right\|^2_{C[0,T]} + \left\|\frac{d^{j+1} v(t)}{dt^{j+1}}\right\|^2_{L_2(0,T)}\right].$$

The last multiplier is bounded because $v \in H^{k+2}(0,T)$, which completes the proof of Proposition 2. ■

4. Estimation of parameters

Exact values of parameters to be estimated are: $\gamma_P^0 = 0.1$, $\gamma_B^0 = 0.01$, $\mu^0 = 0$, $K^0 = 5$. The other parameters are:

$$S = [0,1] \times [0,1], \quad S_P = [0.4, 0.6] \times [0.2, 0.5], \quad \rho_P = 1.5, \quad \rho_B = 1, \quad v(t) = 2\sin^2 5t.$$

The observation $\xi_{mn}^0 = \xi^0(t_m, x_{1n}, x_{2n})$ at points $(t_m, x_{1n}, x_{2n}) = (m \cdot 0.01, n \cdot 0.1, 0.1)$, $m = 1, ..., 10$, $n = 1, ..., 5$, is simulated. The usual residual is minimized:

$$J(\pi) = \sum_{m=1}^{10} \sum_{n=1}^{5} \left[\xi(t_m, x_{1n}, x_{2n}) - \xi_{mn}^0\right]^2, \quad \pi \in \mathcal{P}.$$

The linearized increment of the residual is:

$$J(\pi + \delta\pi) \approx \sum_{m=1}^{10} \sum_{n=1}^{5} \left[\xi(t_m, x_{1n}, x_{2n}) + \sum_{l=1}^{4} \delta^l(t_m, x_{1n}, x_{2n})\delta\pi_l - \xi_{mn}^0\right]^2. \tag{14}$$

Here δ^l, $l = 1, .., 4$, are solutions of the variational equations with the parameter set π. The descent direction $\delta\pi$ is found as a minimizing point of the right-hand-side of (14) which is a quadratic form. Then $\pi := (\pi + \lambda\delta\pi)^+$ is set, where λ is a step's length and "+" denotes the projection onto the set $\mathcal{P}$. The computations are repeated for the new value of π.

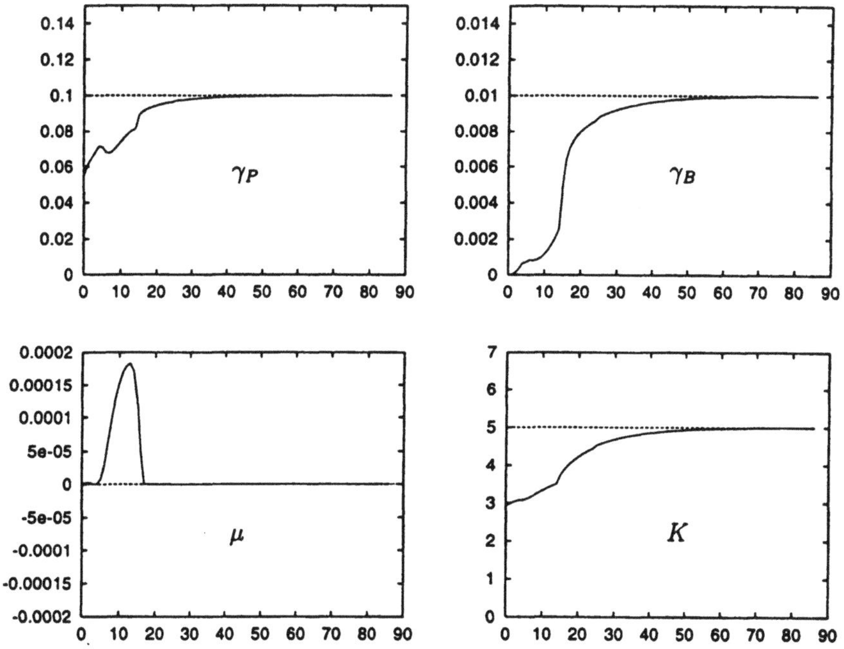

Figure 2:

Figure 2 demonstrates the convergence of the method. The integers under horizontal axis indicate the number of steps of the gradient descent method.

Acknowledgements

The author thanks Prof. Dr. K.-H. Hoffmann for helpful remarks throughout the work.

References

1. H.T.Banks, R.C.Smith, and Y. Wang, Smart Material Structures: Modeling, Estimation and Control, Wiley, Chichester (1996).
2. L.D.Landau and E.M.Lifschitz, Elastizitätstheorie, Akademie - Verlag, Berlin (1975).
3. G.A.Maugin, Continuum Mechanics of Electromagnetic Solids, North-Holland series in Applied Mathematics and Mechanics, 33,J.D.Achenbach, B.Budiansky, W.T.Koiter, H.A.Lauwerier, and L.Van Wijngaarden, (Eds.), North-Holland, Amsterdam-New York-Oxford-Tokyo (1987).
4. H.F.Tiersten, Int. J. Engng Sci., 9, 587–604 (1971).
5. J.Zelenka, Piezoelectric Resonators and Their Applications, Studies in Electrical and Electronic Engineering, 24, Elsevier, Amsterdam-Oxford-New York-Tokyo (1986).
6. J.L.Lions and E.Magenes, Non-homogeneous Boundary Value Problems and Applications, I, Berlin - Heidelberg-New York, Springer Verlag (1972).
7. J. Simon, Ann. Mat. Pura Appl., (IV 146), 65–96 (1987).
8. J. Wloka, Partielle Differentialgleichungen, Stuttgart, B.G. Teubner (1982).

8th Int. Coll. on Differential Equations, pp. 97–104
D. Bainov (Ed)

Efficient and Accurate Local Time Stepping Algorithm for Multi-Rate Problems

LILUN CAO AND JIANPING ZHU
Department of Mathematics and Statistics, NSF Engineering Research Center, Mississippi State University, Mississippi State, MS 39762, USA

Abstract. A multi-rate problem is decribed by a system of coupled partial differenrial equations with different time scales associated with different equations in the system. The numerical solutions to such systems are usually calculated using a time step determined by the most restrictive time scale in the system for stability and accuracy considerations. We demonstrate in this paper that this time step could be excessively small and unnecessary in many situations, and discuss a more efficient time integration method that uses different time steps for different equations depending on their time scales. Numerical results will be presented to demonstrate significant improvement in computational efficiency.

1. INTRODUCTION

Many engineering and physical applications described by coupled systems of partial differential equations involve significantly different time scales. For example, in a coupled mechanical and electrical system, the state variables describing the electrical components of the system could have a much shorter time scale, or a much faster rate of change, than that of the state variables describing the mechanical components in the system. Also in the study of reaction and diffusion processes involving multiple chemical components, the diffusion and reaction rates for different chemical components can be significantly different. A typical equation system describing such a process is the following reaction-diffusion equation

$$\underline{u}_t = D\underline{u}_{xx} + \underline{f}(x, t, \underline{u}) \tag{1}$$

where $\underline{u}$ is a vector of p state variables, D is a diagonal matrix with the diffusion coefficients of different components of $\underline{u}$ on the diagonal, and $\underline{f}(x, \underline{u})$ is a nonlinear reaction term that depends on the space and state variables. To compute numerical solutions, we have to discretize the spatial and temporal dimensions using a grid $x_i = i\Delta x,\ i = 0, \cdots, M$, $t_n = n\Delta t,\ n = 1, \cdots, N$. For simplicity, we have assumed that a uniform grid is used here for both x and t. If an explicit time integration algorithm, such as the forward time central space algorithm(FTCS), is used to compute the numerical solutions at different time steps, we will have the following formula for the computation:

$$\underline{u}_i^{n+1} = \underline{u}_i^n + \frac{\Delta t}{\Delta x^2} D\delta_x^2 \underline{u}_i^n + \Delta t \underline{f}(x_i, t_n, \underline{u}_i^n), \tag{2}$$

$$i = 1, \cdots, M-1, \qquad n = 0, 1, \cdots, N-1,$$

where, to save space, we have introduced the second order central difference operator δ_x^2 to approximate the second order spatial directive:

$$\delta_x^2 \underline{u}_i^n = \underline{u}_{i-1}^n - 2\underline{u}_i^n + \underline{u}_{i+1}^n.$$

It is well known that the stability criterion for the FTCS algorithm requires that

$$\Delta t \leq \frac{\Delta x^2}{2d_{ii}}, \quad i = 1, \cdots, p, \tag{3}$$

where d_{ii}'s are the diagonal elements in matrix D representing different diffusion coefficients for different chemical components. It is clear that if there is one equation in (1) that has a large diffusion coefficient, all the equations in (1) would have to be integrated using the smallest time step given by (3) in order to maintain stability.

On the other hand, if we use an implicit time integration algorithm, such as the Crank-Nicolson scheme(C-N), to compute the numerical solution of (1) at different time steps, we will have the following formula for the computation:

$$\frac{\underline{u}_i^{n+1} - \underline{u}_i^n}{\Delta t} = D[\frac{\delta_x^2 \underline{u}_i^{n+1}}{2\Delta x^2} + \frac{\delta_x^2 \underline{u}_i^n}{2\Delta x^2}] + \frac{1}{2}(\underline{f}_i^{n+1} + \underline{f}_i^n), \tag{4}$$

$$i = 1, \cdots, M-1, \qquad n = 0, \cdots, N-1.$$

Since the computations can not be carried out point by point, Eq. (4) is equivalent to the following:

$$\mathbf{A}\mathbf{u}^{n+1} = \mathbf{B}\mathbf{u}^n + \mathbf{b}^{n+1}$$

where $\mathbf{u}^n = \{\underline{u}_1^n, \underline{u}_2^n, \cdots, \underline{u}_{M-1}^n\}^T$ and $\mathbf{b}^{n+1} = \{\frac{1}{2}(\underline{f}_1^{n+1} + \underline{f}_1^n), \cdots, \frac{1}{2}(\underline{f}_{M-1}^{n+1} + \underline{f}_{M-1}^n)\}^T$. The matrix $\mathbf{A}$ is block-tridiagonal with $I + 2rD$ in the main diagonal and $-rD$ in the super and sub-diagonals, where $r = \frac{\Delta t}{\Delta x^2}$. The matrix $\mathbf{B}$ has the similar structure with $I - 2rD$ in the main diagonal and rD in the super and sub-diagonals. It is also well-known that the C-N algorithm is unconditionally stable, so there is no restriction like (3) for the step size Δt. However, if we want to calculate accurate transient solutions, the time step Δt can not be too big. Otherwise we will not be able to catch the details of the fastest changing state variables. In this case, again, the time step size is dictated by the accuracy requirement to calculate the state variable that changes the fastest, even if other variables change at much slower rates.

To improve the computational efficiency, we will discuss the local time stepping algorithm in this paper. The original local time stepping algorithm was developed for imrpoving the efficiency of explicit time integration algorithms [1–7]. The basic idea is to use a time step size determined by the local stability criterion based on the local grid spacing of each grid points, instead of using a uniform time step determined by the smallest grid spacing for all grid points. A detailed complexity analysis was given in [2] and the truncation analysis was given in [1]. This approach is very efficient for non-uniform grids in which only a small number of grid points are clustered to resolve fast-changing solutions in space. In this paper, we will generalize the local time stepping algorithm to multi-rate problems. The basic idea is to use different time steps for different partial differential equations characterized by different time scales. A small time step can be used for the integration of a PDE with a faster rate of change, while a large time step can be used to treat equations with slower rates of change.

2. LOCAL TIME STEPPING FOR EXPLICIT ALGORITHMS

To simplify the discussion, we use the following system of two PDEs and the FTCS algorithm to demonstrate the use of local time stepping with explicit algorithms:

$$u_t = \alpha u_{xx} + f(x,t,u,v) \tag{5}$$

$$v_t = \beta v_{xx} + g(x,t,u,v)$$

where $\beta >> \alpha$, f and g are nonlinear. The generalization to other explicit algorithms and systems with more equations is straightforward.

Applying the FTCS algorithm to Eq. (5), we will have the following scheme for computing the solutions

$$\frac{u_i^{n+1} - u_i^n}{\Delta t} = \alpha \frac{\delta_x^2 u_i^n}{\Delta x^2} + f(x_i, t_n, u_i^n, v_i^n), \tag{6}$$

$$\frac{v_i^{n+1} - v_i^n}{\Delta t} = \beta \frac{\delta_x^2 v_i^n}{\Delta x^2} + g(x_i, t_n, u_i^n, v_i^n),$$

$$i = 1, \cdots, M-1, \quad n = 0, \cdots, N-1,$$

or equivalently

$$u_i^{n+1} = u_i^n + \alpha \frac{\Delta t}{\Delta x^2} \delta_x^2 u_i^n + \Delta t f(x_i, t_n, u_i^n, v_i^n),$$

$$v_i^{n+1} = v_i^n + \beta \frac{\Delta t}{\Delta x^2} \delta_x^2 v_i^n + \Delta t g(x_i, t_n, u_i^n, v_i^n),$$

$$i = 1, \cdots, M-1, \quad n = 0, \cdots, N-1,$$

where the same time step Δt is used for computing both u and v. Since $\beta >> \alpha$, the size of Δt from stability criterion given by (3) is actually dictated by

$$\Delta t \leq \frac{\Delta x^2}{2\beta}$$

which means an unnecessarily small time step has been used for the computation of u.

To improve the computational efficiency, we can use different time steps for different equations based on the stability requirment of each equation:

$$u_i^{n+1} = u_i^n + \alpha \frac{\Delta t_1}{\Delta x^2} \delta_x^2 u_i^n + \Delta t_1 f(x_i, n\Delta t_1, u_i^n, v_i^n), \tag{7}$$

$$v_i^{n+1} = v_i^n + \beta \frac{\Delta t_2}{\Delta x^2} \delta_x^2 v_i^n + \Delta t_2 g(x_i, n\Delta t_2, u_i^n, v_i^n),$$

where

$$\Delta t_1 \leq \frac{\Delta x^2}{2\alpha}, \quad \Delta t_2 \leq \frac{\Delta x^2}{2\beta}.$$

Note that u and v in the two equations are not defined at the same time level because two different time steps are used for the computation. Since the source terms that do not contain spatial derivatives do not affect the stability of a time integration algorithm, it is obvious from Eq. (7) that the computation will be stable with two different time steps Δt_1 and Δt_2. For the convenience of algorithm implementation, we can assume that $\Delta t_1 = m\Delta t_2$ and the scheme given by (7) becomes

$$u_i^{n+1} = u_i^n + \alpha\frac{\Delta t_1}{\Delta x^2}\delta_x^2 u_i^n + \Delta t_1 f(x_i, n\Delta t_1, u_i^n, v_i^n), \tag{8}$$

$$v_i^{n+\frac{l+1}{m}} = v_i^{n+\frac{l}{m}} + \beta\frac{\Delta t_1}{m\Delta x^2}\delta_x^2 v_i^{n+\frac{l}{m}} + \frac{\Delta t_1}{m}g(x_i, (n+\frac{l}{m})\Delta t_1, u_i^{n+\frac{l}{m}}, v_i^{n+\frac{l}{m}}),$$
$$i = 1,\cdots,M-1, \quad l = 0,\cdots,m-1.$$

It is clear from (8) that for each step Δt_1 in calculating u_i^{n+1}, there are m substeps for calculating $v_i^{n+\frac{l}{m}}$ and we have to evaluate $g_i^{n+\frac{l}{m}}$ using $u_i^{n+\frac{l}{m}}$. Since u is calculated using a larger time step Δt_1, the values $u_i^{n+\frac{l}{m}}$ is undefined. A straightforward way to handle this would be to set

$$u_i^{n+\frac{l}{m}} = u_i^n, \quad i = 1,\cdots,M-1, \quad l = 1,\cdots,m-1, \quad n = 0,\cdots,N-1. \tag{9}$$

We will then have

$$u_i^{n+1} = u_i^n + \alpha\frac{\Delta t_1}{\Delta x^2}\delta_x^2 u_i^n + \Delta t_1 f(x_i, n\Delta t_1, u_i^n, v_i^n) \tag{10}$$

$$v_i^{n+\frac{l+1}{m}} = v_i^{n+\frac{l}{m}} + \beta\frac{\Delta t_1}{m\Delta x^2}\delta_x^2 v_i^{n+\frac{l}{m}} + \frac{\Delta t_1}{m}g(x_i, (n+\frac{l}{m})\Delta t_1, u_i^n, v_i^{n+\frac{l}{m}})$$
$$i = 1,\cdots,M-1, \quad l = 0,\cdots,m-1.$$

The question is whether the accuracy of the original time integration algorithm is significantly affected by using (9) in the right hand side term for the computation of $v_i^{n+\frac{l+1}{m}}$ in (10).

Since the FTCS algorithm is first order accurate in time and second order accurate in space, the truncation error term in (6) is $O(\Delta t) + O(\Delta x^2)$. Since $\Delta t_1 = m\Delta t_2$, the two equations in (8) have the same order of accuracy asymptotically. By substituting (9) into the right hand side of the second equation in (8), we have

$$\begin{aligned}
&\frac{\Delta t_1}{m}g(x_i, (n+\frac{l}{m})\Delta t_1, u_i^n, v_i^{n+\frac{l}{m}}) \\
&= \frac{\Delta t_1}{m}g(x_i, (n+\frac{l}{m})\Delta t_1, u_i^{n+\frac{l}{m}} + u_i^n - u_i^{n+\frac{l}{m}}, v_i^{n+\frac{l}{m}}) \\
&= \frac{\Delta t_1}{m}g(x_i, (n+\frac{l}{m})\Delta t_1, u_i^{n+\frac{l}{m}}, v_i^{n+\frac{l}{m}}) + \frac{\Delta t_1}{m}C_1(u_i^n - u_i^{n+\frac{l}{m}}) \\
&= \frac{\Delta t_1}{m}g(x_i, (n+\frac{l}{m})\Delta t_1, u_i^{n+\frac{l}{m}}, v_i^{n+\frac{l}{m}}) + \frac{\Delta t_1}{m}C_1 \cdot C_2\Delta t_1
\end{aligned}$$

where we have assumed that $|g_u| \le C_1$ and $|u_t| \le C_2$. It is clear that the additional error due to the use of (9) is also $O(\Delta t_1)$ if the right hand side function g_u and u_t are bounded. This can be extended to a general explicit time integration algorithm. The approximation given by (9) can be interpreted as an interpolation of order 0. If the time integration algorithm used is accurate of order q in time, then an interpolation of order $q-1$ for $u_i^{n+\frac{l}{m}}$ is needed to maintain the asymptotic order of accuracy of the original algorithm. For example, if the time integration algorithm is second order accurate in time, then we need to use

$$u_i^{n+\frac{l}{m}} = \frac{l}{m} u_i^{n+1} + \frac{m-l}{m} u_i^n + O(\Delta t^2) \tag{11}$$

to approximate $u_i^{n+\frac{l}{m}}$ for computing $v_i^{n+\frac{l+1}{m}}$. For practical implementation, we can first finish the computation for u_i^{n+1}, and then start the computation for $v_i^{n+\frac{l+1}{m}}$ using (10).

When a qudratic or higher order interpolation in the form of

$$u_i^{n+\frac{l}{m}} = \sum_{j=-q+2}^{1} \alpha_j u^{n+j}$$

has to be used for a time integration algorithm, there will not be enough data at the beginning of the computation to construct the interpolation. It can be shown that we can still maintain the asymptotic order of accuracy of the original time integration algorithm by starting with a linear interpolation at $n = 1$, then a qudratic interpolation at $n = 2$, and gradually increasing the order of interpolation to the required order of $q - 1$. After $q - 1$ time steps, we can maintain this interpolation for all the subsequent time steps.

3. LOCAL TIME STEPPING FOR IMPLICIT ALGORITHMS

Traditionally, local time stepping has been used in combination with only explicit time integration algorithms for solving time dependent PDEs on nonuniform grids. This is mainly because with implicit algorithms, the computation for solutions at different spatial points is not independent to each other. A system of coupled linear or nonlinear algebraic equations must be solved at each time step, which prevents the use of different time steps at different spatial grid points. It is interesting to note that when we extend the idea of local time stepping to the use of different time steps for different equations in some coupled systems, the idea can be readily applied to implicit time integration algorithms.

To demonstrate how local time stepping can be used with implicit algorithms to improve the computational efficiency, let us apply the Crank-Nicolson algorithm to Eq. (5):

$$\frac{u_i^{n+1} - u_i^n}{\Delta t} = \frac{\alpha}{2}\left[\frac{\delta_x^2 u_i^{n+1}}{\Delta x^2} + \frac{\delta_x^2 u_i^n}{\Delta x^2}\right] + \frac{1}{2}[f(x_i, t_{n+1}, u_i^{n+1}, v_i^{n+1}) + f(x_i, t_n, u_i^n, v_i^n)], \tag{12}$$

$$\frac{v_i^{n+1} - v_i^n}{\Delta t} = \frac{\beta}{2}\left[\frac{\delta_x^2 v_i^{n+1}}{\Delta x^2} + \frac{\delta_x^2 v_i^n}{\Delta x^2}\right] + \frac{1}{2}[g(x_i, t_{n+1}, u_i^{n+1}, v_i^{n+1}) + g(x_i, t_n, u_i^n, v_i^n)],$$

which can also be written in a matrix form as

$$Aw^{n+1} = Bw^n + b^{n+1} \tag{13}$$

where matrices A and B are block tridiagonal with a dimension of $2 \times (M - 1)$ and $w = (u, v)^T$. Since we have assumed that β is significantly larger than α, it is reasonable to expect that the variable v will change much faster than u. The time step Δt used for computing v may be unnecessarily small for computing u with the same accuracy. Note that the rational for using different time steps here is different from that with explicit algorithms. It is the accuracy requirement, not stability requirement, that dictates the size of time steps here.

In order to facilitate the use of different time steps for different equations when an implicit algorithm is used, we need to decouple Eq. (12). This can be accomplished by treating the diffusion terms implicitly and the nonlinear reaction terms explicitly:

$$\frac{u_i^{n+1}-u_i^n}{\Delta t} = \frac{\alpha}{2}\left[\frac{\delta_x^2 u_i^{n+1}}{\Delta x^2}+\frac{\delta_x^2 u_i^n}{\Delta x^2}\right]+\frac{3}{2}f(x_i,t_n,u_i^n,v_i^n)-\frac{1}{2}f(x_i,t_{n-1},u_i^{n-1},v_i^{n-1}), \tag{14}$$

$$\frac{v_i^{n+1}-v_i^n}{\Delta t} = \frac{\beta}{2}\left[\frac{\delta_x^2 v_i^{n+1}}{\Delta x^2}+\frac{\delta_x^2 v_i^n}{\Delta x^2}\right]+\frac{3}{2}g(x_i,t_n,u_i^n,v_i^n)-\frac{1}{2}g(x_i,t_{n-1},u_i^{n-1},v_i^{n-1}).$$

It is clear from (14) that the two equations are now decoupled at the new time level t_{n+1}. We can then use different time steps for these two different equations. Similar to the sitution of explicit algorithms, we can assume that $\Delta t_1 = m\Delta t_2$ and obtain the following scheme

$$\frac{u_i^{n+1}-u_i^n}{\Delta t_1} = \frac{\alpha}{2\Delta x^2}[\delta_x^2 u_i^{n+1}+\delta_x^2 u_i^n]+\frac{3}{2}f(x_i,t_n,u_i^n,v_i^n)-\frac{1}{2}f(x_i,t_{n-1},u_i^{n-1},v_i^{n-1}), \tag{15}$$

$$\frac{v_i^{n+\frac{l+1}{m}}-v_i^{n+\frac{l}{m}}}{m\Delta t_1} = \frac{\beta}{2\Delta x^2}[\delta_x^2 v_i^{n+\frac{l+1}{m}}+\delta_x^2 v_i^{n+\frac{l}{m}}]+\frac{3}{2}g(x_i,t_{n+\frac{l}{m}},u_i^{n+\frac{l}{m}},v_i^{n+\frac{l}{m}})$$
$$-\frac{1}{2}g(x_i,t_{n+\frac{l-1}{m}},u_i^{n+\frac{l-1}{m}},v_i^{n+\frac{l-1}{m}}).$$

The terms $u_i^{n+\frac{l}{m}}$ and $u_i^{n+\frac{l-1}{m}}$ can be handled in a similar way as in the case of explicit algorithms. Since The C-N algorithm is second order accurate in both time and space, the truncation error in (14) is $O(\Delta t^2)+O(\Delta x^2)$. By substituting a linear interpolation (11) into the right hand side of the second equation in (15), we have

$$g(x_i,t_{n+\frac{l}{m}},\frac{l}{m}u_i^{n+1}+\frac{m-l}{m}u_i^n,v_i^{n+\frac{l}{m}}) = g(x_i,t_{n+\frac{l}{m}},u_i^{n+\frac{l}{m}}+O(\Delta t^2),v_i^{n+\frac{l}{m}})$$
$$= g(x_i,t_{n+\frac{l}{m}},u_i^{n+\frac{l}{m}},v_i^{n+\frac{l}{m}})+O(\Delta t^2)$$

which shows that the original order of accuracy is maintained. The matrix form of (15) is

$$A_1u^{n+1} = B_1u^n+b_1^n, \qquad n=0,\cdots,N-1 \tag{16}$$

$$A_2v^{n+\frac{l+1}{m}} = B_2v^{n+\frac{l}{m}}+b_2^{n+\frac{l}{m}}, \qquad n=0,\cdots,N-1, \quad l=0,1,\cdots,m-1,$$

where A_1 and A_2 are tri-diagonal matrices of order $M-1$. There are several advantages using this local time stepping approach:

- Since the diffusion terms are treated implicitly in (14), it can be shown that the local time stepping algorithm (15) is still unconditionally stable.
- Different time steps can be used for different equations depending on the accuracy requirement, thereby improving the computational efficiency. Furthermore, for most two or three-dimensional problems, it is more efficient to solve two systems of algebraic equations of order $M-1$, like those in (16), than to solve a single system of order $2\times(M-1)$ in (13).
- For reaction-diffusion equations with linear diffusion and nonlinear reaction terms, this approach does not require linearization as the traditional C-N algorithms does.

Table 1: Maximum error at $T = 0.1$. $\frac{\Delta t_2}{\Delta x^2} = \frac{1}{2}$.

Δx	m = 1	m = 2	m = 4	m = 8	m = 16
0.1	0.7590e-3	0.7592e-3	0.7593e-3	0.7593e-3	0.7595e-3
0.05	0.1909e-3	0.1910e-3	0.1910e-3	0.1910e-3	0.1910e-3
0.01	0.7630e-5	0.7630e-5	0.7630e-5	0.7631e-5	0.7633e-5

4. NUMERICAL EXPERIMENTS

In this section, we will present three numerical examples to demonstrate the use of local time stepping for solving partial differential equations with multiple time scales.
Example 1: In this example, the following system of equations is solved using the FTCS algorithm with local time stepping:

$$u_t = \frac{1}{4}u_{xx} + 0.5v - u + f(x,t) \tag{17}$$
$$v_t = 4v_{xx} - v(v-1) - u + g(x,t)$$

where $f(x,t) = \frac{13}{16}e^{-\frac{t}{4}}\sin(\frac{x}{2}) - 0.5e^{-4t}\sin(2x)$ and $g(x,t) = 11e^{-4t}\sin(2x) + e^{-8t}\sin^2(2x) + e^{-\frac{t}{4}}\sin(\frac{x}{2})$. The true solutions are $u = e^{-\frac{t}{4}}sin(\frac{x}{2})$ and $v = e^{-4t}sin(2x)$. Two different time steps Δt_1 and Δt_2 are used for the two equations in (17) separately, with $\Delta t_1 = m\Delta t_2$. The stability criteria allows a maximum value of $m = 16$. Table 1 shows the error $max|w_i^* - w_i^c|$ at $T = 0.1$, where $w_i^* = (u_i^*, v_i^*)^T$ and $w_i^c = (u_i^c, v_i^c)^T$ are the exact solution and the computed solution of point i at the final time level, respectively. It is clear that with m up to the maximum value of 16, the computed solutions are just as accurate as those computed using a uniform small time step ($m = 1$).
Example 2. In this example, we solve the following system of equations using the C-N algorithm with local time stepping:

$$u_t = \frac{1}{4}u_{xx} + u(u-v) + f(x,t) \tag{18}$$
$$v_t = 4v_{xx} + u^2(1-v) + g(x,t)$$

where

$$f(x,t) = \frac{1}{4}\cos(\frac{t}{4})\sin(\frac{x}{2}) + \frac{1}{16}\sin(\frac{t}{4})\sin(\frac{x}{2}) - \sin(\frac{t}{4})\sin(\frac{x}{2})[\sin(\frac{t}{4})\sin\frac{x}{2} - \sin(4t)\sin(2x)],$$

$$g(x,t) = 4\cos(4t)\sin(2x) + 16\sin(4t)\sin(2x) - (1 - \sin(4t)\sin(2x))\sin^2\frac{t}{4}\sin^2\frac{x}{2}.$$

The true solution is $u = \sin\frac{t}{4}\sin\frac{x}{2}$ and $v = \sin 4t \sin 2x$. Table 2 shows the maximum errors at $T = 0.1$ defined the same way as in Example 1. For the coupled method, the two equations in (18) are discretized using the C-N algorithm and the final system is linearized using Newton's method. For the decoupled method, since the nonlinear terms in (18) are treated explicitly, there is no need for linear linearization. As shown in Table 2, the accuracy of the decoupled method is the same as the coupled method. But the savings in CPU time is significant. Table 3 shows the timing results of coupled C-N method, decoupled C-N method with the same time step, and decoupled method using $m\Delta t$ for the first equation and Δt for the second equation. The accuracy for all cases are comparable. We can see from the table that by decoupling the system of equations, the CPU time reduces from 67.5 seconds to 7.6 seconds. By using larger time steps for the first equation, we can further reduce the CPU time to 4.5 seconds.

Table 2: Maximum error at $T = 0.1$. C-N method.

Methods	$\Delta x = 0.1, \Delta t = 0.01$	$\Delta x = 0.05$, $\Delta t = 0.005$	$\Delta x = 0.01$, $\Delta t = 0.001$
Coupled	0.4110e-3	0.1034e-3	0.4140e-5
Decoupled	0.4107e-3	0.1034e-3	0.4139e-5

Table 3: Timing results of C-N method ($\Delta x = 0.01$, $\Delta t = 0.0001$)

Coupled	Decoupled	$m = 2$	$m = 4$	$m = 8$	$m = 20$
67.5	7.6	6.0	5.2	4.7	4.5

5. CONCLUSIONS

From the truction analysis and numerical experiments presented in this paper, it is clear that we can use local time stepping algorithms to solve systems of nonlinear diffusion-reaction equations with multiple time scales and linear diffusion terms. By decoupling those equations in a system, and using different time steps for different equations, we can signifcantly improve the computational efficiency of existing numerical algorithms.

REFERENCES

1. L. Cao and J. Zhu, IMACS Lecture Notes in Computational and Applied Mathematics, (1998) (in press).

2. W. Zhong, X. Zhuang and J. Zhu, Applied Mathematics and Computation, (1998) (in press).

3. W. Zhong, J. Zhu and X. Zhong, Computer Methods in Applied Mechanics and Engineering, 130, 163-178 (1996).

4. A. H. van den Boogaard, R. de Borst and P. A. J. van den Bogert, Communications in Numerical Methods in Engineering, 10, 837-844 (1994).

5. K. Chen, M. J. Baines and P. K. Sweby, J. Computational Physics, 105, 324-332 (1993).

6. P. Smolinski, Computer Methods in Applied Mechanics and Engineering, 94, 25-34 (1992).

7. X. D. Zhang, J. Y. Trepanier, M. Reggio and R. Camarero, AIAA Journal. 32, 1926-1929 (1994).

8th Int. Coll. on Differential Equations, pp. 105–110
D. Bainov (Ed)

Improving Laguerre's Method to Cope with Symmetry Pitfalls in Polynomial Root-Finding

TIEN CHI CHEN
United College, the Chinese University of Hong Kong, Shatin, Hong Kong
(Present Address: 6566 Tam O'Shanter Drive, San Jose, CA 95120 USA)
email: tcchenuccu@aol.com

Abstract. Symmetrically positioned zeros attract distant iterates yet expel nearby ones, causing poor convergence. An effective countermeasure, used successfully for the *C*luster-*A*dapted *M*ethod (CLAM), has now been tailored for Laguerre's method, producing excellent results in limited tests.

Keywords: Iterative root-finding, symmetry, symmetric cluster, total expulsion, local expulsion, rebound, CLAM, Laguerre's method.

AMS Classification: 65H05

1. INTRODUCTION

Despite its long history, research in iterative polynomial root-finding continues to yield surprises, especially as the role of distance and symmetry is beginning to be understood[1][2]. We shall see that common methods are insensitive to physical distances but vulnerable to expulsion by symmetric (sub-)clusters of zeros. This paper is an attempt to bring Laguerre's method to the high level of robustness and effectiveness as the CLAM (*C*luster-*A*dapted *M*ethod) package.

2. PRELIMINARIES

2.1. Notations and conventions

Our aim is to approximate one or more of the zeros, with associated multiplicities, of a monic polynomial, given only its coefficients. The polynomial is $P(z) = \Sigma b_i z^i = \Pi(z-\alpha_j)^{m_j}$ with $b_N = 1$, $\Sigma i = \Sigma m_j = N$. The centroid of the zeros, weighted by the multiplicities, is $C \equiv -b_{N-1}/Nb_N = \Sigma m_j\alpha_j/\Sigma m_j$. We shall refer often to the symmetric cluster polynomial $S_\lambda(z; c, R, M) \equiv [(z-c)^\lambda - R^\lambda]^M$, with integer fanout of λ and M-fold zeros [2].

During computation we shall need the effective polynomial degree n, and the local centroid c; these are set initially to N and C respectively. A quantity, say q, referring to the ν-th iteration is designated as $q^{(\nu)}$, and where no ambiguity arises, the superscript is omitted; we also use $q*, q**$ as abbreviations of $q^{(\nu+1)}$ and $q^{(\nu+2)}$ respectively.

The principal value of a fractional power is taken as follows, with $-\pi < \theta \leq \pi$:

$$(a+ib)^{u/v} = (re^{i\theta})^{u/v} = |r|^{u/v}e^{i\theta u/v} = \rho e^{i\phi} = x+iy.$$

The $(\nu+1)$-st iteration uses the multiplicity estimate m to generate the new iterate

$$\begin{aligned}
z* &= z - G(\mu;m,n)/s_1 \\
\text{with} \quad G(\mu;m,n) &= m \quad \text{(Newton's method)} \\
&= 2m\mu/(m+\mu) \quad \text{(Halley's method)} \\
&= n/\{1+[(n/\mu-1)(n/m-1)]^{1/2}\} \quad \text{(Laguerre's method)} \\
&= n(1-Q^{m/n})/(1-Q) \qquad \text{(CLAM[2])} \\
\text{where} \quad s_k &\equiv [(-1)^{k-1}/(k-1)!]d^k \ln P(z)/dz^k = \Sigma m_j/(z-\alpha_j)^k \\
\mu &\equiv s_1^2/s_2 = 1/[1-PP''/P'^2] \\
\text{and} \quad Q &= (n/\mu-1)/(n/m-1)
\end{aligned}$$

we also demand $n \neq 0, m$ or μ, lest the two third-order methods lose their fractional powers and degenerate into second order. We shall consider no further the methods by Newton and Halley because of their problems with even just one reflective symmetry axis. Most well-known is the inability to find a complex zero of a real polynomial using a real guess, but the failure persists, except for roundoff errors, despite coordinate translation, rotation and/or stretching [1].

For accuracy, our numerical results were actually obtained in terms of $(1/s_1)$, $PP''/P'^2 = (1-1/\mu)$, and $(1-Q) = n[(1-m)+mPP''/P'^2]/(n-m)$. Using the STSC APL9 package with double-precision floating-point arithmetic on an IBM-compatible personal computer, convergence is said to occur when $E \equiv |P(z)/b_N z^N| \leq 2 \cdot 10^{-14}$.

2.2. Multiplicity estimates and initial guesses

A computed m steers the iterates, and the best choices should bring the iterates near zero-clusters, then home in on one of the component zeros; in doing so m should be allowed to vary with iterate position, and need not match the true multiplicity of the target zero, except for the final few iterations.

For CLAM we define

$$w = \lfloor 1/2 + n/(|n/\mu - 1| + 1) \rfloor;$$

$$\text{then } m = m(z,\nu) = w \text{ iff } (n > w \geq 1) \text{ and } [(\nu \bmod 10) \neq 9], \text{ else } m = 1.$$

As z approaches, say, α_1, then w approaches m_1; and if $P(z) = S_\lambda(z;C,R,M)$, with z lying on the circle passing through all zeros, then w also gives the correct multiplicity M. The reversion to unity every 10 iterations is an escape mechanism designed to break out of persistent misassignments, if any. The same algorithm is used for Laguerre's method in Examples 1-3 below, though minor fine-tuning is suggested in Section 5.

An excellent initial guess for all methods is $z^{(0)} \equiv C + [-P(C)]^{1/N}/b_N$. It is a zero of $S_\lambda(z;C,R,N/\lambda)$ when the multiplicity N/λ is an odd integer, most notably unity. For general polynomials this guess is embedded within the collection of zeros without becoming C itself, unless C is itself a zero.

3. EXPULSION BY SYMMETRIC (SUB-)CLUSTERS OF ZEROS

3.1. Total repulsion

Starting with an adequate initial guess, and navigated by recomputed m at every iteration, one can expect rapid convergence if the zero collection does not possess high symmetry. Otherwise extra measures may be needed, and Laguerre's method is at a relative disadvantage.

Example 1. For $P_1(z) = z^{19} - 1 \equiv S_{19}(z; 0, 1, 1)$, we use $z^{(0)} = 1 + \max|b_i| = 2$, a real point on the Cauchy circle containing all zeros; the nearest zero to it is $\alpha_1 = 1$. Laguerre's method with $n = N$, $m = 1$ throughout gives $z^{(1)} = 0.049$; however, $z^{(2)}$ overflows. In fact, for real $z^{(0)}$ outside the interval $(0.736, 1.311]$, the iterates fluctuate about α_1 with increasing amplitude until overflow, though convergence occurs for the real guess within the interval. CLAM on the other hand, gives α_1 in just one iteration.

Comment. The centroid of symmetric clusters, while an attractor for distant guesses, also repels nearby ones. $S_\lambda(z; C, R, M)$ has λ symmetry axes intersecting at C; inside the cluster, $|s_1|$ and μ are both very small. Laguerre's method, with $n = N$ and arbitrary m, handles the cases with $\lambda \leq 4$ quite well, but if $\lambda \geq 5$ with z outside a roughly annular region enclosing all zeros, its iterates ping-pong between the outer and the inner regions, skipping the annular region, each inward pass getting closer to the centroid, and each outward pass farther away from C, until eventual overflow. The phenomenon with all zeros acting in concert to fling the iterate far away from the convex hull containing all zeros shall be called *total expulsion*, and the polynomial responsible could be $S_\lambda(z; C, R, M)$ or products of symmetric cluster polynomials sharing the same centroid. Virtually all methods, most notably Newton's method, are vulnerable. Because μ/s_1 is also small, Halley's method produce miniscule iterate displacements, which could be misinterpreted as convergence[1]. CLAM is unique in being immune to total expulsions: starting almost anywhere, it reaches a zero of S_λ in just one iteration, except for roundoff errors.

3.2. Symmetric subclusters and local expulsions

Example 2. $P_2(z)$ has zeros $(\pm 1 \pm 0.01 \pm 0.01i)$, forming a dumbell-structure with two 4-th degree symmetric subclusters each of radius $0.01\sqrt{2}$ with local centroids at ± 1. Using CLAM with $z^{(0)} = 200$, and the computed m for each iteration, the iterate becomes attracted by one of the subcluster centroids ($z^{(7)} = 1.00007$). However, the iterate which has entered one subcluster (with multiplicity 4) is immediately rebounded from it, towards the centroid of the opposite subcluster. It took 24 iterations to converge to $(-0.99 + 0.01i)$.

Comment. Iterates from all methods, including CLAM, are vulnerable to expulsion by local zero subclusters with symmetry. To an iterate inside, the subcluster is effectively transparent, as it contributes little to $|s_1|$ and $|\mu|$. The next iterate is expelled from the prior host subcluster, towards zeros outside which provide the only major attraction. This *local expulsion* would be tolerable if accompanied by a large reduction in $|P|$; but more often than not, $|P|$ may even increase by orders of magnitude, undoing the work of many prior iterations. The new iterate, or its progenies, could rebound again and again from subclusters, even the same ones, possibly without end.

4. CLAM

4.1. CLAM and the symmetric cluster polynomial

Not only is CLAM immune from the total expulsion by symmetric clusters, it converges to one of its zeros in one iteration. The reason is, the formula is derivable by matching the value and the first two derivatives of $P(z)$ by $S_\lambda(z; C, R, M)$. Indeed as symmetry plays such an important role in convergence, the multiplicity guess m and the initial guess are all based on properties of S_λ.

4.2. Expulsion countermeasures designed for CLAM

Notwithstanding, on applying the raw CLAM formula iterates can still fall prey to local expulsion.

Typically, $z \equiv z^{(\nu)}$, outside the subcluster, is attracted by a local subcluster centroid. The multiplicity estimate $m \equiv m^{(\nu)} > 1$ brings $z* \equiv z^{(\nu+1)}$ inside the cluster; then local expulsion occurs, putting $z** \equiv z^{(\nu+1)}$ outside. This entire process of iterate entry followed immediately by exit is called *rebounding*; and the available prior history points the way to its detection and correction.

In using the CLAM package, rebounding is considered detected if

$$m \equiv m^{(\nu)} > 1$$

with either $|P(z**)/P(z*)| \geq 20$ (Symptom 1)

and/or $|P(z**)/P(z*)| \geq 0.75$, with $[|z** - z*|/|z* - c| > m\pi/n]$ (Symptom 2)

Our eXPulsion Correction algorithm, or XPC, assumes the local cluster to be $S_m(z;z*,R,1)$ with $r = [-P(z*)]^{1/m}$. A point on the assumed periphery of the cluster, away from the banished $z**$, is taken to be the new $z**$, as follows:

$$\begin{aligned} (z**)_{\text{new}} &= z* + |R| e^{i\phi} \\ \text{with} \quad \phi &= \tan^{-1}\{[\Im(z*) - \Im(z**)]/[\Re(z*) - \Re(z**)]\} \end{aligned}$$

with $(m, z*)$ becoming the new (n, c) until the next rebound, if any.

In addition, one could force m to unity to prevent deep entry of the next iterate into subclusters, but at the risk of slow convergence and wrong ultimate multiplicity identification. This choice, taken once every 10 iterations, has been built into the CLAM multiplicity estimate.

Example 3. For the problem in Example 2, rebounding is detected and XPC duly taken. Then monotonic convergence occurs, converging in only three more iterations to the zero $(1.01-0.01i)$,for a total of 11 iterations. The computed initial guess $(0.9239+0.3827i)$ reduces the iteration count to 5, with XPC applied at the second iteration.

4.3. Iterating with CLAM

Equipped with rebound countermeasures and steered by the multiplicity estimate, the CLAM package has never failed to converge to a zero in extensive testing. Convergence in global iterations, starting from almost anywhere in the complex plane (except at isolated points), has been demonstrated experimentally; using the initial guess, in the overwhelming majority of cases convergence is reached in fewer than 10 iterations using the initial guess.

When more than one zero is sought, conventional deflation is made: the coefficients of the deflated polynomial, namely P_{new} = the polynomial part of $[P(z)/(z-\alpha)^m]$ replaces those of P(z).

During the last few iterations before a zero is found, the multiplicity estimates tend to switch from m_1 (say) to unity; to compensate, the zero just found is taken as the initial guess after every deflation. To find all zeros of a polynomial, the average cost is only about five iterations per zero; this is because once the problem symmetry of the zeros is broken through the extraction of a zero, subsequent iterations tend to encounter fewer expulsions. Also, the last two zeros require but one iteration.

5. RECONSIDERING LAGUERRE'S METHOD

5.1. Comparisons with CLAM

Like CLAM, Laguerre's method is also classically third-order in converging towards a zero, say α_1, for arbitrary $n \neq m = m_1$. It enjoys a lower cost per iteration, and where problem symmetry is not pronounced, it usually converges in fewer iterations and with less roundoff error. If all zeros are real, then a real z converges towards a zero monotonically without the need for complex arithmetic.

If $P = [(z-\alpha)(z-\beta)]^m$, the two methods with $n = 2m$ coalesce, and converge to α in one iteration with arbitrary z.

When $n = m_1 \neq m$ the method is of order 1.5, vs. $(1 + m/n)$ in CLAM. Thus its iterates require fewer iterations than CLAM towards the centroid of a distant cluster; a distinct advantage if the cluster has low symmetry. For high symmetry (sub-)clusters, however, this behavior only increases the chance for expulsions. CLAM, on the other hand, experience neither total expulsions, nor local expulsions if $m = 1$.

5.2 Expulsion countermeasures

It appears that most of the supporting algorithms designed for CLAM can be used unchanged for Laguerre's method, but clearly the expulsion countermeasures devised for CLAM cannot be applied in unchanged form. The assumption of $m > 1$ is particularly untenable; as the iterate $z*$ can be just within the subcluster while z is just outside, even with $m = 1$. The algorithm for estimating the cluster radius is likewise open to doubt; in Example 1, for instance, the total multiplicity of the responsible cluster is actually $n = N$, and the proper R is $[-P(C)]^{1/n}$.

For Laguerre's method we use the CLAM schemes for detection and correction as a basis, but can no longer exclude the $(m = 1)$ case. We also add the condition that, if $m = 1$, then n and c should remain *unchanged.* The inclusion of the $m = 1$ case not only fills an important gap in local expulsion handling; it solves incidentally perhaps the most important total expulsion problem as well, for $m = 1$ is the default assignment whenever the computed m appears unreasonable. Here the new iterate $z**$ is placed on the circle centered at C with radius $R = |[-P(z^{(\nu)})]^{1/N}|$, away from the old $z^{(\nu+2)}$, similar to an automatic invocation of the initial guess algorithm, though using an assumed centroid and a different azimuthal angle. Nevertheless, provision for total expulsion with $m \neq 1$ is still lacking.

Because of a greater propensity for expulsions, the multiplicity estimate is set to 1 every 5 iterations, whenever $\nu \bmod 5 = 4$. XPC for consecutive iterations is disallowed, also for the first iteration, for want of exploitable prehistory.

6. PERFORMANCE COMPARISONS

Example 4. For the prior example $P_1 = (z^{19} - 1)$ the improved Laguerre method with $z^{(0)} = 1.5$ converges in 6 iterations, calling for XPC during the second iteration. ($z^{(0)} = 2.0$

leads to almost irretrievable overflow, even though the overflowed $z**$ is to be replaced anyway.) CLAM converges in *one* iteration. For P_2, with $\{\alpha_j\} = (\pm 1 \pm 0.01 \pm 0.01i)$, Laguerre's method takes 9 iterations with XPC at the 5th; CLAM takes 11, with XPC at the 7th.

Example 5. To find all zeros of P_3 with 10 real simple zeros $\{0.1, 0.2, 0.3, \ldots 0.9, 1.0\}$. Laguerre's method takes 30 iterations, averaging 3 iterations (9 horners) per zero, while CLAM takes 33 iterations, or 3.3 iterations/zero. For unrestricted initial guesses this problem is troublesome for CLAM; for example, with $z^{(0)} = 2i$, CLAM falls into a trap with $m = 7$, and requires the enforced multiplicity setting of $m^{(9)} = 1$ to converge to a zero in 14 iterations. Laguerre's method with its enforced multiplicity setting takes 7 iterations; adopting the same setting, CLAM takes XPC at the 9th iteration, and takes 15 iterations to converge.

Example 6. To find all zeros of $P_4 = (z-3)^4(z^2+9)$, an example from Atkinson's textbook, Muller's method takes 108 horners, or 18 horners per zero[3]. Both Laguerre's method and CLAM take only 3 iterations, or 1.5 horners per zero. All zeros of $P_5 = (z-1-i)^2[(z-4)^2+9](z-3.999+3i)$ take Ralston and Rabinowitz' modified Newton method 30.3 horners/zero and Traub-Jenkins' method 16.6 horners/zero [4]. CLAM takes 10 iterations (6 horners/zero), and Laguerre's method takes only 8 iterations (4.8 horners/zero), and is slightly more accurate. XPC was not needed for this unsymmetric problem.

Example 7. CLAM takes 69 iterations to find all zeros of P_6 with $\{\alpha_j\} = (\pm 1 \pm 0.01 \pm i \pm 0.01i)$, averaging 12.94 horners/zero, plus 3 XPCs. Laguerre's method takes 68 iterations (12.75 horners/zero), plus 5 XPCs.

7. SUMMARY AND CONCLUSIONS

The CLAM formula, supported by devised initial guesses, multiplicity estimates and expulsion countermeasures, usually finds a zero in just a few iterations. Laguerre's method is far more vulnerable to expulsions by symmetrical collection of zeros, but, reinforced by countermeasures for *both* total and local expulsions, it now appears to be at least a match for CLAM in versatility and efficiency. Starting with a lower cost per iteration, it often takes fewer iterations, and appears to be slightly more accurate, though may encounter more expulsions. More testing is needed, however, before the improved Laguerre method can be used with full confidence.

References

1. Tien Chi Chen, ACM Signum Newsletter, 32, 40-43 (January-April 1997).

2. Tien Chi Chen, in: Lecture Notes in Computer Science 1196, Numerical Analysis and Applications, (Proc. First International Workshop (**WNAA'96**), Rousse, Bulgaria, June 1996), L. Vulkov, J. Wasniewski and P. Yalamov, (Eds.), Springer, (Berlin 1997) pp. 98-105.

3. Kendall E. Atkinson, An Introduction to Numerical Analysis, Wiley, New York (1978) p. 87.

4. A. Ralston and P. Rabinowitz, A First Course in Numerical Analysis, 2nd Ed., McGraw-Hill, New York (1978) pp. 391-395.

8th Int. Coll. on Differential Equations, pp. 111–117
D. Bainov (Ed)

A Nonlinear Wave Equation with Jumping Nonlinearity

Q-HEUNG CHOI
Department of Mathematics, Inha University, Incheon 402-751, Korea

TACKSUN JUNG
Department of Mathematics, Kunsan N. University, Kunsan 573-701, Korea

Abstract. We investigate multiplicity of solutions $u(x,t)$ for a piecewise linear perturbation $-(bu^+ - au^-)$ of the one-dimensional wave operator $u_{tt} - u_{xx}$ under Dirichlet boundary condition on the interval $(-\frac{\pi}{2}, \frac{\pi}{2})$ and periodic condition on the variable t. Our concern is to investigate multiplicity of solutions of (0.4) when the nonlinearity crosses two eigenvalues and the source term f is generated by two eigenfunctions ϕ_{00}, ϕ_{10}.
Keywords: Multiplicity of solutions, eigenvalue, eigenfunction
1991 AMS Subject Classification: 35L20

0. Introduction

We investigate multiplicity of solutions $u(x,t)$ for a piecewise linear perturbation $-(bu^+ - au^-)$ of the one-dimensional wave operator $u_{tt} - u_{xx}$ under Dirichlet boundary condition on the interval $(-\frac{\pi}{2}, \frac{\pi}{2})$ and periodic condition on the variable t,

$$u_{tt} - u_{xx} + bu^+ - au^- = f(x,t) \quad \text{in} \quad (-\frac{\pi}{2}, \frac{\pi}{2}) \times R, \tag{0.1}$$

$$u(\pm\frac{\pi}{2}, t) = 0, \tag{0.2}$$

$$u \text{ is } \pi - \text{periodic in } t \text{ and even in } x \text{ and } t. \tag{0.3}$$

When a string with nonuniform density vibrates up and down, the upward restoring coefficient and the downward restoring coefficient of it are different. Hence it happens a nonlinear perturbation in a wave equation. Here we assumed that the upward restoring coefficient and the downward one in the vibrating of the string are constant and they are different.

We let L the wave operator, $Lu = u_{tt} - u_{xx}$. Then the eigenvalue problem for $u(x,t)$

$$Lu = \lambda u \quad \text{in} \quad (-\frac{\pi}{2}, \frac{\pi}{2}) \times R$$

with (0.2) and (0.3), has infinitely many eigenvalues

$$\lambda_{mn} = (2n+1)^2 - 4m^2 \qquad (m, n = 0, 1, 2, \cdots)$$

and corresponding normalized eigenfunctions $\phi_{mn}(m, n \geq 0)$ given by

$$\begin{aligned} \phi_{0n} &= \frac{\sqrt{2}}{\pi} \cos(2n+1)x \text{ for } n \geq 0, \\ \phi_{mn} &= \frac{2}{\pi} \cos 2mt \cos(2n+1)x \text{ for } m > 0, n \geq 0. \end{aligned}$$

We note that all eigenvalues in the interval $(-9, 9)$ are given by

$$\lambda_{21} = -7 < \lambda_{10} = -3 < \lambda_{00} = 1 < \lambda_{11} = 5.$$

Let Q be the square $[-\frac{\pi}{2}, \frac{\pi}{2}] \times [-\frac{\pi}{2}, \frac{\pi}{2}]$ and H the Hilbert space defined by

$$H = \left\{ u \in L^2(Q) : \quad u \text{ is even in } x \text{ and } t \quad \right\}.$$

Then the set of eigenfunctions $\{\phi_{mn}\}$ is an orthonormal base in H. Hence equation (0.1) with (0.2) and (0.3) is equivalent to

$$Lu + bu^+ - au^- = f \quad \text{in} \quad H. \tag{0.4}$$

Our concern is to investigate multiplicity of solutions of (0.4) when the nonlinearity $-(bu^+ - au^-)$ crosses finite eigenvalues and the source term f is generated by two eigenfunctions ϕ_{00}, ϕ_{10}.

1. A Variational Reduction Method

We investigate multiplicity of solutions $u(x,t)$ for a piecewise linear perturbation $-(bu^+ - au^-)$ of the one-dimensional wave operator $u_{tt} - u_{xx}$ with the nonlinearity $-(bu^+ - au^-)$ crosses two eigenvalues. We suppose that $-5 < a < -1$ and $3 < b < 7$. Under this assumption, we have a concern with a relation between multiplicity of solutions and source terms of a nonlinear wave equation

$$Lu + bu^+ - au^- = f \quad \text{in} \quad H. \tag{1.1}$$

Here we suppose that f is generated by two eigenfunctions ϕ_{00} and ϕ_{10}, that is, $f = s_1\phi_{00} + s_2\phi_{10} (s_1, s_2 \in \mathbb{R})$.

Let V be the two dimensional subspace of H spanned by $\{\phi_{00}, \phi_{10}\}$ and W be the orthogonal complement of V in H. Let P be an orthogonal projection H onto V. Then every element $u \in H$ is expressed by

$$u = v + w,$$

where $v = Pu$, $w = (I - P)u$. Hence equation (1.1) is equivalent to

$$Lw + (I - P)(b(v + w)^+ - a(v + w)^-) = 0, \tag{1.2}$$

$$Lv + P(b(v + w)^+ - a(v + w)^-) = s_1\phi_{00} + s_2\phi_{10}. \tag{1.3}$$

We look on (1.2) and (1.3) as a system of two equations in the two unknowns v and w.

Lemma 1.1. *For fixed $v \in V$, (1.2) has a unique solution $w = \theta(v)$. Furthermore, $\theta(v)$ is Lipschitz continuous in terms of v.*

Proof. We use the contraction mapping theorem. Let $\delta = \frac{1}{2}(a + b)$. Rewrite (1.2) as

$$(-L - \delta)w = (I - P)(b(v + w)^+ - a(v + w)^- - \delta(v + w)),$$

or equivalently,

$$w = (-L - \delta)^{-1}(I - P)g_v(w), \tag{1.4}$$

where

$$g_v(w) = b(v + w)^+ - a(v + w)^- - \delta(v + w).$$

Since

$$|g_v(w_1) - g_v(w_2)| \le |b - \delta||w_1 - w_2|,$$

we have

$$\|g_v(w_1) - g_v(w_2)\| \le |b - \delta|\|w_1 - w_2\|,$$

where $\| \ \|$ is the L^2 norm in H. The operator $(-L - \delta)^{-1}(I - P)$ is a self adjoint compact linear map from $(I - P)H$ into itself. The eigenvalues of $(-L - \delta)^{-1}(I - P)$ in W are $(\lambda_{mn} - \delta)^{-1}$, where $\lambda_{mn} \ge 7$ or $\lambda_{mn} \le -5$. Therefore its L^2 norm is $\max\{\frac{1}{7-\delta}, \frac{1}{5+\delta}\}$. Since $|b - \delta| < \min\{7 - \delta, 5 + \delta\}$, it follows that for fixed $v \in V$, the right hand side of (1.4) defines a Lipschitz mapping W into itself with Lipschitz constant $\gamma < 1$. Hence, by the contraction mapping principle, for given $v \in V$, there is a unique $w \in W$ which satisfies (1.2).

Also, it follows, by the standard argument principle, that $\theta(v)$ is Lipschitz continuous in terms of v. ∎

By Lemma 1.1, the study of the multiplicity of solutions of (1.1) is reduced to the study of the multiplicity of solutions of an equivalent problem

$$Lv + P(b(v + \theta(v))^+ - a(v + \theta(v))^-) = s_1\phi_{00} + s_2\phi_{10} \tag{1.5}$$

defined on the two dimensional subspace V spanned by $\{\phi_{00}, \phi_{10}\}$.

While one feels instinctively that (1.5) ought to be easier to solve, there is the disadvantage of an implicitly defined term $\theta(v)$ in the equation. However, in our case, it turns out that we know $\theta(v)$ for some very important v's.

If $v \ge 0$ or $v \le 0$, then $\theta(v) \equiv 0$. For example, let us take $v \ge 0$ and $\theta(v) = 0$. Then equation (1.2) reduces to

$$L0 + (I - P)(bv^+ - av^-) = 0$$

which is satisfied because $v^+ = v$, $v^- = 0$ and $(I-P)v = 0$, since $v \in V$.

Since the subspace V is spanned by $\{\phi_{00}, \phi_{10}\}$ and $\phi_{00}(x,t) > 0$ in Q, there exists a cone C_1 defined by

$$C_1 = \left\{ v = c_1\phi_{00} + c_2\phi_{10} \mid c_1 \geq 0, |c_2| \leq \frac{c_1}{\sqrt{2}} \right\}$$

so that $v \geq 0$ for all $v \in C_1$ and a cone C_3 defined by

$$C_3 = \left\{ v = c_1\phi_{00} + c_2\phi_{10} \mid c_1 \leq 0, |c_2| \leq \frac{|c_1|}{\sqrt{2}} \right\}$$

so that $v \leq 0$ for all $v \in C_3$.

Thus, even if we do not know $\theta(v)$ for all $v \in V$, we know $\theta(v) \equiv 0$ for $v \in C_1 \cup C_3$.

Now, we define a map $\Phi : V \to V$ given by

$$\Phi(v) = Lv + P(b(v+\theta(v))^+ - a(v+\theta(v))^-), \quad v \in V. \tag{1.6}$$

Then Φ is continuous on V and we have the following lemma.

Lemma 1.2. *$\Phi(cv) = c\Phi(v)$ for $c \geq 0$.*

We investigate the images of the cones C_1 and C_3 under Φ. First we consider the image of the cone C_1. If $v = c_1\phi_{00} + c_2\phi_{10} \geq 0$, we have

$$\begin{aligned} \Phi(v) &= L(v) + P(b(v+\theta(v))^+ - a(v+\theta(v))^-) \\ &= c_1\lambda_{00}\phi_{00} + c_2\lambda_{10}\phi_{10} + b(c_1\phi_{00} + c_2\phi_{10}) \\ &= c_1(b+\lambda_{00})\phi_{00} + c_2(b+\lambda_{10})\phi_{10}. \end{aligned}$$

Thus the images of the rays $c_1\phi_{00} \pm \frac{c_1}{\sqrt{2}}\phi_{10} (c_1 \geq 0)$ can be explicitly calculated and they are

$$c_1(b+\lambda_{00})\phi_{00} \pm \frac{c_1}{\sqrt{2}}(b+\lambda_{10})\phi_{10} \qquad (c_1 \geq 0).$$

Therefore Φ maps C_1 onto the cone R_1 in the right half-plane of V, where

$$R_1 = \left\{ d_1\phi_{00} + d_2\phi_{10} \mid d_1 \geq 0, |d_2| \leq \frac{1}{\sqrt{2}}\left(\frac{b+\lambda_{10}}{b+\lambda_{00}}\right) d_1 \right\}.$$

Second we consider the image of C_3. If $v = -c_1\phi_{00} + c_2\phi_{10} \leq 0$, we have

$$\begin{aligned} \Phi(v) &= L(v) + P(b(v+\theta(v))^+ - a(v+\theta(v))^-) \\ &= Lv + P(av) \\ &= -c_1\lambda_{00}\phi_{00} + c_2\lambda_{10}\phi_{10} - ac_1\phi_{00} + ac_2\phi_{10} \\ &= -c_1(\lambda_{00}+a)\phi_{00} + c_2(\lambda_{10}+a)\phi_{10}. \end{aligned}$$

Thus the images of the rays $-c_1\phi_{00} \pm \frac{c_1}{\sqrt{2}}\phi_{10}$ $(c_1 \geq 0)$ can be explicitly calculated and they are

$$-c_1(\lambda_{00}+a)\phi_{00} \pm \frac{c_1}{\sqrt{2}}(\lambda_{10}+a)\phi_{10} \qquad (c_1 \geq 0).$$

Thus Φ maps the cone C_3 onto the cone

$$R_3 = \left\{ d_1\phi_{00} + d_2\phi_{10} \mid d_1 \geq 0,\ |d_2| \leq \frac{1}{\sqrt{2}} \left(\frac{\lambda_{10}+a}{\lambda_{00}+a} \right) d_1 \right\},$$

which is in the right half-plane of V and $R_1 \subset R_3$, since $-5 < a < -1$ and $3 < b < 7$.

Last we investigate the images of the cones C_2 and C_4 under Φ, where

$$C_2 = \{c_1\phi_{00} + c_2\phi_{10} \mid c_2 \geq 0,\ |\frac{c_1}{\sqrt{2}}| \leq c_2\},$$

$$C_4 = \{c_1\phi_{00} + c_2\phi_{10} \mid c_2 \leq 0,\ |\frac{c_1}{\sqrt{2}}| \leq |c_2|\}.$$

To investigate the images of the cones C_2, C_4, we need the following lemma.

Lemma 1.3. *For every $v = c_1\phi_{00} + c_2\phi_{10}$ in V, there exists a constant $d > 0$ such that*

$$(\Phi(v), \phi_{00}) \geq d|c_2|.$$

Lemma 1.3 means that the image of Φ is contained in the right half-plane of V. That is, $\Phi(C_2)$ and $\Phi(C_4)$ are the cones in the right half-plane of V. The image of C_2 under Φ is the cone containing

$$R_2 = \left\{ d_1\phi_{00} + d_2\phi_{10} \mid d_1 \geq 0, -\frac{1}{\sqrt{2}} \left(\frac{\lambda_{10}+a}{\lambda_{00}+a} \right) d_1 \leq d_2 \leq \frac{1}{\sqrt{2}} \left(\frac{\lambda_{10}+b}{\lambda_{00}+b} \right) d_1 \right\}$$

and the image of C_4 under Φ is the cone containing

$$R_4 = \left\{ d_1\phi_{00} + d_2\phi_{10} \mid d_1 \geq 0, -\frac{1}{\sqrt{2}} \left(\frac{\lambda_{10}+b}{\lambda_{00}+b} \right) d_1 \leq d_2 \leq \frac{1}{\sqrt{2}} \left(\frac{\lambda_{10}+a}{\lambda_{00}+a} \right) d_1 \right\}.$$

We note that all the cones R_2, R_3, R_4 contain the cone R_1. Also R_3, R_2 contain the cone $R_2 \backslash R_1$, and R_3, R_4 contain the cone $R_4 \backslash R_1$. Hence we have the theorem.

Theorem 1.1. *Suppose $-5 < a < -1$ and $3 < b < 7$. Let $f = s_1\phi_{00} + s_2\phi_{10}$. Then we have :*

(i) If f belongs to the interior $IntR_1$ of R_1, (1.1) has a positive solution, a negative solution, and at least two sign changing solutions.

(ii) If f belongs to the boundary ∂R_1 of R_1, (1.1) has a positive solution, a negative solution, and at least one sign changing solution.

(iii) If f belongs to $Int(R_3 \backslash R_1)$, (1.1) has a negative solution and at least one sign changing solution.

(iv) If f belongs to ∂R_3, (1.1) has a negative solution.

Remark. *If $f = s_1\phi_{00} + s_2\phi_{10}$ and $s_1 < 0$, then (1.1) has no solution. Also, if $f = s_1\phi_{00} + s_2\phi_{10}$ and $s_1 = 0$, $s_2 \neq 0$, then (1.1) has no solution.*

Proof. If we consider the inner product

$$\left(Lu + bu^+ - au^-, \phi_{00} \right) = (f, \phi_{00}),$$

then the right hand side is s_1 and the left hand side is nonnegative. Hence we have the conclusion. ∎

Remark. *With Theorem 1.1, we can not claim that if f does not belong to R_3 then (1.1) has no solution.*

2. A Region without Solution

In this chapter, we conclude the region of source terms without solution. We assume that $-5 < a < -1, 3 < b < 7$ and f is generated by $\{\phi_{00}, \phi_{10}\}$, that is, $f = s_1\phi_{00}+s_2\phi_{10}(s_1, s_2 \in \mathbb{R})$. We consider a semilinear beam equation

$$Lu + bu^+ - au^- = f \quad \text{in} \quad H. \tag{2.1}$$

The study of the map $\Phi : V \to V$ defined in (1.6) will give a powerful theorem to conclude the region of source terms without solution. We consider the restriction $\Phi|_{C_i}(1 \leq i \leq 4)$ of Φ to the cone C_i. Let $\Phi_i = \Phi|_{C_i}$, i.e.,

$$\Phi_i : C_i \to V.$$

It follows from Lemma 1.2 that $\Phi_1 : C_1 \to R_1$ and $\Phi_3 : C_3 \to R_3$ are bijections.

Now we study the restrictions Φ_2 and Φ_4. Let $i = 2, 4$. Let γ be a simple path in R_i. We investigate the inverse image $\Phi_i^{-1}(\gamma)$ of γ and conclude the region of source terms that (1.1) has no solution. We note that $\Phi_i(C_i)$ contains R_i.

Lemma 2.1. *Let $i = 2, 4$. Let γ be any simple path in R_i with end points on ∂R_i, where each ray (starting from the origin) in R_i intersects only one point of γ. Then the inverse image $\Phi_i^{-1}(\gamma)$ of γ is a simple path in C_i with end points on ∂C_i, where any ray (starting from the origin) in C_i intersects only one point of this path.*

Proof. We note that $\Phi_i^{-1}(\gamma)$ is closed since Φ is continuous and γ is closed in V. Suppose that there is a ray (starting from the origin) in C_i which intersects two points of $\Phi_i^{-1}(\gamma)$, say, p, $\alpha p\, (\alpha > 1)$. Then by Lemma 1.2,

$$\Phi_i(\alpha p) = \alpha\Phi_i(p),$$

which implies that $\Phi_i(p) \in \gamma$ and $\Phi_i(\alpha p) \in \gamma$. This contradicts that each ray (starting from the origin) C_i intersects only one point of γ.

We regard a point p in the plane V as a radius vector. For a point v in V, we define the argument $\arg p$ of p by the angle from the positive ϕ_{00}-axis to p.

We claim that $\Phi_i^{-1}(\gamma)$ meets all ray in C_i, starting from the origin. In fact, if not, $\Phi_i^{-1}(\gamma)$ is disconnected in C_i. Since $\Phi_i^{-1}(\gamma)$ is closed and meets at most one point of any ray in C_i, there are two points p_1 and p_2 in C_i such that $\Phi_{-1}(\gamma)$ does not contain any point p with

$$\arg p_1 < \arg p < \arg p_2.$$

On the other hand, if we let l the segment in C_i with end points p_1 and p_2, then $\Phi_i(l)$ is a path in R_i, where $\Phi_i(p_1)$ and $\Phi_i(p_2)$ belong to γ. Choose a point q in $\Phi_i(l)$ such that $\arg q$ is between $\arg\Phi_i(p_1)$ and $\arg\Phi_i(p_2)$. Then there exist a point q' in γ such that $q' = \beta q$ for some $\beta > 0$ since γ is a simple path. But $\Phi_i^{-1}(q')$ satisfies

$$\arg p_1 < \arg\Phi_i^{-1}(q') < \arg p_2,$$

which is a contradiction. This completes the lemma. ∎

With Lemma 2.1, we have the following theorem, which is very important to investigate the multiplicity of solutions of a semilinear wave equation when the source term varies.

Theorem 2.1. *For $1 \le i \le 4$, the restriction Φ_i maps C_i onto R_i. Therefore, Φ maps V onto R_3. In particular, Φ_1 and Φ_3 are bijective.*

The above theorem implies Theorem 1.1. Furthermore, we conclude the region of source terms in which (1.1) has no solution.

Theorem 2.2. *Under the same condition as in Theorem 1.1, if f does not belong to the cone R_3, then equation (1.1) has no solution.*

Acknowledgements

The research of the first author (Q-Heung Choi) is supported by GARC-KOSEF and Inha University Research Fund.

The research of the second author (Tacksun Jung) is supported in part by BSRI Program BSRI-97-1436.

References

1. H. Amann, Math. Z., 127-166 (1979).
2. A. Ambrosetti and G. Prodi, A Primer of Nonlinear Analysis, Cambridge University Press, Cambridge Studies in Advanced Math., **No. 34**, (1993).
3. Q.H. Choi and T. Jung, J. Differential Equations, 117, 390-410 (1995).
4. Q.H. Choi and T. Jung, Studia Math., 120, 259-270 (1996).
5. A.C. Lazer and P.J. McKenna, J. Math. Anal. Appl., 107, 371-395 (1985).
6. J. Schröder, Operator Inequalities, Academic Press, New York (1980).

8th Int. Coll. on Differential Equations, pp. 119–125
D. Bainov (Ed)

Regular Solutions and Stability for the Nonlinear Thermoviscoelastic System with Nonlinear Damping on the Boundary

A. T. COUSIN AND C. L. FROTA
Departamento de Matemática, Universidade Estadual de Maringá, Agencia Postal UEM, 87.020-900, Maringá - PR, Brasil

N. A. LAR'KIN
The Institute of Theoretical and Applied Mechanics, Novosibirsk - 90. 630090, Russia

Abstract. We prove existence and uniqueness of regular global solutions for the one-dimensional thermoviscoelastic system. Moreover, the stability in time of such solution is obtained.

1. INTRODUCTION

We prove the existence, uniqueness and the asymptotic stability of solutions to the problem

$$(1.1)\qquad \begin{cases} u_{tt} - u_{xxt} - \dfrac{\partial}{\partial x}(F(u_x)) + \theta_x = f, \\ \theta_t - \theta_{xx} + u_{xt} = g, \end{cases} \qquad \text{in } Q = (0,1) \times (0,T)$$

with the nonlinear boundary conditions

$$(1.2)\qquad u|_{x=0} = 0,\ \{u_x + |u_t|^\rho u_t\}_{x=1} = 0,\quad \theta|_{x=0} = \theta|_{x=1} = 0;\quad t \in (0,T)$$

and initial data

$$(1.3)\qquad u(x,0) = u^0(x),\quad u_t(x,0) = u^1(x),\quad \theta(x,0) = \theta^0(x),\ x \in (0,1),$$

where F is a real, continuously differentiable function; and $\rho \geq 0$.

This problem describes the longitudinal deformations of a one-dimensional homogeneuos viscoelastic medium, combined with thermal effects, fixed at $x = 0$ and unilaterally stressed at $x = 1$. Here u is the displacement and θ is the variation of the temperature.

A linear system of thermoelasticity was studied by Dafermos [1] where he proved existence and uniqueness of solutions which were asymptoticaly stable. Elliot and Tang Qi [2] proved the existence result in the one-dimensional case when unilateral boundary conditions were considered. Munoz Rivera and Racke [3] studied smoothing and decay properties for the nonlinear system of thermoelasticity.

For the linear system of thermoviscoelasticity with linear boundary conditions, Munoz Rivera [4] proved that the energy associated to the system goes to zero exponentially in time. In the framework of nonlinear thermoviscoelasticity, Shibata [5] proved the global in time existence of smooth solutions to the mixed problem with small initial data and the Dirichlet boundary conditions.

In our case, we consider the nonlinear boundary condition (1.2) that models some physical problems and has a clear meaning, see Greenberg and Li Ta Tsien [6], Cousin and Lar'kin [7].

The decay of solutions for the quasilinear wave equation with viscosity was analyzed by Nakao in [8]. He proved the algebraic decay of solutions to the mixed problem with the Dirichlet boundary conditions.

In the present work, employing the Faedo-Galerkin method, we established existence and uniqueness results for (1.1)-(1.3) and studied the asymptotic behavior of solutions using Nakao's Lemma [8].

In the first part of this paper we solved an auxiliary problem that did not contain nonlinear terms. Suitable a priori estimates allowed us to pass to the limits and to construct Galerkin's approximations. Then we employed standard arguments: proved local in t existence theorem and extended local solutions to any finite T.

To prove the energy decay, we put $f, g \equiv 0$ and used ideas of Nakao [8].

2. LOCAL SOLUTIONS

We use notations

$$(u,v)(t) = \int_0^1 u(x,t)v(x,t)dx, \quad \|u(t)\|^2 = (u,u)(t);$$

$H^m(0,1)$ is the Sobolev space $W_2^m(0,1)$, see Lions-Magenes [9]; $H_1(0,1)$ is the subspace of $H^1(0,1)$: $H_1(0,1) = \{z \in H^1(0,1)\ ;\ z(0) = 0\}$ with the norm $|z|_{H_1(0,1)} = \|z_x\|$; $|\cdot|$ denotes the absolute value; and $\|\cdot\|_\alpha$ denotes the norm of $L^\alpha(0,1)$.

To prove the decay of solutions, we will need the following Lemma of Nakao [8].

Lemma 2.1. *Let $\phi(t)$ be a nonnegative function on $[0,\infty)$ satisfying*

$$\sup_{t\le\tau\le t+1} \phi(\tau)^{1+r} \le K_0(\phi(t) - \phi(t+1)), \; K_0 > 0$$

for some $r \ge 0$. Then it follows that

$$\phi(t) \le \begin{cases} C\phi(0)(1+t)^{-\frac{1}{r}}, & \text{if } r > 0, \\ C\phi(0)e^{-\lambda t}, \; \lambda > 0, & \text{if } r = 0. \end{cases}$$

Our starting point is to study solutions to the following problem

$$\begin{aligned} u_{tt} - u_{xxt} + \theta_x &= p, \\ \theta_t - \theta_{xx} + u_{xt} &= q \end{aligned} \qquad \text{in } Q \tag{2.1}$$

with conditions (1.2), (1.3), where $p(x,t)$ and $q(x,t)$ are given functions.

Theorem 2.1. *Let $p, q \in L^2(Q)$; $u^0, u^1 \in H^2(0,1) \cap H_1(0,1)$ and $\theta^0 \in H_0^1(0,1)$.*

Then there exists a unique solution $(u(x,t), \theta(x,t))$ to (2.1),(1.2),(1.3) such that

$$\begin{aligned} &u \in C(0,T; H^2(0,1) \cap H_1(0,1)), \; \theta \in C(0,T; H_0^1(0,1) \cap H^2(0,1)), \\ &u_t \in L^\infty(0,T; H_1(0,1)) \cap L^2(0,T; H^2(0,1)), \\ &u_{tt}, u_{xxt}, \theta_t, \theta_{xx} \in L^2(Q). \end{aligned}$$

Proof: Putting $z(x,t) = u_t(x,t)$ and $r(x,t) = \theta_t(x,t)$ in (2.1), we employ the Faedo-Galerkin procedure to construct approximate solutions. Let $\{\omega_j\}$ and $\{\overline{\omega}_j\}$ be basis of $H^2(0,1) \cap H_1(0,1)$ and $H_0^1(0,1)$ respectivelly. Then the approximate solutions will be sought in the form

$$z^N(x,t) = \sum_{j=1}^{N} g_j^N(t)\omega_j(x), \; r^N(x,t) = \sum_{j=1}^{N} h_j^N(t)\overline{\omega}_j(x),$$

where g_j^N, h_j^N are solutions to the problem

$$(z_t^N, w_j)(t) + (z_x^N, w_{jx})(t) + (\rho+1)|z^N(1,t)|^\rho z_t^N(1,t)w_j(1)$$

$$+(\int_0^t r_x^N(\tau)d\tau, \omega_j) = (h, w_j)(t), \tag{2.2}$$

$$(r^N, \overline{\omega}_j)(t) + (\int_0^t r_x^N(\tau)d\tau, \overline{\omega}_{jx})(t) + (z_x^N, \overline{\omega}_j)(t) = (q, \overline{\omega}_j)(t), \tag{2.3}$$

$$g_j^N(0) = (u^1, w_j), \quad j = 1, \cdots, N. \tag{2.4}$$

The functions $g_j^N(t), h_j^N(t)$ $(j = 1, \cdots, N)$ are defined in some interval $(0, T_N)$. Uniform in N estimates will permit us to extend solutions to the interval $(0,T)$ and pass to the limit as $N \to \infty$.

Multiplying equations (2.2) by $g_{jt}^N(t)$ and (2.3) by $h_j^N(t)$ respectivelly, we come after some calculations to the inequality

$$\|z_x^N(t)\|^2 + \|\int_0^t r_x^N(\tau)d\tau\|^2 + \int_0^t \|z_\tau^N(\tau)\|^2 d\tau + \int_0^t \|r^N(\tau)\|^2 d\tau \le C_0, \; t \in (0,T). \tag{2.5}$$

Therefore, we can pass to the limit in (2.2)-(2.4), as $N \to \infty$, to obtain

$$u_t - u_{xx} + \int_0^t \theta_x(\tau)d\tau = \int_0^t p(\tau)d\tau,$$

$$\theta - \int_0^t \theta_{xx}(\tau)d\tau + u_x = \int_0^t q(\tau)d\tau,$$

$$u|_{x=0} = 0, \; \{u_x + |u_t|^\rho u_t\}_{x=1} = 0, \; \theta|_{x=0} = \theta|_{x=1} = 0.$$

This implies that u_{xx}, $\int_0^t \theta_{xx}(\tau)d\tau \in L^2(Q)$. In turn, taking the derivative with respect to t, we get

$$u_{tt} - u_{xxt} + \theta_x = p,$$
$$\theta_t - \theta_{xx} + u_{xt} = q,$$

whence $u_{xxt}, \theta_{xx} \in L^2(Q)$.

The uniqueness result can be established in the usual way hence the proof of Theorem 2.1 is completed. ■

Next we prove existence of local in t solutions.

Theorem 2.2: *Let $F \in C^1(\mathbf{R}), u^1 \in H_1(0,1), u^0 \in H^2(0,1) \cap H_1(0,1),\ \theta^0 \in H_0^1(0,1)$. Then for each $f, g \in L^2(Q)$ there exist $T_0 > 0$ and a unique pair $(u(x,t), \theta(x,t))$*

$$\begin{aligned} &u \in C(0,T_0; H^2(0,1) \cap H_1(0,1)),\ \theta \in C(0,T_0; H_0^1(0,1) \cap H^2(0,1)), \\ &u_t \in L^\infty(0,T_0; H_1(0,1)) \cap L^2(0,T_0; H^2(0,1)), \\ &u_{tt}, u_{xxt}, \theta_t, \theta_{xx} \in L^2(0,T_0; L^2(0,1)) \end{aligned} \tag{2.6}$$

which is a solution to the problem (1.1)-(1.3) in $Q_0 = (0,1) \times (0,T_0)$.

Let us define the ball $W_{R\Gamma}$ by

$$W_{R\Gamma} = \{\eta \in C([0,\Gamma]; H_1(0,1) \cap H^2(0,1))\ /\ \|\eta_x(t)\|^2 + \|\eta_{xx}(t)\|^2 \le R^2,\ t \in [0,\Gamma]\}.$$

For each $v \in W_{R\Gamma}$, we define $u = S(v)$ where (u,θ) is the solution of the system

$$u_{tt} - u_{xxt} + \theta_x = f + \frac{\partial}{\partial x} F(v_x), \tag{2.7}$$

$$\theta_t - \theta_{xx} + u_{xt} = g \tag{2.8}$$

with the initial and boundary conditions (1.2), (1.3).

Lemma 2.1: *For a sufficiently big R, there exists $\Gamma > 0$ such that S maps $W_{R\Gamma} \to W_{R\Gamma}$.*

Proof: From inclusions (2.4) and the hypothesis on F, we have for fixed $R < \infty$

$$|F'(v_x(x,t))| \le \max_{s \in \overline{Q}} |F'(s)|,\ \|F'(v_x)v_{xx}(t)\| \le C(R) \text{ in } Q,$$

where the constant $C(R)$ depends only on R.

Multiplying (2.7) by $-u_{xxt}$, (2.8) by θ_t and integrating with respect to x over (0,1), we come to the inequality

$$\|u_x(t)\|^2 + \|u_{xx}(t)\|^2 + \|\theta_x(t)\| + \|\theta(t)\|^2 \le C_1 + C_2 R^2 t,$$

where C_1, C_2 are positive constants that depend on $u^0, u^1, \theta^0, f, g, R$ and do not depend on t.

Now, let us take $R^2 = 2C_1,\ \Gamma = \frac{1}{2C_2}$. Then

$$\|u_x(t)\|^2 + \|u_{xx}(t)\|^2 + \|\theta_x(t)\| + \|\theta(t)\|^2 \le R^2,\ t \in [0,\Gamma]. \ \blacksquare$$

Lemma 2.2: *There exist $T_0 \in (0,\Gamma)$ and $0 < \gamma < 1$ such that the inequality holds*

$$\|u_t - \tilde{u}_t\|^2_{L^2(0,T_0;L^2(0,1))} + \|u_x - \tilde{u}_x\|^2_{L^2(0,T_0;H_1(0,1))} + \int_0^\Gamma \| \int_0^\tau (\theta_x - \tilde{\theta}_x)(s)ds\|^2 d\tau$$
$$+\|\theta - \tilde{\theta}\|^2_{L^2(0,T_0;L^2(0,1))} \le \gamma \|v_x - \tilde{v}_x\|^2_{L^2(0,T_0;H_1(0,1))},$$

for every $v, \tilde{v} \in W_{R\Gamma}$ and $u = S(v),\ \tilde{u} = S(\tilde{v})$.

Proof: Let $v, \tilde{v} \in W_{R\Gamma}$; $u = S(v), \tilde{u} = S(\tilde{v})$. Then the pair $Z = u - \tilde{u}$ and $Y = \theta - \tilde{\theta}$ is a solution to the problem

$$\begin{aligned} &Z_{tt} - Z_{xxt} + Y_x = \tfrac{\partial}{\partial x} F(v_x - \tilde{v}_x), \\ &Y_t - Y_{xx} + Z_{xt} = 0, \end{aligned} \tag{2.10}$$

$$
\begin{aligned}
&Z|_{x=0} = 0,\ \{Z_x + |u_t|^\rho u_t - |\tilde{u}_t|^\rho \tilde{u}_t\}_{x=1} = 0, \\
&Y|_{x=0} = Y|_{x=1} = 0,\ Z(x,0) = Z_t(x,0) = Y(x,0) = 0.
\end{aligned} \tag{2.11}
$$

Integrating both equations in (2.10) with respect to t, multiplying the first result by $-Z_{xx}$ and the second result by Y, we come to

$$
\|Z_t\|^2_{L^2(0,\Gamma;L^2(0,1))} + \|Z_x\|^2_{L^2(0,\Gamma;H_1(0,1))} + \int_0^\Gamma \| \int_0^\tau Y_x(s)ds\|^2 d\tau + \|Y\|^2_{L^2(0,\Gamma;L^2(0,1))}
$$

$$
\leq C_3(R)\Gamma \|v_x - \tilde{v}\|^2_{L^2(0,\Gamma;H_1(0,1))},
$$

where $C_3(R)$ is a positive constant that depends on R. Putting $T_0 = \frac{1}{2C_3(R)}$, we prove Lemma 2.2. ■

Proof of Theorem 2.2: Let $u^{(n)}, \theta^{(n)}$, $n = 1, \cdots$, be such that

$$
\begin{aligned}
&u_{tt}^{(n)} - u_{xxt}^{(n)} + \theta_x^{(n)} = f + \frac{\partial}{\partial x} F(u_x^{(n-1)}), \\
&\theta_t^{(n)} - \theta_{xx}^{(n)} + u_{xt}^{(n)} = g, \\
&u^{(n)}\Big|_{x=0} = 0,\ \left\{u_x^{(n)} + |u_t^{(n)}|^\rho u_t^{(n)}\right\}_{x=1} = 0, \\
&\theta^{(n)}\Big|_{x=0} = 0,\ \theta^{(n)}\Big|_{x=1} = 0, \\
&u^{(n)}(x,0) = u^0(x),\ u_t^{(n)}(x,0) = u^1(x),\ \theta^{(n)}(x,0) = \theta^0.
\end{aligned}
$$

According to Lemmas 2.1 and 2.2, for T_0 sufficiently small, all $u^{(n)}(x,t)$ belong to the ball W_{RT_0} and satisfy the inequalities

$$
\|u_x^{(n)}(t)\|^2_{H_1(0,1)} + \|\theta^{(n)}(t)\|^2_{H_0^1(0,1)} \leq R^2,\ n \in \mathbf{N},\ t \in [0, T_0], \tag{2.12}
$$

$$
\|u_t^{(n+1)} - u_t^{(n)}\|_{L^2(0,T_0;L^2(0,1))} + \|u_x^{(n+1)} - u_x^{(n)}\|^2_{L^2(0,T_0;H_1(0,1))} + \|\theta^{(n+1)} - \theta^{(n)}\|^2_{L^2(0,T_0;L^2(0,1))}
$$

$$
\leq \int_0^{T_0} \| \int_0^\tau (\theta_x^{(n+1)}(s) - \theta_x^{(n)}(s))ds\|^2 d\tau \leq \gamma \|u_x^{(n)} - u_x^{(n-1)}\|^2_{L^2(0,T_0;H_1(0,1))},\ n \in \mathbf{N}. \tag{2.13}
$$

We can extract subsequences $(u^{(n)}, \theta^{(n)})$ which converge to (u, θ) strongly in

$$
L^2(0,T_0;H_1(0,1) \cap H^1(0,T_0;L^2(0,1)) \times L^2(0,T_0;H_0^1(0,1)).
$$

Hence, the existence is proved. The uniqueness can be proved in a standard way. ■

3. GLOBAL SOLUTIONS

Theorem 3.1. *Let u^0, u^1, θ^0 satisfy all the conditions of Theorem 2.2. If $F \in C^1(\mathbf{R})$ with $F'(s) \geq 0$, then for every $f, g \in L^2(Q)$ and for all finite $T > 0$ there exists in $Q = (0,1) \times (0,T)$ a unique solution to (1.1)-(1.3) satisfying (2.2).*

Proof: Taking into account Theorem 2.2, it is sufficient to prove global a priori estimates permitting to continue a local solution to any finite interval $(0,T)$.

Multiplying equations (1.1) by u_t and θ respectivelly, we obtain

$$
\|u_t(t)\|^2 + \|\theta(t)\|^2 + |u_t(1,t)|^{\rho+2} + \int_0^t \|u_{x\tau}(\tau)\|^2 d\tau + \int_0^t \|\theta_x(\tau)\|^2 d\tau \leq K_1. \tag{3.1}
$$

In turn, multiplying equations (1.1) by $-u_{xx}$ and $-\theta_{xx}$ respectivelly, we come to

$$(3.2)\qquad \|u_{xx}(t)\|^2 + \|\theta_x(t)\|^2 + \int_0^t \|\theta_{xx}(\tau)\|^2 d\tau \le K_2.$$

At last, multiplying equations (1.1) by $-u_{xxt}$ and $-\theta_{xx}$ respectivelly, we get

$$(3.3)\qquad \|u_{xt}(t)\|^2 + \|\theta_x(t)\|^2 + \int_0^t \|\theta_{xx}(\tau)\|^2 d\tau + \int_0^t \|u_{xx\tau}(\tau)\|^2 d\tau \le K_3.$$

The estimates (3.1)-(3.3) allow us to extend the local solutions of Theorem 2.2 to any finite interval (0,T). By Theorem 2.2, this solution is uniquely defined. ■

4. STABILITY OF SOLUTIONS

Theorem 4.1: *Let $(u(x,t),\theta(x,t))$ be a solution to (1.1)-(1.3) with $f \equiv 0, g \equiv 0$ and $F(s) = s + |s|^{\alpha-2}s, \alpha \ge 2$. Then there exist positive constants λ, C such that*

$$E(t) \le CE(0)e^{-\lambda t},\ t \ge 0,$$

where

$$(4.1)\quad E(t) = \frac{1}{2}\|u_t(t)\|^2 + \frac{1}{2}\|u_x(t)\|^2 + \frac{1}{\alpha}\|u_x(t)\|_\alpha^\alpha + \frac{\rho+1}{\rho+2}|u_t(1,t)|^{\rho+2} + \frac{1}{2}\|\theta(t)\|^2,\ t \ge 0.$$

Proof: Multiplying equations (1.1) by u_t and θ respectivelly, we obtain

$$(4.2)\qquad \frac{d}{dt}E(t) + \|u_xt(t)\|^2 + \|\theta_x(t)\|^2 + |u_t(1,t)|^{\rho+2} + |u_t(1,t)|^\beta = 0,$$

then

$$E(t) + \int_0^t \left\{|u_xT(\tau)\|^2 + \|\theta_x(\tau)\|^2 + |u_\tau(1,\tau)|^{\rho+2} + |u_\tau(1,\tau)|^\beta\right\} d\tau = E(0) < \infty,$$

where $\beta = \rho\alpha - \rho + \alpha$.

It means that the energy is bounded and decreasing. Integrating (4.1) over $[t, t+1]$, $t > 0$, we have

$$\int_t^{t+1} \|u_{x\tau}(\tau)\|^2 d\tau + \int_t^{t+1} \|\theta_x(\tau)\|^2 d\tau + \int_t^{t+1} \left[|u_\tau(1,\tau)|^{\rho+2} + |u_\tau(1,\tau)|^\beta\right] d\tau =$$

$$= E(t) - E(t+1) \equiv D^2(t).$$

It follows by Mean Value Theorem, that there exist $t_1 \in [t, t+\frac{1}{4}]$ and $t_2 \in [t+\frac{3}{4}, t+1]$ such that

$$(4.3)\qquad \|u_{xt}(t_i)\|^2 + \|\theta_x(t_1)\|^2 + |u_t(1,t_i)|^{\rho+2} + |u_t(1,t_i)|^\beta \le 4D^2(t),\ i = 1, 2.$$

Multiplying equations (1.1) by u and θ repectivelly, integrating over $[t_1, t_2]$ and using (4.3), we have

$$(4.4)\quad \int_{t_1}^{t_2} \left[\|u_x(\tau)\|^2 + \|u_x(\tau)\|_\alpha^\alpha + \|\theta_x(\tau)\|^2\right] d\tau \le C(D^2(t) + \varepsilon_1 E(t) + \varepsilon_2 \sup_{t\le\tau\le t+1} E(\tau)),$$

where $\varepsilon_1, \varepsilon_2$ are arbitrary positive numbers.

From (4.1), (4.4) by the Mean Value Theorem, there is $t^* \in [t_1, t_2]$ such that

$$E(t^*) \leq C(D^2(t) + \varepsilon_1 E(t) + \varepsilon_2 \sup_{t \leq \tau \leq t+1} E(\tau)).$$

Integrating (4.2) from t to t^*, we find

$$E(t) \leq \int_{t_1}^{t_2} E(\tau)d\tau + D^2(t) \leq C(D^2(t) + \varepsilon_1 E(t) + \varepsilon_2 \sup_{t \leq \tau \leq t+1} E(\tau)).$$

Choosing $\varepsilon_1, \varepsilon_2$ sufficiently small, we get

$$\sup_{t \leq \tau \leq t+1} E(\tau) \leq CD^2(t).$$

Using Nakao's Lemma, we prove Theorem 4.1. ■

Acting in the similar manner, we can prove

Theorem 4.2. *Let $(u(x,t), \theta(x,t))$ be a solution to (1.1)-(1.3) with $f \equiv 0, g \equiv 0$, and $F(s) = |s|^{\alpha-2}s$, $\alpha > 2$. Then there exist positive constants C and $r < 2$ such that the following inequality holds*

$$E(t) \leq C(1+t)^{\frac{r}{r-2}},\ t \geq 0,$$

where $E(t)$ is the energy given by

$$E(t) = \frac{1}{2}\|u_t(t)\|^2 + \frac{1}{\alpha}\|u_x(t)\|_\alpha^\alpha + \frac{\rho+1}{\rho+2}|u_t(1,t)|^{\rho+2} + \frac{1}{2}\|\theta(t)\|^2,\ t \geq 0,$$

$r = \min\{\frac{\alpha}{\alpha-1}, 2\frac{\rho+2}{\beta}\}$. ■

ACKNOWLEDGEMENTS

The third author (N. A. Lar'kin) is supported by CNPq-Brasil as a Visiting Professor at the State University of Maringá.

REFERENCES

1. C. M. Dafermos, J. Differential Equations, 6, 71-86 (1969).
2. C. M. Elliott and Tang Qi, Nonlinear Analysis, 23, 883-898 (1994).
3. J. E. Munoz Rivera and R. Racke, SIAM J. Math.Anal., 26, 1547-1563 (1995).
4. J. E. Munoz Rivera, Mat. Aplic. Comp., 11, 45-71 (1992).
5. Y. Shibata, Math. Methods Appl. Sci., 18, 871-895 (1995).
6. A. T. Cousin and N. A. Lar'kin, Nonlinear Analysis, (in press).
7. J. M. Greenberg and Li Ta Tsien, J. Differential Equations, 52, 66-75 (1984).
8. M. Nakao, Math. Z., 219, 289-299 (1995).
9. J. L. Lions and E. Magenes, Problems aux Limites Non Homogènes et Applications, Tome I, Dunod, Paris (1968).

8th Int. Coll. on Differential Equations, pp. 127–136
D. Bainov (Ed)

Analytical Solutions of the Navier-Stokes Equation for the Laminar and Turbulent Boundary Layer, Over a Flat Plate

LEONARD D'ATTORRE
33655 Scotty's Cove Dr., Dana Point, California 92629, USA

Abstract. The PDE of Navier-Stokes is implemented with an analytic model to represent the turbulent variable stress in the BL. The new PDE in non divergence form is transformed into a nonlinear ODE in terms of similarity variable. Also are shown two analytic solutions that satisfy it. The solutions share the properties of the inverse functions. The present approach, with a single analytical form can describe the BL regimes, from pure laminar to fully turbulent.

Introduction

Consider an, incompressible steady flow, moving along a flat surface in the x-direction of the main flow, with constant velocity U far from the plate.

Near the plate surface, the turbulent shear stress is represented by, τ, also by u, the velocity component along x and by v, perpendicular to the plate, in addition, the pressure and density ρ, are considered as constants. The flow velocity components and the turbulent shear stress, are assumed to be represented by their statistical averages that satisfy the hydrodynamic equations of conservation of mass $\mathrm{div} V = 0$, and momentum, the Navier-Stokes, namely, in two dimensions read:

$$(1) \qquad \left(\frac{\delta u}{\delta x}\right) + \left(\frac{\delta v}{\delta y}\right) = 0, \qquad u\left(\frac{\delta u}{\delta x}\right) + v\left(\frac{\delta v}{\delta y}\right) = \frac{\delta(\tau/\rho)}{\delta y}.$$

By hypothesis the turbulent stress is represented by the expression: $\tau/\rho = \nu_t(du/dy)$, in which, $\nu_t(y)$ represents the coefficient of eddy kinematic viscosity, and assumed to varies with y.

Introducing the stream function ψ, such that: $u = \delta\psi/\delta y = \psi_y$, $-v = \delta\psi/\delta_x$ the mass conservation, reduced to $\psi_{xy} - \psi_{yx} = 0$, and the Navier-Stokes equation becomes:

$$\psi_y \cdot \psi_{yx} - \psi_x \cdot \psi_{yy} = \frac{\delta(\nu_t du/dy)}{\delta y}.$$

Calling: $\dot{\nu}_t = d\nu/dy$, by expression of the Navier-Stokes equation finally reads:

$$(2) \qquad \psi_y \cdot \psi_{yx} - \psi_x \cdot \psi_{yyy} + \dot{\nu}_t \cdot \psi_{yy}.$$

This partial differential equation is assumed to model the turbulent BL with variable eddy viscosity given as function of y, and to be called solution, the function, ψ, that satisfy it.

Definitions and transformations

$$\psi = \nu_t F, \quad Z = \frac{y}{x}, \quad u_t = u \cdot a_t, \quad a_t = a \cdot y^p, \quad \dot{\nu}_t = \nu_t \cdot \frac{p}{y}, \quad P_j = p - j$$

in which: ν_t, the coefficient of kinematic turbulent viscosity, is expressed by the factors: ν, coefficient of laminar (molecular) viscosity and nondimensional, a_t given in terms of the power, p, of the y, variable. Lets introduce the following derivatives:

$$F' = \frac{dF}{dZ}, \qquad F'' = \frac{dF'}{dZ}, \qquad F''' = \frac{dF''}{dZ},$$

$$\frac{dZ}{dy} = \frac{1}{x}, \qquad \frac{dZ}{dx} = -\frac{y}{x^2}, \qquad \frac{d^2Z}{dxdy} = -\frac{1}{x^2}$$

to simplify the notation, lets introduce the following parameters:

$$G_1 = (p), \qquad G_2 = (-P_1), \qquad G_3 = (1),$$

$$G_4 = (2p), \qquad G_5 = (4p), \qquad G_6 = (3pP_1 + 2p^2), \qquad G_7 = (2pP_1^2).$$

The equation (2) reads:

$$(3) \qquad Z^3(G_1FF'' + G_2F'^2 + G_3F''') + Z^2(G_4FF' + G_5F'') + Z(G_6F') + G_7F = 0.$$

Consequently the original PDE on ψ, and x, y is transformed into a nonlinear Ordinary DE, on terms of new variable, Z, and the F, function and their derivatives, in conjunction of a set of constant parameters.

I. Solution of the ODE

Solutions of the ODE (3) are similarity solutions.

First assuming the function F to be of the form: $F = AZ^m$, being $A =$ const. The derivatives of F are: $F' = Fm/Z$, $F'' = FmM_1/Z^2$, $F''' = FmM_1M_2$, and: $M_j = m - j$. When those values are replaced into equation (3) we obtain:

$$(4) \qquad Z \cdot F(G_1mM_1 + G_2m^2 + G_4m) + (G_3mM_1M_2 + G_5mM_1 + G_6m + G_7) = 0.$$

The convenient choice of the value of: $m = -1$, then, $M_1 = -2$ and $M_2 = -3$ permits to eliminate the F, and Z, from equation (4), namely after replacing the values of $G's$, becomes the following:

$$(5) \qquad A = 2p^2 - 7p + 6.$$

This important equation (5) relates the value of the constant A, that defined the solution function F, with the parameter, p, exponent in the assumed law of turbulent kinematic coefficient of eddy viscosity is instructive to observe that when, $p = 0$, $A = 6$, then could represent constant viscosity, either laminar or turbulent BL. Consequently also are obtained the following:

$$F = \frac{A}{Z}, \qquad F' = -\frac{A}{Z^2}, \qquad F'' = \frac{2A}{Z^3}, \qquad F''' = -\frac{6A}{Z^4}. \tag{6}$$

The velocities are: $u = \psi_y = \nu_t(F \cdot p/y + F'1/x)$, $v = -\psi_x = \nu_t \cdot F'(-y/x^2)$. Also:

$$u = \nu_t \cdot A \cdot \frac{P_1}{xZ^2} = \nu_t \cdot A \cdot P_1 \cdot \frac{x}{y^2}, \qquad v = \nu_t \cdot \frac{A}{xZ} = \nu_t \cdot A \cdot \frac{y}{y^2}. \tag{7}$$

Since: $\nu_t = \nu \cdot a_t$, and calling, local Reynolds number, the value $R = Ux/\nu$. The velocity components equation (7), when are non dimensionalized by the reference velocity U, permits to write

$$\frac{u}{U} = a_t \cdot A \cdot \frac{P_1}{RZ^2}, \qquad \frac{v}{U} = a_t \cdot \frac{A}{RZ}. \tag{8}$$

From equation (7), is also obtain the important ratio:

$$\frac{v}{u} = \frac{Z}{P_1}. \tag{9}$$

Law of defect-velocity-law of the wall

From equation (7), $u = \nu_t A \cdot P_1 x/y^2$ and when $\nu_t = \text{const}$, the derivative of, $u \cdot du/dy = -2\nu_t \cdot A \cdot P_1 \cdot x/y^3$, now following a customary procedure in BL analysis, for values near the surface define the parameters: $u = u_\tau$, $y = y_\tau$ and permits to write:

$$u_\tau = \nu_t \cdot A \cdot P_1 \cdot \frac{x}{y_\tau^2}, \qquad \frac{du}{dy} = -2 \cdot u_\tau \left(\frac{1}{y}\right) \tag{10}$$

when the last in integrated between: U. and u. and δ, and y,

$$\frac{U - u}{u_\tau} = 2 \cdot \ln\left(\frac{y}{\delta}\right). \tag{11}$$

Formula established empirically by von Karman and Prandtl. Moreover when equation (10) is integrated between u, and 0, and y, y_τ, to obtain: $u/u_\tau = 2 \cdot \ln(y/y_\tau)$, using customary definitions, like: $\tau/\rho = u_\tau^2 = \nu \cdot du/dy$, and call: $y_\tau = \nu/u_\tau$ to have finally:

$$\frac{u}{u_\tau} = 2\ln\left[\left(\frac{u_\tau}{\nu}\right) \cdot y\right]. \tag{12}$$

Representing, the law of the wall, also being established empirically.

Lets introduce some useful parameters. Assume the value of the total velocity U_{BL}, in the BL, to be:

$$\begin{gathered} U_{BL}^2 = u^2 + v^2 = u^2\left[1 + \left(\frac{v}{u}\right)^2\right], \\ \text{and defining: } k_u^2 = \left[1 + \left(\frac{v}{u}\right)^2\right], \quad \text{and } U_{BL} = k_u \cdot U. \end{gathered} \tag{13}$$

Also introducing the parameter k_1, relating the unperturbed main stream U, with the BL velocity U_{BL} such that: $U = k_1 \cdot U_{BL}$, consequently

$$U = k_1 \cdot k_u \cdot u. \tag{14}$$

From equation (13): $v/u = (k_u^2 - 1)^{1/2}$ into equation (12) becomes:

$$Z = P_1(k_U^2 - 1)^{1/2} \tag{15}$$

as is customary for BL experimental analyst to defined, friction, eddy, Reynolds number, for example: $R_{ed} = u_\tau \cdot \delta / \nu_t$.

Introducing equation (4), namely: $u = u_t \cdot A \cdot P_1 \cdot / x Z^2$ and equation (15) is obtained: $u \cdot y = \nu_t \cdot A/(k^{2|u01})^{1/2}$, near the wall $u = u_\tau$, the friction velocity and making $y = \delta$, the BL thickness, actually the eddy Reynolds number becomes:

$$R_{ed} = \frac{A}{(k_u^2 - 1)^{1/2}}. \tag{16}$$

Velocity distribution in the BL

Expressions for the velocity components of the BL already are shown in equation (8), the equations satisfy the ODE, the equation (3), but not the boundary condition at the wall, since instead of $u = 0$, it becomes and infinity or a singularity.

Since the basic solution, after the introduction of certain parameters, proved to be useful, in the following, a procedure is developed, based in new definitions and parameters.

Introducing the variable χ, and when $y = \delta$, the BL thickness, $Z_1 = \delta/x$ and $\chi = Z/Z_1$, after introducing those values into equation (8), results

$$\frac{u}{U} = \frac{a_t A P_1}{R Z_1^2 \chi^2}, \qquad \frac{v}{U} = \frac{a_t A}{R Z_1 \chi}. \tag{17}$$

From equation (9) results:

$$\frac{1+\chi}{\chi^2} = \left(\frac{u}{U}\right) \cdot \frac{R Z_1^2 + R Z_1 Z}{a_t A P_1}. \tag{18}$$

Defining the functions:

$$E(\chi) = \frac{1+\chi}{\chi^2}; \qquad F'(\chi) = \frac{u}{U}; \qquad H = (R Z_1^2 + R Z_1 Z) a_t A P_1. \tag{19}$$

The same equation (19) is recast as: $E(\chi) = H F'(\chi)$, and also:

$$E(\chi) = G \cdot \overline{F'^2(\chi)} \cdot F'(\chi)^{-1} \tag{20}$$

in which the $\overline{F'^2(\chi)}$ is taken as the mean value of the following integral:

$$\overline{F'^2(\chi)} = \left(\frac{1}{\varsigma_1} - \varsigma_0\right) \cdot \int_0^1 F'^2(\chi) d\chi. \tag{21}$$

From equation (14), when the average integral is interpreted in terms of the parameters k_1 and k_u, the equation (20) results in: $\overline{F'^2(\chi)} = \overline{1/k_1^2 d_u^2}$ then when replace into equation (20) results:

$$F'(\chi) = \frac{u}{U} = \left(H \cdot \frac{\chi}{k_1^2 k_u^2}\right) \cdot \left(\frac{\chi}{1} + \chi\right). \tag{22}$$

When: $\chi = 0$, $u/U = 0$; $\chi = 1$, $u/U = 1$, consequently equation (22) satisfies the boundary conditions when: $(H\chi/k_1^2 k_u^2) = 2$ for $\chi = 1$, also $Z = Z_1$, from equation (19) moreover, since $H = 2RZ_1^2/a_t A \cdot P_1$, the following results; finally in

$$R \cdot \frac{Z_1^2}{a_t} \cdot A \cdot P_1 \cdot k_1^2 k_u^2 = 1 \tag{23}$$

and

$$F' = \frac{2\chi}{1} + \chi. \tag{24}$$

The last expression can be transformed yet, permits to write with good approximation, consequently:

$$F'(\chi) = \frac{u}{U} = \frac{2\chi}{(1+\chi)^{\chi}}. \tag{25}$$

Moreover, a more general expression can be generated; by introducing the exponent N :

$$\frac{u}{U} = \left[\frac{2\chi}{(1+\chi)^{\chi}}\right]^N. \tag{26}$$

The parameter N, to be used as a constant, to correlated the BL, flow regimes, namely: $N = 1$ for laminar; $N = 1/7$ for fully turbulent; $N = 1/2$ for transitional regime $id \cdot id \cdot N = 3/4$; $N = 2$, and $N = 3$, incipient separation.

Boundary layer thickness, and coefficient of friction

To start from equation (23) and equation (15), can be written:

$$Z_1^2 = \frac{a_t \cdot A \cdot P_1 \cdot k_1^2 \cdot k_u^2}{R}, \qquad Z_1^3 = P_1^3 (k_u^2 - 1)^{3/2}$$

in order to correlate Z_1 to the fifth power, as is customary in BL analysis, from the product of last two, we can obtain:

$$\frac{\delta}{x} = Z_1 = \frac{K_t}{R^{1/5}}, \qquad K_t = \left[a_t \cdot P_1^4 \cdot k_1^2 k_u^2 (k_u^2 (k_u^2 - 1)^{3/2}\right]^{1/5} \tag{27}$$

the expression of turbulent BL thickness.

With regard to the coefficient of friction, lets refer to the following definitions: $\tau/\rho = C_f \cdot \frac{1}{2}\rho u^2 = \nu_T \cdot du/dy$ and $C_f = (2\nu_t/U)d(u/U)/dy = (\nu/Ux)2a_t \cdot x \cdot d(u/U)/dy$

$$C_f = \left(\frac{2a_t \cdot x}{R}\right) \cdot \frac{d(u/U)}{dy}, \qquad y = \hat{Y} \tag{28}$$

the velocity derivative, from equation (27)

$$\frac{d(u/U)}{dy} = \left(\frac{N}{xZ_1}\right) \cdot \left[\frac{2\chi}{(1+\chi)^{\chi}}\right]^{N-1} \cdot [2] \quad \text{for } \chi = 0. \tag{29}$$

From equation (23): $1/RZ_1 = a_t \cdot A \cdot P_1 k_1^2 k_u^2$, when this form, and equation (33) are introduced into equation (28), the following is obtained for turbulent BL

$$C_f = \left[\frac{4Z_1}{A} \cdot P_1 k_1^2 k_u^2\right] \cdot N \cdot \left[\frac{2\chi}{(1+\chi)^{\chi}}\right]^{N-1}. \tag{30}$$

The last equation (31) also can be applied to laminar BL flow by making $N = 1$, the corresponding exponent of the viscosity law, $p = 0$ then $P_1 = 1$ to have for laminar BL, the following expression after equation (5) for $A = 6$ shows:

$$C_f = 4\frac{Z_1}{6k_1^2 k_u 2}. \tag{31}$$

The boundary conditions on equation (32) are non conflicting the T. von Karman concept of the existence of a "laminar sub-layer".

The expressions (32) and (31) show the functional dependence of the coefficient of friction by the BL thickness, since $Z_1 = \delta/x$.

From equation (28), the value of $\delta/x = K_t/R^{1/5}$ when substituted in equation (31) for

$$\chi = \hat{\chi}, \qquad C_f = \left(\frac{1}{R^{1/5}}\right) \cdot \left(4\frac{K_t}{A} \cdot P_1 k_1^2 k_u^2\right) \cdot N \cdot \left[\frac{2\hat{\chi}}{(1+\hat{\chi})^{\hat{\chi}}}\right]^{N-1} \tag{32}$$

final expression for the turbulent BL, coefficient of friction in this equation the value of $\hat{\chi} = \hat{Z}/Z_1$, $\hat{Z} = \hat{y}/x$, $\hat{y}$, is the thickness of the laminar sub-layer.

Integral averages-parameters

Assuming the flow moving in the region, two times the BL thickness. Introducing a region of uniform velocity U_L different from the distant uniform flow U. Also introduce the parameter $k_L = U_L/U$. From equations (16) and (17): $k_L \cdot U = k_L \cdot k_1 \cdot k_u \cdot u = (k_1/k_u)(k_L \cdot k_u/k_1)$, and then:

$$k_L = \frac{k_1}{k_u} \tag{33}$$

and balance of mass require:

$$2\delta \cdot U = \delta \cdot U_1 + \delta \cdot U_{BL}, \quad 2U = U_L + U_{BL} = k_L U + U_{BL}, \quad k_L = 2 - \frac{U_{BL}}{U}. \tag{34}$$

The integral average of the velocity U_{BL}

$$\frac{\bar{U_{BL}}}{U} = \frac{1}{\delta}\int_0^{\delta} \chi^N dy \tag{35}$$

after the turbulent velocity law, equation (29) is approximated by $\chi^N = (y/\delta)^N$. As an example for $N = 1/7$, corresponding to fully turbulent BL, from equation (37)

$$(36) \qquad \frac{U_{BL}}{U} = \left(\frac{1}{\delta^{8/7}}\right) \cdot \delta^{8/7} \cdot /(\frac{8}{7}) = \frac{7}{8}$$

or, $U = 8/7 U_{BL}$ then: $k_1 = 8/7$ also:

$$(37) \qquad k_L = \frac{9}{8} \qquad \text{and} \qquad k_u = \frac{k_1}{k_1} = \frac{8/7}{9/8}$$

$.v/u = (k_u^2 - 1)^{1/2} = \chi Z/P_1$, then $\chi Z = P_1(k_u^2 - 1)^{1/2} = \chi Z/P_1$, also: $Z_1 = P_1(k_u^2 - 1)^{1/2}$ from

$$u = \frac{v}{(k_u^2 - 1)^{1/2}} = \frac{\nu_t \chi F}{x(k_u^2 - 1)^{1/2}}, \qquad \nu_t = \nu a_t, \qquad u = \frac{\nu a_t \phi_1 \chi e^{1-\chi}}{x Z_1 (k_u^2 - 1)^{1/2}}.$$

The (41) formulas, represents average parameters for the turbulent BL. Similarly, the laminar BL parameter, can be determined: In general, moreover take the integral average of the linear and nonlinear values and have the following integrals:

$$I_{NL} - \int_0^1 \left[\frac{2\chi}{(1+\chi)^\chi}\right] d\chi = \left[\frac{\chi^2}{(1+\chi)^{1/2}\chi} \cdot d\chi\right]_0^1 = \frac{1}{2^{1/2}};$$

$$I_L = \int_0^1 \chi \cdot d\chi = \frac{1}{2} \frac{u}{U(I_{NL} - I_L)} = \frac{u}{U}(k_U - 1)$$

or:

$$(38) \qquad \left[\frac{1}{2^{1/2}} - \frac{1}{2}\right] + 1 = k_u \qquad \text{for} \qquad k_1 = 2.$$

II. Exponential solution of the ODE

Beginning with equation (3). Introducing:

$$(39) \qquad F = b\exp(mZ), \qquad F' = mF, \qquad F'' = m^2 F, \qquad F''' = m^3 F.$$

Choice of $m = -1/Z_1$ and calling: $\chi = Z/Z_1$, is obtained:

$$Z_a t F[\chi^3(G_1 + G_2) - \chi^2 G_4] = \chi^3 G_3 - \chi^2 G_5 + \chi G_6 - G_7$$

and, $F = b \cdot \exp(-\chi)$, Z_1 is the value of Z when $y = \delta$, (BL thickness) by calling:

$$\phi(\chi) = \frac{\chi^3 G_3 - \chi^2 G_5 + \chi G_6 - G_7}{\chi^3(G_1 + G_2) - \chi^2 G_4},$$

$$F = \frac{\phi(\chi)}{Z_1}, \qquad \phi(1) = \phi_1, \qquad F_1 = \frac{\phi_1}{Z_1}.$$

Results:

$$(40) \qquad \phi(\chi) = \frac{\chi^3 - 4p\chi^2 + p(5p-3)\chi - 2pP_1^2}{\chi^3 + 2p\chi^2}.$$

Since: $\psi = \nu_t F$, $Z = y/x$, $\nu_t = \nu a y^p m d d \nu_t/dy = \nu a y^p$, $d\nu_t/dy = \nu \cdot p/y$, $F = be^{-\chi}$ for: $\chi = 1$, $b = eF_1$, $F_1 = \phi_1/Z_1$ and results:

(41)
$$F = \frac{\phi_1}{Z_1} \cdot \exp(1 - \chi).$$

The derivatives of the stream function determine the velocity components:

$$\psi_y = u, \qquad \psi_x = v, \qquad u = \nu_t F \left[\frac{p}{y} - \frac{1}{xZ_1}\right] \qquad \text{also if:} P_{k_2} = p - k_2$$

$$u = \frac{\nu_t F P_{k_2}}{Y}, \quad \text{and } v = \frac{\nu_t F \chi}{x} \text{ consequently: } \frac{v}{u} = \frac{Z\chi}{P_{k_2}}$$

from: $u = v/(k_u^2 - 1)^{1/2} = \nu_t \chi F/x(k_u^2 - 1)^{1/2} u = \nu a_t \phi_1 \chi e^{1-\chi}/xZ_1(k_u^2 - 1)^{1/2}$ at the edge of the $BL \cdot \chi = 1 \cdot u = U$. Consequently: $U = \nu a_t \phi_1 / xZ_1(K_u^2 - 1)^{1/2}$ and finally the u, velocity can be expressed by the ratio:

(42)
$$\frac{u}{U} = \chi e^{1-\chi}.$$

The turbulent BL velocity component, can be correlated by a more general expression:

(43)
$$\frac{u}{U} = (\chi e^{1-\chi})^N$$

the value of N, to represent different flow regimes $N = 1$ for pure laminar flow, $N = 1/7$, $1/8$ for pure turbulent then intermediate values for different degrees of transition.

Assuming that turbulent flow BL correlated to the fifth powers, the turbulent BL thickness, have the expression:

(44)
$$Z_1 = \frac{\delta}{x} = \frac{K_t}{R^{1/5}}, \qquad K_t = [a_t \phi_1 k_1^2 k_u^2 p_1^4 (k_u^2 - 1)^{3/2}]^{1/5}.$$

For the coefficient of friction C_f based on the relation: $C_f \cdot \frac{1}{2}\rho U^2 = \varepsilon_t [du/dy]_{\hat{y}} \cdot \hat{y}$, is the thickness of the laminar sub-layer. Moreover:

$$\frac{\varepsilon_t}{\rho} = \nu_t = \nu a_t. \qquad C_f = \frac{2a_t \nu}{U} \cdot \left[\left(\frac{du}{U}\right) dy\right]_{\hat{y}}.$$

The velocity gradient becomes:

(45)
$$\left[\frac{du}{dy}\right]_{\hat{y} = (1/xZ_1) \cdot [N\chi^{N-1}(1-\chi)e^N(1-\chi)]}$$

and near the plate results:

$$\left[\frac{(du/U)}{dy}\right]_{\hat{y}} = \left(\frac{1}{xZ_1}\right) \cdot [N\chi^{N-1} \cdot e^N]$$

consequently:

(46)
$$C_f = \frac{2a_T \nu}{U} \cdot \left(\frac{1}{xZ_1}\right)[N\chi^{N-1}e^N] = \frac{2a_t Z_1}{RZ_1^2}[N\chi^{N-1}e^N].$$

As a particular solution of the turbulent flow, can be derived the corresponding to the laminar, from the velocity law taken $N = 1$ reads: $u/U = \chi e^{1-\chi} = 1$ integrated by parts to obtained the mass flux $I = -(1+\chi)e \cdot e^{-\chi} + e \cdot C$, when $\chi = 0$, $I(0) = 0$, for $\chi = 1$, $C = 1 : I = e - (1+\chi)e^{1-\chi}$, $\chi = 0 I(1) = 0$, $\chi = 1 I(1) = e - 2$. For the second derivative, $I'' = dI'/d\chi = (1-\chi)e^{1-\chi}$, the boundary conditions: $\chi = 1 I''(1) = 0$ developing the exponential in series and taking two terms:

$$I'' = 1 + (1-\chi) - \chi(1-\chi) \qquad \text{then} \qquad \chi = 0, \quad I'' = 2.$$

For the third derivative, let, defined $I'' = ye^{y} = f(y)$, $y = 1 - \chi$

$$f(y_1) = y_1 e^{y_1}, \qquad Z_1 = \frac{\delta}{x} = \frac{K_1}{R^{1/2}}, \qquad K_1 = [\phi_1 k_1^2 k_u^2]^{1/2} \tag{47}$$

the corresponding coefficient of friction becomes:

$$C_f = \frac{2Z_1}{\phi_1 k_1^2 k_u^2}\left[\frac{(du/U)}{d\chi}\right], \qquad \left[\frac{(du/U)}{d\chi}\right]_0 = 2,$$

$$C_F = \frac{4K_1}{R^{1/2}\phi_1 k_1^2 k_u^2}. \tag{48}$$

The expressions developed here resembles the similar obtained from the first solution, and suggest that at least for the laminar case, the constants: A and ϕ_1 are equivalent. Moreover from the first solution the value of the mass flow function, $I = \int_0^1 2\chi/(1+\chi)^{\chi} \cdot d\chi = (1/2)^{1/2}$ the corresponding value, from the second solution, is $I = e - 2$, they are within 1%. It is instructive to introduce the Blasius parameters: $\eta = ZR^{1/2}$, and $f = \int u/U \cdot d\eta = R^{1/2} \cdot Z_1 1 I$; integrated over the BL thickness:

$$f_1 = K_1 I_1 = \frac{\phi_1}{Z_4}, \quad \text{also } K_1^2 = Z_1^2 R = \phi_1 k_1^2 k_u^2, \quad \phi_1 = \frac{K_1^2}{k_1^2 k_u^2}. \tag{49}$$

Theoretical parameters

First is necessary to establish the value of the function ϕ_1, from equation (40), after $\chi = 1$, becomes:

$$H = 1 - 2p, \qquad \phi_1 = \frac{1 + p[H(p-4) - 5]}{H}. \tag{50}$$

Using the results of laminar solution, equation (50), $\phi_1 = 6$, expression (51) for $p = 0$, gives one. The final expression reads:

$$H = 1 - 2P, \qquad \phi_1 = 5 + \frac{1 + p[H(P-4) - 5]}{H}. \tag{51}$$

For the fully turbulent BL, the velocity exponent $N = 1/7$, the values of the nondimensional parameters are listed in the table, in terms of p, values the exponent of the law of turbulence for both solutions.

$N = 1/7$, $y_7 = 8/7$, $k_u = (8/7)/(9/8)$, $K_t = 0/39$

		SOL-I				SOL-II			
P	$\hat{\chi}$	A	K_f	a_t	R_{eddy}	ϕ_1	K_f	a_t	R_{eddy}
7/2	0.05	6	0.0752	0.0044	35	6	0.076	0.0044	35
1/7	0.22	5.41	0.0754	0.38		4.849	0.078	0.36	
−1/7	0.10	7.04	0.079	0.087		6.925	0.08	0.088	
0	0.15	6	0.075	0.1725	35	6	0.075	0.172	35

The value $p = 0$, correspond to the assumption of constant eddy viscosity.

For the laminar BL, the choice of parameters, require: $N = 1, p = 0$, $A = 6$, $P - 1 = 1$ the original ODE, equation (3) since $G_2 = G_3 = 1$, reduces to:

$$Z^3(F'^2 + F'') = 0, \qquad (F'^2 + F''') = 0$$

from equation (9a), when $A = 6$, becomes $F'^2 = 36/Z^4$, moreover also: $\frac{1}{2}F \cdot F'' = 36/Z^4$, consequently: $(F \cdot F'' + 2F''') = 0$ expression known as Blasius equation, that basically govern the laminar BL. The velocity function is given by equation (27), when $N = 1$, namely: $u/U = 2\chi/(1+\chi)^\chi$. For the expressison of the BL thickness, when $a_t = 1$, from equation (24): and equation (42), $k_1^2 \cdot k_u^2 = 6$, $Z_1^2 = 6k_1^2k_u^2/R$, or: $Z_1 = 6/R^{1/2}$, $\delta/x = K_1/R^{1/2}$, and for the coefficient of friction, from equation (32), $C_f = 4Z_1/Ak_1^2k_\mu^2 = (4/6)(1/R^{1/2}) = K_f/R^{1/2}$, $K_1 = 6$, $K_f = 4/6$. The results are remarkable close to the accepted values.

References

1. L. D'Attorre, Turbulent boundary layer, an analytic solution of the nonlinear differential equation for the flat plate, Joint Mathematics meetings, January 8–11, 1997, San Diego, California.

2. L. D'Attorre, A closed form solution for the laminar boundary layer, Blasius equation: $F \cdot f_2'' f''' = 0$, for the flat plate, Annual meeting SIAM, AMS, San Diego, California, 1994.

3. L. D'Attorre, Comressible Time Dependent Navier-Stokes Two Dimensional Flow Equations, Finite Difference Solution, TRW IR & D Project, 1978.

8th Int. Coll. on Differential Equations, pp. 137-144
D. Bainov (Ed)

Solving Singular Integral Equations

J.H. DE KLERK
Dept of Mathematics & Applied Mathematics, Potchefstroom University for CHE, Private Bag X6001, Potchefstroom, South Africa

Abstract. Integral equations (particularly strongly singular equations) can be solved effectively by means of an L_1-approximation technique. In this discussion attention is paid to some numerical aspects of this method.
Keywords: Integral equation, approximation

1. Introduction

Integral equations are encountered in many mathematical models and a variety of numerical methods for obtaining solutions to integral equations has been proposed during the last few decades.

In some of these methods different kinds of orthogonal polynomials have been used to solve these equations (among others, that of Razzaghi *et al.* [1]). The basis of these methods is the reduction of the integral equation to a set of algebraic equations by means of these orthogonal polynomials. In other methods, successive approximations (among others, that of Hanna & Brown [2]) or iterative methods (for example, that of Kleinman & Van den Berg, [3]) are used.

Still another strategy for solving integral equations is a technique based on an L_p-approximation. By reformulating the problem of solving an integral equation as a continuous L_p-approximation problem, it could be solved numerically by means of a minimization technique, thus obtaining an approximate solution to the original problem. In many of the problems in my own studies specific attention has been paid to the case of the L_1-approximation, and this norm will feature in the present discussion. This method is discussed in detail in De Klerk [4] and De Klerk, Eyre & Venter [5].

The purpose of this paper is to discuss some aspects of this technique, especially concerning the case of (a) the parameter of the integral equation tending towards one of the eigenvalues of the problem (with the kernel taking the form of a Pincherle-Goursat kernel); (b) deviations in the original data (discrete noise) as well as the case of poorly rounded off data (continuous noise); (c) bifurcation points, that is, the situation where two or more different solutions to the problem lie very close together; (d) solving strongly singular integral equations.

2. The problem and its reformulation as an approximation problem

2.1. The general problem

The general problem under consideration in this discussion is the Fredholm integral equation (either linear or non-linear) of the first and second kind, namely:

$$\lambda \int_a^b K(x,t,\phi(t))dt = \psi(x), \tag{1}$$

and

$$\phi(x) - \lambda \int_a^b K(x,t,\phi(t))dt = \psi(x), \tag{2}$$

with $x \in [a,b]$.

In these equations it will be assumed that the driving function ψ and the kernel K is known, while the function ϕ is to be found. The kernel need not be continuous.

2.2. The general continuous L_p-approximation problem

The general continuous L_p-approximation problem can be formulated as follows: Let a function f, defined on a real interval X, be given, together with an approximation function $F(A)$ which depends on the n parameters $A = (a_1, a_2, \ldots, a_n)$ in R_n. For the continuous L_p-approximation problem one then has to find a vector $\hat{A} \in R_n$ such that

$$\int_X |f(x) - F(\hat{A},x)|^p dx \leq \int_X |f(x) - F(A,x)|^p dx,$$

for all $A \in R_n$.

2.3. A reformulation of the problem as an approximation problem

The problem of solving the integral equation as a continuous L_p-approximation problem can now be reformulated as follows: Assume that the solution function ϕ of the integral equation can be approximated by the function $F(A)$. Substitution of this function in equation (2) leads to

$$F(A,x) - \int_a^b K(x,t,F(A,t))dt \approx \psi(x), \qquad x \in [a,b].$$

The approximation problem in the L_p-norm then becomes

$$\min_A \int_a^b \left| F(A,x) - \int_a^b K(x,t,F(A,t))dt - \psi(x) \right|^p dx,$$

where $F(A,x)$ is defined on R_n with respect to A and on $[a,b]$ with respect to x. If the integral

$$J(A) = \int_a^b \left| F(A,x) - \int_a^b K(x,t,F(A,t))dt - \psi(x) \right|^p dx,$$

is considered as a function $J : R_n \to R_1$, depending on A only, the approximation problem can be viewed as an unconstrained minimization problem in the n variables $a_1, \ldots, a_n$. Therefore J only has to be minimized with respect to these variables. In this way the L_p-approximation problem (and therefore also the integral equation) can be solved numerically as a minimization problem.

2.4. The procedure

The approximate solution of the integral equation is obtained from the coefficients $\hat{A}$ resulting from the minimization problem. The two major steps in the numerical technique are:

- A numerical search procedure to calculate the coefficients $\hat{A}$; and
- a numerical integration procedure to evaluate the L_p-norm.

  The following comments about these procedures are in order:

1. The minimization is performed using standard optimization algorithms (for example UMPOL in the IMSL Library and the downhill simplex method according to Nelder and Mead (see Press *et al.* [6])). Although the simplex method is not very fast, it has the advantage that it is very robust and does not need derivatives.
2. The numerical integration procedure used for evaluating integrals over absolute values of functions (as in the case of the L_1-norm which was specifically studied) is performed via a technique developed by colleagues Laurie and Venter [7, 8]. For this technique the standard adaptive integration routine QAGE (from the QUADPACK subroutine package of Piessens *et al.* [9]) was modified to allow for the efficient evaluation of integrals of the general form $\int_a^b |f(x)|dx$. Using the function values available during the integration process, the zeros of the function are found and used as subdivision points.

3. The sensitivity of the technique

Due to the fact that the technique of solving an integral equation by means of an L_1-approximation method is based on a whole series of different numerical techniques (among others an approximation process, an integration of an integrand consisting of an absolute value, and a minimization process) the sensitivity of the technique could be questioned. Therefore, to provide an answer to the question of sensitivity, the following problems were considered:

- The case of the parameter tending towards one of the eigenvalues of the problem, with the kernel taking the form of a Pincherle-Goursat kernel;
- the case of poorly rounded off data (continuous noise); and
- the case of bifurcation points, that is, the situation where two or more different solutions to the problem lie very close together.

In equations (1) and (2) above the constant parameter λ will be assumed to be known. As will become clear in due course, the value of λ may play a vital role in some of the cases under discussion.

3.1. The parameter λ tending towards one of the eigenvalues of the problem

Consider a kernel $K(x,t)$ of the Pincherle-Goursat type, that is,

$$K(x,t) = \sum_{k=1}^{n} M_k(x)N_k(t),$$

where $\{M_1(x), \dots, M_n(x)\}$ and $\{N_1(t), \dots, N_n(t)\}$ are two sets of linearly independent functions.

In this case the Fredholm equation of the second kind has in general one and only one solution. The only exception is when the parameter λ takes the value of one of the eigenvalues of the kernel, in which case there is in general no solution (Tricomi [10]).

In the following example we consider a case which exhibits the efficiency of the technique quite clearly.

Example 1

Consider the Fredholm integral equation of the second kind

$$y(x) - \lambda \int_0^1 (x+t)y(t)\,dt = \frac{5x}{6} - \frac{1}{9}.$$

We set

$$y(x) \equiv F(A,x) = a_0 + a_1 x + a_2 x^2 + a_3 x^3,$$

so that one has to compute the parameters $a_0, \dots, a_3$ such that

$$J(A) = \int_0^1 \left| a_0 + a_1 x + a_2 x^2 + a_3 x^3 - \right.$$

$$\left. -\lambda \int_0^1 (x+t)(a_0 + a_1 t + a_2 t^2 + a_3 t^3)\,dt - \frac{5x}{6} + \frac{1}{9} \right| dx$$

is minimized.

The eigenvalues of this problem are found to be $\lambda \approx 0.928\,203$ and $\lambda \approx -12.928\,203$ (to six decimal places). Using the L_1-approximation method and letting λ tend to 0.928, the following results were obtained:

λ	$y(x)$, analytically calculated	$y(x)$, numerically calculated
0.3333	x	$0.0000 + 1.0000x$
0.8	$1.0606 + 2.8030x$	$1.6060 + 2.8030x$
0.9	$5.8120 + 11.0256x$	$5.8116 + 11.0251x$
0.92	$20.6573 + 36.7371x$	$20.6588 + 36.7398x$
0.928	$844.6970 + 1464.0151x$	$844.7736 + 1464.1480x$

It is clear that the algorithm efficiently obtains the solutions to this problem, even in the case of λ differing by as little as 2.03×10^{-3} from one of the eigenvalues.

3.2. Deviations in the original data

An especially interesting and difficult problem with regard to integral equations is the case when some of the data has been contaminated by noise (Rall [11]). In many engineering problems, the solution function ϕ in equation (2) corresponds with some actual physical phenomenon that is observed or measured, the operator K in (2) corresponds with the action of the measuring instrument, while ψ in (2) represents the set of observations or data on which the calculation of ϕ is to be based. What concerns us, is the case of noise occurring in the observed values of ψ.

Example 2

Consider the Fredholm integral equation of the second kind

$$u(x) - \frac{1}{2}\int_0^1 tu(t)dt = \psi(x),$$

with

$$\psi(x) = \frac{3}{2}e^x - \frac{1}{2}xe^x - \frac{1}{2}.$$

The solution to this problem can be determined analytically to be

$$u(x) = \frac{3}{2}e^x - \frac{1}{2}xe^x - \frac{1}{3}e + 1.$$

We will show that the solution of the problem with $\psi(x)$ substituted by $\psi_d(x)$, is in fact influenced to some extent by the contamination spread over the whole interval.

If we select

$$u(x) \equiv F(A, x) = a_0 + a_1 x,$$

we then have to compute the parameters a_0, a_1, such that

$$J(A) = \int_0^1 \left| \frac{3}{2}e^x - \frac{xe^x}{2} - 0.5 - \frac{3}{4}a_0 + \frac{a_1}{6} - a_1 x \right| dx$$

is minimized. (It is sufficient to consider only a linear approximation, because, as will become clear in due course, the influence of the contamination can already be seen in the linear case.)

Using the algorithm to minimize $J(A)$, with $\psi(x)$ calculated as accurately as possible, we found the best solution to this problem in the L_1-sense for the above-mentioned choice of $F(A, x)$ to be

$$F(\hat{A}, x) = 1.550\,471 + 1.232\,180x.$$

It can be noted that in the calculation of this result, the values of the partial derivatives of J with respect to a_0 and a_1, are respectively found to be 5.4×10^{-7} and 3.1×10^{-7}. This assures one that this value is correct and can in fact be used as a basis for further comparisons. When one now considers the function $\psi_d(x)$ instead of $\psi(x)$, the following results are obtained:

$d = 9: \quad F(\hat{A}, x) = 1.550\,470 + 1.232\,181x,$

$d = 6: \quad F(\hat{A}, x) = 1.550\,463 + 1.232\,191x.$

It is immediately clear on the one hand that the technique has to some extent been influenced by the spread of noise over the whole interval. On the other hand, however, one also notices that the influence in the results is more or less of the same order as the original contamination and therefore one can still consider these results positively.

3.3. Bifurcation points

A phenomenon that occurs in certain types of integral equations is the presence of bifurcation points in the solution function ϕ of the integral equation when considered as a function of the parameter λ. (See Tricomi [10], who apparently is one of the first authors to mention such a case, and also Rall [11].) The consequence of this phenomenon is that the number of solutions may vary due to a change in the value of λ. Therefore, starting with a given value of λ for which the different solutions are known and letting λ tend towards such a bifurcation point, one may ask whether – and for how far – the technique will still be able to discriminate between these different solutions.

Example 3

Consider the following non-linear Fredholm integral equation of the second kind:

$$\phi(x) - 10\,000\lambda \int_0^1 \phi^3(t)\,dt = 0.01.$$

Calculating the solution to this problem analytically, it can be shown that the number of finite solutions for different values of λ is spread as follows:

$\lambda < 0.0$	1 solution
$0.0 < \lambda < 4/27$	3 solutions
$\lambda = 4/27$	2 solutions
$\lambda > 4/27$	1 solution

We choose

$$\phi(x) \equiv F(A,x) = a_0 + a_1 x,$$

for this case, and have to calculate the parameters a_0 and a_1 such that

$$\begin{aligned} J(A) &= \int_0^1 \left| a_0 + a_1 x - 10\,000\lambda \int_0^1 [a_0 + a_1 t]^3 \, dt - 0.01 \right| dx \\ &= \int_0^1 \left| a_0 + a_1 x - 10\,000\lambda \left[a_0^3 + 1.5a_0^2 a_1 + a_0 a_1^2 + 0.25 a_1^3 \right] - 0.01 \right| dx \end{aligned}$$

is minimized.

In the following calculations, we choose, among others, $\lambda = 0.148\,148$, this value being very close to the bifurcation value of $4/27$ ($= 0.148\,148\ldots$).

Using the algorithm to minimize $J(A)$, for the best solution of this problem in the L_1-norm and for the above-mentioned choice of $F(A,x)$, one finds that (a) the second parameter, a_1, is equal to zero, at least as far as 6 decimal places, and (b) the first parameter, a_0, takes the following values, as given in the last column, for the stated values of λ – first for the interval $0.0 < \lambda < 4/27$ and then for $\lambda > 4/27$.

λ	a_0, analytically calculated	a_0, numerically calculated
0.12	-0.032957; 0.012155; 0.020802	-0.032957; 0.012155; 0.020802
0.14	-0.030765; 0.013276; 0.017489	-0.030765; 0.013276; 0.017489
0.148	-0.030013; 0.014733; 0.015281	-0.030013; 0.014733; 0.015281
0.148 148	-0.030000; 0.014991; 0.015009	-0.030000; 0.014991; 0.015009
0.15	-0.029835;	-0.029835; (0.014907)

It is clear that the algorithm exactly (in any case, to six decimal places) obtained all the solutions to the problem, even in the close case of $\lambda = 0.148\,148$, differing by as little as 1.5×10^{-6} from the bifurcation value. (In fact, in the last case, it was even somewhat over-enthusiastic in the sense that it also found a value which was not a solution, but which was due to a local minimum of the approximation problem.) The solutions definitely depend on the choice of starting values, but, as can be seen, all values could be found.

4. Strongly singular integral equations

The analysis of cracks in brittle solids like ceramics and ceramic composites is presently a topic of much interest (Hori & Nemat-Nasser [12]). Analysis of a broad class of problems of this kind leads to the following strongly singular integral equation:

$$f(x) = g(x) + \mathop{=\!\!\!\!\!\!\int}\nolimits_a^b k(x,t) \frac{f(t)}{(t-x)^2} dt, \quad a < x < b.$$

The integral in this integral equation is to be understood in the finite part integral or Hadamard sense (Bertram & Ruehr [13]).

The approximation technique was also applied to this type of integral equation - as was reported on at the Second International Conference on Dynamic Systems and Applications held at Morehouse College, Atlanta, USA in 1995 (De Klerk [14]). To conclude, I would like to give the following example showing the performance of this method in the case of a strongly singular integral equation.

Example 4

Consider the finite part integral equation (Bertram & Ruehr [13]),

$$f(x) - \neq\!\!\!\!\int_0^1 \frac{f(t)}{(t-x)^2}dt = \sin\pi x - [-\pi \sin\pi x\{Si(\pi(1-x)) + Si(\pi x)\}$$
$$+\pi\cos\pi x\{Ci(\pi x) + Ci(\pi(1-x))\}],$$

where $Si(x)$ and $Ci(x)$ are respectively the sine and cosine integral function

$$Si(x) = \int_0^x \frac{\sin u}{u}du \quad \text{and} \quad Ci(x) = \int_x^\infty \frac{\cos u}{u}du,$$

with solution, $f(x) = \sin\pi x$. Setting as approximate solution,

$$f(x) \approx F(A,x) = \frac{A_2x^2 + A_1x + A_0}{B_2x^2 + B_1x + 1},$$

the approximation problem becomes,

$$\min_{A,B} \int_0^1 \left| M(x) - \neq\!\!\!\!\int_0^1 N(t,x)dt + P(x) \right| dx,$$

where

$$M(x) = \frac{A_2x^2 + A_1x + A_0}{B_2x^2 + B_1x + 1}, \qquad N(t,x) = \frac{A_2t^2 + A_1t + A_0}{(B_2t^2 + B_1t + 1)(t-x)^2},$$

$$P(x) = -\sin\pi x - \pi\sin\pi x[Si\{\pi(1-x)\} + Si(\pi x)] + \pi\cos\pi x[Ci(\pi x) + Ci\{\pi(1-x)\}].$$

The strategy used here was to calculate the inner integral by means of Gauss quadrature. It follows that

$$F(A,x) = \frac{-3.212833x^2 + 3.212833x + 0.000099}{0.793135x^2 - 0.793136x + 1}.$$

The approximation in this case is very good. The sine and cosine integral functions have been calculated by means of an infinite series approximation. Fortunately, using only a few terms in the series, an accuracy of up to 6 decimal places was obtained.

5. Conclusion

It has been shown previously that integral equations can be solved effectively by means of an L_1-approximation technique. In this discussion the purpose was to pay attention to some details concerning this method. We also wished to pay attention to the case of strongly singular integral equations. By means of these examples it has now been shown that this technique can be used successfully in more difficult cases.

References

1. M. Razzaghi, M. Razzaghi and A. Arabshahi, Applied Mathematics and Computation, 40, 215–224 (1990).

2. O.T. Hanna and L.F. Brown, Chemical Engineering Science, 46, 2749–2753 (1991).

3. R.E. Kleinman and P.M. van den Berg, Radio Science, 26, 175–181 (1991).

4. J.H. de Klerk, Applied Mathematics and Computation, 62, 171–182 (1994).

5. J.H. de Klerk, D. Eyre and L.M. Venter, Applied Mathematics and Computation, 72, 285-300 (1995).

6. W.H. Press, B.P. Flannery, S.A. Teukolsky and W.T. Vetterling, Numerical Recipes in Pascal: The art of scientific programming, Cambridge University Press, Cambridge (1989).

7. D.P. Laurie and A Venter, (Technical note T/15. Potchefstroom University, Potchefstroom, 1993).

8. A Venter and D.P. Laurie, BIT, 36, 387-394 (1996).

9. R. Piessens, E. de Doncker-Kapenga, C.W. Überhuber and D.K. Kahaner, QUADPACK – A Subroutine Package for Automatic Integration, Springer-Verlag, Berlin (1983).

10. F.G. Tricomi, Integral Equations, Interscience Publishers, New York (1957), pp. 55ff and 161ff.

11. L.B. Rall, in: Numerical Solutions of Integral Equations, L.M. Delves and J. Walsh (eds.), Clarendon Press, Oxford (1974) pp. 207–222.

12. M. Hori and S. Nemat-Nasser, SIAM Journal of Applied Mathematics, 50, 716-725 (1990).

13. B. Bertram and O. Ruehr, Journal of Computational and Applied Mathematics, 41, 163-173 (1992).

14. J.H. de Klerk, in: Proceedings of Dynamic Systems and Applications, II, G.S. Ladde and M. Sambandham, (eds), Dynamic Publishers, Atlanta (1996) pp. 315–322.

8th Int. Coll. on Differential Equations, pp. 145-152
D. Bainov (Ed)

Asynchronous Multisplitting Methods with Flexible Communication for Pseudolinear P.D.E.

DIDIER EL BAZ
LAAS du CNRS, 7, avenue du Colonel Roche, 31077 Toulouse Cedex 4, France

PIERRE SPITERI
ENSEEIHT-IRIT/LIMA, 2, rue Camichel, 31071 Toulouse Cedex 4, France

JEAN CLAUDE MIELLOU
LCS, Université de Franche Comté, 16, Route de Gray, 25030 Besançon Cedex, France

Abstract. We consider asynchronous iterations whereby every processor can have access to the current value of the components of the iteration vector updated by any other processor, i.e. parallel schemes of computation which allow exchanges between processors of partial updates resulting from updating phases in progress. Application of asynchronous iterations with flexible communication to the solution of nonlinear boundary value problems via multisplitting methods is presented.
Keywords: flexible communication, asynchronous iterations, parallel computing, multisplitting technics, boundary value problems.

1. INTRODUCTION

In this paper we consider asynchronous multisplitting methods with flexible communication. The originality of this class of parallel iterative methods is that updating phases can use data which are issued from computations in progress. This leads to a better coupling between communication and computation. Asynchronous iterations with flexible communication are parallel schemes of computation whereby every processor can have access to the current value of the components of the iteration vector updated by any other processor. In particular, they allow the exchange between processors of partial updates which result from updating phases in progress. In contrast with previous studies (see [1]–[8]), the methods presented here permit one to consider block iterative schemes with delayed access to block subvectors, the components of which are relative to different iteration numbers or intermediate computations not explicitly labelled by an iteration number. Another important feature of this paper is the use of mappings which approximate the solution (see [9]).

Asynchronous iterations with flexible communication were introduced in [10] in order to speed up the convergence of parallel iterative algorithms in the context of M-functions. This new model of parallel asynchronous iterations was applied with success to subdomain methods for the solution of partial differential equations (see [11]) and approximate relaxation methods for network flow problems (see [12] and [13]). We note that computational results on shared memory and distributed memory multiprocessors are presented in detail in [12], [13], and [11]. Application of asynchronous iterations with flexible communication to the solution of nonlinear boundary value problems via multisplitting methods is presented in this paper; this includes in particular, discrete analogs of the multisubdomain Schwarz method (see [14]) and two-stage multisplitting methods (see [14]–[17], and [18]).

In Section 2 we present the main problem, give some background material on multisplitting and study the properties of the extended nonlinear problem. Section 3 deals with parallel asynchronous multisplitting methods with flexible communication.

2. PSEUDOLINEAR PROBLEM AND FORMAL MULTISPLITTING

We consider the solution of the following pseudolinear problem:

$$Ax^* + \varphi(x^*) = 0, \tag{2.1}$$

where $A \in L(R^n)$ is an M-matrix, $\varphi : R^n \to R^n$ is a continuous diagonal maximal monotone operator. Let the following regular splittings of matrix A :

$$A = M^l - N^l, \quad l = 1, ..., m, \tag{2.2}$$

where $(M^l)^{-1} \geq 0$ and $N^l \geq 0$. Let $F^l : R^n \to R^n$, $l = 1, ..., m$ be the fixed point mappings associated with problem (2.1) and defined by:

$$F^l(x) = y \text{ such that } M^l y = N^l x - \varphi(x). \tag{2.3}$$

A formal multisplitting associated with problem (2.1) is defined by the collection of fixed point problems (see [14]):

$$x^* - F^l(x^*) = 0, \; l = 1, ..., m. \tag{2.4}$$

Let $E = (R^n)^m$ and consider the following block-decomposition of E:

$$E = \prod_{l=1}^{m} E_l,$$

where $E_l = R^n$. Each subspace E_l is endowed with the natural (or component-wise) partial ordering associated with the cone $K_l = R^n_+$ of vectors with nonnegative components in R^n. Let $X \in E$. We have the following block-decomposition of X :

$$X = \{X_1, \ldots, X_l, \ldots, X_m\} \in \prod_{l=1}^{m} E_l.$$

Definition 1. The extended fixed point mapping $T : E \to E$ associated with the formal multisplitting is given as follows (see [14]):

$$T(X) = Y, \text{ such that } Y_l = F^l(Z_l) \text{ with } Z_l = \sum_{k=1}^{m} W_{lk} X_k, \; l = 1, ..., m, \tag{2.5}$$

where W_{lk} are nonnegative diagonal weighting matrices satisfying for all $l \in \{1, \ldots, m\}$

$$\sum_{k=1}^{m} W_{lk} = I,$$

I being is the identity matrix in $L(E_l)$. Let the following block-decomposition of the mapping T :

$$T(X) = \{T_1(X), \ldots, T_l(X), \ldots, T_m(X)\} \in \prod_{l=1}^{m} E_l.$$

The extended fixed point mapping T is associated with the following extended nonlinear problem:

$$a(X^*) = 0, \tag{2.6}$$

where the mapping $a : E \to E$ is given by:

$$a(X) = A^e X + \varphi^e(X), \tag{2.7}$$

the function $\varphi^e : E \to E$ being the extended monotone perturbation operator and for all $l \in \{1, \ldots, m\}$:

$$A_l^e X = M^l X_l - N^l \sum_{k=1}^{m} W_{lk} X_k. \tag{2.8}$$

In the sequel $a_l(X_1, \ldots, X_{l-1}, Y_l, X_{l+1}, \ldots, X_m)$ will also be denoted by $a_l(Y_l\,; X)$.

Proposition 1. Let the above assumptions hold. The mapping a is a continuous surjective M-function.

Proof: We recall that an M-function is off-diagonally antitone and inverse isotone (see [19] and [20]). It follows from (2.8) and the definition of matrices W_{lk} that the off-diagonal entries of the matrix A^e are nonpositive since A is an M-matrix. Let $U = \{u, \ldots, u\}$ be the vector of E with block-components equal to the eigenvector u associated with the greatest eigenvalue of the Jacobi matrix $D^{-1}(D - A)$, where D denotes the diagonal part of matrix A and consider the Jacobi matrix J derived from the extended system (2.6). For all $l \in \{1, \ldots, m\}$, we have:

$$J_l u = (D^e)_l^{-1} D_l^e u - (D^e)_l^{-1} M^l u + (D^e)_l^{-1} N^l \sum_{k=1}^{m} W_{lk} u =$$

$$(D^e)_l^{-1}(D_l^e - M^l + N^l)u = D^{-1}(D - A)u = \rho(D^{-1}(D - A))u.$$

It follows from [8] that the Jacobi matrix J is contracting relative to a weighted maximum norm and the result follows from Proposition 2.4.17 in [19]. The continuity and surjectivity of a follows from the continuity and maximal monotonicity of φ.

It follows from Proposition 1 that

$$\begin{cases} \text{For all } l \in \{1, \ldots, m\} \text{ and all } X \in E, \text{ the mapping: } \; Y_l \to a_l(Y_l, X), \\ \text{is a continuous surjective } M\text{-function of } E_l \text{ onto } E_l, \end{cases} \tag{2.9}$$

(see [20, theorem 3.5]). Moreover it follows from Proposition 1 that

$$\begin{cases} \text{for all } l \in \{1, \ldots, m\} \text{ and } X \in E \text{ the problem: } \; a_l(Y_l\,; X) = 0, \\ \text{has a unique solution } Y_l. \end{cases} \tag{2.10}$$

It follows from the above assumptions that T is isotone on E (see [9]).

Definition 2. A vector $Y \in R_+^{nm}$ is an a-supersolution (an a-subsolution) if $a(Y) \geq 0$ $(a(Y) \leq 0)$.

We note that the above multisplitting corresponds to the discrete analog of the multisubdomain Schwarz method.

We also note that the above results are still valid in the case where φ is a multivalued diagonal maximal monotone operator.

3. ASYNCHRONOUS MULTISPLITTING METHODS WITH FLEXIBLE COMMUNICATION

In Section 2 we have presented a collection of fixed point mappings associated with the exact solution of the problem (2.1). In this Section we consider fixed point mappings associated with approximate solutions of the problem (2.1). Such mappings were introduced in [9] and are used in this paper for the solution of problem (2.1) via parallel asynchronous iterative methods with flexible communication.

Throughout the paper we adopt the following notation for order intervals. Let $X_l, Y_l \in E_l$, be such that $X_l \leq Y_l$; then an order interval in E_l is denoted by:

$$\langle X_l, Y_l \rangle_l = \{X_l' \in E_l \mid X_l \leq X_l' \leq Y_l\}.$$

Similarly, let $X, Y \in E$, be such that $X \leq Y$; then an order interval in E is denoted by:

$$\langle X, Y \rangle = \{X' \in E \mid X \leq X' \leq Y\}.$$

Definition 3. Let a be an M-function and T defined by (2.5). Then the mapping T^a with components T_l^a is an a-supermapping associated with T if for all $l \in \{1, \ldots, m\}$ and Y element of the domain: $\{Y \in E \mid a_l(Y) \geq 0\}$, we have $T_l^a(Y) \leq Y_l$, $a_l(T_l^a(Y), Y) \geq 0$, and $T_l^a(Y) \neq Y_l$ if $T_l(Y) \neq Y_l$.

Remark 1. We can analogously define a-submappings associated with T by reversing the inequalities.

The class of asynchronous iterations presented in this paper allows flexible communication between processors. We introduce first the concepts of steering and sequence of delays.

Definition 4. A steering of components of the iteration vector is a sequence $\{s(p)\}$ also denoted by $\mathcal{S}$, where $p \in N$ the set of natural integers, which satisfies

$$\text{for all } p \in N, \quad s(p) \in \{1, \ldots, m\}. \tag{3.1}$$

Definition 5. A sequence of maximal delays in the access to the updated components of the iteration vector, is denoted by $\{r(p)\}$ (or $\mathcal{R}$) and satisfies

$$r(p) = \{r_1(p), \ldots, r_l(p), \ldots, r_m(p)\} \in N^m, \text{ for all } p \in N,$$

where for all $l \in \{1, \ldots, m\}$ and $p \in N$, $0 \leq r_l(p) \leq p$ and $r_l(p) = 0$ if $l = s(p)$.

We note that the latest condition has been introduced in order to simplify the presentation of the algorithms. This assumption is more restrictive than the one used in the classical model, however it is natural since it implies that each processor has access to its own working space without delays.

We introduce the functions $\rho_l : N \to N, l \in \{1, \ldots, m\}$, defined by

$$\rho_l(p) = p - r_l(p),$$

which satisfy for all $l \in \{1, \ldots, m\}$ and $p \in N$:

$$0 \leq \rho_l(p) \leq p, \tag{3.2}$$

$$\rho_l(p) = p, \text{ if } l = s(p). \tag{3.3}$$

In the sequel, we will denote by $\rho(p)$ the mapping from N^m into itself, with components $\rho_l(p), l \in \{1, \ldots, m\}$.

We introduce now the sequence $\{K_l^p\}, p \in N$ defined by

$$K_l^p = \{k \in N \mid s(k) = l, 0 \leq k < p\}, l \in \{1, \ldots, m\}, p \in N. \tag{3.4}$$

The elements of the set K_l^p correspond to the iteration numbers between 0 and p at which the l-th block-component of the iteration vector is updated. We have

(3.5) $\left\{K_l^p\right\}$ is a denumerable sequence of finite elements of the set of parts of N.

Moreover $\left\{K_l^p\right\}$ is nested:

$$K_l^p \subseteq K_l^{p+1}. \tag{3.6}$$

Definition 6. Asynchronous multisplitting iterations with flexible communication associated with the a-supermapping T^a are sequences $\{Y^p\}$ defined recursively as follows. For all $p \in N$,

$$\begin{cases} Y_l^{p+1} = T_l^a(\tilde{Y}^p), & \text{if } l = s(p), \\ Y_l^{p+1} = Y_l^p, & \text{if } l \neq s(p), \end{cases} \tag{3.7}$$

where $\tilde{Y}^p$ is implicitly defined by

$$\tilde{Y}^0 = Y^0, \text{ an } a\text{-supersolution} \tag{3.8}$$

and

$$\tilde{Y}^p \in \langle Y^p, Y^{\rho(p)} \wedge \tilde{Y}^q \rangle, \text{ if } p \geq 1, \tag{3.9}$$

the block-components of the vector $Y^{\rho(p)} \in E$ are the subvectors $Y_l^{\rho_l(p)}, l \in \{1, \ldots, m\}$, and $q = \max_{k \in K_{s(p)}^p} k$. We note that $\tilde{Y}^p \in \langle Y^p, Y^{\rho(p)} \rangle$ if $K_l^p = \emptyset$, where $\emptyset$ denotes the empty set.

Remark 2. Asynchronous iterations defined by (3.7)-(3.9) are general iterative methods whereby iterations are carried out in parallel by m processors in arbitrary order and

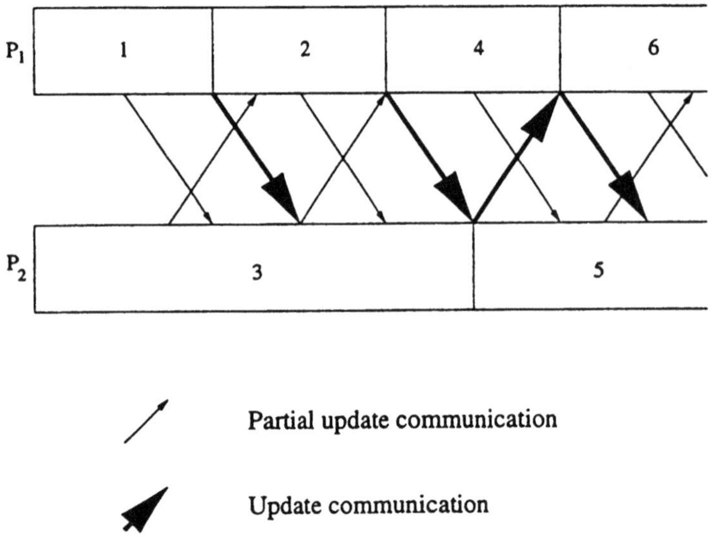

Figure 1: Asynchronous iterations with flexible communication

without any synchronization. Flexible communication between processors is the main originality of this class of algorithms. Indeed we note that the block-components $\tilde{Y}_j^p$ of the iteration vector used at the time of the updating of Y_l^{p+1} are taken anywhere in the interval $\langle Y_j^p, Y_j^{\rho(p)} \wedge \tilde{Y}_j^q \rangle_j$, where $\tilde{Y}_j^q$ denotes the value used for the latest update of the block-component Y_l and $Y_j^{\rho(p)}$ models the non-deterministic behavior of the iterative scheme. The introduction of order intervals permits us to take into account data coming from computations which are in progress and which are not explicitly labelled by an iteration number (see figure 1 where P_1 and P_2 denote respectively processor 1 and 2 and the white boxes which are labelled by an iteration number represent updating phases). We note that computations overlap communications. Thus, there is a better coupling between communication and computation and we may expect a faster convergence particularly in the case of monotone convergence. Moreover each communication of a block subvector is not necessarily associated with a given iteration number. The exchanged values of the components of each block subvector can be relative to different iteration numbers.

Proposition 2. Let a be a continuous surjective M-function, T the fixed point mapping defined by (2.5), T^a an a-supermapping associated with T, $Y^0 \in E$ an a-supersolution, $\mathcal{S}$, and $\mathcal{R}$ a steering and a sequence of delays, respectively. Then, the asynchronous multisplitting iteration $\{Y^p\}$ given by (3.7)-(3.9) is well defined and satisfies:

$$Y^* \leq \cdots \leq Y^{p+1} \leq Y^p \leq \cdots \leq Y^0,$$

and $\lim_{p\to\infty} Y^p = Y^*$, where Y^* is an a-supersolution of problem (2.6).

Proof: The result follows from Proposition 1 in Section 2 and Proposition 6 in [10].

We now introduce two classes of a-supermappings for which we have convergence properties.

Definition 7. The mapping T^a with components T_l^a defined on the domain $\{Y \in E \mid a_l(Y) \geq 0\}$ is an a-supermapping of the first kind associated with T if T^a is an a-supermapping and there exists an a-supermapping T'^a with components $T_l'^a$ defined on the domain $\{Y \in E \mid a_l(Y) \geq 0,\}$ such that for all $l \in \{1, \ldots, m\}$ and Y element of the domain $\{Y \in E \mid a_l(Y) \geq 0\}$, we have

$$T_l'^a(Y) \in \langle T_l^a(Y), Y_l \rangle_l \tag{3.10}$$

and

$$Y^p \downarrow Y^*, p \to \infty \text{ implies } T_l'^a(Y^p) \downarrow T_l'^a(Y^*), p \to \infty \text{ for all } l \in \{1, \ldots, m\}, \tag{3.11}$$

where the notation $Y^p \downarrow Y^*, p \to \infty$ means that

$$Y^* \leq \cdots \leq Y^{p+1} \leq Y^p \leq \cdots \leq Y^0 \text{ and } \lim_{p \to \infty} Y^p = Y^*.$$

Remark 3. The relation (3.11) can be interpreted as a property of continuity of T'^a related to the partial ordering.

Proposition 3. Let the assumptions of Proposition 2 hold. Let T^a be an a-supermapping of the first kind associated with T and assume that the steering $\mathcal{S}$ satisfies:

$$\text{for all } l \in \{1, \ldots, m\}, \text{ the set } \{p \in N \mid s(p) = l\} \text{ is infinite} \tag{3.12}$$

and

$$\text{for all } l \in \{1, \ldots, m\}, \quad \lim_{p \to \infty} \rho_l(p) = +\infty. \tag{3.13}$$

Then, the asynchronous multisplitting iteration $\{Y^p\}$ defined by (3.7)-(3.9) satisfies $Y^p \downarrow X^*$, where X^* is the unique solution of problem (2.6).

Proof: The result follows from Proposition 1 in Section 2 and Proposition 7 in [10].

Hypothesis (3.12) assumes that no subvector is abandonned for ever in the updating process.

Hypothesis (3.13) assumes that the old values of a subvector are abandonned as the computation progresses.

Definition 8. The mapping T^a with components T_l^a is an a-supermapping of the second kind associated with T if T^a is an a-supermapping and for all $l \in \{1, \ldots, m\}$ and Y element of the domain: $\{Y \in E \mid a_l(Y) \geq 0\}$, there exists $\delta > 0$ such that

$$\|Y_l - T_l^a(Y)\|_l \geq \delta \|Y_l - T_l(Y)\|_l, \tag{3.14}$$

where $\|.\|_l$ is a norm defined in E_l.

Proposition 4. Let the assumptions of Proposition 2 hold and assume that T^a is an a-supermapping of the second kind associated with T and (3.12) and (3.13) hold. Then, the asynchronous multisplitting iteration $\{Y^p\}$ defined by (3.7)-(3.9) satisfies $Y^p \downarrow X^*$, where X^* is the unique solution of problem (2.6).

Proof: The result follows from Proposition 1 in Section 2 and Proposition 8 in [10].

Parallel asynchronous multisplitting methods with flexible communication can be applied for example to the solution of the following problem. Find v such that

$$(5.6)\qquad \begin{cases} -\Delta v + e^{cv} = f & \text{everywhere in } \Omega, \\ v \mid \Omega = 0, \end{cases}$$

where Ω is an open domain of R^2 (or R^3), c is positive real number, and we consider discretization using five-point difference equations for the Laplace operator. Inner iteration can be for example a conjugate gradient method used with an LU incomplete preconditionning.

References

1. G. M. Baudet, J.A.C.M. 25, 226-244 (1978).

2. D. P. Bertsekas and D. El Baz, SIAM J. on Control and Optimization, 25, 74-85 (1987).

3. D. P. Bertsekas and J. Tsitsiklis, Parallel and Distributed Computation, Numerical Methods. Prentice Hall, Englewood cliffs (1989).

4. D. Chazan, D. and W. Miranker, Linear Algebra Appl., 2, 199-222 (1969).

5. D. El Baz, SIAM J. Numer. Anal., 27, 136-140 (1990).

6. D. El Baz, Computational Optimization and Applications, 5, 187-205 (1996).

7. M. N. El Tarazi, Num. Math., 39, 325-340 (1982).

8. J.C. Miellou, R.A.I.R.O. R1, 55-82 (1975).

9. J. C. Miellou, in: Parallel Algorithms and Architectures M. Cosnard et al. (Eds) North-Holland, Amsterdam (1986) p. 85-96.

10. J. C. Miellou, D. El Baz and P. Spiteri, Mathematics of Computation, (1997).

11. P. Spiteri, J.C. Miellou and D. El Baz, IRIT report 95-17-R (1995).

12. D. El Baz, P. Spiteri, J. C. Miellou and D. Gazen, Journal of Parallel and Distributed Computing, 38, 1-15 (1996).

13. D. El Baz, D. Gazen, J. C. Miellou and P. Spiteri, Calculateurs Parallèles, 8, 393-410 (1996).

14. J. Bahi, J. C. Miellou and K. Rhofir, Numerical Algorithms (1997).

15. R. Bru, V. Migallon, J. Penadés and D. Szyld, ETNA, 3, 24-38 (1995).

16. A. Frommer and D. Szyld, Num. Math., 69, 141-153 (1994).

17. M. Jones and D. Szyld, Numer. Linear Algebra Appl., 3, 113-124 (1996).

18. D. O'Leary and R. White, SIAM J. on Algebraic and Discrete Mathematics, 6, 630-640 (1985).

19. J. M. Ortega and W. C. Rheinboldt, Iterative Solution of Nonlinear Equations in Several Variables, Academic Press, New York (1970).

20. W. C. Rheinboldt, J. Math. Anal. and Appl., 32, 274-307 (1970).

21. J. Bahi and J. C. Miellou, Parallel Computing, 19, 511-523 (1993).

8th Int. Coll. on Differential Equations, pp. 153-160
D. Bainov (Ed)

On the Approximation of an Infinite-Dimensional Linear Programming Problem

M. H. FARAHI
School of Mathematics, University of Mashad, P.O.Box 1159-91775, Mashad, Iran

Abstract. The boundary control of a linear wave equation with an integral performance criterion and fixed final states is considered. The problem is first modified to one in which a linear form is minimized over a set of positive Radon measures satisfying linear constraints. Then the minimum value of the objective function can be estimated by solving this measure problem when only a finite number of constraints are satisfied. The new measure nonclassical problem can be recast as a discrete infinite-dimensional linear programming problem. The next object of this paper is to study of two different approximation methods for the solution of the infinite-dimensional system to give rise an upper bound as well as a lower bound for the solution of this infinite-dimensional linear programming problem which is in fact the minimum value of the objective function.
Keywords: Hyperbolic equation, Optimal control, Measure theory, Linear programming.

1 Introduction

In Farahi et al [1], we considered a control system whose evolution in time was described by the function $(x,t) \to Y(x,t)$, defined on $(0,S) \times (0,T)$, with $T, S > 0$, satisfying the wave equation

$$Y_{tt}(x,t) = a^2 Y_{xx}(x,t), \tag{1}$$

where a is the velocity of wave propagation and $Y(x,t)$ describes the variation of the oscillation at point x and time t. The initial conditions were

$$\begin{aligned} Y(x,0) &= f(x) \\ Y_t(x,0) &= h(x), \end{aligned} \tag{2}$$

for the given functions $f(x)$ and $h(x)$, and the boundary conditions were

$$\begin{aligned} Y(0,t) &= u(t) \\ Y(S,t) &= u(t), \end{aligned} \tag{3}$$

where $t \in [0,T] \longrightarrow u(t) \in R$ is called the control function. We assumed the control u to be *admissible*, that is, be a Lebesgue measurable function on $[0,T]$ and
(a) $u(t) \in [-1,1]$ a.e. for $t \in [0,T]$.

(b) The solution of the partial differential equation (1) corresponding to the initial conditions (2) and the boundary conditions (3) satisfies the terminal conditions

$$\begin{aligned} Y(x,T) &= g_1(x) \\ Y_t(x,T) &= g_2(x), \end{aligned} \tag{4}$$

where $g_1(x)$, $g_2(x) \in L_2(0,S)$. Our control problem consisted of finding an admissible control $u(\cdot)$ which minimizes the functional

$$I[u(\cdot)] = \int_0^T f_0[t, u(t)]dt, \tag{5}$$

where $f_0 \in C(\Omega)$, the space of continuous functions on $\Omega=[0,T] \times [-1,1]$, with the topology of uniform convergence.
Some problem arose when we considered the classical problem (1)-(5). The set of admissible controls may be empty; the functional I in (5) measuring the performance of the system may not achieve its minimum in this set; if the set of admissible controls is non empty and a minimizing u for I does exists, it may be difficult to characterize it; there are no comprehensive ways of estimating numerically the optimal control.
To overcome these troubles associated with the classical control problem (1)-(5), we metamorphose the problem, this change is done by employing methods developed in Rubio [2], and the mathematical framework of Butkovskiy [3].

2 Modified Control Problem

We consider the solution of (1), (2), and (3) in the sense defined by Butkovskiy [3]:

$$Y(x,T) = Y_1(x,T) + Y_2(x,T),$$

where

$$Y_1(x,T) = \sum_{n=1}^{\infty}(A_n \cos\frac{\pi naT}{S} + B_n \sin\frac{\pi naT}{S}) \sin\frac{\pi nx}{S},$$

$$A_n = \frac{2}{S}\int_0^S f(\xi)\sin\left(\frac{\pi n\xi}{S}\right)d\xi,$$

$$B_n = \frac{2}{Sna}\int_0^S h(\xi)\sin\left(\frac{\pi n\xi}{S}\right)d\xi,$$

and

$$Y_2(x,T) = \int_0^T K(x, T-t)u(t)dt,$$

where

$$K(x,t) = \sum_{n=1}^{\infty} c_n \sin\frac{\pi nx}{S}\sin\frac{\pi nat}{S},$$

and $c_n = \frac{2a}{S}[1-(-1)^n]$.
Since $g_1(x) \in L_2(0,S)$, and $g_2(x) \in L_2(0,S)$, they possess a half-range Fourier series

$$g_1(x) = \sum_{n=1}^{\infty} a_n \sin\frac{\pi na}{S}$$

$$g_2(x) = \sum_{n=1}^{\infty} b_n \sin \frac{\pi n a}{S}.$$

Hence, the control problem is reduced to finding a measurable control function $u(t) \in [-1, 1]$ for $t \in [0, T]$ which satisfies

$$\int_0^T \psi_n(t, u(t))\, dt = \alpha_n, \tag{6}$$

and minimizes the functional (5), where in (6) ψ_n and α_n are indicated functions. Now we proceed to enlarge the set U, the set of admissible controls, as in Rubio [2]. (1) For a fixed control function $u(\cdot)$ the mapping

$$\Lambda_F : F \longrightarrow \int_0^T F[t, u(t)]dt, \qquad F \in C(\Omega)$$

defines a positive linear functional on $C(\Omega)$.

(2) There exists a unique positive Radon representing measure μ on Ω such that

$$\int_0^T F[t, u(t)]dt = \int_\Omega F d\mu = \mu(F), \qquad F \in C(\Omega). \tag{7}$$

Now, we replace the original minimization problem by one in which we will find the minimum of $\mu(f_0)$ over a set Q of positive Radon measures on Ω. These measures are required to have certain properties which are abstracted from the definition of admissible controls. First from (7),

$$|\mu(F)| \leq T \sup_\Omega |F(t, u)|;$$

hence,

$$\mu(1) \leq T.$$

Next, the measures in Q must satisfy an abstracted version of equation (6),

$$\mu(\psi_n) = \alpha_n, \qquad n = 1, 2, 3, \ldots.$$

Note that this is possible since $\psi_n \in C(\Omega)$. Finally, suppose that $G \in C(\Omega)$ does not depend on u, that is,

$$G(t, u_1) = G(t, u_2), \text{ for all, } t \in [0, T],\ u_1,\ u_2 \in [-1, 1],$$

then, measures in Q must satisfy,

$$\int_\Omega G d\mu = \int_0^T G[t, u]dt = a_G,$$

where u is an arbitrary number in the set $[-1, 1]$, and a_G is the integral of $G(\cdot, u)$ over $[0, T]$, independent of u. Let $\mathcal{M}^+(\Omega)$ be the set of all positive Radon measures on Ω. Then the set Q is defined as a subset of $\mathcal{M}^+(\Omega)$:

$$Q = S_1 \cap S_2 \cap S_3$$

where,

$$S_1 = \left\{\mu \in \mathcal{M}^+(\Omega) : \mu(1) \leq T\right\}$$

$$S_2 = \left\{\mu \in \mathcal{M}^+(\Omega) : \mu(\psi_n) = \alpha_n\right\}$$

$$S_3 = \left\{\mu \in \mathcal{M}^+(\Omega) : \mu(G) = a_G, G \in C(\Omega), \text{independent of } u\right\}$$

Now if we topologize the space $\mathcal{M}(\Omega)$ by the weak*-topology, it can be seen that S_1 is compact, S_2 and S_3 are closed, therefore Q is a closed subset of the compact set S_1, and hence is compact. The functional $I : Q \longrightarrow R$ defined by

$$I(\mu) = \int_\Omega f_0 d\mu = \mu(f_0) \in R, \qquad \mu \in Q, \tag{8}$$

is a linear continuous functional on a compact set Q, therefore attains its minimum, and thus the measure-theoretical problem, which consists of finding the minimum of the functional (8) over the subset Q of $\mathcal{M}^+(\Omega)$ possesses a minimizing solution μ^*, say, in Q. Thus the optimal control problem changes to the following linear programming problem:

$$\text{Minimize} \qquad \mu : f_0 \longrightarrow \mu(f_0)$$

subject to

$$\begin{aligned} \mu(\psi_n) &= \alpha_n, & n &= 1, 2, ..., \\ \mu(G_r) &= a_{G_r}, & r &= 0, 1, ..., G_r = t^r \end{aligned} \tag{9}$$

where ψ_n, α_n, G_r, a_{G_r} are indicated functions of t, u, and the Fourier coefficients of the given functions $f(x)$, $h(x)$, $g_1(x)$, and $g_2(x)$.

By considering the measure-theoretical nonclassical problem (9), it is possible, therefore, to solve a modified type of optimal control problem. This problem is actually an infinite-dimensional linear programming problem. In this paper we approximate the solution of the problem (9) by a truncated linear system; only a finite number M_1 of the functions ψ_n will be considered. Also, only a finite number of functions G will be considered, in fact these functions will be denoted by $G_i = t^i$, $i = 1, 2, ..., M_2$. So the new linear programming problem is as follows:

$$\text{Minimize} \qquad \mu : f_0 \longrightarrow \mu(f_0)$$

subject to

$$\begin{aligned} \mu(\psi_n) &= \alpha_n, & n &= 1, 2, ..., M_1 \\ \mu(G_r) &= a_{G_r}, & r &= 0, 1, ..., M_2 \end{aligned} \tag{10}$$

In the following we will consider the maximum value of

$$|\mu^*(f_0) - \mu^*_{Q(M_1,M_2)}(f_0)| \tag{11}$$

where $\mu^*(f_0)$ is the optimal value of the primal problem (9) and $\mu^*_{Q(M_1,M_2)}(f_0)$ is the minimum of the problem (10); $Q(M_1, M_2)$ is the set of measures in $\mathcal{M}^-(\Omega)$ satisfying (10).

Many authors have proposed and studied this kind of approximation, that is, approximating an infinite-dimensional linear programming problem by a finite-dimensional one. It appears that the first attempt was done by Vershik and Temel't [4], Temel't [5] and Vershik [6], and then followed by many others, we mention only Dahleh and Pearson [7], Mendlovitz [8] and Staffans [9]-[10].

3 Approximation to the optimal measure

In this section we are going to construct a method to estimate the value of $\mu_Q^*(f_0)$ by $\mu^*_{Q(M_1,M_2)}(f_0)$ when M_1 and M_2 both are finite positive integers. In this construction we follow two basic steps to calculate the lower-upper bounds of $\mu^*(f_0)$
(i) It is assumed that the first function appearing in the set of second equalities in (9) and (10) is $G_0 = 1$ and as we defined before $a_{G_0} = \int_0^\pi dt = T = \pi$. Assume that in (10), $F_1 = 1, F_2 = \psi_1, ..., F_{M_1+1} = \psi_{M_1}, F_{M_1+2} = G_1, ..., F_{M_1+M_2+1} = G_{M_2}$, thus $Q(M_1, M_2)$ is the set of positive Radon measures on $\Omega = [0, \pi] \times [-1, 1]$ satisfying

$$\mu(F_i) = \alpha_i, \quad \alpha_1 = \pi, i = 1, 2, ..., M_1 + M_2 + 1.$$

Proposition 1 Let the functions $F_1, F_2, ..., F_{M_1+M_2+1}$ be continuous on a compact Hausdorff topological space Ω, with $F_1(Z) = 1, Z \in \Omega$, and let $Q(M_1, M_2)$ be the set of positive Radon measures on Ω, $Q(M_1, M_2) = \{\mu \in \mathcal{M}^+(\Omega) : \mu(F_i) = \alpha_i, i = 1, 2, ..., M_1 + M_2 + 1\}$ with $\alpha_1 > 0$. Then if $Q(M_1, M_2)$ is nonempty, it is a compact convex subset of $\mathcal{M}^+(\Omega)$, and if $\mu^*_{Q(M_1,M_2)}$ is an extremal point of $Q(M_1, M_2)$, then it is of the form

$$\mu^*_{Q(M_1,M_2)} = \sum_{i=1}^{M_1+M_2+1} a_i\delta(Z_i), \quad Z_i \in \Omega, a_i \geq 0$$

where $\delta(Z_i)$ is the atomic measure with support the singleton set $\{Z_i\}$.
So using Proposition 1, the minimization problem (10) changes to the following one:

$$\text{Minimize} \quad \mu_{Q(M_1,M_2)}(f_0) = \sum_{i=1}^{M_1+M_2+1} a_i f_0(Z_i), \quad Z_i \in \Omega, a_i \geq 0$$

subject to

$$\begin{aligned} \sum_{i=1}^{M_1+M_2+1} a_i &= \pi = \alpha_1 \\ \sum_{i=1}^{M_1+M_2+1} a_i\psi_j(Z_i) &= \alpha_{j+1}, \quad j = 1, 2, ..., M_1 \\ \sum_{i=1}^{M_1+M_2+1} a_i G_r(Z_i) &= \alpha_{G_r}, \quad r = 1, 2, ..., M_2. \end{aligned} \tag{12}$$

This problem sets up a semi-infinite linear programming problem; the Z_i's are in the set Ω but there are only a finite number of constraints, the solution of (12) gives us $\mu^*_{Q(M_1,M_2)}(f_0)$ which is a suboptimal solution, i.e., it is a lower bound for the primal solution $\mu_Q^*(f_0)$.
(ii) To find an upper bound for $\mu_Q^*(f_0)$, first we have,

Proposition 2 Let ω be a countable dense subset of Ω. Given $\epsilon' > 0$, a measure $\nu \in \mathcal{M}^+(\omega)$ can be found such that

$$\begin{aligned} |\mu^*_{Q(M_1,M_2)}(f_0) - \nu(f_0)| &< \epsilon' \\ |\mu^*_{Q(M_1,M_2)}(F_i) - \nu(F_i)| &< \epsilon', \\ i = 1, 2, ..., M_1 + M_2 + 1 & \end{aligned}$$

and

$$\nu = \sum_{k=1}^{M_1+M_2+1} a_k \delta(Z_k), \quad Z_k \in \omega.$$

By Proposition 2 and Theorem III.1 of Rubio [2] our linear programming problem consists of minimizing the linear form

$$\mu_{Q(M_1,M_2)}(f_0) = \sum_{i=1}^{N} a_i f_0(Z_i)$$

over the set of coefficients $a_i \geq 0$, subject to

$$\begin{aligned} &\textstyle\sum_{i=1}^{N} a_i = \pi = \alpha_1 \\ &\textstyle\sum_{i=1}^{N} a_i \psi_j(Z_i) = \alpha_{j+1}, \quad j = 1, 2, ..., M_1 \\ &\textstyle\sum_{i=1}^{N} a_i G_r(Z_i) = a_{G_r}, \quad r = 1, 2, ..., M_2, \end{aligned} \tag{13}$$

where $Z_i \in \omega$; a dense subset of Ω. Define $\omega_N = \{Z_k, k = 1, 2, ..., N\}$, then $\omega_N \subseteq \omega$. Now let λ be the following transformation

$$\lambda : N \times N \longrightarrow N$$

defined by $\lambda(i, j) = H(j-1) + i$, $i = 1, 2, ..., H$, $j = 1, 2, ..., K$, where H and K are two positive integers, and let the set $\omega_M = \{Z_l, l = 1, 2, ..., M = HK\} \subset \omega$ be the set of points that are defined as

$$Z_l = Z_{\lambda(i,j)} = Z(t_i, u_j) = \left(\frac{\pi}{2H} + \frac{(i-1)\pi}{H}, -1 + \frac{1}{K} + \frac{2(j-1)}{K}\right),$$

$$i = 1, 2, ..., H, \quad j = 1, 2, ..., K$$

where $l = H(j-1) + i$; it is clear that $\omega = \cup \omega_M$ and for sufficiently large H and K, the set ω_M will contain the set ω_N.

Now we assume that M_1 and M_2 tend to infinity, it is easy to prove that (13) changes to the following one

$$\text{Minimize} \sum_{i=1}^{\infty} a_i f_0(Z_i), \quad Z_i \in \omega, \ a_i \geq 0$$

subject to

$$\begin{aligned} &\textstyle\sum_{i=1}^{\infty} a_i = \pi = \alpha_1 \\ &\textstyle\sum_{i=1}^{\infty} a_i \psi_j(Z_i) = \alpha_{j+1}, \quad j = 1, 2, ... \\ &\textstyle\sum_{i=1}^{\infty} a_i G_r(Z_i) = a_{G_r}, \quad r = 1, 2, ... \end{aligned} \tag{14}$$

Then the solution of (14), if it exists, is $\mu_Q^*(f_0)$. The new linear programming problem (14) is an infinite-dimensional one. To find an upper bound for the solution of this problem, we restrict the variables in (14) while the constraints are still infinite, i.e., we consider the following problem

$$\text{Minimize} \sum_{i=1}^{N'} a_i f_0(Z_i), \quad Z_i \in \omega, \ a_i \geq 0$$

subject to

$$\begin{aligned} \textstyle\sum_{i=1}^{N'} a_i &= \pi = \alpha_1 \\ \textstyle\sum_{i=1}^{N'} a_i\psi_j(Z_i) &= \alpha_{j+1}, \quad j = 1, 2, \ldots \\ \textstyle\sum_{i=1}^{N'} a_i G_r(Z_i) &= a_{G_r}, \quad r = 1, 2, \ldots. \end{aligned} \tag{15}$$

Although it is clear in principle what we should do, but still two major obstacles remain: How do we solve the truncated problems (12) and (15)?
Let us first discuss about the solution of the semi-infinite problem (12). We have chosen here a method due to Glashoff and Gustafson (see [11]) to solve such a problem. The most difficult problem encountered in using this method was that of finding an initial solution from which one can start the method and find successive values towards the minimum. This was done here by means of finite-dimensional linear program, obtained by discretization.
The situation is more complicated for the problem (15). First we change (15) to its dual, and then we restrict once again the variables in the final problem, so the new problem is as follows:

$$\text{Maximize } \sum_{i=1}^{M} \beta_i \alpha_i$$

subject to

$$\sum_{i=1}^{M} \beta_i F_i(Z_j) \leq f_0(Z_j), \quad j = 1, 2, \ldots, N' \tag{16}$$

where the F_i's are the known functions in term of ψ_i's and G_r's. The solutions of this final problem (16) are lower bound for the solution of (15) (since (16) is a maximization problem). Now let the number of variables M in (16) increases, for larger value of M the value of objective function in (16) increases; if we assume that the value of the objective function of (16) in this case is $I_{M,N'}(f_0)$, then by increasing M we determine successive values of $I_{M,N'}(f_0)$. Since the sequence $\{I_{M,N'}(f_0), M = 1, 2, \ldots\}$ converges to a superoptimal solution of (15), one can estimate, by increasing M, an upper bound for $\mu^*(f_0)$ to any desired accuracy.
By using the both solutions of (12) and (15) at the same time, we have access to an upper bound as well as a lower bound of the optimal value $\mu_Q^*(f_0)$. To illustrate the method, we consider the following example.

Example 1 Consider the problem (1)-(4) where in (2) $f(x) \equiv 0$, $h(x) \equiv 0$ and in (4) $g_1(x) = \sin(x)$, $g_2(x) \equiv 0$ and in (5), $f_0 = u^2$, $t \in [0, \pi]$, $x \in [0, \pi]$, $u \in [-1, 1]$. We assume $M_1 = 6$, $M_2 = 10$.

To start the algorithm and find a suboptimal solution one needs a basic set from which can be iterated toward the minimum. Here we find a basic set by means of a finite-dimensional linear program obtained by discretization. We cover the set $\Omega = [0, \pi] \times [-1, 1]$ with a grid by taking $m_1 + 1$ and $m_2 + 1$ points along the t-axis, and u-axis each separately in the ordered mentioned. These points are equidistant of $\frac{\pi}{m_1}$ and $\frac{2}{m_2}$ respectively. Now Ω is divided into $N = m_1 \cdot m_2$ number of equal volume rectangles Ω_j, $j = 1, 2, \ldots, N$, and $Z_j = (t_j, u_j)$ is chosen as the center of Ω_j. We assume here $m_1 = 10$, $m_2 = 10$, so $N = 100$. Then a finite-dimensional simplex program is run, with 100 variables and, of course, 16 constraints. The minimum obtained for the objective function is 0.807839 with only 16 variables that are nonzero. From these variables we could find the basic set, then the simplex algorithm of Glashoff and Gustafson is applied by using this basic set, and after several iteration the value of 0.7853981 is obtained as the best improvement for the value of objective function. So

$$\mu^*_{(6,10)}(f_0) = 0.7853981.$$

We apply then the method introduced in this section for the estimation of a superoptimal solution and find the number 0.8202272. So

$$0.785398 < \mu^*(f_0) < 0.8202272.$$

References

1. M. H. Farahi, J. E. Rubio and D. A. Wilson, Int. J. Control, 63, 833-848 (1996).
2. J. E. Rubio, Control and Optimization: The Linear Treatment of non Linear Problems, Manchester University Press, England (1986).
3. A. G. Butkovskiy, Distributed Control Systems, American Elsevier Publishing Company, Inc., (1969).
4. A. M. Vershik and V. Temel't, Siberian Maths, 9, 591-601 (1968).
5. V. Temel't, Kibernetika, 5, 80-87 (1969).
6. A. M. Vershik, Russian Mathematical Surveys, 25, 117-124 (1971).
7. M. A. Dahleh and J. B. Pearson, IEEE Transaction on Automatic Control, 33, 722-731 (1988).
8. M. A. Mendlovitz, Systems & Control Letters, 12, 461-463 (1989).
9. O. J. Staffans, JOTA, 70, 173-189 (1991).
10. O. J. Staffans, SIAM J. Control and Optimization, 31, 747-779 (1994).
11. K. Glashoff and S. A. Gustafson, Linear Optimization and Approximation, Springer-Verlag, Berlin (1983).

8th Int. Coll. on Differential Equations, pp. 161-167
D. Bainov (Ed)

Partial Hölder Continuity Results for Solutions of non Linear non Variational Elliptic Systems with Strictly Controlled Growth

LUISA FATTORUSSO AND GIOVANNA IDONE
Università degli Studi di Reggio Calabria, Dipartimento di Ingegneria Elettronica e Matematica Applicata, Via Graziella, Località Feo di Vito, Reggio Calabria, Italy

Abstract. In a bounded open set Ω we consider the non linear non variational system

$$a(x, u, Du, H(u)) = b(x, u, Du)$$

where $a(x, u, Du, H(u))$ and $b(x, u, Du)$ are vectors in R^N, $N \geq 1$, measurable in x and continuous in the other variables. We prove that if $u \in H^2(\Omega)$ is a solution in Ω to above system, if $b(x, u, Du)$ has a strictly controlled growth, if $a(x, u, \mu, \xi)$ is of class C^1 in ξ and satisfies the condition (A) and, together with $\dfrac{\partial a}{\partial \xi}$, certain continuity conditions, then the vector Du is partially Hölder continuous for every exponent $\alpha < 1 - \frac{n}{p}$.
Keywords: *Hölder continuity, Sobolev spaces, elliptic system.*

1. Introduction

In this work we study the partial Hölder continuity for solutions of second order non linear non variational elliptic systems of type

$$a(x, u, Du, H(u)) = b(x, u, Du)$$

with stricty controlled growth. This result is similar to the one proved in this case of "linear growth" by M. Marino - A. Maugeri for the parabolic systems in [1] and by S. Campanato for the elliptic case in [2].

In his work Campanato obtains, for any n, the partial Hölder continuity for the vector Du.

He further shows, for $n \leq 4$, the global Hölder continuity for a solution u and, for $n = 2$, the global Hölder continuity for the gradient Du.

Campanato supposes that the vector $a(x, u, p, \xi)$ satisfies the following condition:

(A) there exist three positive constants α, γ and δ with

$$\gamma + \delta < 1$$

such that, $\forall u \in R^N, \forall p \in R^{nN}, \forall \eta, \tau \in R^{n^2N}$ and for almost every $x \in \Omega$ we have

$$\left\| \sum_i \tau_{ii} - \alpha \left[a(x, u, p, \tau + \eta) - a(x, u, p, \eta) \right] \right\|_N^2 \leq \gamma \|\tau\|^2 + \delta \left\| \sum_i \tau_{ii} \right\|_N^2 .$$

With this condition (A) the vector $a(x,u,p,\xi)$ is elliptic and satisfies the non linear condition of Cordés.

Let Ω be a bounded open subset of R^n, $n > 4$, of class C^2 and let $x = (x_1, x_2, \dots, x_n)$ be a generic point in it.

In what follows we denote by $\mu = (\mu_1, \mu_2, \dots, \mu_n)$, $\mu_i \in R^N$, N integer ≥ 1, a generic vector of R^{nN} and by $\xi = \{\xi_{ij}\}$, $i,j = 1,2,\dots,n$, $\xi_{ij} \in R^N$ a generic element of R^{n^2N}.

We consider in Ω the following second order non linear non variational system

$$a(x,u,Du,H(u)) = b(x,u,Du) \tag{1}$$

where $a(x,u,\mu,\xi)$ and $b(x,u,\mu)$ are vectors in R^N, measurable in x, continuous in (u,μ,ξ) and (u,μ) respectively, satistying the conditions:

$$a(x,u,\mu,0) = 0. \tag{2}$$

(A) there exist three positive constants α, γ and δ with $\gamma + \delta < 1$, such that, $\forall u \in R^N$, $\forall \mu \in R^{nN}$, $\forall \eta, \tau \in R^{n^2N}$ and for almost every $x \in \Omega$ we have

$$\left\| \sum_{i=1}^{n} \tau_{ii} - \alpha \left[a(x,u,\mu,\tau+\eta) - a(x,u,\mu,\eta) \right] \right\|^2 \leq \gamma \|\tau\|^2 + \delta \left\| \sum_{i=1}^{n} \tau_{ii} \right\|^2 \tag{3}$$

there exists a constant c such that, $\forall u \in R^N$, $\forall \mu \in R^{nN}$ and for almost every $x \in \Omega$ we have

$$\|b(x,u,\mu)\| \leq c\left(f(x) + \|u\|^\alpha + \|\mu\|^\beta\right)$$

with $f \in L^2(\Omega)$ and with

$$1 \leq \alpha < \frac{n}{n-4}, \qquad 1 \leq \beta < \frac{n}{n-2}.$$

We shall denote by:

$B(x^0,\sigma) = \{x \in R^n : \|x - x^0\| < \sigma\}$

$u_\sigma = u_{B(x^0,\sigma)} = \displaystyle\fint_{B(x^0,\sigma)} u\,dx = \frac{\int_{B(x^0,\sigma)} u\,dx}{meas B(x^0,\sigma)}$

$H^{S,p}\left(\Omega, R^N\right)$, s integer≥ 0, $p \in [1,+\infty)$, the usual Sobolev spaces.

By a solution to the system (1) we mean a vector $u \in H^2\left(\Omega, R^N\right)$ satisfying (1) for almost all $x \in \Omega$.

To achieve the partial Hölder regularity of the gradient we need to obtain the local L^q – *regularity*, $q > 2$, of the derivaties $D_{ij}u$ and suitable estimates.

2. Local L^q-regularity of the matrix $H(u)$

First of all we shall show a "Caccioppoli's type inequality", from which we will be able to derive, by means of a lemma of Gehring-Giaquinta-G.Modica, a L^q – *local* regularity result for the matrix $H(u)$; this result, of interest in itself, is preliminary to the study of the partial Hölder continuity of the vector Du.

Under assumption (2), (3) and (A), taking into account the Miranda-Talenti estimate and the theorem of Poincarè, we have the following Caccioppoli estimate

$$\int_{B(x^0,\sigma)} \|H(u)\|^2 dx \le c\sigma^{-2} \int_{B(x^0,2\sigma)} \|Du - (Du)_{2\sigma}\|^2 dx \tag{4}$$
$$+ c \int_{B(x^0,2\sigma)} \|b(x,u,Du)\|^2 dx.$$

Now, setting $F(x) = \|H(u)\|^{\frac{2n}{n+2}}$, $G(x) = \|b(x,u,Du)\|^{\frac{2n}{n+2}}$ the estimate (4) can be written in the following manner

$$⨍_{B(x^0,\sigma)} F^{\frac{n+2}{n}} dx \le \left(⨍_{B(x^0,2\sigma)} F\, dx \right)^{\frac{n+2}{n}} + c \left(⨍_{B(x^0,2\sigma)} G^{\frac{n+2}{n}} dx \right).$$

From this and by a lemma of Gehring-Giaquinta-G.Modica we deduce that:

— *If $u \in H^2(\Omega, R^N)$ is a solution to the system (1) and if the assumption (2),(3) with $f \in L^{q_0}(\Omega)$, $2 < q_0 \le 2^*$ (where 2^* is the Sobolev exponent) and (A) hold, then there exists $\bar q$*

$$2 < \bar q \le q_0 \wedge \frac{2n}{\alpha(n-4)} \wedge \frac{2^*}{\beta},$$

such that $\forall q \in [2, \bar q)$, $u \in H^{2,q}_{loc}(\Omega, R^N)$ and $\forall B(x^0, 2\sigma) \subset\subset \Omega$, with $\sigma < 1$, one has

$$\left(⨍_{B(x^0,\sigma)} \|H(u)\|^q dx \right)^{\frac{1}{q}} \le k \left\{ \left(⨍_{B(x^0,2\sigma)} \|H(u)\|^2 dx \right)^{\frac{1}{2}} + \left(⨍_{B(x^0,\sigma)} \|b\|^q dx \right)^{\frac{1}{q}} \right\} \tag{5}$$

where K does not depend on σ.

From (5), evaluting the last term of the right hand side, we have the following result:

— *If $u \in H^2(\Omega, R^N)$ is a solution to the system (1) and if the assumptions (2),(3) with $f \in L^p(\Omega)$, $p > n$, and (A) hold, then there exists $\bar q$* with

$$2 < \bar q \le \frac{2n}{\alpha(n-4)} \wedge \frac{2^*}{\beta},$$

such that $\forall q \in [2, \bar q)$, $\forall B(x^0, 2\sigma) \subset\subset \Omega$, with $\sigma < 1$, it results:

$$\left(\int_{B(x^0,\sigma)} \|H(u)\|^q dx \right)^{\frac{1}{q}} \le c\sigma^{n\left(\frac{1}{q}-\frac{1}{2}\right)} \left[\Phi\left(u, x^0, 2\sigma\right) \right]^{\frac{1}{2}} \tag{6}$$

where c does not depend on σ and

$$\Phi\left(u, x^0, \sigma\right) = \sigma^{\xi} + \int_{B(x^0,\sigma)} \|u\|^{\frac{2n}{n-4}} dx + \int_{B(x^0,\sigma)} \|Du\|^{\frac{2n}{n+2}} dx + \int_{B(x^0,\sigma)} \|H(u)\|^2 dx$$

with

$$\xi = n\left(1 - \frac{2}{p}\right).$$

Now introducing the polynomial vector $P = (P_1, P_2, \dots, P_N)$ of degree ≤ 1, such that

$$\int_{B(x^0,\sigma)} D^{\alpha}(u - P) = 0 \qquad \forall \alpha, |\alpha| \le 1$$

where

$$D^\alpha = D_1^{\alpha_1} D_2^{\alpha_2} \dots D_n^{\alpha_n}, \qquad \alpha = (\alpha_1, \alpha_2, \dots, \alpha_n), |\alpha| = \alpha_1 + \alpha_2 + \dots + \alpha_n,$$

α_i integer ≥ 0, taking into account that

$$\int_{B(x^0,\tau\sigma)} \|u\|^{\frac{2n}{n-4}} \leq c(\tau\sigma)^n \|P\|^{\frac{2n}{n-4}} + \int_{B(x^0,\tau\sigma)} \|u - P\|^{\frac{2n}{n-4}} dx$$

we obtain the following fundamental estimate, needed to achieve the partial Hölder continuity of the vector Du.

— *If $u \in H^2\left(B\left(x^0, \sigma\right), R^N\right)$ with $\sigma \in (0,1)$ is a solution to the system (1) and if the assumptions (2),(3) with $f \in L^p(\Omega), p > n$, and (A) hold, then there exists $\bar{q}$*

$$2 < \bar{q} \leq \frac{2n}{\alpha(n-4)} \wedge \frac{2^*}{\beta}$$

such that $\forall q \in [2, \bar{q})$, $\forall \tau \in (0,1)$ one has

$$\left(\int_{B(x^0,\tau\sigma)} \|u\|^{\frac{2n}{n-4}} + \|Du\|^{\frac{2n}{n-2}}\right) dx \leq c\tau^n \left(\int_{B(x^0,\sigma)} \|u\|^{\frac{2n}{n-4}} + \int_{B(x^0,\sigma)} \|Du\|^{\frac{2n}{n-2}} dx\right) +$$

$$c\sigma^{2\left(1-\frac{2}{q}\right)} \int_{B(x^0,\sigma)} \|H(u)\|^2 dx \tag{7}$$

where c does not depend on σ and τ.

3. Partial Hölder continuity of the vector Du

Let $u \in H^2\left(\Omega, R^N\right)$ be a solution to the non variational system

$$a(x, u, Du, H(u)) = b(x, u, Du) \tag{8}$$

where $a(x, u, \mu, \xi)$ and $b(x, u, \mu)$ are vectors of R^N, with the following properties:

(i) $b(x, u, \mu)$ is measurable in x, continuous in (u, μ) and such that $\forall u \in R^N, \forall \mu \in R^N$ and for x a.e. in Ω

$$\|b(x, u, \mu)\| \leq c\left\{f(x) + \|u\|^\alpha + \|\mu\|^\beta\right\},$$

with $f \in L^p(\Omega), p > n$ and with

$$1 \leq \alpha < \frac{n}{n-4}, \qquad 1 \leq \beta < \frac{n}{n-2};$$

(ii) $a(x, u, \mu, \xi)$ is continuous in (x, u, μ), of class C^1 in ξ, with derivaties $\dfrac{\partial a}{\partial \xi_{ij}}$ uniformly continuous and bounded in $\Omega \times R^N \times R^{nN} \times R^{n^2N}$ and such that

$$a(x, u, \mu, 0) = 0,$$

(A) there exist three positive constants α, γ and δ with $\gamma + \delta < 1$ such that $\forall u \in R^N, \forall \mu \in R^{nN}, \forall \eta, \tau \in R^{n^2N}$ and for almost every $x \in \Omega$:

$$\left\|\sum_{i=1}^n \tau_{ii} - \alpha\left[a(x, u, \mu, \tau + \eta) - a(x, u, \mu, \eta)\right]\right\|^2 \leq \gamma \|\tau\|^2 + \delta \left\|\sum_{i=1}^n \tau_{ii}\right\|^2,$$

(B) there exists a non negative function $\omega(t)$, defined for $t \geq 0$, continuous, bounded, concave, non decreasing, with $\omega(0) = 0$, such that $\forall x, y \in \Omega, \forall u, v \in R^N, \forall \mu, \bar{\mu} \in R^{nN}$ and $\forall \xi, \tau \in R^{n^2N}$:

$$\|a(x,u,\mu,\xi) - a(y,v,\bar{\mu},\xi)\| \leq \omega\left(d^2(x,y) + \|u-v\|^2 + \|\mu-\bar{\mu}\|^2\right) \cdot \|\xi\|,$$

$$\left\|\frac{\partial a(x,u,\mu,\xi)}{\partial \xi} - \frac{\partial a(x,u,\mu,\tau)}{\partial \xi}\right\| \leq \omega\left(\|\xi-\tau\|^2\right).$$

The first result obtained is the following:

— *If $u \in H^2\left(\Omega, R^N\right)$ is a solution to the system (8) and if the assumptions (i) and (ii) hold, then,* $\forall B(x^0,\sigma) \subset\subset \Omega$, with $\sigma < 2, \forall \tau \in (0,1)$ and $\forall \varepsilon \in (0, n-2-2(\frac{1}{p}+\frac{1}{q})]$ *where* $q \in (2, \bar{q})$, it results

$$\begin{aligned} \Phi\left(u,x^0,\tau\sigma\right) \leq\ & A\Phi\left(u,x^0,\sigma\right)\left\{\tau^\lambda + \sigma^{\left(1-\frac{2}{q}\right)} + \left[\omega\left(c\sigma^{2-n}\Phi\left(u,x^0,\sigma\right)\right)\right]^{1-\frac{2}{q}}\right\} + \\ & A\Phi\left(u,x^0,\sigma\right)\left\{\left[\omega\left(\int_{B(x^0,\sigma)} \|H(u) - (H(u))_\sigma\|^2 d\sigma\right)\right]^{1-\frac{2}{q}}\right\} \end{aligned} \tag{9}$$

where $\lambda = n\left(1-\frac{2}{p}\right) - \varepsilon$, *and*

$$\Phi\left(u,x^0,\sigma\right) = \sigma^\xi + \int_{B(x^0,\sigma)} \left[\|u\|^{\frac{2n}{n-4}} dx + \|Du\|^{2^*} + \|H(u)\|^2\right] dx \tag{10}$$

with

$$\xi = n\left(1-\frac{2}{p}\right).$$

To achieve the estimate (9) we use the following techinique: let us fix the ball $B(x^0,\sigma)$, with $\sigma < 1$, such that $B(x^0, 2\sigma) \subset\subset \Omega$ and let us

$$\frac{\partial \tilde{a}(x,u,\mu,\eta)}{\partial \xi_{ij}} = \left\{\int_0^1 \frac{\partial a^h(x,u,\mu,t\eta)}{\partial \xi_{ij}^k} dt\right\} \qquad h,k = 1,2,\ldots,N$$

$$\frac{\partial \tilde{a}(x,u,\mu,\eta)}{\partial \xi} = \left\{\frac{\partial \tilde{a}(x,u,\mu,\eta)}{\partial \xi_{ij}}\right\} \qquad i,j = 1,2,\ldots,n.$$

In $B(x^0,\sigma)$ the system (8) can also be written in the following form

$$\sum_{i,j=1}^n \frac{\partial \tilde{a}\left(x^0, u_\sigma, (Du)_\sigma, (H(u))_\sigma\right)}{\partial \xi_{ij}} D_{ij}u = B_1 + B_2 + b(x,u,Du)$$

where $B_1 = a\left(x^0, u_\sigma, (Du)_\sigma, H(u)\right) - a(x,u,Du,H(u))$ and

$$B_2 = \sum_{i,j=1}^n \left(\frac{\partial \tilde{a}\left(x^0,u_\sigma,(Du)_\sigma,(H(u))_\sigma\right)}{\partial \xi_{ij}} - \frac{\partial \tilde{a}\left(x^0,u_\sigma,(Du)_\sigma,H(u)\right)}{\partial \xi_{ij}}\right) D_{ij}u.$$

Setting w to be the solution in $B(x^0,\sigma)$ to the elliptic Dirichlet problem

$$\begin{cases} w \in H_0^2\left(B(x^0,\sigma), R^N\right) \\ \sum_{i,j=1}^n \frac{\partial \tilde{a}\left(x^0,u_\sigma,(Du)_\sigma,(H(u))_\sigma\right)}{\partial \xi_{ij}} D_{ij}w = B_1 + B_2 \end{cases}$$

it results, in $B(x^0,\sigma)$ $u = w+v$ where $v \in H^2\left(B(x^0,\sigma), R^N\right)$ is the solution to the linear system

$$\sum_{i,j=1}^{n} \frac{\partial \tilde{a}\left(x^0, u_\sigma, (Du)_\sigma, (H(u))_\sigma\right)}{\partial \xi_{ij}} D_{ij} v = b(x, u, Du).$$

We obtain (9), since $u = w + v$, in virtue of the estimate for v and w, taking into account the fundamental estimate (7) and using an algebraic lemma.

Now let us set:

$$\mathcal{B}_1 = \left\{x \in \Omega : \lim_{\sigma\to 0}{}'' \fint_{B(x,\sigma)} \|H(u) - (H(u))_\sigma\|^2 dy > 0\right\}$$

$$\mathcal{B}_2 = \left\{x \in \Omega : \lim_{\sigma\to 0}{}' \sigma^{2-n}\Phi(u, x, \sigma) > 0\right\}.$$

By a property of the Lebesgue integral we have

$$meas\mathcal{B}_1 = 0$$

and taking into account a theorem of Giusti we obtain

$$\mathcal{H}_{n-2}(\mathcal{B}_2) = 0$$

where $\mathcal{H}_\gamma$ is the $\gamma-$dimensional Hausdorff measure.

Hence the set $\mathcal{B}_1 \cup \mathcal{B}_2$ has measure zero.

Take $x \in \Omega \backslash \mathcal{B}_1 \cup \mathcal{B}_2$, we have

$$\lim_{\sigma\to 0} \fint_{B(x,\sigma)} \|H(u) - (H(u))_\sigma\|^2 dx = 0$$

$$\lim_{\sigma\to 0} \sigma^{2-n}\Phi(u, x, \sigma) = 0.$$

It follows from these facts that for every fixed $\varepsilon \in \left(0, 1 - \frac{n}{p}\right)$, it is possible to associate to every $x^0 \in \Omega\backslash\mathcal{B}_1 \cup \mathcal{B}_2$ a ball $B(x^0, R_{x^0}) \subset \Omega\backslash\mathcal{B}_2$ and a positive number σ_ε such that, $\forall t \in (0,1)$ and $\forall y \in B(x^0, R_{x^0})$

$$\Phi(u, y, t\sigma_\varepsilon) \le (1 + A)\, t^{n\left(1-\frac{2}{p}\right)-2\varepsilon}\Phi(u, y, \sigma_\varepsilon)$$

and hence

$$H(u) \in L^{2,n\left(1-\frac{2}{p}\right)-2\varepsilon}\left(B\left(x^0, R_{x^0}\right), R^{n^2N}\right), \quad Du \in \mathcal{L}^{2,n\left(1-\frac{2}{p}\right)-2\varepsilon+2}\left(B\left(x^0, R_{x^0}\right), R^{nN}\right).$$

where

$$L^{2,n\left(1-\frac{2}{p}\right)-2\varepsilon}{}_{\left(B(x^0,R_{x^0}),R^k\right)} \quad k \text{ integer} > 0,$$

denotes the space of those functions $v \in L^2\left(B(x^0, R_{x^0}), R^k\right)$ such that

$$\sup_{x\in B\left(x^0,R_{x^0}\right),\rho>0} \left\{\rho^{-\left[n\left(1-\frac{2}{p}\right)-2\varepsilon\right]} \int_{B(x,\rho)} \|v\|^2 dx\right\} < +\infty.$$

$\mathcal{L}^{2,n\left(1-\frac{2}{p}\right)-2\varepsilon+2}\left(B(x^0, R_{x^0}), R^{nN}\right)$ denotes the space of those functions

$$v \in L^2\left(B\left(x^0, R_{x^0}\right), R^{nN}\right)$$

for which

$$\sup_{x\in B(x^0,R_{x^0}),\rho>0}\left\{\rho^{-[n(1-\frac{2}{p})-2\varepsilon+2]}\int_{B(x,\rho)}\left\|v-v_{B(x,\rho)}\right\|^2dx\right\}<+\infty.$$

If $v\in\mathcal{L}^{2,n(1-\frac{2}{p})-2\varepsilon+2}\left(B\left(x^0,R_{x^0}\right),R^{nN}\right)$, then v is $\left(1-\frac{n}{p}\right)$ –Hölder continuous in $B\left(x^0,R_{x^0}\right)$.

Now, if $Du\in\mathcal{L}^{2,n(1-\frac{2}{p})-2\varepsilon+2}\left(B\left(x^0,R_{x^0}\right),R^{nN}\right)$, then there exists a set $\mathcal{B}_0$, with $meas\mathcal{B}_0=0$, closed in Ω, with

$$\mathcal{B}_2\subset\mathcal{B}_0\subset\mathcal{B}_1\cup\mathcal{B}_2$$

such that

$$Du\in C^{0,\alpha}\left(\Omega\backslash\mathcal{B}_0,R^{nN}\right),\qquad\forall\alpha<1-\frac{n}{p}.$$

References

1. M. Marino and A. Maugeri, Rend. Matem. Series VII, Roma, 13, (1993).

2. S. Campanato, Rend. Matem. Series VII, Roma, 10, (1990).

8th Int. Coll. on Differential Equations, pp. 169-178
D. Bainov (Ed)

A Structured Approach to Toeplitz and Toeplitz-like Linear Systems with Applications to (semi) Elliptic Problems

GIUSEPPE FIORENTINO
Dipartimento di Matematica, Università di Pisa, Italy

Abstract. We survey a structured approach for solving Toeplitz and Toeplitz-like linear systems. We emphasize how matrix algebras and generating functions can be used as valuable tools to devise efficient sequential and parallel iterative solvers. The applicability of this approach falls beyond the strictly Toeplitz area covering, for instance, many computations arising from the numerical treatment of boundary value problems.

Keywords: CG preconditioning, Algebraic multigrid, Elliptic problems.

1 Introduction

A matrix $\mathbf{A}_n \in C^{n\times n}$ is said to be *Toeplitz* when its entries are constant along each diagonal, i.e., $(\mathbf{A}_n)_{ij} = a_{i-j}$. Since this feature "captures" the time/space shift invariance of many problems in pure and applied mathematics, it is not surprising that Toeplitz matrices and operators arise in many applicative fields such as partial differential equations, Markov chains and signal processing.

The important feature that allows to compute the product of a Toeplitz matrix with a vector by means of FFT's has driven more attention on iterative methods whose complexity is dominated by such convolutions. Among these, major attention has been devoted to the preconditioned conjugated gradient (PCG) and to the multigrid. Both methods allow an approximation of the solution within a given error bound in a number of iterations independent of the matrix size n and thus leading to an overall complexity of $O(n \log n)$ operations.
This occurs provided that a good preconditioner or a good projector is found.

In this paper we underline how matrix algebras and the generating functions provide a unifying approach to the study of sequential and parallel iterative solvers whose asymptotical complexity is optimal, i.e., the cost of the product of the coefficient matrix with a vector.

2 Spectral properties of Toeplitz and τ matrices

In this section we introduce our main tools and report some useful theorems on Toeplitz matrices, τ matrices and generating functions.

2.1 Toeplitz matrices and generating functions

Let $a(x)$ be a real-valued continuous function defined on $[-\pi, \pi]$ and let

$$a_k = \frac{1}{2\pi}\int_{-\pi}^{\pi} a(\theta)e^{ik\theta}d\theta, \quad k \text{ integer},$$

be the Fourier coefficients of $a(x)$. Under such assumptions, it is possible to define a class of Toeplitz matrices $\{\mathbf{A}_n \equiv T_n(a)\}_n$ whose entries are given by $(\mathbf{A}_n)_{i,j} = a_{i-j}$. The function $a(x)$ is called the *generating function* of the class.

The spectral properties of the the matrices in $\{T_n(a)\}_n$ are well described by the generating function as shown by the following theorem due to Szegö.

Theorem 1 *[1] Assume that $\{\mathbf{A}_n\}_n$ is generated by the continuous function $a(x)$ and let $m_a = \min_{[0,\pi]} a(x)$ and $M_a = \max_{[0,\pi]} a(x)$. If we call $\lambda_i^{(n)}$ the eigenvalues of $\mathbf{A}_n$, the relation $\lim_{n\to\infty} \frac{1}{n}\sum_{i=1}^{n} F(\lambda_i^{(n)}) = \frac{1}{\pi}\int_0^{\pi} F(a(x))dx$ holds for every continuous test function $F : [m_a, M_a] \to \Re$.* ∎

Moreover, if $a(x) \in L^1(I)$ then the set $\{\lambda_i^{(n)} : n = 1, \ldots, \infty;\ n = 1, \ldots, n\}$ is dense in $[m_a, M_a]$; in particular, the sets $\{\lambda_i^{(n)}\}_{i=1}^{n}$ and $\{a(\frac{i\pi}{n+1})\}_{i=1}^{n}$ are *equally distributed* [1]. The following theorem shows us how the extreme eigenvalues tend to the extreme values of the range of the generating function $a(x)$.

Theorem 2 *[1] For all n the spectrum of $\mathbf{A}_n$ is contained in the interval $[m_a, M_a]$ and moreover $m_a = \inf_{n\in\mathcal{N}} \min_{i=1,\ldots,n} \lambda_i^{(n)}$, $M_a = \sup_{n\in\mathcal{N}} \max_{i=1,\ldots,n} \lambda_i^{(n)}$.* ∎

However, provided that $a(x)$ is nonnegative, the matrices $\mathbf{A}_n$ are positive definite for all n even when $m_a = 0$, see [2].

2.2 τ matrices

For any given integer n, we define τ_n as the matrix algebra generated by

$$\mathbf{H}_n = \begin{bmatrix} 0 & 1 & & \\ 1 & \ddots & \ddots & \\ & \ddots & \ddots & 1 \\ & & 1 & 0 \end{bmatrix}.$$

Any matrix $\mathbf{A} \in \tau_n$ can be therefore obtained as a polynomial in $\mathbf{H}_n$ of degree at most $n-1$ and therefore shares the same set of eigenvectors $\mathbf{v}^{(i)}$ of $\mathbf{H}_n$:

$$(\mathbf{v}^{(i)})_j = \sqrt{\frac{2}{n+1}}\sin\frac{ij\pi}{n+1} \qquad i = 1, \ldots, n, \quad j = 1, \ldots, n.$$

Now, if we denote with $\mathbf{F}_n$ the matrix whose columns are the eigenvectors of $\mathbf{H}_n$, for any $\mathbf{A} \in \tau_n$ we obtain $\mathbf{A} = \mathbf{F}_n \mathrm{diag}(\lambda_1, \ldots, \lambda_n)\mathbf{F}_n$, where the λ_i's are the eigenvalues of $\mathbf{A}$. It is worth noting that such diagonalization can be performed with $O(n \log n)$ operations by means of the fast sine transform; it follows that the product of a τ matrix with a vector can be computed with the same computational cost.

Similar results also hold for the two-level (block) τ algebra $\tau_{N,M}$ generated by the matrices $\mathbf{H}_N \otimes \mathbf{I}_M$ and $\mathbf{I}_N \otimes \mathbf{H}_M$ where $\otimes$ denotes the Kronecker product. Also in this case, any matrix in $\tau_{N,M}$ can be diagonalized by $\mathbf{F}_N \otimes \mathbf{F}_M$ whose columns are the common eigenvectors of the class.

2.3 τ approximations for Toeplitz matrices

There is a close relationship between symmetric Toeplitz matrices and τ matrices. Given the symmetric Toeplitz matrix $\mathbf{T}$ generated by $a(x) = a_0 + 2\sum_{k=1}^{n-1} a_k \cos(kx)$, it is possible to obtain a τ approximation $\mathbf{A} = \mathbf{T} - H(\mathbf{T})$ where $H(\mathbf{T})$ is the Hankel operator having $(a_2, \ldots, a_{n-1}, 0, 0)$ on the first row and $(0, 0, a_{n-1}, \ldots, a_2)$ on the last one. This approximation, besides being structural in the case of banded Toeplitz matrices since $\mathbf{A}$ is obtained through a low rank correction, is effective in spectral terms: $\mathbf{A}$ and $\mathbf{T}$ share the same generating function. Moreover, in the τ case the eigenvalues of the τ approximation $\mathbf{A}$ are known exactly: these are obtained by sampling the function given by the first n terms of the Fourier series for $a(x)$ on the regular grid $\{x_i = \frac{i\pi}{n+1} : i = 1, \ldots, n\}$. A similar approach can be used to obtain a two-level τ approximation for any block symmetric Toeplitz matrix.

3 PCG preconditioning for Toeplitz matrices

It is well known that the convergence of the PCG strongly depends on the distribution of the eigenvalues of the preconditioned matrix. In [3], O. Axelsson and G. Lindskög prove that when the eigenvalues are clustered we obtain a superlinear convergence after a number of iterations equal to the number of the "outliers".

Given a Toeplitz matrix $\mathbf{A}$, we need to find a preconditioner $\mathbf{P}$ such that:

a) $\mathbf{P}$ is so "close" to $\mathbf{A}$ that the spectrum of $\mathbf{P}^{-1}\mathbf{A}$ is clustered around 1 except for a constant number of outliers;

b) the solution a linear system involving $\mathbf{P}$ is not more expensive than computing a matrix-vector product.

These two issues are clearly conflicting since when a matrix is too close to $\mathbf{A}$ it also requires a comparable computational effort to be inverted. Moreover, since the overall cost per iteration of the unpreconditioned CG on Toeplitz matrices only requires $O(n \log n)$ operations, if we want to keep the same computational cost per iteration we get a tight upper bound to the arithmetic cost required by point **b**.

A way to meet point **b** is then to look for preconditioners within matrix algebras like the circulant class or the τ class. This choice typically offers a better knowledge of how close we may choose the approximation and the possibility to use a uniform and efficient ($O(n \log n)$ steps) algorithm to solve the preconditioning system. This approach, together with the insight deriving from the generating functions, also provides a good way to meet point **a**.

3.1 Some spectral results

In many works the computational behavior of iterative methods for solving symmetric positive definite Toeplitz systems has been analyzed. Efficient preconditioning techniques have been proposed assuming that the class $\{\mathbf{A}_n\}_n$ of nested Toeplitz matrices generated by $a(x)$ belongs to the *Wiener class*: i.e., the Fourier coefficients satisfy the assumption $\sum_{k=0}^{\infty} |a_k| < +\infty$.

If the matrix $\mathbf{A}_n$ is ill-conditioned, such as when the generating function $a(x)$ of $\mathbf{A}_n$ has a root in $[0, \pi]$, it is necessary to find a preconditioner $\mathbf{P}_n$ such that the spectrum of $\mathbf{P}_n^{-1}\mathbf{A}_n$ is "moved away from zero". The following theorem shows a way to accomplish this task.

Theorem 3 *[2] Let $\mathbf{A}_n$ and $\mathbf{P}_n$ be Toeplitz matrices generated by nonzero continuous functions $a(x)$ and $p(x)$. If there exist positive constants $m < M$ such that*

$$m = \inf_{x\in[0,\pi]} \frac{a(x)}{p(x)} \le \frac{a(x)}{p(x)} \le \sup_{x\in[0,\pi]} \frac{a(x)}{p(x)} = M,$$

then all the eigenvalues of $\mathbf{P}_n^{-1}\mathbf{A}_n$ lie in the open interval (m, M).

In particular, $\kappa_2(\mathbf{P}_n^{-\frac{1}{2}}\mathbf{A}_n\mathbf{P}_n^{-\frac{1}{2}}) \le M/m$. ∎

The following theorem investigates the density of eigenvalues of preconditioned Toeplitz matrices and gives us an insight about how well the ratio $r(x) = a(x)/p(x)$ describes the eigenvalues of $\mathbf{P}_n^{-1}\mathbf{A}_n$.

Theorem 4 *[4] Let $\mathbf{P}_n$ and $\mathbf{A}_n$ satisfy the hypothesis of theorem 3, let $m(\cdot)$ denote the Lebesgue measure and assume that $m\{x : p(x) = 0\} = 0$; then the union $\bigcup_{n\in\mathcal{N}} \sigma(\mathbf{P}_n^{-1}\mathbf{A}_n)$ of the spectra of $\mathbf{P}_n^{-1}\mathbf{A}_n$ is dense on the interval $[m, M]$.* ∎

3.2 Preconditioners for nonnegative generating functions

The main consideration that arises from the previous section is that it is possible to handle Toeplitz matrices and their preconditioners in terms of their generating functions; this reduces the problem of finding a semi-inverse of $\mathbf{A}_n$ to the approximation of $a(x)$ by means of trigonometric polynomials. In the following we introduce some approximation techniques to provide fast methods.

Dealing with the zeros of $a(x)$. Since the eigenvalues of $\mathbf{P}_n^{-1}\mathbf{A}_n$ are evenly distributed in the range of $r(x)$, in order to obtain a convergence rate independent of the dimension we must avoid the cases where the ratio $r(x)$ tends to zero or to infinity. Whenever $a(x) = 0$ for some $\hat{x} \in [0, \pi]$, in order to obtain

$$m \le r(x) \le M \qquad \forall x \in [0, \pi], \tag{1}$$

for some positive and finite m and M, we must choose as minimal preconditioner the matrix generated by a polynomial $p(x)$ having a unique zero in $\hat{x}$ of the same order, i.e. such that for a positive constant c we have $\lim_{x\to\hat{x}} r(x) = c$.

When the matrices $\mathbf{A}_n$ are banded or, in the general case, when the order of the zero of $a(x)$ is even, a good choice for $p(x)$ is given by a suitable power of the trigonometric polynomial $p_{\hat{x}}(x) = 2\cos\hat{x} - 2\cos x$. The polynomial $p_{\hat{x}}(x)$ has a single zero in $\hat{x}$ and is associated with a tridiagonal Toeplitz (and τ) matrix $\mathbf{P}_{\hat{x}}$. From theorem 3, we obtain that this choice ensures condition (1).

Convergence acceleration. In order to obtain a faster convergence, we may "enrich" the minimal preconditioner $p(x)$ by means of a convolution with a low degree trigonometric approximation of $r(x)$ to obtain a better approximation of $a(x)$.

The most natural way to approximate $r(x)$ is to use its representation by means of trigonometric series, this provides an efficient 2-norm approximation/interpolation technique.

When $a(x)$ is even so is $r(x)$, therefore $r(x)$ can be well represented by as a trigonometric series of cosines. Moreover, when $a(x)$ is "regular", say it is in the Wiener class, we are also guaranteed of a fast convergence of the series.

The method we introduced in [4] proceeds as follows: first we sample $r(x)$ at $m+1$ points $x_j = k\pi/m$ for $k = 0, \ldots, m$ then we build the trigonometric polynomial

$$r_s(x) = \frac{A_0}{2} + \sum_{j=1}^{s} A_j \cos(jx), \tag{2}$$

for some $s < m$, here the values A_j are given by

$$A_j = \frac{1}{m}[r(0) + (-1)^j r(\pi) + 2\sum_{k=1}^{m-1} r(x_j)\cos(kx_j)].$$

The series (2), when $s = m$, is the interpolating polynomial of $r(x)$ at the Chebyshev points x_j while for $s < m$ we obtain the best approximating polynomial for these points.

Applying this method to the case $\mathbf{A}_{200}$ generated by $a(x) = x^2$ leads to very good results: we use the minimal preconditioner $p(x) = 2 - 2\cos(x)$ and sample the function $x^2/p(x)$ in two hundred points using only the very first A_j's. In the following table we report the error reductions in ∞–norm as the band of the preconditioner increases: Table

	Degree of $r_s(x)$			
Step	1	2	3	4
2	9.132 e-03	3.886 e-03	1.586 e-03	6.606 e-04
4	9.375 e-05	1.520 e-05	1.741 e-06	6.747 e-07
6	9.912 e-07	5.370 e-08	3.597 e-09	7.896 e-10
8	9.647 e-09	1.480 e-10	6.018 e-12	6.485 e-13
10	1.198 e-10	4.396 e-13	6.455 e-14	2.914 e-14

Table 1: CG preconditioning by series truncation.

1 shows the fast convergence of the method even for small values of s.

R. Chan and P. Tang have proposed another method to to shrink the range (m, M) of $r(x)$: they compute $p^*(x)$ minimizing $\|(a-p)/a\|_\infty$ over all trigonometric polynomials of a fixed degree. This approach gives an a priori bound on the number of iterations required for the convergence of the PCG method but does not lead to dramatic improvements over the proposed techniques.

Good preconditioning techniques have also been adopted for the solution of block Toeplitz systems with Toeplitz blocks. In this case, if we use an algebraic multigrid [5] for the solution of the preconditioned system then we obtain a computational cost per iteration bounded by $O(nm\log(nm))$ ops, i.e., the cost of a bivariate fast Fourier transform. Moreover, since the resulting required number of iteration is constant with respect to the matrix dimensions n and m we obtain that the overall cost is also bounded by $O(nm\log(nm))$ ops.

It is worth noting that in the block case there is no alternative to the band Toeplitz preconditioning since the approaches based on block matrix algebras do not ensure a convergence speed independent of the matrix dimensions.

4 Algebraic multigrid for Toeplitz matrices

Another interesting application of the joint use of structured matrices and generating functions is the definition of a fully algebraic multigrid for Toeplitz matrices whose convergence rate is independent of the dimension of the matrix even in case of asymptotic ill–conditioning [6, 5, 7].

Outline of the method. Consider the linear system $\mathbf{A}_n\mathbf{x} = \mathbf{b}$ and an iterative method $x^{(j+1)} = \mathcal{S}(\mathbf{x}^{(j)})$ given by

$$\mathbf{x}^{(j+1)} = \mathbf{S}\mathbf{x}^{(j)} + \mathbf{b}_1 \tag{3}$$

where $\mathbf{S} = \mathbf{I} - \mathbf{M}^{-1}\mathbf{A}_n$ and $\mathbf{b}_1 = \mathbf{M}^{-1}\mathbf{b}$ for some matrix $\mathbf{M}$. Given a matrix $\mathbf{p} \in \mathcal{C}^{n\times k}$, with $k < n$, a general Two Grid Method (TGM) can be defined by the following iteration matrix [6]

$$TGM(\mathbf{S}, \mathbf{p}) = \left[\mathbf{I} - \mathbf{p}^T\left(\mathbf{p}\mathbf{A}\mathbf{p}^T\right)^{-1}\mathbf{p}\mathbf{A}\right]\mathbf{S}^{\nu}, \tag{4}$$

where the expression within square brackets concerns the "coarse grid correction" and depends on the projection operator $\mathbf{p}$ while the term $\mathbf{S}^{\nu}$ consists in applying ν times the "smoothing iteration" $\mathbf{S}$.

4.1 Multigrid for Tau Matrices

We consider a linear system $\mathbf{A}\mathbf{x} = \mathbf{b}$ where $\mathbf{A} \in \tau_n$ is generated by $a(x)$. If we use as "smoothing iteration" the relaxed method defined by the iteration matrix $\mathbf{S} = \mathbf{I} - \mathbf{A}/\bar{a}$ where $\bar{a} = \max_{x\in[0,\pi]} a(x)$, we find that $\mathbf{S} \in \tau_n$ is generated by $s(x) = 1 - a(x)/\bar{a}$ and that $\rho(\mathbf{S}) < 1$.

When the generating function $a(x)$ is bounded and the condition number is also bounded for any n, it means that there exists $\alpha > 0$ such that $a(x) \geq \alpha$ for all $x \in [0, \pi]$. In this case the iterative method (3) already provides a method whose convergence speed is independent of the dimension.
On the other hand, when the system is asymptotically ill-conditioned then there must be a $\hat{x} \in [0, \pi]$ such that $a(\hat{x}) = 0$. In this case the iterative method (3) achieves a poor approximation in the subspace generated by the eigenvectors of $\mathbf{A}$ whose eigenvalues are close to zero; it is therefore necessary to solve the problem in this subspace with a suitable coarse grid correction.

In order to to obtain a projector from an arbitrary matrix $\mathbf{P} \in \tau_n$, we use the *cutting matrix* [6] $\mathbf{T} \in \Re^{k\times n}$, defined by $(\mathbf{T})_{i,j} = \delta_{2i,j}$. Given a matrix $\mathbf{P} \in \tau_n$ generated by $p(x)$, a projector $\mathbf{p} \in \Re^{k\times n}$ is given by $\mathbf{p} = \mathbf{T} \cdot \mathbf{P}$. The function $p(x)$, as we will see, sets the weights for the frequencies in the projector, that is, the spectral behavior of $\mathbf{P}$ selects the subspace where the problem is projected.

Spectral properties. In [6, 5] by means of a diagonalization technique we obtain the generating function $\lambda(x)$ of the whole TGM (4):

$$\lambda(x) = \frac{p^2(x)a(x)s^{\nu}(\pi - x) + p^2(\pi - x)a(\pi - x)s^{\nu}(x)}{p^2(x)a(x) + p^2(\pi - x)a(\pi - x)}. \tag{5}$$

When $\mathbf{A}_n$ is positive definite, $\lambda(x)$ is just a weighted mean of the values that $s(x)$ assumes at the *mirror points* x and $\pi - x$; in terms of spectral radius it means that the TGM is never worse than the iterative method (3) alone.

When $a(x)$ is not smaller than a positive constant, it is possible to choose $p(x)$ such that the convergence of the iterative method, still independent of the dimension, is further improved. Equation (5), rewritten in the form

$$\lambda(x) = \frac{s^{\nu}(\pi - x) + w(x)s^{\nu}(x)}{1 + w(x)}, \tag{6}$$

points out how the ratio

$$w(x) = \frac{p^2(\pi - x)a(\pi - x)}{p^2(x)a(x)},$$

depending on the free function $p(x)$, determines the weights to associate to the terms $s(x)$ and $s(\pi - x)$. Our goal is to find a $p(x)$ that globally minimizes $\lambda(x)$.
When $s(x)$ takes its maximum value, say at $\hat{x}$, we must force $r(\hat{x})$ to zero to obtain a value smaller than $s^\nu(\hat{x})$. Let $s(\hat{x}) = \max_{x\in[0,\pi]} s(x)$, if we assume that $s(\pi - \hat{x}) < s(\hat{x})$, in order to minimize $\lambda(\hat{x})$ we impose $w(\hat{x}) = 0$.
Now, if we assume that the previous (nonsymmetry) relation also holds for $p(x)$, the last requirement is equivalent to ask for $p(\pi - \hat{x}) = 0$; this can be easily accomplished using the trigonometric polynomial $p_{\pi-\hat{x}}(x) = 2\cos\hat{x} + 2\cos x$ associated to a tridiagonal τ matrix.

TGM **as a solver.** When $a(\hat{x}) = 0$ for some $\hat{x}$, without any hypotheses on $p(x)$, the outlined *TGM* is useless since the mean in (5) is eluded. In this case we need that $\lim_{x\to\hat{x}} r(x) = 0$, that, under the hypothesis of nonsymmetry of the roots, it is equivalent to ask for

$$\lim_{x\to\hat{x}} \frac{p^2(\pi - x)}{a(x)} = 0.$$

In other terms, $\pi - \hat{x}$ should be a root of $p(x)$ of order not less than a half of the order of the zero $\hat{x}$ of $a(x)$. Such a polynomial is easily obtained: let $p_{\pi-\hat{x}}(x)$ be the base polynomial with a zero in $\pi - \hat{x}$, then it suffices to set $p(x) = p^{\beta}_{\pi-\hat{x}}(x)$ where β is the smallest integer i such that

$$\lim_{x\to\hat{x}} \frac{(x - \hat{x})^{2i}}{a(x)} < +\infty.$$

When the order of zero of $a(x)$ is even, equation (5) turns to a real mean yielding $s^\nu(\pi - \hat{x}) < \lambda(\hat{x}) < s^\nu(\hat{x})$.

4.2 Algebraic multigrid for Toeplitz matrices

In [6, 5] we generalize the outlined multigrid for a class of interesting Toeplitz problems including matrices arising from self-adjoint differential problems with constant coefficients. For these matrices we approximately know the distribution of the eigenvalues and the structure of the eigenvectors, this information is enough to define a fast algebraic multigrid method.

We consider linear systems of the form $\mathbf{A}\mathbf{x} = \mathbf{b}$ where $\mathbf{A}$ is a symmetric Toeplitz matrix generated by some function $a(x)$; according to [6], we use two "smoothers", called Pre- and Post-smoother, defined by the relaxed methods

$$\mathbf{x}^{(j+1)} = \mathbf{x}^{(j)} - \left(\mathbf{A}\mathbf{x}^{(j)} - \mathbf{b}\right)/\theta_i, \quad i = 1, 2.$$

More precisely, setting $\theta = \|\mathbf{A}\|_2$, the pre-smoother $\mathcal{S}_1$ is defined by iteration matrix $\mathbf{S}_1 = \mathbf{I} - \mathbf{A}/\theta$ while the post-smoother $\mathcal{S}_2$ is obtained from $\mathbf{S}_2 = \mathbf{I} - 2\theta/\mathbf{A}$.

Thus $TGM(\mathbf{S}_1, \mathbf{S}_2, \mathbf{p})$ is a multi-iterative method where the pre-smoother, the coarse grid correction and the post-smoother approach the solution in complementary spaces.

As an example, we consider again the Toeplitz matrix generated by $a(x) = x^2$ and since $a(0) = 0$ with order two, we choose as projector $\mathbf{p} = \mathbf{T}\cdot\mathbf{P}$ where $\mathbf{P}$ is generated by $p_\pi(x)$.

In table 2 we report the results of the first 10 iterations of our method applied to a system of size 1000 with $\theta = \pi^2$. the parameters **pre** and **post** denote the number of the pre- and post-smoothing steps made for each global *TGM* iteration; the errors are given in ∞−norm.
Though the second method is more effective, both are convergent and lead to iteration matrices whose spectral radii are smaller than one and very close to the theoretical values: 0.5 and 0.33 respectively.

Step	Pre and Post smoother iterations 2 0	1 1
2	0.12837 e+00	0.63175 e-01
4	0.38976 e-01	0.75145 e-02
6	0.11360 e-01	0.89100 e-03
8	0.32907 e-02	0.10812 e-03
10	0.95533 e-03	0.14174 e-04

Table 2: Algebraic multigrid with $a(x) = x^2$.

5 Structured solvers for elliptic problems

In this section we apply our approach to the solution of positive definite linear systems arising from the discretization of elliptic operators. These systems are commonly characterized by condition numbers that grow with the dimension of the matrices. An example is given by the Poisson problem on $\Omega = (0,1)$:

$$\begin{cases} -\left(a(x)u_x\right)_x = f & \text{on } \Omega, \\ u = 0 & \text{on } \partial\Omega. \end{cases} \tag{7}$$

Discretizing (7) by means of finite differences on a uniform grid with n points, leads to a linear system $\mathbf{Ax} = \mathbf{f}$ whose condition number grows as $O(n^2)$ even in the simplest case where $a(x) \equiv 1$. This difficulty, together with the sparsity of such matrices, makes iterative methods more attractive than direct ones especially when multi-dimensional problems are considered or when we are interested in parallel computation. In general, when we deal with differential problems like (7), we are given a class α of matrices associated with the differential operator. In order to efficiently apply a PCG method we have to deal again with two main issues:

c) choose a suitable class β of matrices "close" enough to α and whose elements are easy to invert;

d) devise a suitable projection operator $\mathcal{P} : \alpha \longmapsto \beta$ to obtain the "best" approximation $\mathbf{P} \in \beta$ for any given $\mathbf{A} \in \alpha$.

Again, a good way to meet point **c** is to look for preconditioners within matrix algebras. Point **d** is delicate, since when $a(x)$ is not a constant we need to reduce the "informational content" of the matrix to meet requirement **c**; for this purpose some averaging schemes proved to be useful tools.

In [8] we propose the use of the τ algebra as the approximating class and a projection operator that uses an averaging scheme similar to the one found in [9]. This allows us to obtain a fast convergence even in case of strong ill-conditioning arising from nonnegative coefficients in the differential operator.

5.1 Some spectral results

Given a grid size n and setting $a_i = a(\frac{i}{n+1})$, equation (7) can be discretized by means of centered differences to obtain a symmetric linear system $\mathbf{Ax} = \mathbf{f}$ where

$$\mathbf{A} = \left[-a_{i-\frac{1}{2}},\ a_{i-\frac{1}{2}} + a_{i+\frac{1}{2}},\ -a_{i+\frac{1}{2}}\right] \equiv \begin{bmatrix} a_{\frac{1}{2}} + a_{\frac{3}{2}} & -a_{\frac{3}{2}} & \\ -a_{\frac{3}{2}} & \ddots & \ddots \\ & \ddots & \ddots \end{bmatrix}. \tag{8}$$

Following the approach found in [9], we use an averaging scheme to compute the quantities $c_1 = -\frac{1}{n-1}\sum_{j=1}^{n-1} a_{j+\frac{1}{2}}$ and $c_0 = -2c_1(n-1)/n + (a_{\frac{1}{2}} + a_{n+\frac{1}{2}})/n$ that define the "best" τ approximation for $\mathbf{A}$: $\tau_{\mathbf{A}} = [c_1, c_0, c_1]$. We find that $\tau_{\mathbf{A}}$ is asymptotically close to the well known matrix $c_1[-1,2,-1]$; in fact, $\tau_{\mathbf{A}} = c_1[-1,2,-1] + \mathbf{E}$ with $\|\mathbf{E}\|_\infty \in O(1/n)$. Therefore, if we ignore the correction $\mathbf{E}$ and set $\tau_{\mathbf{A}} = c_1[-1,2,-1]$, it is easy to verify that when $0 < a_{min} \leq a(x) \leq a_{max}$, both $\mathbf{A}$ and $\tau_{\mathbf{A}}$ are positive definite and that the following theorem holds.

Theorem 5 *[8] The matrices*

$$\mathbf{A} - a_{min}[-1,2,-1] \text{ and } a_{max}[-1,2,-1] - \mathbf{A}$$

are positive semidefinite ∎

It follows [8] that $a_{min} \leq \rho\left(\tau_{\mathbf{A}}^{-1}\mathbf{A}\right) \leq a_{max}$ and and that the resulting PCG method achieves a convergence speed independent of the dimension of $\mathbf{A}$.

Another interesting remark is that $\mathbf{A}$ is approximately a scaling of the matrix $\Delta = [-1,2,-1]$, i.e., setting $\mathbf{D}_{\mathbf{A}} = \text{diag}(a_{i-\frac{1}{2}} + a_{i+\frac{1}{2}})$ we have $\mathbf{D}_{\mathbf{A}}^{-1}\mathbf{A} = \Delta + \mathbf{F}$ with $\|\mathbf{F}\|_\infty \in O(1/n)$. Consequently, a better symmetric positive definite approximation of $\mathbf{A}$, to be used as a preconditioner, is given by

$$\tau_{\mathbf{A}}^* = \mathbf{D}_{\mathbf{A}}^{\frac{1}{2}}\Delta\mathbf{D}_{\mathbf{A}}^{\frac{1}{2}}.$$

In this case we find a good clustering property for the spectrum of $\tau_{\mathbf{A}}^{*\,-1}\mathbf{A}$ and a superlinear convergence of this PCG is observed in numerical tests.

Similar results also hold for two dimensional elliptic problems of the form

$$-(a(x,y)u_x)_x - (a(x,y)u_y)_y = f(x,y) \tag{9}$$

with homogeneous boundary conditions, see [8, 10].

A numerical example. As an example, we consider a 1–dimensional problem like (7) with $a(x) = x^2$ and discretize it as in (8) on a uniform grid with $n = 200$ points. In table 3 we report the 2-norm of the errors at each iteration of the method. For comparison, we report the results obtained using 4 different preconditioners: the identity matrix $\mathbf{I}_n$, the diagonal part $\mathbf{D}_{\mathbf{A}}$ of $\mathbf{A}$, the τ approximation $\tau_{\mathbf{A}}$ and the improved τ approximation $\tau_{\mathbf{A}}^* = \mathbf{D}_{\mathbf{A}}^{\frac{1}{2}}\tau_{\mathbf{A}}\mathbf{D}_{\mathbf{A}}^{\frac{1}{2}}$. It should be noted that while the basic preconditioners

	Preconditioner			
Step	$\mathbf{I}_n$	$\mathbf{D}_{\mathbf{A}}$	$\tau_{\mathbf{A}}$	$\tau_{\mathbf{A}}^*$
2	9.318 e-01	8.019 e-01	8.322 e-01	1.259 e-04
4	9.191 e-01	6.823 e-01	7.680 e-01	1.152 e-09
6	9.091 e-01	6.377 e-01	6.738 e-01	1.077 e-14
8	8.969 e-01	6.447 e-01	6.180 e-01	3.886 e-15
10	8.820 e-01	5.939 e-01	6.074 e-01	9.881 e-15

Table 3: Equation (7) with $a(x) \equiv x^2$.

are not effective, their combination yields a very fast method. For more examples, also in the two dimensional case, see [8, 10].

6 Conclusions

An approach based on structured matrices and generating functions for solving Toeplitz and Toeplitz–like systems offers a unified framework to the search for suitable preconditioners or projector operators. The generating functions allows one to treat difficult linear properties in terms of simple functional terms while matrix algebras provide efficient tools to keep the overall complexity as low as possible. In fact, each iteration of our PCG method, in the Toeplitz case, costs $O(n \log n)$ ops, that is, the cost of computing a fast transform of size n. When **A** is banded, this cost can be lowered to $O(ln)$ using the algebraic multigrid method, where l is the bandwidth and c is a constant. The number of iteration is asymptotically constant due to the reduction of the condition number from $O(n^2)$ to $O(1)$.

References

1. U. Grenander and G. Szegö, Toeplitz Forms and Their Applications. Second Edition, Chelsea, New York (1984).

2. R. Chan, IMA J. Numer. Anal., 11, 333-345 (1991).

3. O. Axelsson and G. Lindskog, Numer. Math., 52, 499-523 (1986).

4. F. Di Benedetto, G. Fiorentino and S. Serra, Comput. Math. Appl., 25, 33-45 (1993).

5. G. Fiorentino and S. Serra, SIAM J. Sci. Comp., 17, 1068-1081 (1996).

6. G. Fiorentino and S. Serra, Calcolo, 28, 283-305 (1991).

7. G. Fiorentino and S. Serra, in: Proc. of the conf. "Algebraic Multilevel Iteration Methods with Applications", Nijmegen (The Netherlands), (1996).

8. G. Fiorentino and S. Serra, in: Proc. 2nd IMACS conf. on "Iterative Methods in Linear Algebra", Blagoevgrad (Bulgaria), (1995).

9. R. Chan and T. Chan, J. Numer. Lin. Algebra Appl., 1, 77-101 (1992).

10. G. Fiorentino and S. Serra, Comput. and Math. with Appl., 32, 61-68 (1996).

8th Int. Coll. on Differential Equations, pp. 179-185
D. Bainov (Ed)

Finding Voxels Shared by a Ray and a Linear Octree

I. GARGANTINI AND J-Y LIU
Department of Computer Science, London, Ontario, Canada N6A 5B7
e-mail: irene@csd.uwo.ca; http://www.csd.uwo.ca/faculty/irene.html

Abstract. The voxels shared by a ray - encoded as a sequence of octal codes - and a linear octree, representative of a scene in 3D space, can be calculated by comparing, left-to-right, the corresponding pairs of digits of ray and object's node. A preliminary comparison of both codes, interpreted as integers, allows to establish whether or not an intersection is at all possible. The algorithm automatically skips all the empty spaces in the given 3D-scene.

Key Words: Linear octrees, Intersection ray-object, Ray casting, Visualization.

AMS Subject Classification: 68U05

1. INTRODUCTION

Casting rays to mimic the penetration of light is one of the most powerful methods used to reveal the internal structures of the objects which form a given 3D-scene. The latter can be defined mathematically (as in a CAD-system) or acquired through some sensory devices (CAT scanners, for instance). Transparency, i.e. the attenuation of light as it travels through an object, can be implemented using several approaches. Most of these require the evaluation of the volume elements common to both ray and 3D-scene. We address this central problem in a linear octree environment, where ray and objects are represented as sets of (suitably) sorted octal codes. Even at the price of repeating well-known topics, we feel that an introduction to locational codes and linear octrees is necessary to the understanding of our new approach.

2. ENCODING: ONE DIGIT FOR POINTER AND LOCATION

We represent the recursive subdivision of a given $2^n \times 2^n \times 2^n$ raster by a *tree* constructed as follows: nonterminal *nodes* correspond to subsets containing free or occupied voxels and have 8 children; terminal nodes or leaves correspond to maximal subsets of voxels with the same attribute, labelled *white* or *black*. *Arcs* represent the successive subdivisions in terms of cubes of equal size, called *octants*; the height of the thus generated octree is at most n. See, for instance, [1]. The recursive subdivision described above can also be given a different but equally useful interpretation by linking the path from the octree root to a node with the (x, y, z) coordinates of the northern, western, frontal vertex of the octant corresponding to that node. This can be achieved by encoding each octant with the scheme given below; by juxtaposing the digit of a child-octant to the right of that of

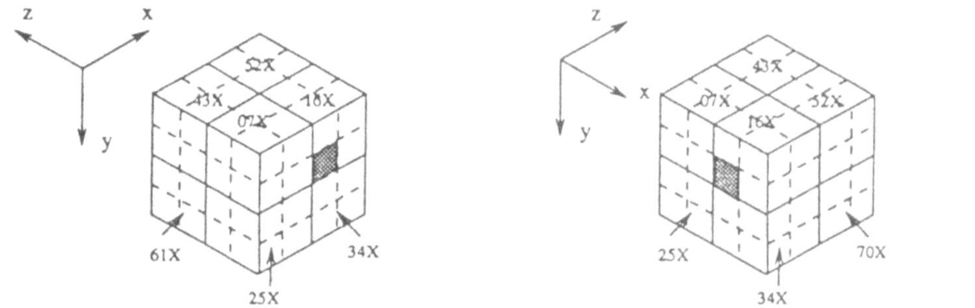

a. viewed from the top-front-left direction **b. viewed from the top-front-right direction**

Figure 1: Object identified by octal codes

the father-octant; and by interlacing the binary digits representing the (x, y, z) vertex to form an octal digit, i.e. a digit in base 8 (see, for instance, [2]): 0 for front-north-west (FNW), 1 for FNE, and so on, through FSW, FSE, BNW, BNE, BSW, and BSE in order.

Let $x = (a_{n-1} \dots a_1 a_0), y = (b_{n-1} \dots b_1 b_0)$, and $z = (c_{n-1} \dots c_1 c_0)$ be the binary representations of x, y, and z. The octal code's digit of weight 8^i is given by $(c_i b_i a_i)$. For example, the shaded voxel in Figure 1a, which has the coordinates $(4, 3, 2) = (100_2, 011_2, 010_2)$, is represented by the octal code $(001_2) \times 8^2 + (110_2) \times 8^1 + (010_2) \times 8^0 = 162_8$.

The size of a $(2^m \times 2^m \times 2^m)$-octant can be encoded by the logarithm (base 8) of the number of voxels it contains, so that a complete representation is given by the pair (octal-code, m). A short-hand for this is to replace the least significant m digits of the code with a non-numerical symbol, such as d for 'don't care'. By encoding an octree arc (point) with the *same* digit as the child-octant it points to, the path from the root to a leaf of size m coincides with the octal-code for that leaf and the value $n - m$ is the length of that path: the rth digit from the left in the octal code corresponds to the rth subdivision, $r = 1, 2, \dots, n - m$.

3. SORTING THE OCTAL CODES

One of the advantages of the representation of an object by octal codes is that most of the visibility problems can be solved by simply looking at the coordinates of the VPN-vector in relation to the octal codes. The VPN-vector is indicative of the direction along which the viewer is positioned. Let the origin of the raster be located at the front, top, left corner of the raster and the VPN $= (x_1, y_1, z_1)$, with $x_1 y_1 z_1 \neq 0$. A ray cast from infinity in the direction determined by the VPN will encounter the nodes of an octree according to the priority specified in Table 1 [3]. Note that the VPN's coordinates are set in the correct mathematical context; reversing their sign gives the direction along which the viewer is located.

Looking at Table 1 we see that, given the octree sorted for $x_1 > 0, y_1 > 0, z_1 > 0$ (first row), the linear octree for $x_1 < 0, y_1 < 0, z_1 < 0$ (last row) is obtained by reversing the order; similarly for rows 2 & 7, 3 & 6, and 4 & 5. This means that the preprocessing to put the linear octree in position to be seen from any directions is reduced to one half. The order initially given may not necessarily be one of the valid priority orders, but by taking the reversals into account, we need only to apply a radix-sort at most four times. Object in Figure 1a is seen from the top-front-left-direction $(x_1, y_1, z_1 > 0)$; that in Figure 1b from the top-front-right-direction ($x_1 < 0; y_1, z_1 > 0$).

x	y	z	Priority Order	x	y	z	Priority Order
>0	>0	>0	0,1,2,4,3,5,6,7	<0	<0	<0	7,6,5,3,4,2,1,0
<0	>0	>0	1,0,3,5,2,4,7,6	>0	<0	<0	6,7,4,2,5,3,0,1
>0	<0	>0	2,3,0,6,1,7,4,5	<0	>0	<0	5,4,7,1,6,0,3,2
<0	<0	>0	3,2,1,7,0,6,5,4	>0	>0	<0	4,5,6,0,7,1,2,3

Table 1: Visiting the octal nodes according to the viewing direction

4. THE RAY TRAVERSAL

To reduce the number of intersections, it is imperative for the ray to step over all empty regions. While this approach has been used in several instances [4, 5, 6, 7, 8, 9, 10], our technique differs in the use we make of the locational code to predict how far a ray freely travels between objects or from the raster's border to the first object encountered.

Assume that the ray has travelled up to V, where V is the voxel with locational code $c_{n-1}\dots c_0$ and that the object list has been (orderly) traversed up to LO, where LO is the node with grouping factor gf, $0 \leq gf \leq n-2$. In terms of the locational code's expansion, this means that each of the digits $d_{gf-1}, d_{gf-2}, \dots, d_0$ assumes every value from 0 to 7, and that LO consists of 8^{gf} voxels. We express this by underlying these 'don't care' combinations: $d_{n-1}d_{n-2}\dots d_{gf}\underline{d_{gf-1}\dots d_0}$ or $d_{n-1}d_{n-2}\dots d_{gf}\underbrace{\underline{d\dots d}}_{gf\ times}$. Now we have

Proposition 1. If $c_i \neq d_i, i = n-1, n-2, \dots, gf$, then no intersection between ray-voxel V and node LO is possible.

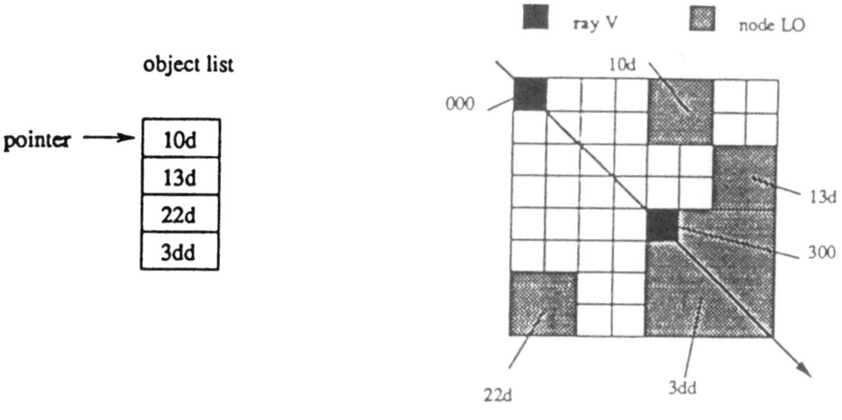

Figure 2: Ray casting in a 2D scene

If the test expressed by Proposition 1 fails, we have three cases. Each refers to the $(n-g)$-th subdivision, identified by the g-th octal digit. We will explain these cases in 3D, but all figures will be drawn in two dimensions.

Case 1: $c_g < d_g$, with $\{c_i = d_i,\ i = n-1, n-2, \dots, g+1, \quad gf \leq g < n-1$ or $g = n-1\}$.
The current node LO is in front of the ray and will be the first node the ray may encounter. As $c_g < d_g$, we also know that the octant $(c_{n-1}c_{n-2}\dots c_g)$ of size 8^g (to which V belongs)

is empty. We advance the ray by the entire line segment crossing the empty octant and rename the first voxel, outside the latter, with V. For example, in Figure 2 the first ray voxel is V='000' and the first node of the object list is LO='10d'. Since c_2 is less than d_2 and g (g=2) greater than gf (gf = 1), we advance the ray to exit the empty node '0dd' and move it to '300', the new value of V.

Case 2: $c_g > d_g$, with $\{c_i = d_i, i = n-1, \ldots, g+1, \quad gf \leq g < n-1 \text{ or } g = n-1\}$. Here the ray has passed node LO. Before comparing the ray with a new node, we try to exclude all other possible object nodes, which the ray has also passed by. We proceed as follows: Let LO' be the successor of LO in the object list, with locational code

$$d'_{n-1}d'_{n-2}\ldots d'_{gf'}d'_{gf'-1}\ldots d'_0.$$

If $d_i = d'_i$, for $i = n-1, n-2, \ldots, g$, we know that node LO' has also been passed by. We then move the pointer to the object list to the successor of LO' and keep doing so until, for the digits of the current LO and LO', we have $d_i \neq d'_i$, where i can take any of the $n-1, n-2\ldots, g$ values. If this takes place, we update LO to LO'. For example, in Figure 2, V is '300' and LO is '10d'. Since c_2 is greater than d_2 and g greater than gf, we first get LO' whose octal code is '13d'. As d_2 equals d'_2, this means that node LO' has been passed by too. The new LO' is '22d'. As this time d_2 differs from d'_2, we have to compare V='300' with LO='22d'.

Case 3: $c_g = d_g$, with $\{i = n-1, n-2, \ldots, g, \quad 0 \leq g < gf\}$. In this case voxel V is travelling inside the object. If gf is 0, LO and the ray share precisely that voxel V. Otherwise, V pierces the object formed of the entire octant $(d_{n-1}d_{n-2}\ldots d_{gf}\underbrace{d\ldots d}_{gf\ times})$ of size 8^{gf}. In Figure 2, when V is '300' and LO is '3dd', as c_2 equals d_2 and g (g = 1) is less than gf (gf = 2), we have an intersection. The voxels met by the ray are '300', '303', '330', '333'.

Algorithm 'Traversal' is given below. It finds the intersections between a linear octree and a ray, using the *Digital Differential Analyzer (DDA)* algorithm coupled with the approach given in [6] to intersect the ray with the principal planes representing the faces of the encountered octant.

```
Data structure:
LO                      /* octal code of linear octree */
gf-oct                  /* grouping factor of current node of linear octree */
V                       /* octal code of ray-voxel */
S-RAY                   /* set of voxels common to ray and object */
initialize
g=Resolution-1          /* raster is of size 8^Resolution; */
evaluate the intersection (V) of the ray with the raster border
  while (ray is still within the raster OR linear octree has unvisited elements) do
    /* 'switch' decides if the current part of the ray is in front of an object
     (digit==1), inside an object (digit==0), or past the object (digit==-1) */
   switch (digit (g. V, LO))
     case 1:     /* V is in front of LO; the g-th digit of V is less than that of LO */
     advance_ray (V, g)             /* let ray skip the empty octant of size 8^g */
     get new V                      /* new V is the successor of the exit voxel */
     case -1:    /* V is past LO; the g-th digit of V is greater than that of LO */
     advance_octree (LO, g)         /* get new LO */
     case 0:     /* V is inside LO */
     insert_voxels (S-RAY, V, g)    /* insert all shared voxels into S-RAY */
```

```
        advance pointer in linear octree  /* get next node of linear octree */
        get new LO and gf-oct
        g=Resolution-1                    /* reset g */
      end switch
    end while
digit (g, V, LO)            /* find first octal digit for which octal code V and
                             * LO differ; start from leftmost position */
        c = V xor LO                            /* operator exclusive or */
        while ((c >> g) == 0)                   /* shift c of g positions to the right */
                g = g-1
        end while
        if (g < gf-oct)   return 0
        mask = 1 << g                           /* shift 1 of g positions to the left */
        if ((V and mask) < (LO and mask))   return 1
        else              return -1
end
advance_ray (V, g)                  /* the ray still navigates in an empty
                                     * octant: advance V by this octant of size 8^g */
    V-fnw = V and ( ~ 0 << g)       /* evaluate front-north-west corner of the octant,
                                     * ~ 0 is the inverse of each '0'_bit */
    V-size = 8^g                    /* the size of the current octant */
    evaluate the octant's plane from which the ray exits and the ray-plane intersection (V)
end
advance_octree (LO, g)              /* the ray passes the object: advance LO until
any of the digits, from (n-1) to g, of previous and current node differs */
    LO-org = LO                     /* keep original LO to compare with new LO */
    advance LO                      /* and get a new node */
    while (((LO-org xor LO) >> g) == 0)
/* test if any of the code digits, from (n-1) to g, of the new LO has changed using bitwise
'exclusive or' and shifting */
        advance LO by one node
    end while
end
insert_voxels (S-RAY, V, g)         /* insert all shared voxels into S-RAY */
    ...    ...
    this part is the same as advance_ray()
    ...    ...
    insert all the voxels in the above octant into S-RAY
end
```

In Algorithm Traversal each ray goes through the linear octree only once. The comparison of the ray and the linear octree terminates when either the ray exits or the linear octree has been completely visited. Advancing the ray is carried out in time proportional to αK, where α is a positive constant and $K (K < 2^n)$ is the number of nodes (empty or full) which are pierced by the ray. Advancing the pointer in the linear octree is carried out in time proportional to N, where N is the total number of nodes in the scene. The amount of computation involved in Algorithm Traversal is $\beta(N + \alpha K)$, where $0 < \beta \leq 1$.

5. THE POTENTIAL OF THE ALGORITHM

As we keep in memory the original 3D-scene sorted according to the four principal directions (and, implicitly, also to their diagonally opposite) it will be possible to spawn a secondary ray. Assume that the surface's voxel hit by a primary ray is (x_0, y_0, z_0), with octal encoding 'aaa', and the ray bends according to some reflection vector R. We could first evaluate the outer normal to (x_0, y_0, z_0) and analyze along which of the eight principal directions the reflection takes place; then find the voxel with encoding 'aaa' in the linear octree sorted according to the direction of the reflection vector R. This voxel is the starting point of the secondary ray. Note that all nodes preceding 'aaa' would not need to be examined: algorithm given in Section 4 can then be reapplied.

Algorithm Traversal had been tested on simple solids, on a voxel-to-voxel comparison, to verify the correct intersections of a ray with the given 3D-scene. To give the reader an idea of the algorithm, we have visualized the trajectory of each ray cast into the 3D scene as follows. Assume that a ray piercing screen's pixel 'a' intersects M voxels while another ray, piercing screen's pixel 'b' intersects M' voxels with $M > M'$. We assign to each pixel an intensity of $(2^n - M)^2/2^{2n}$ on a scale of [0,255], resulting in 'a' being darker than pixel 'b'. The first 3D scene consists of six *solid* cones touching at their apices, digitized in a $(2^8 \times 2^8 \times 2^8)$-raster, to create 26,170 octal nodes. A projection from the (top-front-left)-direction is shown in Figure 3a, where all the 256 grey levels have been used. One can see that the darkest areas correspond to the superposition of the two pairs of cones, corresponding to the intersections with the largest values of M. A second 3D-scene, simulating *a solid* glass, consists of a cone, a cylinder and a 10-voxels-thick ellipsoid for a total of 20,621 octal nodes. Viewed from the center-top (Figure 3b), one can see that the longest trajectories are generated when the rays pierce the object from the glass' rim downwards. Note also that the superposition of the glass' stem and basis is highlighted correctly.

(a)

(b)

Figure 3: Visualization of composite solids

6. CONCLUSION AND FUTURE WORK

A new algorithm to find all the voxels shared by a ray and a linear octree has been presented. Its two major components consist of skipping the empty spaces and deciding whether or not an intersection can occur before actually doing any evaluation. Both these components are based on comparing the octal encodings of ray and object's nodes.

The algorithm lends itself to a simple visualization showing how deep a ray penetrates an object. It is currently being extended to incorporate transparency.

References

1. H.H. Chen and T.S. Hang, Comp. Vision, Graphics, and Image Proc., 43, 409-431 (1988).
2. I.Gargantini, Computer Graphics and Image Processing, 20, 265-274 (1982).
3. I. Gargantini, T.R. Walsh and O.L. Wu, IEEE Computer Graphics & Applications, 12-21 (1986).
4. I. Gargantini and H.H. Atkinson, Computer Graphics Forum, 112, 199-210 (1993).
5. I. Gargantini and J.H.G. Redekop, in: Proc. COMPUGRAPHICS'95, (1995) Alvor, Portugal, pp. 278-284.
6. A.S. Glassner, IEEE Computer Graphics & Applications, 15–22 (1984).
7. M. Levoy, ACM Transactions on Graphics, 9, 245-261 (1990).
8. H. Samet, Computer Graphics, 13, 445-460 (1989).
9. J. Spackman, and P. Willis, Computers & Graphics, 15, 185-194 (1991).
10. N. Stolte and R. Caubet, Computer Graphics Forum, 14, C383-C394 (1995).

8th Int. Coll. on Differential Equations, pp. 187-193
D. Bainov (Ed)

Existence and Uniqueness Theorem for an Oblique Derivative Problem for Nonlinear Elliptic Discontinuous Operators in the Plane

SOFIA GIUFFRE'
Dipartimento di Matematica, Università di Catania, Viale "A. Doria" 6, 95125 Catania, Italy

Abstract. The oblique derivative problem is studied for a class of nonlinear discontinuous differential operators in the plane. Making use of Campanato's theory of near operators, we prove existence and uniqueness of strong solutions to the problem under consideration.
Keywords: *Nonlinear discontinuous elliptic operators, oblique derivative problem*

1. Introduction

Let $\mathcal{A}(x, H(u))$ be an operator with $u\colon\ \Omega \subset \mathbb{R}^2 \to \mathbb{R}$ and $H(u) = \{D_{ij}u\}_{i,j=1,2}$ being the Hessian matrix of u. Our main goal is to study the following regular oblique derivative problem:

$$(1.1)\qquad \begin{cases} \mathcal{A}(x, H(u)) - \lambda u = f(x) \in L^2(\Omega) & \text{almost everywhere in } \Omega, \\ \partial u/\partial \ell = 0 & \text{on } \partial\Omega, \end{cases}$$

where the number λ is greater than zero.

Let us point out that the linear planar problem (1.1) with $\lambda = 0$, i.e., the case when $\mathcal{A}(x,\xi) = \sum_{i,j=1}^{2} a_{ij}(x)\xi_{ij}$ has been well studied by Talenti (cf. [1]), who established a $W^{2,2}(\Omega)$ solvability assuming $a_{ij}(x)$ to be bounded and measurable functions. Concerning the multi-dimensional linear oblique derivative problem, there is a classical result of C. Miranda ([2]) who assumed the coefficients $a_{ij}(x)$ to belong to the Sobolev space $W^{1,n}(\Omega)$ (n is the dimension of the space and $\Omega \subset \mathbb{R}^n$). Recently, the result of Miranda has been improved by G. Di Fazio and D.K. Palagachev ([3]) who assumed the coefficients of the elliptic differential operator to belong to the space VMO of functions with vanishing mean oscillations. Their results have been successfully applied in the study of oblique derivative problem for quasilinear elliptic operators with VMO principal coefficients and general boundary operator (see [4] and [5]).

In the case of Dirichlet problem for general nonlinear operators $\mathcal{A}(x, u, Du, H(u))$ satisfying Campanato's condition, we should mention the papers of Palagachev [6] and L. Softova [7] where strong ($W^{2,2}(\Omega)$) solvability has been derived in the cases of subquadratic and quadratic growth with respect to the gradient Du, respectively.

The general aim of the present paper is to prove existence and uniqueness of strong solutions to the planar problem (1.1), assuming $\mathcal{A}(x,\xi)$ to satisfy Caratheodory's condition and a suitable ellipticity condition introduced by S. Campanato (see [8]). We should point out that, in view of the Caratheodory character of $\mathcal{A}(x,\xi)$, a general enough discontinuity with to respect to $x \in \Omega \subset \mathbb{R}^2$ is allowed (in fact, we need only measurability with respect to x). Concerning the dependence on the second derivatives $H(u)$ in $\mathcal{A}(x, H(u))$, it is subordinated to an ellipticity condition which ensures the possibility to apply Campanato's theory of "near mappings" in the study of (1.1).

2. Main results

Let $\Omega \subset \mathbb{R}^2$ be a bounded and open set with a C^2-smooth boundary $\partial\Omega$. We assume that $\partial\Omega$ is a closed curve and let

$$\begin{cases} x_1 = x_1(\varphi) \\ x_2 = x_2(\varphi) \end{cases}$$

be the normal parameterization of $\partial\Omega$, with φ being a curvilinear parameter, $\varphi \in [0, L]$. Considering the unit outward normal $\boldsymbol{n} = (X_1, X_2)$ to $\partial\Omega$ with $X_1 = x_2'(\varphi)$, $X_2 = -x_1'(\varphi)$, we denote by $\kappa(\varphi)$ the mean curvature of $\partial\Omega$ at the point $(x_1(\varphi), x_2(\varphi))$, i.e., $\kappa = x''_1 x_2' - x_1' x''_2$.

Further, let $\ell = (Y_1(\varphi), Y_2(\varphi))$ be a unit vector field of class $C^1(\partial\Omega)$, such that $Y_1(L) = Y_1(0)$ and $Y_2(L) = Y_2(0)$. Denoting by θ the angle between the field ℓ and the normal $\boldsymbol{n}$, we always assume that $\sum_{i=1}^{2} X_i(\varphi)Y_i(\varphi) = \cos\theta > 0$.

Our aim will be to study the following *regular* oblique derivative problem:

$$\begin{cases} \mathcal{A}(x, H(u)) - \lambda u = f(x) \in L^2(\Omega) & \text{a.e. in } \Omega, \\ \partial u/\partial \ell = 0 & \text{on } \partial\Omega, \end{cases}$$

with $\lambda > 0$.

We will prove existence and uniqueness of a strong solution to this problem. That is, a function $u(x)$ belonging to the Sobolev class $W^{2,2}(\Omega)$ of functions which are twice weakly differentiable with $L^2(\Omega)$-summable derivatives up to order 2.

We impose the following assumptions:

1) *the function $\mathcal{A}(x, \xi)$ is measurable in $x \in \Omega$ for all $\xi \in \mathbb{R}^4$ and continuous in ξ for almost all x, and it satisfies Campanato's* A) *condition (see* [8]*): there exist $\alpha, \gamma, \delta > 0$, $\gamma + \delta < 1$, such that for all $\xi, \tau \in \mathbb{R}^4$ and for almost all $x \in \Omega$ we have*

$$\left| \sum_{i=1}^{2} \xi_{ii} - \alpha \left[\mathcal{A}(x, \xi + \tau) - \mathcal{A}(x, \tau) \right] \right| \le \gamma \|\xi\|_4 + \delta \left| \sum_{i=1}^{2} \xi_{ii} \right|,$$

where the symbol $\|\cdot\|_4$ states for the usual Euclidean norm in $\mathbb{R}^4$;

2) $\mathcal{A}(x, 0) = 0$;

3) $\kappa \le \dfrac{d\theta}{d\varphi} \le 0$, *with* $\dfrac{d\theta}{d\varphi} - \kappa \not\equiv 0$.

We observe that it follows from the hypothesis 3) that

$$-1 \le \frac{\theta(L) - \theta(0)}{2\pi} \le 0,$$

since $\int_0^L \kappa(\varphi)d\varphi = -2\pi$. The above relation means that the angle θ decreases when φ increases. We assume that $\theta(L) = \theta(0) - \pi$ or $\theta(L) = \theta(0) - 2\pi$ and thus it follows that the unit vector ℓ makes a half turn or a complete turn around the normal $\boldsymbol{n}$ when the point $(x_1(\varphi), x_2(\varphi))$ makes a complete round in a counter-clockwise direction on $\partial\Omega$.

We shall prove the following theorems:

Theorem 2.1. *Under the assumptions* 1), 2) *and* 3) *the problem*

$$\begin{cases} \mathcal{A}(x, H(u)) - \lambda u = f(x) \in L^2(\Omega) & a.e. \ in \ \Omega, \\ \partial u/\partial \ell = 0 & on \ \partial\Omega \end{cases}$$

admits a unique solution $u \in W^{2,2}(\Omega)$.

Theorem 2.2. *Under the same assumptions as above, there exists a number* $q_0 > 2$ *such that for each* $q \in [2, q_0)$ *and each* $f(x) \in L^q(\Omega)$, *the problem*

$$\begin{cases} \mathcal{A}(x, H(u)) - \lambda u = f(x) \in L^q(\Omega) & a.e. \ in \ \Omega, \\ \partial u/\partial \ell = 0 & on \ \partial\Omega \end{cases}$$

admits a unique solution $u \in W^{2,q}(\Omega)$.

3. Preliminary results

Now, we recall some definitions and auxiliary results which will be used in the proofs of Theorems 2.1 and 2.2. We start with the definition of near operators.

Definition 3.1. *Let* $\mathcal{B}$ *be a set and* $\mathcal{B}_1$ *a real Banach space. Consider two mappings A and B defined on* $\mathcal{B}$ *with values in* $\mathcal{B}_1$. *The mapping A is said to be* **near** *B if there exist two positive constants* α *and* K, $0 < K < 1$, *such that for all* u, $v \in \mathcal{B}$ *we have*

$$\left\| B(u) - B(v) - \alpha\left[A(u) - A(v)\right]\right\|_{\mathcal{B}_1} \leq K\left\|B(u) - B(v)\right\|_{\mathcal{B}_1}.$$

The next theorem is proved by Campanato in [8]:

Theorem 3.1. *The mapping* $A\colon \ \mathcal{B} \to \mathcal{B}_1$ *is injective, or surjective, or bijective if and only if it is near to a mapping* $B\colon \ \mathcal{B} \to \mathcal{B}_1$ *which is injective, or surjective, or bijective, respectively. Moreover, there is the estimate*

$$\left\|B(u) - B(0)\right\|_{\mathcal{B}_1} \leq \frac{\alpha}{1-K}\left\|A(u) - A(0)\right\|_{\mathcal{B}_1}.$$

We recall also the definition of monotone operators:

Definition 3.2. *Let* $\mathcal{B}$ *be a set and* $\mathcal{B}_1$ *a Hilbert space, and suppose A and B to be two mappings from* $\mathcal{B}$ *into* $\mathcal{B}_1$. *The mapping A is said to be* **monotone** *with respect to B if for all* u, $v \in \mathcal{B}$ *we have*

$$\Big(A(u) - A(v), B(u) - B(v)\Big)_{\mathcal{B}_1} \geq 0$$

where $(\cdot,\cdot)_{\mathcal{B}_1}$ *is the scalar product in* $\mathcal{B}_1$.

In order to prove the strong solvability of the problem (1.1), we shall make use of the following theorem (see [8]):

Theorem 3.2. *Let A, B and C be mappings from* $\mathcal{B}$ *into* $\mathcal{B}_1$ *and suppose that A is near B with constants* α *and* K, *and C is monotone with respect to B.*

Then the operator $A + C$ *is near* $B + \alpha C$ *with the same constants* α *and* K.

We will need also the following Miranda–Talenti estimate (cf. [2] and [1]):

Lemma 3.1. *Let $\Omega \subset \mathbb{R}^2$ be a bounded and open set with C^2-smooth boundary and assume* $\frac{d\theta}{d\varphi} - \kappa \geq 0$.

Then

$$\left\|H(u)\right\|_{L^2(\Omega)} \leq \left\|\Delta u\right\|_{L^2(\Omega)} \qquad (\Delta \equiv \textit{ Laplace's operator})$$

for any $u \in W^{2,2}(\Omega)$ *such that* $\frac{\partial u}{\partial \ell} = const$ *on* $\partial\Omega$.

4. Proofs of the results

At first we observe that Campanato's A) condition ensures that the mapping $\mathcal{A}(x, H(u))$ is near to the Laplacian Δu both considered as mappings from

$$\mathcal{B} = \overline{\left\{u \in C^2(\overline{\Omega}) : \ \frac{\partial u}{\partial \ell} = \text{const} \quad \text{on } \partial\Omega\right\}}^{W^{2,2}(\Omega)}$$

into $\mathcal{B}_1 = L^2(\Omega)$. Later, since the problem

$$\begin{cases} \Delta u - \lambda u = f(x) \in L^2(\Omega) & \text{a.e. in } \Omega, \\ \partial u/\partial \ell = 0 & \text{on } \partial\Omega \end{cases}$$

admits a unique solution $u \in W^{2,2}(\Omega)$, it follows that $\Delta u - \lambda u$ is bijective mapping from $\mathcal{B}$ onto $\mathcal{B}_1$. Therefore, in view of Theorem 3.1, to derive unique solvability in $W^{2,2}(\Omega)$ of the problem

$$\begin{cases} \mathcal{A}(x, H(u)) - \lambda u = f(x) & \text{a.e. in } \Omega, \\ \partial u/\partial \ell = 0 & \text{on } \partial\Omega, \end{cases}$$

it is enough to prove that the mapping $\mathcal{A}(x, H(u)) - \lambda u$ is near to the operator $\Delta u - \lambda u$. On the other hand, bearing in mind the nearness of $\mathcal{A}(x, H(u))$ to Δu and Theorem 3.2, we will achieve that $\mathcal{A}(x, H(u)) - \lambda u$ is near to $\Delta u - \alpha\lambda u$ if we just prove that $-\lambda u$ is a monotone operator with respect to Δu.

Thus way, it is enough to prove that

$$-\lambda \int_\Omega u \Delta u \, dx_1 dx_2 \geq 0.$$

For this goal, we make use of the identity

$$u\Delta u = \left[\frac{\partial}{\partial x_1}(u u_{x_1}) + \frac{\partial}{\partial x_2}(u u_{x_2})\right] - \left(u_{x_1}^2 + u_{x_2}^2\right)$$

and the Green–Gauss formula, in order to achieve:

$$\begin{aligned} -\lambda \int_\Omega u \Delta u \, dx_1 dx_2 &= \lambda \int_\Omega \left(u_{x_1}^2 + u_{x_2}^2\right) dx_1 dx_2 - \lambda \int_{\partial\Omega^+} u u_{x_1} dx_2 + \lambda \int_{\partial\Omega^+} u u_{x_2} dx_1 \\ &= \lambda \int_\Omega \left(u_{x_1}^2 + u_{x_2}^2\right) dx_1 dx_2 - \lambda \int_0^L u u_{x_1} X_1 d\varphi - \lambda \int_0^L u u_{x_2} X_2 d\varphi \end{aligned}$$

(here $\partial\Omega^+$ is the contour $\partial\Omega$ runed in the positive direction).

Let us consider the term

$$-\lambda\int_0^L u\left[u_{x_1}X_1+u_{x_2}X_2\right]d\varphi.$$

We have on $\partial\Omega$ the system

$$\begin{cases}u_{x_1}Y_1+u_{x_2}Y_2=0\\ u_{x_1}x_1'+u_{x_2}x_2'=du/d\varphi.\end{cases}$$

Recalling the assumption $Y_1x_2'-x_1'Y_2=\cos\theta\neq 0$, we derive

$$u_{x_1}=\frac{\begin{vmatrix}0 & Y_2\\ \frac{du}{d\varphi} & x_2'\end{vmatrix}}{\cos\theta}=-\frac{Y_2}{\cos\theta}\frac{du}{d\varphi}\quad\text{and}\quad u_{x_2}=\frac{\begin{vmatrix}Y_1 & 0\\ x_1' & \frac{du}{d\varphi}\end{vmatrix}}{\cos\theta}=\frac{Y_1}{\cos\theta}\frac{du}{d\varphi}.$$

This way, one obtains

$$\begin{aligned}-\lambda\int_0^L u\left[u_{x_1}X_1+u_{x_2}X_2\right]d\varphi &= -\lambda\int_0^L u\left[-\frac{Y_2X_1}{\cos\theta}+\frac{Y_1X_2}{\cos\theta}\right]\frac{du}{d\varphi}d\varphi\\ &= -\frac{\lambda}{2}\int_0^L\frac{du^2}{d\varphi}\left[-\frac{Y_2X_1}{\cos\theta}+\frac{Y_1X_2}{\cos\theta}\right]d\varphi.\end{aligned}$$

On the other hand, $\sin\theta=Y_2X_1-X_2Y_1$ and therefore

$$-\lambda\int_0^L u\left[u_{x_1}X_1+u_{x_2}X_2\right]d\varphi=\frac{\lambda}{2}\int_0^L\frac{du^2}{d\varphi}\operatorname{tg}\theta d\varphi=\frac{\lambda}{2}\left[u^2\operatorname{tg}\theta\right]\Big|_0^L-\frac{\lambda}{2}\int_0^L\frac{u^2}{\cos^2\theta}\frac{d\theta}{d\varphi}d\varphi.$$

Since $x_1(0)=x_1(L)$, $x_2(0)=x_2(L)$, $Y_1(0)=Y_1(L)$, $Y_2(0)=Y_2(L)$, $x_1'(0)=x_1'(L)$ and $x_2'(0)=x_2'(L)$, the first term above is equal to zero.

Therefore, we obtain

$$-\lambda\int_\Omega u\Delta u dx_1dx_2=\lambda\int_\Omega\left(u_{x_1}^2+u_{x_2}^2\right)dx_1dx_2-\frac{\lambda}{2}\int_0^L\frac{u^2}{\cos^2\theta}\frac{d\theta}{d\varphi}d\varphi.$$

Since we have assumed $\dfrac{d\theta}{d\varphi}\leq 0$, it follows $-\lambda\int_\Omega u\Delta u dx_1dx_2\geq 0$, that is, the operator $-\lambda u$ is really monotone with respect to the Laplacian Δu.

Remembering the observations already done above, we conclude that the problem (1.1) admits a unique solution $u\in W^{2,2}(\Omega)$. This completes the proof of Theorem 2.1. ■

Remark. Let us consider the non-homogeneous problem

$$\begin{cases}\mathcal{A}(x,H(u))-\lambda u=f(x)\in L^2(\Omega) & \text{a.e. in }\Omega,\\ \partial u/\partial\ell=g(x) & \text{on }\partial\Omega\end{cases}\tag{4.1}$$

with $g(x) \in W^{1/2,2}(\partial\Omega)$ (for the definition of the fractional Sobolev space $W^{1/2,2}(\partial\Omega)$ see [9]).

Since the unit vector field ℓ is nowhere tangential to the boundary we have $\cos\theta > 0$, and therefore there exists a function $v \in W^{2,2}(\Omega)$ such that $\partial v/\partial \ell = g(x)$ on $\partial\Omega$. Consider now the problem

$$\text{(4.2)} \qquad \begin{cases} \mathcal{A}(x, H(w) + H(v)) - \lambda w = f(x) + \lambda v \in L^2(\Omega) & \text{a.e. in } \Omega, \\ \partial w/\partial \ell = 0 & \text{on } \partial\Omega. \end{cases}$$

We already know that the operator $\mathcal{A}(x,\xi)$ satisfies Campanato's A) condition, that is, there exist $\alpha,\ \gamma,\ \delta > 0,\ \gamma + \delta < 1$ such that for each $\xi,\ \tau' \in \mathbb{R}^4$ and for almost all $x \in \Omega$ one has

$$\left|\sum_{i=1}^{2} \xi_{ii} - \alpha\left[\mathcal{A}(x,\xi+\tau') - \mathcal{A}(x,\tau')\right]\right| \le \gamma \|\xi\|_4 + \delta \left|\sum_{i=1}^{2} \xi_{ii}\right|.$$

Therefore, choosing $\tau' = \tau + H(v)$, $\tau \in \mathbb{R}^4$, we derive

$$\left|\sum_{i=1}^{2} \xi_{ii} - \alpha\left[\mathcal{A}(x,\xi+\tau+H(v)) - \mathcal{A}(x,\tau+H(v))\right]\right| \le \gamma \|\xi\|_4 + \delta \left|\sum_{i=1}^{2} \xi_{ii}\right|,$$

which means that the mapping $\mathcal{A}(x,\xi + H(v))$ also satisfies the condition A). Now, in view of Theorem 2.1, we conclude that there exists a unique solution $w \in W^{2,2}(\Omega)$ of the problem (4.2). Setting $u = w + v$, we obtain that it belongs to $W^{2,2}(\Omega)$ and solves the problem

$$\begin{cases} \mathcal{A}(x, H(u)) - \lambda u = f(x) \in L^2(\Omega) & \text{a.e. in } \Omega, \\ \partial u/\partial \ell = \partial v/\partial \ell & \text{on } \partial\Omega, \end{cases}$$

that is, the problem (4.1).

Proof of Theorem 2.2. Consider the problem

$$\text{(4.3)} \qquad \begin{cases} \mathcal{A}(x, H(u)) - \lambda u = f(x) \in L^q(\Omega) & \text{a.e. in } \Omega, \\ \partial u/\partial \ell = 0 & \text{on } \partial\Omega. \end{cases}$$

Since $f \in L^q(\Omega)$ and $q > 2$, it follows $f \in L^2(\Omega)$ and Theorem 2.1 asserts that there exists a unique solution $u \in W^{2,2}(\Omega)$ of the problem (4.3). Then, by virtue of the Sobolev embedding theorem, we have $\lambda u + f \in L^q(\Omega)$, $q > 2$. Rewriting (4.3) as

$$\begin{cases} \mathcal{A}(x, H(u)) = f(x) + \lambda u(x) \equiv f^*(x) \in L^q(\Omega) & \text{a.e. in } \Omega, \\ \partial u/\partial \ell = 0 & \text{on } \partial\Omega, \end{cases}$$

Theorem 2.2 of [10] asserts that there exists a number $q_0 > 2$ such that for each $q \in [2, q_0)$ the solution u belongs to $W^{2,q}(\Omega)$. This proves Theorem 2.2. ■

References

1. G. Talenti, Boll. Un. Mat. Ital. (3), 22, 505–526 (1967).

2. C. Miranda, Ann. Mat. Pura Appl., 63, 353–386 (1963).

3. G. Di Fazio and D.K. Palagachev, Comment. Math. Univ. Carolinae, 37, 537–556 (1996).

4. G. Di Fazio and D.K. Palagachev, Bull. Austral. Math. Soc., 53, 501–513 (1996).

5. A. Maugeri and D.K. Palagachev, Forum Math., 9, (1997), (in print).

6. D.K. Palagachev, Le Matematiche, XLVIII, 311–321 (1993).

7. L. Softova, Z. Anal. Anwend., 17, 1–9 (1998).

8. S. Campanato, Ann. Mat. Pura Appl., 167, 243–256 (1994).

9. R. Adams, Sobolev Spaces, Academic Press, New York (1975).

10. S. Giuffrè, Comm. Appl. Anal., 2, 585–594 (1998).

11. M. Chicco, Rend. Ist. Mat. Univ. Trieste, 4, 85–94 (1972).

8th Int. Coll. on Differential Equations, pp. 195–202
D. Bainov (Ed)

Special Functions in Fractional Relaxation-Oscillation and Fractional Diffusion-Wave Phenomena

RUDOLF GORENFLO
Department of Mathematics and Computer Science, Free University of Berlin, Arnimallee 2-6, D-14195 Berlin, Germany
gorenflo@math.fu-berlin.de

FRANCESCO MAINARDI
Department of Physics, University of Bologna, Via Irnerio 46, I-40126 Bologna, Italy
mainardi@bo.infn.it

HARI M. SRIVASTAVA
Department of Mathematics and Statistics, University of Victoria, Victoria, British Columbia V8W 3P4, Canada
hmsri@uvvm.uvic.ca

Abstract: The basic processes of relaxation, diffusion, oscillations, and wave propagation have been generalized by several authors by introducing fractional derivatives in the governing (ordinary or partial) differential equations. This leads to superslow or intermediate processes that, in mathematical physics, we may refer to as fractional phenomena. Our analysis of these phenomena, carried out by means of fractional calculus and Laplace transforms, leads to certain special functions in one variable of Mittag-Leffler and Wright type. In this note these special functions are investigated systematically as relevant cases of the general class of H-functions.
Key Words and Phrases: *Fractional derivatives, Laplace transforms, Mittag-Leffler functions, Green's functions, Wright functions, Fox H-functions.*
1991 Mathematics Subject Classification. Primary 26A33, 33E20, 45J05, 70J99; Secondary 33C60.

1. Introduction and Preliminaries

In recent years considerable interest has been shown in the so-called *fractional calculus*, which allows one to consider integration and differentiation of any order, not necessarily integer. Such interest has been stimulated by the applications that this calculus finds in different areas of physics and engineering, possibly including fractal phenomena.

It is the purpose of this presentation to provide an approach to fractional calculus that is as simple as possible. Here we adopt the following definition for the *fractional derivative of order* $\alpha > 0$ of a causal function $f(t)$ (*i.e.*, $f(t) = 0$ for $t < 0$):

$$\frac{d^\alpha}{dt^\alpha} f(t) := \begin{cases} f^{(n)}(t) & \text{if} \quad \alpha = n \in \mathbb{N}_0 := \mathbb{N} \cup \{0\} \quad (\mathbb{N} := \{1,2,3,\cdots\}), \\ \dfrac{1}{\Gamma(n-\alpha)} \displaystyle\int_0^t \frac{f^{(n)}(\tau)}{(t-\tau)^{\alpha+1-n}}\, d\tau & \text{if} \quad n-1 < \alpha < n \quad (n \in \mathbb{N}), \end{cases} \tag{1}$$

where $f^{(n)}(t)$ denotes the usual derivative of order n and Γ is the Gamma function. We apply the above notion in order to generalize some basic topics of classical mathematical physics, which are treated by simple, linear, ordinary or partial, differential equations.

We start by considering the *fractional* ordinary differential equation:

$$\frac{d^\alpha u}{dt^\alpha} + \omega^\alpha u(t;\alpha) = 0 \qquad 0 < \alpha \le 2, \tag{2}$$

where ω is a positive constant with the dimension T^{-1} as a frequency, and the field variable $u(t;\alpha)$ is assumed to be a causal function of time. According to the cases $0 < \alpha \le 1$ and $1 < \alpha \le 2$, the equation (2) can be referred to as the *fractional relaxation* or the *fractional oscillation* equation, respectively.

We also consider the following *fractional* partial differential equation:

$$\frac{\partial^{2\beta} u}{\partial t^{2\beta}} = D\,\frac{\partial^2 u}{\partial x^2} \qquad (-\infty < x < \infty;\ 0 < \beta \le 1), \tag{3}$$

where D denotes a positive constant with the dimension $L^2\,T^{-2\beta}$, and the field variable $u = u(x,t;\beta)$ is assumed to be a causal function of time with $u(\mp\infty,t;\beta) = 0$. It should be noted that here, for the sake of convenience, we agree to denote the order of the time derivative as 2β rather than α. As before, we consider two cases, namely $0 < \beta \le \frac{1}{2}$ and $\frac{1}{2} < \beta \le 1$, for which the equation (3) can be referred to as the *fractional diffusion* or the *fractional wave-propagation* equation, respectively.

The solutions of the fractional differential equations (2) and (3) will be obtained by applying the classical technique involving the Laplace transform. In this respect, using the following notation for the Laplace transform pairs:

$$f(t) \div \tilde{f}(s) = \mathcal{L}\{f(t) : s\} := \int_0^\infty e^{-st} f(t)\,dt \qquad \left(t \in \mathbb{R}^+;\ \ s \in \mathbb{C}\right), \tag{4}$$

we recall here the Caputo rule [1]:

$$\mathcal{L}\left\{\frac{d^\alpha}{dt^\alpha} f(t) : s\right\} = s^\alpha \tilde{f}(s) - \sum_{k=0}^{n-1} s^{\alpha-1-k} f^{(k)}(0+) \qquad (n-1 < \alpha \le n). \tag{5}$$

An extensive treatment of fractional calculus and its applications may be found in the works of (for example) Samko *et al.* [2] (see also Miller and Ross [3]). For more details about the results presented here, one can refer to the earlier works [4]–[11] (see also [12] for a similar application of fractional derivatives).

2. The Fractional Relaxation-Oscillation Equation

According to our plan outlined in Section 1, we begin by considering the *fractional* ordinary differential equation (2) distinguishing the cases $0 < \alpha \le 1$ (*fractional relaxation*) and $1 < \alpha \le 2$ (*fractional oscillation*). In the former case, Equation (2) must be equipped with a single initial condition, say $u(0+;\alpha) = u_0$, and in the latter with two initial conditions, say $u(0+;\alpha) = u_0$ and $\dot{u}(0+;\alpha) = v_0$. We assume that $v_0 \equiv 0$ in order to ensure the continuous dependence of the solution of (2) on the parameter α also in the transition from $\alpha = 1-$ to $\alpha = 1+$.

The solution of (2) can be obtained by applying the Laplace transform technique and using the rule (5) with $n = 1, 2$; in both cases, since $v_0 \equiv 0$, the image solution turns out to be

$$\tilde{u}(s;\alpha) = u_0\,\frac{s^{\alpha-1}}{s^\alpha + \omega^\alpha}. \tag{6}$$

In order to invert (6), we can use the series expansion theorem or the complex Bromwich formula; we thus obtain

$$u(t;\alpha) = u_0 \, E_\alpha[-(\omega t)^\alpha], \tag{7}$$

where E_α denotes the Mittag-Leffler function [13, Chapter 18]. In the complex z-plane this function is defined by the following series and integral representations:

$$E_\alpha(z) := \sum_{n=0}^{\infty} \frac{z^n}{\Gamma(\alpha n+1)} = \frac{1}{2\pi i}\int_{Ha} \frac{\zeta^{\alpha-1} e^\zeta}{\zeta^\alpha - z}\, d\zeta \qquad (\alpha>0;\;\; z\in\mathbb{C}), \tag{8}$$

where Ha denotes the Hankel path, a loop which starts from $-\infty$ along the lower side of the negative real axis, encircles the circular disk $|\zeta| \le |z|^{1/\alpha}$ in the positive sense and ends at $-\infty$ along the upper side of the negative real axis.

From the series representation in (8) we can easily recognize that $E_\alpha(z)$ is an *entire function* (of order $1/\alpha$), which provides a simple generalization of the exponential function ($\alpha = 1$).

The most interesting properties of the Mittag-Leffler function $E_\alpha(z)$ are associated with its asymptotic expansions as $z \to \infty$ in various sectors of the complex z-plane and with its connection with the Laplace integral [13, Chapter 18]. For the present purposes, our interest in this function is limited to negative argument and $0 < \alpha \le 2$. Taking, for the sake of simplicity, $\omega \equiv 1$, and putting

$$e_\alpha(t) := E_\alpha(-t^\alpha) = \sum_{n=0}^{\infty} (-1)^n \frac{t^{\alpha n}}{\Gamma(\alpha n+1)}, \tag{9}$$

the main properties are given below:

(i) For $0 < \alpha \le 1$, $e_\alpha(t)$ turns out to be *positive* and *completely monotonic*, *i.e.*,

$$(-1)^n \frac{d^n}{dt^n} e_\alpha(t) > 0 \qquad (t > 0;\;\; 0 < \alpha \le 1). \tag{10}$$

(ii) For $1 < \alpha < 2$, $e_\alpha(t)$ exhibits an odd numbers of zeros, while for $\alpha = 2$, $e_2(t) = \cos t$ exhibits an infinite number of zeros.

(iii) For $0 < \alpha < 1$ and $1 < \alpha < 2$, the following asymptotic expansion is valid:

$$e_\alpha(t) \sim -\sum_{n=1}^{\infty} (-1)^n \frac{t^{-\alpha n}}{\Gamma(1-\alpha n)} \qquad (t\to\infty;\;\; 0<\alpha<1;\;\; 1<\alpha<2). \tag{11}$$

Other relevant properties of $e_\alpha(t)$ can be derived in detail for $0 < \alpha \le 2$, directly from its representation as a Laplace inversion integral (Bromwich formula), without detour on the general theory of Mittag-Leffler functions in the complex plane. The essential step consists in decomposing $e_\alpha(t)$ into two parts according to $e_\alpha(t) = f_\alpha(t) + g_\alpha(t)$, as indicated below. The part $f_\alpha(t)$, present for α not an integer, turns out to be a *completely monotonic* function, which tends to zero as t tends to infinity; the other part $g_\alpha(t)$ turns out to be of *oscillatory* character if $1 < \alpha < 2$, but identically vanishing if $0 < \alpha < 1$.

We separately discuss the cases: **(a)** $0 < \alpha < 1$, **(b)** $1 < \alpha < 2$, recalling that for $\alpha = 1, 2$ we know that $e_\alpha(t)$ as elementary function, namely, $e_1(t) = \exp(-t)$, $e_2(t) = \cos t$. In order to obtain the desired decomposition of $e_\alpha(t)$, we bend the Bromwich path of integration Br into the equivalent Hankel path $Ha(1+)$, a loop which starts from $-\infty$ along the lower side of the negative real axis, encircles the circular disk $|s| = 1$ in the positive sense and ends at $-\infty$ along the upper side of the negative real axis. One obtains, in both cases **(a)** and **(b)**, $e_\alpha(t) = f_\alpha(t) + g_\alpha(t)$ $(t \geq 0)$ with

$$f_\alpha(t) := \frac{1}{2\pi i} \int_{Ha(\epsilon)} e^{st} \frac{s^{\alpha-1}}{s^\alpha + 1} \, ds, \tag{12}$$

where now the Hankel path $Ha(\epsilon)$ denotes a loop constituted by a small circle $|s| = \epsilon$ with $\epsilon \to 0$ and by the two borders of the cut negative real axis, and

$$g_\alpha(t) := \sum_h e^{s'_h t} \operatorname{Res} \left[\frac{s^{\alpha-1}}{s^\alpha + 1} \right]_{s'_h}, \tag{13}$$

where s'_h are the relevant poles of $s^{\alpha-1}/(s^\alpha + 1)$. In fact, the poles turn out to be $s_h = \exp[i(2h+1)\pi/\alpha]$ with unitary modulus; they are all simple, but the relevant ones are only those situated in the main Riemann sheet, *i.e.*, the poles s'_h for which $-\pi < \arg(s'_h) < \pi$.

If $0 < \alpha < 1$, there are no such poles, since for all integers h we have $|\arg(s_h)| = |2h + 1|\pi/\alpha > \pi$; as a consequence, $g_\alpha(t) \equiv 0$; hence $e_\alpha(t) = f_\alpha(t)$ if $0 < \alpha < 1$.

If $1 < \alpha < 2$, then there exist precisely two relevant poles, namely $s'_0 = \exp(i\pi/\alpha)$ and $s'_{-1} = \exp(-i\pi/\alpha) = \bar{s}'_0$, and one obtains

$$g_\alpha(t) = \frac{2}{\alpha} e^{t\cos(\pi/\alpha)} \cos\left[t \sin\left(\frac{\pi}{\alpha}\right)\right] \quad \text{if} \quad 1 < \alpha < 2. \tag{14}$$

We note that this function exhibits oscillations with circular frequency $\sin(\pi/\alpha)$ and with an amplitude which exponentially decays with the rate $|\cos(\pi/\alpha)|$. Of course, in the classical case $\alpha = 2$, we recover the sinusoidal behaviour with unitary frequency.

It is now an exercise in complex analysis to show that the contribution from the Hankel path $Ha(\epsilon)$ as $\epsilon \to 0$ is provided by

$$f_\alpha(t) := \int_0^\infty e^{-rt} K_\alpha(r) \, dr \quad \text{with} \quad K_\alpha(r) = \frac{1}{\pi} \frac{r^{\alpha-1} \sin(\alpha\pi)}{r^{2\alpha} + 2r^\alpha \cos(\alpha\pi) + 1}. \tag{15}$$

This function $K_\alpha(r)$ vanishes identically if α is an integer, it is positive for all r if $0 < \alpha < 1$, negative for all r if $1 < \alpha < 2$. In fact, in (15) the denominator is, for α not an integer, always positive, being $> (r^\alpha - 1)^2 \geq 0$. Hence $f_\alpha(t)$ in Case **(a)** and $-f_\alpha(t)$ in Case **(b)** have the aforementioned property of complete monotonicity, decreasing to zero for $t \to \infty$. We also note that, in order to satisfy the initial condition $e_\alpha(0+) = 1$, we find that

$$\int_0^\infty K_\alpha(r) \, dr = \begin{cases} 1 & \text{if} \quad 0 < \alpha < 1, \\ 1 - \dfrac{2}{\alpha} & \text{if} \quad 1 < \alpha < 2. \end{cases} \tag{16}$$

3. The Fractional Diffusion-Wave Equation

Let us now consider the *fractional* partial differential equation (3) distinguishing the cases $0 < \beta \leq \frac{1}{2}$ (*fractional diffusion*) and $\frac{1}{2} < \beta \leq 1$ (*fractional wave*). In the former case, Equation (3) must be equipped with a single initial condition, say $u(x, 0+; \beta) = f(x)$, while in the latter case with two initial conditions, say $u(x, 0+; \beta) = f(x)$ and $\dot{u}(x, 0+; \beta) = g(x)$. We assume that $g(x) \equiv 0$ in order to ensure the continuous dependence of the solution of (3) on the parameter β also in the transition from $\beta = \frac{1}{2}-$ to $\beta = \frac{1}{2}+$.

Let us now solve the above initial-value problem (or *Cauchy problem*) recurring to the *Green function* $\mathcal{G}_c(x,t;\beta)$ which represents the solution for $f(x) = \delta(x)$. This method allows one to obtain the required solution in the form:

$$u(x,t;\beta) = \int_{-\infty}^{\infty} \mathcal{G}_c(\xi,t;\beta)\, f(x-\xi)\, d\xi. \tag{17}$$

It should be noted that $\mathcal{G}_c(x,t;\beta) = \mathcal{G}_c(|x|,t;\beta)$, since the Green function turns out to be an even function of x.

The Green function $\mathcal{G}_c(x,t;\beta)$ can be determined by using the technique of the Laplace transform. This technique allows us to obtain the transformed function $\widetilde{\mathcal{G}_c}(x,s;\beta)$ by solving the following non-homogeneous differential equation of the second order in x:

$$s^{2\beta}\,\widetilde{\mathcal{G}_c} - \delta(x)\, s^{2\beta-1} = D\,\frac{d^2\,\widetilde{\mathcal{G}_c}}{dx^2}, \tag{18}$$

which eventually yields

$$\widetilde{\mathcal{G}_c}(x,s;\beta) = \frac{1}{2\,\sqrt{D}\,s^{1-\beta}}\, e^{-(|x|/\sqrt{D})\, s^{\beta}}. \tag{19}$$

We are going to show how the inversion of the Laplace transform in (20) leads us to the *auxiliary function* of the *similarity variable*

$$z = \frac{|x|}{\sqrt{D}\, t^{\beta}} \quad \text{when} \quad 0 < \beta < 1. \tag{20}$$

In fact, applying the inverse Bromwich formula to (20), introducing the *similarity variable* z as in (20), and putting $\sigma = st$, we just obtain

$$|x|\mathcal{G}_c(x,t;\beta) = \frac{z}{2} M(z;\beta) \text{ with } M(z;\beta) := \frac{1}{2\pi i}\int_{Br} e^{\sigma - z\sigma^{\beta}}\,\frac{d\sigma}{\sigma^{1-\beta}} \;\; (0<\beta<1), \tag{21}$$

where $M(z;\beta)$ is our *auxiliary function* and Br denotes the Bromwich path. The above definition of $M(z;\beta)$ (*Bromwich representation*) can be analytically continued from $z > 0$ to $\forall z \in \mathbb{C}$, adopting suitable integral and series representations valid in all of $\mathbb{C}$. We thus finally obtain the following *series representation*:

$$M(z;\beta) := \sum_{n=0}^{\infty} \frac{(-z)^n}{n!\,\Gamma[-\beta n + (1-\beta)]} \qquad (0 < \beta < 1). \tag{22}$$

Since the radius of convergence of the power series in (22) can be proven to be infinite for $0 < \beta < 1$, our function turns out to be entire in z.

In the theory of special functions (see, *e.g.*, [13, Chapter 18]), we find an entire function of z, referred to as the *Wright function*, which reads (in our notation)

$$W(z;\lambda,\mu) = \sum_{n=0}^{\infty} \frac{z^n}{n!\,\Gamma(\lambda n + \mu)} := \frac{1}{2\pi i}\int_{Ha} e^{\sigma - z\sigma^{-\lambda}}\,\frac{d\sigma}{\sigma^{\mu}}, \tag{23}$$

where $\lambda > -1$ and $\mu > 0$. We thus recognize that $M(z;\beta) = W(-z;-\beta,\, 1-\beta)$.

We can prove that for $\beta = \frac{1}{q}$, where $q \geq 2$ is a positive integer, the auxiliary function can be expressed as a sum of $q-1$ simpler entire functions. In the particular cases when $q = 2$ and $q = 3$, we find that

$$M\left(z;\tfrac{1}{2}\right) = \frac{1}{\sqrt{\pi}} \sum_{m=0}^{\infty} (-1)^m \left(\frac{1}{2}\right)_m \frac{z^{2m}}{(2m)!} = \frac{1}{\sqrt{\pi}} \exp\left(-\frac{z^2}{4}\right) \tag{24}$$

and

$$\begin{aligned} M\left(z;\tfrac{1}{3}\right) &= \frac{1}{\Gamma\left(\frac{2}{3}\right)}\sum_{m=0}^{\infty}\left(\frac{1}{3}\right)_m \frac{z^{3m}}{(3m)!} - \frac{1}{\Gamma\left(\frac{1}{3}\right)}\sum_{m=0}^{\infty}\left(\frac{2}{3}\right)_m \frac{z^{3m+1}}{(3m+1)!} \\ &= 3^{\frac{2}{3}}\,\mathrm{Ai}\left(\frac{z}{3^{\frac{1}{3}}}\right), \end{aligned} \tag{25}$$

where $(a)_m := a(a+1)\cdots(a+m-1)$ denotes the Pochhammer symbol and Ai denotes the *Airy function*. Thus we recognize that, for $\beta = \frac{1}{2}$, the *Gaussian auxiliary function* of the classical diffusion equation is recovered. Furthermore, always for $\beta = \frac{1}{q}$, it can be proved that our function satisfies the following differential equation of order $q-1$:

$$\frac{d^{q-1}}{dz^{q-1}} M\left(z;\tfrac{1}{q}\right) + \frac{(-1)^q}{q}\, z\, M\left(z;\tfrac{1}{q}\right) = 0, \tag{26}$$

subject to the following $q-1$ initial conditions at $z = 0$, derived from (22):

$$M^{(h)}(0;\beta) = \frac{(-1)^h}{\pi}\Gamma[\beta(h+1)]\sin[\pi\beta(h+1)] \qquad (h = 0, 1, \cdots q-2). \tag{27}$$

We note that, for $q \geq 4$, Equation (26) is akin to the *hyper-Airy differential equation* of order $q-1$. Consequently, in view of the above considerations, the auxiliary function can be considered as a *generalized hyper-Airy function.*

For the fractional diffusion-wave equation (3), only real positive values of z are of interest. As a matter of fact, it turns out that, for $z \in \mathbb{R}^+$, the function $M(z;\beta)$ is positive; in particular, it is *monotonically decreasing* for $0 < \beta < \frac{1}{2}$, while for $\frac{1}{2} < \beta < 1$ it is *unimodal*, *i.e.*, it exhibits only one maximum, $M_0(\beta)$, at a certain point $z_0(\beta)$. It should be noted that, as $\beta \to 1-$, $M_0(\beta) \to \infty$ and $z_0(\beta) \to 1$. The case $\beta = 1$, for which (3) reduces to the classical wave equation, is of course singular, since $M(z;1) \propto \delta(z-1)$.

4. Relationships with the Functions of Fox and Wright

The Mittag-Leffler function $E_\alpha(z)$ defined by (8), which appears in our solution (7), and the function $M(z;\beta)$ defined by (22), which is involved in our solution (17), are related to the Wright function ${}_p\Psi_q$ defined by (*cf.*, *e.g.*, Srivastava *et al.* [14, Chapter 1])

$$\begin{aligned} {}_p\Psi_q \left[\begin{matrix}(\alpha_1, A_1), \cdots, (\alpha_p, A_p); \\ (\beta_1, B_1), \cdots, (\beta_q, B_q);\end{matrix}\ z\right] &:= \sum_{n=0}^{\infty} \frac{\prod_{j=1}^{p}\Gamma(\alpha_j + A_j n)}{\prod_{j=1}^{q}\Gamma(\beta_j + B_j n)}\frac{z^n}{n!} \\ &= H^{1,p}_{p,q+1}\left[-z \,\middle|\, \begin{matrix}(1-\alpha_1, A_1), \cdots, (1-\alpha_p, A_p) \\ (0,1),\ (1-\beta_1, B_1), \cdots (1-\beta_q, B_q)\end{matrix}\right], \end{aligned} \tag{28}$$

where $H^{m,n}_{p,q}$ denotes Fox's H-function defined by [*loc. cit.*]

$$\begin{aligned} &H^{m,n}_{p,q}\left[z \,\middle|\, \begin{matrix}(\alpha_1, A_1), \cdots, (\alpha_p, A_p) \\ (\beta_1, B_1), \cdots, (\beta_q, B_q)\end{matrix}\right] \\ &= \frac{1}{2\pi i}\int_C \frac{\prod_{j=1}^{m}\Gamma(\beta_j - B_j\zeta)\ \prod_{j=1}^{n}\Gamma(1-\alpha_j + A_j\zeta)}{\prod_{j=m+1}^{q}\Gamma(1-\beta_j + B_j\zeta)\ \prod_{j=n+1}^{p}\Gamma(\alpha_j - A_j\zeta)}\, z^\zeta\, d\zeta \end{aligned} \tag{29}$$

$$\left(\alpha_j, \beta_k \in \mathbb{C};\quad A_j, B_k \in \mathbb{R}^+;\quad j = 1, \cdots, p;\quad k = 1, \cdots, q\right),$$

$\mathcal{C}$ being a suitable contour of the Mellin-Barnes type (which, for example, runs from $-i\infty$ to $i\infty$ such that the poles of $\Gamma(\beta_j - B_j\zeta)$ $(j = 1, \cdots, m)$ lie to the right of $\mathcal{C}$ and the poles of $\Gamma(1 - \alpha_j + A_j\zeta)$ $(j = 1, \cdots, n)$ lie to the left of $\mathcal{C}$, *et cetera*). In fact, we have

$$E_\alpha(z) = {}_1\Psi_1 \left[\begin{matrix} (1,1); \\ (1,\alpha); \end{matrix} \, z \right] = H^{1,1}_{1,2} \left[-z \,\middle|\, \begin{matrix} (0,1) \\ (0,1), \ (0,\alpha) \end{matrix} \right] \tag{30}$$

and

$$M(z; -\beta) = {}_0\Psi_1 \left[\begin{matrix} \overline{\hspace{2em}}; \\ (1-\beta, \beta); \end{matrix} \, -z \right] = H^{1,0}_{0,2} \left[z \,\middle|\, \begin{matrix} \overline{\hspace{2em}} \\ (0,1), (\beta,\beta) \end{matrix} \right]. \tag{31}$$

Making use of these last relationships (30) and (31), it is not difficult to apply various known properties of the aforementioned functions of Fox and Wright in order to deduce the corresponding properties of the functions $E_\alpha(z)$ and $M(z;\beta)$ of interest in our present investigation (see also [12]). For example, we have the following results giving the images of the functions $E_\alpha(z)$ and $M(z;\beta)$ in the Laplace transform defined by (4) and the Mellin transform defined by

$$\mathcal{M}\{f(t) : s\} := \int_0^\infty t^{s-1} f(t)\, dt \qquad (t \in \mathbb{R}^+;\ s \in \mathbb{C}), \tag{32}$$

provided that the integral exists (see, for details, Srivastava *et al.* [14, p. 15 *et seq.*]):

$$\mathcal{L}\left\{t^{\lambda-1} E_\alpha(-zt^\mu) : s\right\} = s^{-\lambda} H^{1,2}_{2,2} \left[zs^{-\mu} \,\middle|\, \begin{matrix} (1-\lambda, \mu), \ (0,1) \\ (0,1), \ (0,\alpha) \end{matrix} \right] \tag{33}$$

$$(\Re(s) > 0; \quad \Re(\lambda) > 0; \quad \mu > 0);$$

$$\mathcal{L}\left\{t^{\lambda-1} M(zt^\mu; -\beta) : s\right\} = s^{-\lambda} H^{1,1}_{1,2} \left[zs^{-\mu} \,\middle|\, \begin{matrix} (1-\lambda, \mu) \\ (0,1), \ (\beta,\beta) \end{matrix} \right] \tag{34}$$

$$(\Re(s) > 0; \quad \Re(\lambda) > 0; \quad \mu > 0);$$

$$\mathcal{M}\{E_\alpha(-zt) : s\} = z^{-s} \frac{\Gamma(s)\Gamma(1+s)}{\Gamma(1-\alpha s)} \qquad (\Re(s) > 0); \tag{35}$$

$$\mathcal{M}\{M(zt; -\beta) : s\} = z^{-s} \frac{\Gamma(s)}{\Gamma(1-\beta-\beta s)} \qquad (\Re(s) > 0). \tag{36}$$

In each of our results (33) to (36), we tacitly assume that the various conditions corresponding appropriately to the existence and convergence of the Mellin-Barnes contour integral in (29), which defines the H-function, are satisfied.

Numerous further properties and characteristics of the functions $E_\alpha(z)$ and $M(z;\beta)$ can indeed be derived in the manner which we have illustrated here.

Acknowledgements

The present investigation was supported, in part, by the Research Commission of the Free University of Berlin, CNR and INFN (sez. Bologna), DAAD (Deutscher Akademischer Austausch–Dienst), and the Natural Sciences and Engineering Research Council of Canada.

References

1. M. Caputo, Elasticità e Dissipazione, Zanichelli, Bologna (1969).

2. S.G. Samko, A.A. Kilbas and O.I. Marichev, Fractional Integrals and Derivatives: Theory and Applications, Gordon and Breach, Amsterdam (1993). [English Translation from the Russian edition, Nauka i Tekhnika, Minsk, 1987]

3. K.S. Miller and B. Ross, An Introduction to the Fractional Calculus and Fractional Differential Equations, John Wiley and Sons, New York (1993).

4. R. Gorenflo and F. Mainardi, in: Recent Advances in Applied Mathematics, (Proceedings of an International Workshop held at Kuwait University on May 4-7, 1996), pp. 193-208, Kuwait Foundation for the Advancement of Science, State of Kuwait (1996).

5. R. Gorenflo and F. Mainardi, in: Fractals and Fractional Calculus in Continuum Mechanics, (A. Carpinteri and F. Mainardi, Editors), pp. 223-276. Springer-Verlag, Vienna and New York (1997). [Lecture Notes of the Advanced School held at CISM, Udine, Italy, September 23-27, 1996]

6. R. Gorenflo and R. Rutman, in: Transform Methods and Special Functions (Bankya [Sofia], August 12-17, 1994) (P. Rusev, I. Dimovski, and V. Kiryakova. Editors), pp. 61-81, Science Culture Technology Publishing Company, Singapore (1995).

7. F. Mainardi, Chaos, Solitons and Fractals, 7, 1461-1477 (1996).

8. F. Mainardi, Appl. Math. Lett., 9, 23-28 (1996).

9. F. Mainardi, in: Proceedings of the Seventh International Colloquium on Differential Equations, (Plovdiv, Bulgaria, August 18-23, 1996) (D. Bainov. Editor), pp. 221-229, VSP, Utrecht (1997).

10. F. Mainardi, in: Fractals and Fractional Calculus in Continuum Mechanics, (A. Carpinteri and F. Mainardi, Editors), pp. 291-348, Springer-Verlag. Vienna and New York (1997). [Lecture Notes of the Advanced School held at CISM. Udine, Italy, September 23-27, 1996]

11. F. Mainardi and M. Tomirotti, in: Transform Methods and Special Functions (Bankya [Sofia], August 12-17, 1994) (P. Rusev, I. Dimovski, and V. Kiryakova, Editors), pp. 171-183, Science Culture Technology Publishing Company, Singapore (1995).

12. W.R. Schneider and W. Wyss, J. Math. Phys., 30, 134-144 (1989).

13. A. Erdélyi, W. Magnus, F. Oberhettinger and F.G. Tricomi, Higher Transcendental Functions, Vol. 3, McGraw-Hill, New York (1953).

14. H.M. Srivastava, K.C. Gupta, and S.P. Goyal, The H-Functions of One and Two Variables with Applications, South Asian Publishers, New Delhi and Madras (1982).

15. W.R. Schneider, Exposition. Math., 14, 3-16 (1996).

8th Int. Coll. on Differential Equations, pp. 203-210
D. Bainov (Ed)

Mathematical Modelling of Immunization Programs for Hepatitis A in Bulgaria

DAVID GREENHALGH
Department of Statistics and Modelling Science, University of Strathclyde, Livingstone Tower, 26 Richmond Street, Glasgow G1 1XH, U.K.

Abstract. This paper is concerned with evaluating vaccination programs for infectious diseases. An age-structured model is used as such models have been shown to fit the data better. First of all we describe previous work on age-structured models for common childhood diseases. We then describe our basic model. The spread of the disease under a specified immunization program is described by two coupled partial integro-differential equations with appropriate boundary conditions. We then proceed to discuss our theoretical results for this model. A key parameter is R_ϕ, the reproduction number of the disease under a given vaccination program ϕ. We expect the disease to take off if R_ϕ exceeds one and die out if R_ϕ is less than one. We use this result to evaluate vaccination programs under different mixing assumptions, in particular homogeneous, proportional, associative and symmetric mixing. Our theoretical results show that R_ϕ is maximised by associative mixing, so that more vaccination is required with associative mixing than with any other mixing assumption.

A numerical application of this procedure to evaluate hepatitis A vaccination in Bulgaria under different mixing assumptions is discussed. We use pre-vaccination serological data on hepatitis A in Bulgaria and estimate the minimum vaccination proportions necessary to eradicate the disease with one stage and two stage vaccination policies.

Keywords: Disease, age-structure, partial differential equations, who acquires infection from whom matrix, threshold value, vaccination, hepatitis A.

1. BACKGROUND

This paper is concerned with the evaluation of vaccination programs for diseases spread by direct contact, such as measles, mumps, rubella and hepatitis A. For these diseases a typical individual starts off susceptible, at some stage catches the disease and after a short infectious period becomes permanently immune. Of these diseases measles has previously been the subject of much mathematical modelling attention. Measles is spread by airborne transmission and has an incubation period of 7-18 days and an infectious period of 7-10 days [1]. Measles is very infectious and confers permanent immunity. In the UK it is a notifiable disease and as a consequence the data are very good. A measles vaccine is available which is roughly 95% effective and the current UK policy is to vaccinate all children at 15 months of age with a combined measles, mumps and rubella vaccine. It is not advantageous to vaccinate children much earlier than this as they are then protected by maternal antibodies.

In contrast hepatitis A, which is the disease mainly considered in this paper, is spread by the oral-faecal route. The main modes of transmission are contaminated water supplies and food handling. The incubation period is between 15 and 50 days and the infectious period between 7 and 25 days. The pattern differs between countries. Hepatitis A is most prevalent in Southern European countries such as Greece and Bulgaria, rather than northern countries such as Scandinavia. Vaccines against hepatitis A are available but until recently there was no widespread vaccination against hepatitis A in Bulgaria.

The simplest models assume that the population amongst whom the disease is spreading mix homogeneously. However age-structured models have been shown to fit the data better. In addition vaccination programs are often age-dependent. So in this paper we consider an age-structured model. There are two main sources of data. First, case reports of diseases are often collated by a central source. However these have been shown to suffer from age-related reporting bias. A better source of data is age-serological profiles in which a large sample of the population are tested for antibodies to the disease. A key parameter in our model is the age-dependent force of infection, denoted by $\lambda(a,t)$. This denotes the rate of attack of a single susceptible individual of age a at time t.

The earliest work is due to Collins [2] who examines age-related case incidence reports for common childhood diseases. Meunch [3] fitted a catalytic model using a constant force of infection. This means that $p(a)$, the proportion of individuals who have experienced the disease by age a, is

$$1 - e^{-\lambda a}.$$

This assumes that $p(a)$ has settled down to its equilibrium value. Later Griffiths [4] considered a catalytic type curve with an age-related force of infection so

$$p(a) = 1 - e^{-\int_0^a \lambda(\xi)d\xi}.$$

He fitted a linear force of infection in the age range 0-10 years. This was a better fit to the data than a constant force of infection.

2. BASIC MODEL.

Our basic model is described by Dietz and Schenzle [5]. We denote by $x(a,t)$ the density with respect to age of the number of susceptibles of age a at time t. This means that the number of susceptibles between ages A_1 and A_2 at time t is given by

$$\int_{A_1}^{A_2} x(a,t)da.$$

Similarly $y(a,t,c)$ denotes the density with respect to age a and duration of infectiousness c of the number of infectives at time t. The following set of differential equations describes the progress of the disease:

$$\frac{\partial x}{\partial a}+\frac{\partial x}{\partial t} = -[\lambda(a,t)+\phi(a)+\mu(a)]x(a,t), \tag{1}$$

and
$$\frac{\partial y}{\partial a}+\frac{\partial y}{\partial t}+\frac{\partial y}{\partial c} = -[\gamma(c)+\mu(a)]y(a,t,c),$$

with boundary conditions $x(0,t) = \nu$, $y(0,t,c) = 0$ and $y(a,t,0) = \lambda(a,t)x(a,t)$.

Here $\phi(a)$ is the rate at which susceptibles of age a are vaccinated and $\mu(a)$ is the per capita death rate of individuals of age a (assumed the same for susceptible, infectious and immune individuals). N denotes the total population size and ν the total birth rate. $\gamma(c)$ denotes the rate at which an infectious individual who has had the disease for time c becomes immune. $\kappa b(a,a')/N$ denotes the total rate at which an infected individual of age a' transmits infection to a susceptible of age a. Often this function is expressed as a matrix by dividing the age range up into discrete age classes. This matrix is called the who-acquires-infection-from-whom (WAIFW) matrix. Equations (1) are explained more fully in Dietz and Schenzle [5].

Let $\pi(a)$ be the fraction of individuals surviving to age a. Then

$$\pi(a) = exp\left\{-\int_0^a \mu(\xi)d\xi\right\}.$$

For the total population size to remain constant we must have

$$N = \nu \int_0^L \pi(a)da.$$

The force of infection at time t is defined as

$$\lambda(a,t) = \frac{\kappa}{N}\int_0^L \int_0^{a'} b(a,a')f(c)y(a',t,c)dcda'.$$

Here $f(c)$ denotes the probability that an individual is infectious if he or she has had the disease for time c.

Anderson and May [6-8] have previously discussed vaccination policies against rubella and measles, where the transmission coefficient $b(a,a')$ is age-dependent. They look at different configurations for the WAIFW matrix. Case reports or an age-serological profile were used to estimate the force of infection and the matrix of transmission coefficients. Simulation was used to evaluate vaccination programs for measles. This work has been very influential in the design of UK vaccination programs.

3. THEORETICAL RESULTS

A key parameter is the reproductive number. The reproductive number with steady state vaccination campaign ϕ is defined to be the expected number of secondary cases of disease produced by a single infectious case entering a disease-free population at equilibrium with steady-state vaccination campaign ϕ. For a simple homogeneously mixing model where $b(a,a')$ is independent of a and a' (respectively the ages of the susceptible and infected individuals between whom transmission takes place) and

$$\frac{\kappa b(a,a')}{N} = \beta$$

we have R_0, the basic reproductive number in the absence of vaccination, is given by

$$R_0 = \frac{\beta N}{\mu + \gamma} = \beta\tau_0 N,$$

where $\tau_0 = (\gamma+\mu)^{-1}$ is the average infectious period.

For the two age-class model suppose that the age-range is divided into two disjoint age classes I_1 and I_2, and the function $b(a,a')$ is expressed as a matrix

$$\begin{pmatrix} \beta_{11} & \beta_{12} \\ \beta_{21} & \beta_{22} \end{pmatrix}$$

with $b(a,a') = \beta_{ij}$ if $a \in I_i$ and $a' \in I_j$, $i,j = 1,2$. Suppose that $D = \int_0^T f(c)dc$, the average 'effective', infectious period and N_i is the number of susceptible individuals in age-class i at the disease-free equilibrium. Define

$$\hat{\beta}_{ij} = \frac{\kappa\beta_{ij}}{N}DN_i.$$

$\hat{\beta}_{ij}$ is the expected number of cases in class i produced by one infected class j individual. Then the matrix

$$\begin{pmatrix} \hat{\beta}_{11} & \hat{\beta}_{12} \\ \hat{\beta}_{21} & \hat{\beta}_{22} \end{pmatrix}$$

has spectral radius R_ϕ. (Recall that the spectral radius of a finite dimensional matrix **A** is the largest modulus of the eigenvalues of **A**.)

We expect that if $R_\phi > 1$ then the disease takes off, whereas if $R_\phi < 1$ then the disease dies out. This result will be used to evaluate vaccination programs. We proceed as follows: Given the age-serological profile of the population in the absence of vaccination we use this to estimate $\hat{\lambda}_0(a)$. Suppose that the age range is divided up into n disjoint age classes $I_1, I_2, \ldots I_n$ and

$$\lambda_i = \int_{I_i} \lambda(\xi) d\xi$$

and A_i is the length of the interval I_i for $i = 1, 2 \ldots n$. We define

$$\psi_j = \int_{I_j} \lambda(\xi) d\xi, \quad \Phi(\xi) = \int_0^\xi \phi(\xi) d\xi \quad \text{and} \quad M(\xi) = \int_0^\xi \mu(\xi) d\xi.$$

Then we can use the equations

$$\lambda_i = \sum_{j=1}^{n} \nu \frac{A_i}{N_i} \hat{\beta}_{ij} \psi_j \tag{2}$$

to estimate the quantities $\hat{\beta}_{ij}$, and hence the quantities β_{ij}.

Unfortunately this causes us a problem as it gives n equations in n^2 unknowns and we need to make additional assumptions about the relationships between the β_{ij}. When we have estimated the β_{ij} we can calculate R_ϕ with a given vaccination program ϕ and thus determine whether this vaccination program is sufficient to eliminate the disease.

4. MIXING ASSUMPTIONS

To reduce the equations (2) to n equations in n unknowns, and hence solve them we must make some assumptions about the mixing of the population amongst whom the disease is spreading. We discuss four types of assumption: homogeneous, proportional, associative and symmetric mixing. We have already briefly discussed homogeneous mixing. Proportional mixing assumes that the rate at which a person in group i meets (and transmits disease to) a person in group j is proportional to $b_i b_j$, where $b_1, b_2, \ldots b_n$ are specified constants. This may be suitable if there is a communal meeting place and b_i is the proportion of his or her time that person i spends in the meeting place.

Associative mixing assumes that people mix only with their own age group. Whilst this is a rather extreme assumption it does serve as a benchmark with which to compare other mixing patterns. Also as in practice as much of the spread of common childhood diseases such as measles, rubella and mumps takes place in schools we expect within age group contact rates to be higher than between age group contact rates.

Symmetric mixing means that the contact matrix is symmetric so $\beta_{ij} = \beta_{ji}$ for each i and j. This is often assumed in practice. Note that the total number of contacts between any two groups must be symmetric. For example the total number of contacts of adults with children must equal the total number of contacts of children with adults. Thus symmetric mixing is appropriate if direct contacts between individuals correspond closely to transmission of disease. This is possibly true for measles, mumps and rubella. The procedure for evaluating vaccination strategies under the symmetric mixing assumption is to assume several elements of the contact matrix equal on biological grounds to reduce the number of unknowns to n. We then solve equations (2) for β_{ij}. If the resulting estimates for β_{ij} are all positive we can then proceed to evaluate vaccination strategies. If some are negative the biological assumptions are infeasible and alternative ones must be explored.

The theoretical results which we have obtained can be summarised as follows: Suppose that we are given $\lambda_0(a)$, the force of infection in the absence of vaccination. Then for a fixed vaccination campaign ϕ, R_ϕ is maximised by associative mixing. Moreover more vaccination is required with associative mixing than with any other mixing assumption. R_ϕ is minimised if everyone can catch the disease but it is spread only by those in the youngest age-class. For symmetric mixing

$$R_\phi \geq \left\{\frac{\int_0^L \lambda(\xi)^2 e^{-\Phi(\xi)-M(\xi)} d\xi}{\int_0^L \lambda(\xi) e^{-2\Lambda_0(\xi)-M(\xi)+\Phi(\xi)} d\xi}\right\}^{\frac{1}{2}}.$$

5. APPLICATION TO HEPATITIS A IN BULGARIA

First we outline how $\lambda_0(a)$ is estimated from the age-serological profile. This method is due to Keiding [9]. We suppose that we are given data $(Z_1, \delta_1), (Z_2, \delta_2), \ldots (Z_n, \delta_n)$, ordered by age, where Z_i is the current age of person i and

$$\delta_i = \begin{cases} 1 & \text{if } i \text{ is immunised} \\ 0 & \text{otherwise.} \end{cases}$$

So if Y_i denotes the age of immunisation of i, $\delta_i = I(Z_i > Y_i)$, where $I(E)$ is the standard indicator function of the event E,

$$I(E) = \begin{cases} 1 & \text{if E happens} \\ 0 & \text{otherwise.} \end{cases}$$

We want to find $G(y) = 1 - e^{-\Lambda_0(y)}$. The non-parametric maximum likelihood estimator $\hat{G}$ of G is defined as follows: Let $H(i) = \sum_{j=1}^{i} \delta_i$. Recall that the convex minorant of H is the largest convex function which is everywhere less than H. We define $\hat{G}(z_i)$ to be the left continuous derivative of the convex minorant of H at i, which is a step function. We then use $\hat{G}(z)$ to estimate λ using kernel smoothing by

$$\hat{\lambda}_0(a) = \frac{1}{b}\int_{a-b}^{a+b} K\left(\frac{x-a}{b}\right)\frac{d\hat{G}(x)}{1-\hat{G}(x^-)}.$$

Kernel smoothing has been used to give a smooth estimate. Here K is a symmetric probability density, called the kernel, and b is the bandwidth. Similar estimates can be used to evaluate the mixing matrix and different vaccination strategies. We chose the Epanechnikov kernel, $K(x) = 0.75(1 - x^2)$ on $[-1, 1]$, and handled tail problems by using a truncated Epanechnikov kernel. By trial and error we found that a bandwidth of 15 years was best. In our numerical results we estimated the proportions necessary to be vaccinated to eradicate hepatitis A in Bulgaria using one and two stage vaccination strategies. Five years is a convenient age for the second vaccination as most children start school then. The results show that a wide variation in the minimum vaccination proportions can be obtained by using different mixing assumptions.

For associative mixing the population is divided into four age classes: 0-4 years, 5-14 years, 15-34 years and 35-87 years. These age classes correspond roughly to pre-school children, school-children, young adults with children and older adults. For symmetric mixing the population is divided into the same age-classes as before. Two symmetric configurations of the mixing matrix were examined:

$$\text{I}\quad \begin{pmatrix} b_2 & b_2 & b_3 & b_4 \\ b_2 & b_1 & b_3 & b_4 \\ b_3 & b_3 & b_3 & b_4 \\ b_4 & b_4 & b_4 & b_4 \end{pmatrix} \quad \text{and} \quad \text{II} \quad \begin{pmatrix} b_1 & b_1 & b_1 & b_1 \\ b_1 & b_2 & b_2 & b_2 \\ b_1 & b_2 & b_3 & b_4 \\ b_1 & b_2 & b_4 & b_4 \end{pmatrix}.$$

Configuration I is infeasible but for Configuration II all of the coefficients are positive so it is feasible. From Keiding's age-structured serological data we estimated the values $\bar{\lambda}_1 = 0.034$, $\bar{\lambda}_2 = 0.037$, $\bar{\lambda}_3 = 0.052$ and $\bar{\lambda}_4 = 0.045$. This data was originally collected by Professor G. Frösner, Münich, from Bulgaria in 1964. We found that the values of b_i were in the ratio 3.84 : 4.18 : 7.79 : 5.74. This means that the transmission rates are highest amongst young adults (not schoolchildren) and thus suggests a different mode of transmission than measles, mumps and rubella. Configuration I is the form of transmission matrix suitable for measles, mumps and rubella and the fact that it turns out to be infeasible is also indicative of a different transmission mode. In fact measles, mumps and rubella are all spread via aerosol transmission whereas hepatitis A is spread by the oral-faecal route.

Table I. Elimination vaccination proportions using a one stage vaccination policy under proportional and symmetric mixing.

a_0(years)	Proportional %	Symmetric %
0.0	75.8	73.4
5.0	78.5	76.1
10.0	81.4	79.9
15.0	85.5	84.5
20.0	90.9	93.5
25.0	99.4	****

Table II. Elimination vaccination vaccination proportions using a one stage policy under associative mixing.

a_0(years)	Associative %
0.0	90.5
1.0	90.5
2.0	90.5
3.0	90.5
4.0	90.5
4.3	90.5
4.4	96.3

Table III Elimination vaccination proportions using a two stage vaccination policy under the different mixing assumptions.

$A_1 = 2$ %	$A_2 = 5$ Proportional %	$A_2 = 5$ Symmetric %	$A_2 = 5$ Associative %
0	78.5	76.1	****
10	75.9	73.1	****
20	72.6	69.6	88.1
30	68.4	64.9	86.4
40	62.8	58.6	84.2
50	55.0	49.9	80.1
60	43.2	36.8	76.2
70	23.6	15.0	68.3
80	****	****	52.5
90	****	****	4.5

Table I shows the fraction of the population who need to be vaccinated at a given age a_0 to eliminate hepatitis A in Bulgaria using a one-stage vaccination policy under the proportional and symmetric mixing assumptions. The results are very similar. The asterisks in the symmetric mixing column corresponding to $a_0 = 25$ years indicate that with this assumption it is not possible to achieve elimination by making a single vaccination at $a_0 = 25$ years. Table II shows the corresponding vaccination proportions for associative mixing. Note that these are much higher than the corresponding proportions for both symmetric and proportional mixing and indeed for a_0 greater than 4.38 years it was not possible to achieve elimination with a single vaccination.

Table III shows the fraction of the population who require to be vaccinated using a two stage vaccination policy with vaccinations at ages $A_1 = 2$ and $A_2 = 5$ years. With 0% or 10% vaccinated at age two it is not possible to achieve elimination of hepatitis A under the associative mixing assumption. Similarly with 80% or 90% of children vaccinated at age two hepatitis A will always be eliminated under the proportional and symmetric mixing assumptions, whatever the proportions vaccinated at age five. Note that the elimination proportions requiring vaccination under associative mixing are higher than those under proportional or symmetric mixing as predicted by our theoretical results.

6. SUMMARY

In this paper we have examined an age-structured mathematical model for the evaluation of vaccination campaigns under different mixing assumptions. Such models are suitable for diseases such as measles, mumps, rubella and hepatitis A. We concentrated particularly on associative, proportional and symmetric mixing. A key parameter was the basic reproduction number R_ϕ. We expect the disease to take off if R_ϕ exceeds one and die out if R_ϕ is less than one. We used this result to evaluate vaccination campaigns. Our theoretical results show that for a given vaccination campaign ϕ, R_ϕ is maximised under associative mixing and more vaccination effort is required under this mixing assumption than any other. Finally these results were illustrated with reference to serological data on hepatitis A in Bulgaria. These results are discussed in more detail in Greenhalgh and Dietz [10].

References

1. A.S. Beneson, Control of Communicable Diseases in Man, (Sixteenth Edition), American Public Health Association, Washington D.C., (1990).
2. S.D. Collins, U.S. Public Health Rep., 44, 763-828 (1929).
3. H. Meunch, J. Amer. Statist. Assoc., 29,25-38 (1934).
4. D. Griffiths, Appl. Statist., 23, 330-339 (1974).
5. K. Dietz and D. Schenzle, J. Math. Biol., 22, 117-120 (1985).
6. R.M. Anderson and R.M. May, J. Hygiene, 90, 259-325 (1983).
7. R.M. Anderson and R.M. May, IMA J. Math. Appl. Med. Biol., 1, 233-266 (1984).
8. R.M. Anderson and R.M. May, J. Hygiene, 94, 365-436 (1985).
9. N. Keiding, J. Roy. Statist. Soc. Ser. A, 154, 371-412 (1991).

10. D. Greenhalgh and K. Dietz, Math. Biosci., 124, 9-57 (1994).

8th Int. Coll. on Differential Equations, pp. 211-218
D. Bainov (Ed)

Complete-Integrable Calogero Dynamics for Random Matrix Theories

HIROSHI HASEGAWA
Research Center of Quantum Communications, Tamagawa University, Machida, Tokyo 194, Japan

Abstract. We discuss a matrix formalism of the Lax type for the Calogero-Euler system to establish the full equivalence between the matrix form and the Hamiltonian form of the equations of motion, proving the complete integrability in the sense of solvability by quadrature for a given initial condition.
Keywords: random matrix theories (RMT), level dynamics, symplectic structure, complete integrability

1. Introduction

Random matrix theories (RMT) were initiated around 1962 by Dyson and Mehta after some conceptual development of Wigner's idea to describe complex quantum levels by a statistical means [1],[2]. It's starting point of the formulation is to set up a concrete expression for the probability density function $P_N(x_1, x_2, .., x_N)$ that each x_i of N eigenvalues of an $N \times N$ Hermitian matrix lies in an interval $[x_i, x_i + dx_i)$, $i = 1, 2, ..N$, can be written as $P_N(x_1, x_2, .., x_N)dx_1 dx_2..dx_N$. Two basic hypotheses for determining the function, called *N-level joint distribution*, at the initial stage were the following.

(A) The pertinent Hermitian matrix H is regarded as an arbitrary sample of an ensemble of such matrices characterized only by the symmetry with respect to time-reversal, for which a probability $P(H)dH$ can be defined invariantly of the representation basis, under the specified symmetry, of the H.
(B) There exists a set of elementary sample Hermitians, denoted by $\{H_\alpha\}$ in which H_α's are statistically independent of each other, and any sample H can be expanded such that $H = \sum_\alpha H_\alpha$.

The above two requirements suggest the form of $P_N(H)$ and, consequently, the distribution(probability density function) of the eigenvalues $\{x_i\}_{i=1}^N$ of H, $P_N(x_1, .., x_N)$, through a change of the basis in which H is represented as diagonal. Namely,

$$P(H)dH = Ce^{-\frac{1}{2\sigma^2}\mathrm{Tr}H^2} \prod_\alpha dH_\alpha \quad \alpha = (m,n)_{m\leq n} \text{ and } \nu \text{ (to specify symmetry)} \tag{1.1}$$

$$P_N(x_1, .., x_N)dx_1..dx_N = C_N e^{-\frac{1}{2\sigma^2}(x_1^2+..+x_N^2)} \prod_{i<j} |x_i - x_j|^\nu dx_1..dx_N \tag{1.2}$$

where C and C_N are normalization constants to satisfy $\int PdH = 1$ and $\int P_N(x)dx = 1$, respectively (depending on the symmetry type ν): the extra index ν specifies that the H, an $N \times N$ Hermitian, belongs to

$\nu = 1$ real symmetric type (Gaussian orthogonal ensemble, GOE)

$\nu = 2$ complex hermitian type (Gaussian unitary ensemble, GUE)

$\nu = 4$ quaternion real type (Gaussian symplectic ensemble, GSpE).

It is easily seen that form (1.1) satisfies the two requirements such that

$$\text{(A)} \qquad P(H) = P(U^*HU) \qquad \textit{(unitary invariance)}$$

$$\text{(B)} \quad P(H) = C\prod_m e^{-\frac{1}{2\sigma^2}H_{mm}^2} \prod_{a=1}^{\nu} \prod_{m<n} e^{-\frac{1}{\sigma^2}H_{mn}^{(a)2}} \qquad \textit{(statistical independence)}$$

(Index a in $H_{mn}^{(a)}$ specifies components of H_{mn}: for GUE *real* and *imaginary* part, and for GSpE) four components of a quaternion.

Also in (A), $dH (= \prod_\alpha dH_\alpha) = d(U^*HU)$ holds unless any constraint is imposed other than the symmetry preserving unitarity on the new basis associated to the U. On the other hand, if a new basis is chosen as the N eigenvectors of H so that the U diagonalizes H, a complex factor arises as the Jacobian function associated to $dH = J\prod dx_i \times$ every other $d\phi$'s (ϕ stands for all supplementary coordinates, hidden in (1.2), to x_i's):

$$J = \text{const.} \prod_{i<j} |x_i - x_j|^\nu \tag{1.3}$$

which is called *level repulsion factor.* This function is of fundamental significance in the original RMT, as it provides all the origin of correlations between (quantum) levels, and will be shown to originate from the Riemannian metric tensor for the above coordinate transformation, i.e.

$$dH \longrightarrow (dx_1, .., dx_N; d\phi) \qquad \text{defined in} \qquad \mathcal{U}^*H\mathcal{U} = \text{diag}(x_1 .., x_N). \tag{1.4}$$

It can be observed that the Gaussian trace form for $P(H)$,(1.1), automatically satisfies another condition of equal probability that all the elementary sample matrices $H_\alpha; \alpha = (m < n)$ occur with equal probability, reflecting the uniformity in correlation strength for any pair of eigenvalues chosen from $\{x_i\}$, as manifest in the level repulsion factor (1.3). This is a rather uneasy shortcoming of the classical RMT which makes its applicability narrow.

We aim to reconstruct RMT which is based on hypothsis (A) and (B) but not necessarily with the equal probability principle stated above. To outline the idea, let us put the diagonalization process implied in the latter part of eq.(1.4) in perturbation theory:

$$\mathcal{U}^*H\mathcal{U} = \text{diag}(x_1, .., x_N), \qquad H = H_0 + \lambda V \tag{1.5}$$

where H_0 and V are two fixed Hermitians and λ the parameter of perturbation strength. Then, each of eigenvalues and eigenvectors is an analytic function of λ so that

$$x_i = x_i(\lambda), \ i = 1, .., N; \qquad \mathcal{U} = \mathcal{U}(\lambda). \tag{1.6}$$

Our first task is to write down a deterministic set of ordinary differential equations to pursue the variation of x_i with λ that is regarded as the time variable of dynamics.

Then, two questions arise as to whether this set can be put into a Hamiltonian equation of motion, and as to the nature of the initial-value problem of the differential equation i.e. whether any solution of the problem can be regarded as the diagonalization process of a certain Hermitian of the form (1.5) with H_0 and V prescribed by the only knowledge of the initial data. We answer the questions affirmatively by means of the Calogero-Euler dynamics [3]. For the associated subject of physics (quantum chaos), see Hasegawa [4].

2. Differential equations in level dynamics

It is straightforward to write down the aimed differential equations, starting from the eigenvalue equation $(H_0 + \lambda V)u_i = x_i u_i$ and by taking inner products with other eigenvectors [5]:

$$\frac{dx_i}{d\lambda} = V_{ii}, \qquad \frac{dV_{ii}}{d\lambda} = 2\sum_{j\neq i}\frac{V_{ij}V_{ji}}{x_i - x_j} \qquad (V_{ij}^* = V_{ji})$$

$$\frac{dV_{ij}}{d\lambda} = \sum_{k\neq i,j} V_{ik}V_{kj}\left(\frac{1}{x_i - x_k} + \frac{1}{x_j - x_k}\right) - \frac{V_{ij}(V_{ii} - V_{jj})}{x_i - x_j}, \quad i \neq j.$$

Thus, a closed set of differential equations for $2N + \frac{\nu}{2}N(N-1)$ variables has been obtained; the first N for $\{x_i\}$, and the rest $N + \frac{\nu}{2}N(N-1)$ for the matrix elements of V, the latter number representing the degree of freedom f of the system, provided the system meets the condition of Hamiltonian dynamics. One of the conditions can be expressed by the Liouville theorem that a Hamiltonian flow is divergenceless, that is, when the equations of motion are expressed as $\dot{y}_i = f_i(y)$, then $\sum_i \partial f_i/\partial y_i = 0$ must hold, but it is not the case for the above third equation. However, the situation is changed by adopting a set of new variables

$$f_{ij} \equiv (x_i - x_j)V_{ij} \qquad (f_{ij}^* = -f_{ji}) \tag{2.1}$$

making the above set of equations simplified:

$$\frac{dx_i}{d\lambda} = V_{ii}, \quad \frac{dV_{ii}}{d\lambda} = -2\sum_{j\neq i}\frac{f_{ij}f_{ji}}{(x_i - x_j)^3} \qquad \text{(Calogero type equation of motion)} \tag{2.2}$$

$$\frac{df_{ij}}{d\lambda} = \sum_{k\neq i,j} f_{ik}f_{kj}\left[\frac{1}{(x_i - x_k)^2} - \frac{1}{(x_j - x_k)^2}\right] \qquad \text{(Euler equation of rotation).} \tag{2.3}$$

Now, the dynamical flows associated to V_{ii} and f_{ij} besides x_i are clealy divergenceless, as can be seen from the fact that each right hand side of $\dot{y}$ in the above three equations does not contain y.

3. Hamiltonian formalism

Let us first introduce a Riemannian metric on matrix spaces by defining the distance between two infinitesimally different matrices X and $X + dX$ in terms of the Hilbert-Schmidt norm, i.e.

$$ds^2 = \|dX\|^2 \equiv \mathrm{Tr}\, dX^* dX. \tag{3.1}$$

(This is not a unique definition of metric on matrix spaces which is representation invariant: if a density matrix is allowed to enter the metric which changes covariantly with the representation basis, a variety of metrics may arise. However, it is unique as far as such a density matrix is not allowed [6].) We seek the actual metric tensor for a given Hermitian and its diagonalizing transformation (1.4), namely for $dH =$

$$\mathcal{U}dH_D\mathcal{U}^* + d\mathcal{U}H_D\mathcal{U}^* + \mathcal{U}H_Dd\mathcal{U}^* = \mathcal{U}(dH_D + [\mathcal{U}^*d\mathcal{U}, H_D])\mathcal{U}^* \qquad (d\mathcal{U}^*\mathcal{U} = -\mathcal{U}^*d\mathcal{U}) \quad (3.2)$$

where

$$H_D \equiv \mathrm{diag}(x_1, x_2, .., x_N). \tag{3.3}$$

Inserting this into (3.1), we obtain

$$ds^2 = \mathrm{Tr}dH_D^2 + \mathrm{Tr}[\mathcal{U}^*d\mathcal{U}, H_D]^2 + 2\mathrm{Tr}\mathcal{U}^*d\mathcal{U}[H_D, dH_D].$$

Here, we impose the condition of *simultaneous diagonality* on the differential dH_D, i.e.

$$[H_D, dH_D] = 0, \qquad \text{and} \quad dH_D = \mathrm{diag}(dx_1, dx_2, .., dx_N), \tag{3.4}$$

and then

$$ds^2 = \mathrm{Tr}dH_D^2 + \mathrm{Tr}[H_D, \mathcal{U}^*d\mathcal{U}]^2. \tag{3.5}$$

Hence we have two expressions of the metric of dH in the original fixed representation (called *fixed frame*) and the H-diagonal representation (called *moving frame*):

$$ds^2 = \sum_n (dH_{nn})^2 + 2\sum_{m<n} |dH_{mn}|^2 \qquad \text{(fixed frame)} \tag{3.6}$$

$$= \sum_{i=1} dx_i^2 + 2\sum_{i<j}(x_i - x_j)^2|\Omega_{ij}|^2 \qquad \text{(moving frame)} \tag{3.7}$$

where

$$\Omega \equiv \mathcal{U}^*d\mathcal{U} = -d\mathcal{U}^*\mathcal{U} \tag{3.8}$$

is a matrix-valued (differential) 1-form. Both (3.6) and (3.7) express the Riemannian metric in the respective frame as an $f \times f$ diagonal tensor, and from expression (3.7) the Jacobian function (1.3) is assured.

The Hamiltonian formalism of the present dynamical system may be established by the standard means of a Lagrangean based on the metric (3.7), i.e.

$$T = \frac{1}{2}\sum_i\left(\frac{dx_i}{d\lambda}\right)^2 + \sum_{i<j}(x_i - x_j)^2|A_{ij}|^2 \qquad A_{ij} \equiv \Omega_{ij}/d\lambda \qquad \text{(angular velocity)}. \tag{3.9}$$

Note that Ω can not be a differential of any (matrix-valued) function (see Maurer-Cartan equation below).

conjugate momentum In the fixed frame, the λ- dependence is linear, shown in $H_0 + \lambda V$, for which the matrix velocity is just equal to V. In the moving frame, a complication arises due to the unitary evolution so that

$$P_\alpha = P_{nn}\delta_{\alpha,nn} + (1 - \delta_{\alpha,nn})\frac{P_{mn}}{2}, \qquad \alpha = mn\,(m \leq n);$$

$$P_{nn} = V_{nn},\ P_{mn} = 2V_{mn}\ (m < n) \qquad \text{(fixed frame)} \tag{3.10}$$

$$p_i = \frac{dx_i}{d\lambda}, \qquad M_{ij} = -2(x_i - x_j)^2 A_{ij} \qquad \text{(moving frame)}. \tag{3.11}$$

In general, $2f$ coordinates form a smooth manifold with exterior derivatives, called *symplectic manifold* $\mathcal{M}^{2f}$, where symplectic differential forms d's with symbol $\wedge$ play a basic role [7].

symplectic 1-form $\omega^{(1)} = \sum p_\alpha dq_\alpha$

$$\omega^{(1)} = \mathrm{Tr} V dH = \sum P_{nn} dH_{nn} + \sum_{m<n} P_{mn} dH_{mn} \qquad \text{(fixed frame)} \qquad (3.12)$$

$$= \sum p_i dx_i + \sum_{i<j} M_{ij}\Omega_{ij} \qquad \text{(moving frame)} \qquad (3.13)$$

symplectic 2-form $\omega^{(2)} = \sum dp_\alpha \wedge dq_\alpha = d\omega^{(1)}$

$$(dx \wedge dy = -dy \wedge dx,\ dx \wedge dx = 0,\ (dx \wedge dy) \wedge dz = dx \wedge (dy \wedge dz),\ d(dx) = 0)$$

This is the canonical 2-form by which the canonical Poisson bracket (P.b.) relations are endowed to the $2f$ coordinates, making them canonical coordinates. One of major subjects here is to establish the P.b. relations involving the f noncanonical coordinates $\{M_\alpha\}$, each M_α being conjugate to Ω_α. Our answer can be found in the following Lemma.

Lemma *Let G be the adjoint representation of a Lie algebra on a matrix space, generated by the basis set $\{E_\alpha\}(\Omega = \sum_\alpha \Omega_\alpha E_\alpha)$ satisfying the structure relations*

$$[E_\alpha, E_\beta] = \sum_{\gamma=1}^{f} c^{\gamma}_{\alpha,\beta} E_\gamma. \qquad \alpha, \beta = 1, .., f. \qquad (3.14)$$

There exists an operation $\{\,,\,\}$ on the symplectic manifold $\mathcal{M}^f$; $C^\infty(\mathcal{M}^f) \times C^\infty(\mathcal{M}^f) \longrightarrow C^\infty(\mathcal{M}^f)$, called Berezin *bracket for F and $G \in C^\infty(\mathcal{M}^f)$, given by*

$$\{F, G\} = -\sum_{\alpha,\beta,\gamma} c^{\gamma}_{\alpha,\beta} M_\gamma \frac{\partial F}{M_\alpha}\frac{\partial G}{M_\beta}. \qquad \textit{in particular,} \quad \{M_\alpha, M_\beta\} = -\sum_\gamma c^{\gamma}_{\alpha,\beta} M_\gamma. \quad (3.15)$$

It provides the coadjoint representation G^* of G .

Note that the Berezin bracket [8] is an example of more general form of binary operations $\{\,,\,\}$ on $\mathcal{M}^{2f}$, $\omega^{ij}\partial_i F \partial_j G$ (in the tensor notation) with a skew symmetric, nondegenerate $2f \times 2f$ tensor ω^{ij} ;

$$(\omega^{ij}) = (\omega_{ij})^{-1} \quad \text{where} \quad \omega_{ij} = \text{coefficient tensor of } \omega^{(2)};\ \ d\omega^{(2)} = 0\ (\omega^{(2)} \text{ closed}). \quad (3.16)$$

The closedness of $\omega^{(2)}$ is equivalent to the Jacobi identity satisfied by the operation [8].

proof of Lemma This is by computing $\omega^{(2)} = d\omega^{(1)}$ given by eq.(3.13) in the moving frame: **symplectic 2-form in the moving frame**

$$d\left(\sum_i p_i dx_i + \sum_{\alpha=(i<j)} M_\alpha \Omega_\alpha\right) = \sum_i dp_i \wedge dx_i + \sum_\alpha (dM_\alpha \wedge \Omega_\alpha + M_\alpha d\Omega_\alpha). \qquad (3.17)$$

But we have $d\Omega = -\Omega \wedge \Omega$ which is called Maurer-Cartan equation [9], as can be derived from the definition (3.8) i.e. $d\Omega = -\mathcal{U}^*(d\mathcal{U} \wedge \Omega) = -(\mathcal{U}^* d\mathcal{U}) \wedge \Omega = -\Omega \wedge \Omega$. Hence in matrix forms,

$$(\omega_{ij}) = \begin{pmatrix} 0 & -I \\ I & 0 \end{pmatrix} \bigoplus \begin{pmatrix} C & -I \\ I & 0 \end{pmatrix}$$

a direct sum of two symplectic matrices, $f(=N)$ dimensional elementary one for the vector (dx, dp) and $f(= \frac{1}{2}N(N-1))$ dimensional non-elementary one for the vector (Ω, dM). The inverse of the above matrix yields

$$(\omega^{ij}) = \begin{pmatrix} 0 & I \\ -I & 0 \end{pmatrix} \bigoplus \begin{pmatrix} 0 & I \\ -I & C \end{pmatrix}$$

providing the coefficients of the Berezin bracket from the nonvanishing C part: namely, by inserting $\Omega = \sum \Omega_\alpha E_\alpha$ into the matrix 2-form $\Omega \wedge \Omega$ to get $\mathrm{Tr} M \Omega \wedge \Omega = \frac{1}{2} \sum_{\alpha,\beta} \Omega_\alpha \wedge \Omega_\beta \mathrm{Tr} M [E_\alpha, E_\beta]$, and by virtue of eq.(3.14) the coefficient of $\Omega_\alpha \wedge \Omega_\beta$ in $\mathrm{Tr} M d\Omega$ is equal to $-\sum_\gamma c^\gamma_{\alpha,\beta} M_\gamma$, where $M = \sum M_\gamma E_\gamma$. □

<u>**Hamiltonian function**</u> We have made a natural introduction of the angular momentum M as the conjugate momentum to the angular velocity, as in (3.11): in matrix form

$$M = -2[H_D, [H_D, A]]. \qquad \text{(moving frame)} \qquad (3.18)$$

One can adopt another definition of the angular momentum by omitting the factor 2, thus

$$F = \frac{1}{2} M = -[H_D, [H_D, A]] \equiv (f_{ij}), \qquad \mathrm{Tr} F A = \mathrm{Tr}[H_D, A][H_D, A]. \qquad (3.18')$$

This change of definition only results in a factor 2 modification of the Poisson bracket relations, $c^\gamma_{\alpha,\beta} \to \frac{1}{2} c^\gamma_{\alpha,\beta}$. With this remark, we can now write the Hamiltonian function as

$$\mathcal{H} = \frac{1}{2} \sum_i p_i^2 + \frac{1}{2} \sum_{i \neq j} \frac{|f_{ij}|^2}{|x_i - x_j|^2} \qquad (3.19)$$

and P.b.'s: (a) the canonical P.b. for (x,p)'s, (b) coadjoint structure relations for f's and (c) vanishing P. b. between (x,p)'s and f's (mechanical independence of these two modes) by which equations of motion are written as $\dot{y} = \{y, \mathcal{H}\}$ leading to eqs.(2.2) and (2.3).

It is remarked that the vanishing of P.b.'s between (x,p)'s and (f)'s is owing to the above direct-sum decomposition and stems from *simultaneous diagonality* condition (3.4).

4. Matrix formalism and integrating the equations

Let us recall the so-called Lax type representation of the equations of motion and establish its equivalence with the Hamiltonian formalism (here, we use the notation $\dot{y}$ for $dy/d\lambda$.)

I (a) $\dot{X} = [X, A] + V$, (b) $\dot{V} = [V, A]$, (c) $[X, \dot{X}] = 0$

where $X^* = X$, $V^* = V$, and $A^* = -A \in N \times N$ matrix algebra

II $\dot{x}_i = p_i$, $\dot{p}_i = -2 \sum_{j \neq i} \frac{|f_{ij}|^2}{(x_i - x_j)^3}$, $\dot{f}_{ij} = \sum_{k \neq i,j} f_{ik} f_{kj} \left[\frac{1}{(x_i - x_k)^2} - \frac{1}{(x_j - x_k)^2} \right]$

derived from $\dot{y} = \{y, \mathcal{H}\}$ ($\mathcal{H}$ given by (3.19)) by means of P. b.'s

$$(a) \{x_i, p_j\} = \delta_{ij}, \{x_i, x_j\} = \{p_i, p_j\} = 0, \qquad (b) \{f_\alpha, f_\beta\} = -\frac{1}{2} \sum_\gamma c^\gamma_{\alpha,\beta} f_\gamma,$$

$$\text{and} \qquad (c) \{(x,p)'s, f_\alpha\} = 0.$$

Theorem *Given a set of matrix differerntial equations* I*(a) and* I*(b) under constraint* I*(c), then it is possible to set up a Hamiltonian system* II *by identifying*

$$(x_1, x_2, .., x_N) = \mathrm{diag}(X) \equiv \textit{diagonal part of } X, \qquad (p_1, p_2, .., p_N) = \mathrm{diag}(V),$$

$$\text{and} \qquad (f_{ij}) = [X, V].$$

Conversely, a Hamiltonian system II *can be represented by a set of matrix differential equations* I(a) *and* I(b) *under constraint* I(c) *by identifying*

$$X = \text{diag}(x_1, x_2, .., x_N), \quad V = \text{diag}(p_1, p_2, .., p_N) + (\frac{f_{ij}}{x_i - x_j}), \text{ and } (A_{ij}) = (\frac{-f_{ij}}{(x_i - x_j)^2}).$$

proof I $\rightarrow$ II First, we show that the matrix equations of motion I(a) and I(b) yield another equation for the angular momentum

$$\text{I}(d) \qquad F \equiv [X, V] \qquad \dot{F} = [F, A], \tag{4.1}$$

as $\dot{F} = [\dot{X}, V] + [X, \dot{V}] = [[X, A], V] + [X, [V, A]] = [[X, V], A]$ by virtue of Jacobi identity. Next, we consider the matrix equations I$(a) \sim (d)$ in the X-diagonalized representation, which can be performed by means of a unitary transformation $X \rightarrow \bar{X} = \mathcal{U}^* X \mathcal{U}$, $V \rightarrow \bar{V} = \mathcal{U}^* V \mathcal{U}$ and $A \rightarrow \bar{A} = \mathcal{U}^* A \mathcal{U} + \mathcal{U}^* \dot{\mathcal{U}}$. Since the resulting matrix equations assume the same form as I$(a) \sim (c)$(and also (d) by $\bar{F} = \mathcal{U}^* F \mathcal{U}$), we can regard that X and $\dot{X}$ are diagonal from the outset without loss of generality. Then, taking all the matrix elements of four equations I$(a) \sim (c)$ is seen to yield the system II in accordance with Section 3.

II $\rightarrow$ I (a proof based on Wojciechowski [10]) The stated identification of the three matrices X, V and A enables us to rewrite the Hamiltonian equations of motion in II in terms of a single matrix equation I(b) i.e. $\dot{V} = [V, A]$ (its diagonal part for p_i and its off-diagonal part for f_{ij}). In order to see the other two equations I(a) and I(c), we take the P.b.$\{K, \}$ with both sides of the identity $\{V, \mathcal{H}\} = [V, A]$, where $K \equiv \frac{1}{2}\sum x_i^2$, obtaining $\{X, \mathcal{H}\} - V$ for the left side, and $[X, A]$ for the right side. This assures the equation in I(a), and also I(c) because $(p_1, .., p_N) = \text{diag}(V)$ must hold. □

Our argument in the above **proof** provides a means of integrating the matrix equations and hence the equivalent Hamiltonian equations of motion of the present dynamics as the initial value problem. Namely, a change of the representation frame in terms of a unitary transformation generally leads to a covariant form of equations:

$$\dot{\bar{X}} = [\bar{X}, \bar{A}] + \bar{V}, \quad \dot{\bar{V}} = [\bar{V}, \bar{A}]; \qquad \bar{X} = \mathcal{U}^* X \mathcal{U}, \quad \bar{V} = \mathcal{U}^* V \mathcal{U} \tag{4.2}$$

and

$$\bar{A} = \mathcal{U}^* A \mathcal{U} + \mathcal{U}^* \dot{\mathcal{U}}. \tag{4.3}$$

We abandon the diagonality condition, but require the special frame that $\bar{A} = 0$. Such a frame exists, uniquely for a given matrix A, which is predicted by the unitary transformation $\mathcal{U}$ to satisfy

$$\frac{d\mathcal{U}}{d\lambda} = -A\mathcal{U}. \tag{4.4}$$

This is just the fixed frame in Sec.3, where the matrix equations become $\dot{X} = V$, and $\dot{V} = 0$ i.e. the solution

$$X = X^0 + \lambda V^0, \qquad V = V^0 \quad \text{also} \quad F = [X^0 + \lambda V^0, V^0] = [X^0, V^0] \equiv F^0 \tag{4.5}$$

which are related to the initial data of the Hamiltonian equations of motion:

$$(x_1^0, .., x_N^0) = \text{diag}X^0, \quad (p_1^0, .., p_N^0) = \text{diag}V^0, \quad (f_{ij}^0) = F^0. \tag{4.6}$$

The full solution of the initial value problem must be settled by the determination of the unitary transformation $\mathcal{U}$ subject to eq.(4.4), which is available from solutions of the eigenvalue problem

$$(H^0 + \lambda V^0)u_i = x_i u_i. \tag{4.7}$$

References

1. C.E. Porter, in: Statistical Theories of Spectra: Fluctuations, (Ed.) C.E.Porter, Academic Press, New York (1965).

2. M.L. Mehta, Random Matrices, Academic Press, New York (1991).

3. M.A. Olshanetsky and A.M. Perelomov, Phys. Rep., 94, 314-400 (1981).

4. H. Hasegawa, in: Proc. Intern. 3rd Summer School *"Let's face Chaos through Nonlinear Dynamics"*, (Ed.) M. Robnik *Open Systems and Information Dynamics*, (1997), (in press).

5. T. Yukawa, Phys. Lett. A, 116, 227-229 (1986).

6. D. Petz, Linear Alg. Appl., 244, 81-96 (1996).

7. A.T. Fomenko, Symplectic Geometry, Gordon and Breach Publishers, 2nd ed. (1995).

8. A.T. Fomenko and V.V. Trofimov, Integrable Systems on Lie Algebras and Symplectic Spaces, Gordon Breach Publishers (1988).

9. D.H. Sattinger and O.L. Weaver, Lie Groups and Algebras with Application to Physics, Geometry and Mechanics, Chap.7, Springer Verlag, Berlin (1986).

10. S. Wojciechowski, Phys. Lett. A, 111, 101-103 (1985).

8th Int. Coll. on Differential Equations, pp. 219-225
D. Bainov (Ed)

On Optimal Order of Approximation from Bivariate Spline Spaces

DON HONG
Department of Mathematics, East Tennessee State University, Johnson City, TN 37614, USA

Abstract. In this paper, we discuss the optimal order of approximation from bivariate spline spaces over arbitrary triangulations, and some specific triangulations. We also study the construction of optimal triangulation for the purpose that the C^1 quartic spline space over such a triangulation admits the optimal (full) order of approximation. Some most recent results are sketched on these topics.
Keywords: approximation order, bivariate spline, interpolation, optimal triangulation, scattered data representation.

1 Introduction

Let Ω be a connected polygonal domain in $\mathbb{R}^2$ and Δ be a triangulation of Ω. i.e., Δ is a finite collection of triangles τ in $\mathbb{R}^2$ such that $\cup_{\tau\in\Delta}\tau = \Omega$ and the intersection of any two adjacent triangles in Δ is either a common edge or a common vertex. As usual, for any nonnegative integers k and r, we define the multivariate spline space $S_k^r(\Delta)$ to be the space of all piecewise polynomial functions on Δ which are of total degree at most k and are smooth of order r. More precisely, $f \in S_k^r(\Delta)$ means that $f \in C^r(\Delta)$ and f is a polynomial with degree at most k on each triangle of Δ.

In applications of the spline space $S_k^r(\Delta)$, it is critical to answer the question of how well the splines can approximate classes of smooth functions. In this paper, we investigate approximation properties of the bivariate spline space $S_k^r(\Delta)$. We begin with a definition of the approximation order of a function space.

Definition 1 *The approximation order of a space S of functions on $\mathbb{R}^2$ is defined to be the largest real number ρ for which*

$$dist(f,S) \leq Const\|D^{k+1}f\|\ |\Delta|^\rho \tag{1.1}$$

for any sufficiently smooth function f, with the distance measured in the maximum norm $\|\ \|$, and with the mesh size $|\Delta| := \sup_{\tau\in\Delta} \operatorname{diam}\tau$.

It is clear that the approximation order from the spline space $S_k^r(\Delta)$ cannot be better than $k+1$ regardless of r and is trivially $k+1$ in the case $r = 0$. A great deal of work has been done on the approximation order of the bivariate spaces $S_k^r(\Delta)$ in the last decade. (cf.

[1]–[13] and the references therein). In general, it is well known that the approximation order of $S_k^r(\Delta)$ not only depends on k and r, but also on the geometric structure of the partition Δ. According to the results of finite elements analysis in [14], it was believed in the past that the full approximation order of $\rho = k+1$ can be obtained from the spline space $S_k^r(\Delta)$ only when the degree of the polynomial k is at least $4r+1$. In a more recent work, however, Chui, Hong, and Jia [7] proved that if $k \geq 3r+2$, $S_k^r(\Delta)$ has a stable local basis and therefore, it has optimal approximation order $k+1$. This result is sharp for the three-direction mesh Δ (cf. [15] for $r \leq 3$, [4] for general r), the approximation order is less than $k+1$.

Though the approximation order of $S_4^1(\Delta)$ is at most 3 instead of the optimal order 4 if Δ is a three-direction mesh, the approximation order of $S_4^1(\Delta)$ is optimal provide that Δ if a four-direction mesh (cf. [16]). Therefore, the choice of triangulations plays an important role in working with spline functions. Chui and Hong [6] studied the problem of constructing optimal triangulations. By optimal triangulation, we mean that the space of bivariate splines with degree k and smoothness order r on this triangulation achieves the full order $k+1$ of approximation. This paper is organized as follows. In Section 2, we investigate the approximation property of $S_k^r(\Delta)$ over arbitrary triangulations. In Section 3, we consider the optimal order of approximation from $S_k^r(\Delta)$ over some special triangulations. In the final section, we sketch results on the study of the optimal triangulations.

2 Spline spaces over arbitrary triangulations

In this section, we consider the optimal order of approximation from spline space $S_k^r(\Delta)$ over arbitrary triangulation Δ. It was believed in the past that the optimal (full) approximation order of $\rho = k+1$ can be obtained from the spline space $S_k^r(\Delta)$ only when the degree of the polynomial k is at least $4r+1$ (cf. [14]). However, de Boor and Höllig [3] found that $S_k^r(\Delta)$ already has full approximation order, provided that $k \geq 3r+2$. To achieve this goal the technique used in [3] was to "disentangle the rings" by carefully working with the smoothness conditions around each vertex using the B-nets and the duality. Consequently, they obtained the approximation order without exhibiting an approximation scheme that attains this order.

For applications, one is required to construct an efficient scheme to achieve the full order $k+1$ of approximation. For this purpose, explicit bases for the spaces $S_k^r(\Delta)$ are set up when $k \geq 3r+2$ and an approximation scheme using such bases to achieve the optimal order of approximation was discussed in [17]. The subspace of super splines of smoothness r and degree $\leq k$ with enhanced smoothness order $\mu \geq r$ is defined as

$$S_k^{r,\mu}(\Delta) = \{s \in S_k^r(\Delta) : \ s \in C^\mu \text{ at each vertex of } \Delta\}.$$

Chui and Lai [8] gave a basis of vertex splines for the super spline spaces $S_k^{r,\mu}(\Delta)$ of $S_k^r(\Delta)$, where $\mu \geq r + \lfloor \frac{k-2r-1}{2} \rfloor$ and $k \geq 3r+2$. They also constructed an approximation scheme by using this super vertex spline basis to prove that the super spline subspace of the space $S_k^r(\Delta)$ already achieves the full order of approximation. Nevertheless, as in [17], the estimation constant in (1. 1) obtained by scheme in [8] not only depends on the smallest angle of Δ but also on the measurement of the near-singularity of Δ, i.e., the constant becomes large for near singular vertices (see [7]). This dependence was also observed by de Boor who pointed out in [1] that the original argument presented in [3] does not fully support the conclusion that the estimation constant depends only on the smallest angle in Δ (see also Schumaker [18], p.547). Therefore, it becomes even more significant to construct a scheme and to show that the estimation constant depends only on the smallest angle a of the partition Δ. The study of this topic is presented in [7]. The main step in [7] is to use a new formulation of smoothness conditions of splines presented in [19] to construct a stable locally supported basis of the super spline space $S_k^{r,\mu}(\Delta)$, i.e., to obtain the following.

Theorem 2 *If $k \geq 3r+2$, then there is stable basis $\{B_\ell, \ell = 1, \cdots, \dim(S_k^r(\Delta))\}$, for $S_k^{r,\mu}(\Delta)$ in the sense that there exist two positive constants C_1 and C_2, depending only on k and the smallest angle θ of the triangulation Δ, such that*

$$C_1 \sup_\ell |a_\ell| \leq \left\| \sum_\ell a_\ell B_\ell \right\|_\infty \leq C_2 \sup_\ell |a_\ell|. \tag{2.1}$$

This basis is also local in the sense that, for any ℓ, there exists a vertex u such that the support of B_ℓ

$$\text{supp } B_\ell \subseteq \overline{St}^{\lfloor r/2 \rfloor + 1}(u), \tag{2.2}$$

where the closed star of a vertex v, denoted by $\overline{St}(v) =: \overline{St}^1(v)$, is the union of all the triangles attached to v, and the m-star of v, denoted by $\overline{St}^m(v)$, is the union of all triangles that intersect with $\overline{St}^{m-1}(v)$, $m > 1$.

Now, it becomes not so difficult to construct an interpolation scheme to achieve the optimal approximation order by using the stable local supported basis in Theorem 2. Thus, we have the following.

Theorem 3 *For arbitrary triangulation Δ and $k \geq 3r+2$, the full order $k+1$ of approximation from the space $S_k^r(\Delta)$ can be achieved via a quasi-interpolation scheme using this local super spline basis in Theorem 2 with the estimation constant depending only on k and the smallest angle of the partition Δ.*

Theorem 3 shows that if the k is sufficiently large compared to r, the spline space $S_k^r(\Delta)$ provides the full accuracy expected of piecewise polynomials of degree k. If $k \leq 3r+1$, generally speaking, the space $S_k^r(\Delta)$ does not have full order of approximation. This is shown by some observations on special triangulations. A good challenging case is provided by the space of C^1 cubic splines. For Δ is the three-direction mesh, its approximation order is known to be 3 rather than 4 (cf. [15]), but it is not known what its generic approximation order is. de Boor conjectured that bivariate C^1 cubic spline space has approximation order 0 generically (cf. [2]).

3 Splines over some specific triangulations

In 1980's, much work has been done on various specific triangulations, particularly, three- or four-directional meshes, or cross-cut partitions (see [20]–[21], [16], [10] and the references therein). The two most important special triangulations are obtained from a uniform rectangular partition. The triangulation which results when all northeast diagonals are drawn in is called a type-1 or three-direction mesh, and it is denoted by $\Delta^{(1)}$. The triangulation which results when all northwest diagonals are also drawn in is called a type-2 or four-direction mesh, and it is denoted by $\Delta^{(2)}$. For a survey of the work on dimension problems of spline spaces over some special partitions, see [22] and also [18].

It was shown in [15] that the approximation order of the space $S_3^1(\Delta^{(1)})$ is only three. In general, de Boor and Jia [4] proved the following.

Theorem 4 *For a three-direction mesh $\Delta^{(1)}$, the bivariate spline space $S_k^r(\Delta^{(1)})$ has approximation order at most k for $k \le 3r+1$.*

It is well-known that for a type-2 triangulation $\Delta^{(2)}$, Dahmen and Micchelli [16] proved that the space $S_4^1(\Delta^{(2)})$ arrives at the optimal approximation order 5.

For a type-2 triangulation $\Delta^{(2)}$, Jia [11] proved the following general result by considering the local approximation order provided by the box splines in $S_k^r(\Delta^{(2)})$.

Theorem 5 *The approximation order of $S_k^r(\Delta^{(2)})$ for a four-direction mesh $\Delta^{(2)}$ is $k+1$ if $r \le 1$ and $k \ge r+1$.*

Theorem 5 shows that the space $S_k^1(\Delta^{(2)})$ has full order of approximation if $k \ge 2$. This includes a later result published in [12] on C^1 cubic spline space $S_3^1(\Delta^{(2)})$.

It will be very interesting to find some triangulations Δ, which are somewhat specific but more general than the type-2 triangulation, such that the spline space $S_k^r(\Delta)$ still has full approximation order $k+1$. To achieve the optimal (fifth) order approximation from the space $S_4^1(\Delta)$ of C^1 quartic spline functions on a triangulation Δ, a so-called local Clough-Tocher refinement procedure of an arbitrary triangulation Δ is introduced by Chui and Hong in [5]. There, some triangles are subdivided, using an interior point (such as the centroid of the triangle), into three subtriangles. To avoid introducing some new data points in addition to the vertex set of the triangulation Δ, as in the local Clough-Tocher refinement, we introduce a so-called mixed three-directional patch in [9] and define a particular three-directional mesh $\Delta^{(3)}$. A rectangle with northeast diagonal is called a *NE-rectangle*. Similarly, a rectangle with northwest diagonal is called a *NW-rectangle*. For a triangulation Δ which consists of NE- and NW-rectangles we may call Δ a mixed three-direction mesh and denote it by $\Delta^{(3)}$. We consider the approximation property of the space $S_4^1(\Delta^{(3)})$. Compactly supported basis functions, which are called mixed three-directional elements, for the space $S_4^1(\Delta^{(3)})$ are constructed, and the fact that the space $S_4^1(\Delta^{(3)})$ has optimal-order of approximation is proved. We also construct an explicit interpolation scheme by using the mixed three-directional elements to achieve this optimal-order of approximation. In [9], we obtain the following.

Theorem 6 *For a mixed three-direction mesh $\Delta^{(3)}$ there is a linear interpolating operator $T:\ f \in C^1(\Delta^{(3)}) \mapsto s \in S_4^1(\Delta^{(3)})$ such that $Tp = p$ for any polynomial $p \in \pi_4$ and such that T achieves the optimal order of approximation; that is,*

$$\|Tg - g\| \le C\|g^{(5)}\||\Delta^{(3)}|^5, \text{ for } g \in C^5(\Delta^{(3)}), \tag{3.1}$$

where $|\Delta^{(3)}|$ is the mesh size of $\Delta^{(3)}$.

Therefore, the mixed three-directional mesh $\Delta^{(3)}$ is better than the three-directional mesh in the sense that the corresponding spline space has a higher order of approximation. Also the mixed three-directional mesh $\Delta^{(3)}$ is better than local refinements in the sense that the C^1 quartic spline space achieves the optimal approximation order by using a smaller number of data sites in the interpolation. In comparison, the mixed three-directional mesh using the data only at the intersections of rectangle lines and with the optimal-order 5 can also be achieved by the space $S_4^1(\Delta^{(3)})$. Therefore, the mixed three-directional mesh is also better than the four-directional mesh in this point.

4 Optimal triangulations

In the representation of scattered data by smooth spline functions, one would like to construct an optimal approximation scheme based on the given sample data set. Therefore, one of the most important problems is to find an optimal triangulation of the given sample sites. Here, we use the term optimal triangulation to mean the optimal order of approximation can be achieved over this triangulation with respect to the given order r of smoothness and degree k of the polynomial piece of the spline functions. Now, we consider C^1 quartic spline approximation and provide an algorithm for triangulating any finite arbitrarily scattered sample sites, such that these sample sites are the only vertices of the triangulation and such that, for any discrete data given at these sample sites. Furthermore, the C^1 quartic polynomial spline space on this triangulation has the optimal order of approximation. We begin with a notion of type-O triangulation.

The degree of a vertex v, denoted by $deg(v)$, is the number of edges attached to v in a triangulation Δ. A vertex u will be called a *type-O vertex* of Δ if u satisfies at least one of the following.
(a) u is a boundary vertex of Δ.
(b) $u \in V_I$ with $deg(u) = 4$.
(c) $u \in V_I$ and $deg(u)$ is an odd integer.
(d) $u \in V_I$ and there exists a vertex v of Δ that satisfies either (i) $v \in V_I$ and $deg(v) = 4$ or $deg(v)$ = an odd integer, or (ii) $v \in V_b$, such that $[u, v]$ is a nondegenerate edge of Δ with respect to u.

We will use V_O to denote the collection of all type-O vertices in V. A triangulation of V with only type-O vertices (i.e., $V = V_O$) is called a *type-O triangulation*. The reason for introducing the notion of type-O triangulations is the following (see [6]).

Theorem 7 *Any type-O triangulation Δ admits the 5th order of approximation from $S_4^1(\Delta)$.*

As a consequence of Theorem 7, we have

Corollary 8 *If a triangulation Δ consists only of odd-degree interior vertices, then the spline space $S_4^1(\Delta)$ yields the optimal order of approximation.*

To convert any triangulation to be a type-O triangulation, we introduce a so-called swapping algorithm. Every interior edge e of a triangulation Δ is the diagonal of a quadrilateral Q_e which is the union of two triangles of Δ with common edge e. Following [23], we say that e is a swappable edge if Q_e is convex and no three of its vertices are collinear. If an edge e of a triangulation Δ is swappable, then we can create a new triangulation by swapping the edge. That is, if $v_1, \cdots, v_4$ are the vertices of Q_e ordered in the counterclockwise direction, and if e has endpoints v_1 and v_3, then the swapped edge has endpoints v_2 and v_4. Two vertices in Δ will be called *neighbors* of each other if they are the endpoints of the same edge in Δ. Hence, while v_1 and v_3 are neighbors in the original triangulation Δ, v_2 and v_4 become neighbors in the new triangulation after the edge e is swapped.

For any given set of sample sites, it is clear that, with the exception of those that are collinear, there is a triangulation with these sample sites as its only vertices. Let Δ be a triangulation associated with the given set V, and let V_O be the set of all type-O vertices in Δ. Set

$$\tilde{V} = V \setminus V_O.$$

If $u \in \tilde{V}$, then u and all its neighbors with nondegenerate edges with respect to u must be even-degree vertices with $deg(u) \geq 6$. We can see that, for every interior vertex u with $n := deg(u) \geq 5$, there is a swappable edge $e \in E_u$, the set of edges attached to u.

We can describe the *Swapping Algorithm* for constructing a type-O triangulation $\widehat{\Delta}$, starting with any triangulation Δ, as follows.

Swapping Algorithm

Do while ($\tilde{V} \neq \emptyset$)
Pick any vertex u in $\tilde{V}$ and consider its neighbors.
Pick any neighbor v of u so that the edge $[u, v]$ is swappable.
Swap $[u, v]$, yielding a new edge $[u', v']$.
Form a subset of $\tilde{V}$ by deleting from $\tilde{V}$ all the neighbors w of $w' := u$, v, u', or v', with $[w, w']$ being a nondegenerate edge with respect to w.
Call this subset $\tilde{V}$.
Enddo

Combining the Swapping Algorithm with Theorem 7, we have the following.

Theorem 9 *Every finite set V of sample sites admits an optimal triangulation $\widehat{\Delta}$, such that the C^1 quartic spline space $S_4^1(\widehat{\Delta})$ has the fifth order of approximation.*

Remark. Some special partitions for optimal order of approximation purpose have been discussed in [13] and [24]. In general, the question of how to find an optimal triangulation for the bivariate spline space $S_k^r(\Delta)$ with $k \leq 3r + 1$ is still open. Particularly, to find a type-O triangulation for the C^1 cubic spine space will be a very challenge problem.

Acknowledgment. This research was supported in part by a Research Development Grant from East Tennessee State University. The author is grateful for the support by the Department of Mathematics, the College of Arts and Sciences, and the Office of International Program of ETSU.

References

1. C. de Boor, in: Multivariate Approximation IV, (C. K. Chui, W. Schempp, and K. Zeller, Eds.), Birkhäuser, Basel (1989) pp. 25–30.

2. C. de Boor, in: Computations of Curves and Surfaces, (Dahmen, Gasca, and Micchelli, Eds.), Kluwer, Dordrecht, Netherland (1990) pp. 313–345.

3. C. de Boor and K. Höllig, Math. Z., 197, 343–363 (1988).

4. C. de Boor and R. Q. Jia, J. Approx. Theory, 72, 24–33 (1993).

5. C. K. Chui and D. Hong, Math. Comp., 65, 85–98 (1996).

6. C.K. Chui and D. Hong, SIAM J. Numer. Anal., 34, 1472–1482 (1997).

7. C.K. Chui, D. Hong and R.Q. Jia, Trans. of Amer. Math. Soc., 374, 3301–3318 (1995).

8. C. K. Chui and M. J. Lai, Constr. Approx., 6, 399–419 (1990).

9. D. Hong and R. N. Mohapatra, Computers and Mathematics with Applications, (1998), (in press).

10. R.Q. Jia, Trans. Amer. Math. Soc., 295, 199–212 (1986).

11. R.Q. Jia, Lecture Notes on Multivariate Splines, Department of Mathematics, University of Alberta, Edmonton, Canada (1990).

12. M.J. Lai, Comput. Aided Geometric Design, 11, 215–223 (1994).

13. M.J. Lai and L.L. Schumaker, SIAM Numer. Anal., 34, 905–921 (1997).

14. A. Ženišek, Numer. Math., 15, 283–296 (1970).

15. C. de Boor and K. Höllig, Proc. Amer. Math. Soc., 87, 649–655 (1983).

16. W. Dahmen and C.A. Micchelli, J. Comp. Appl. Math., 10, 255–273 (1984).

17. D. Hong, On Bivariate Spline Spaces Over Arbitrary Triangulations, Master's Thesis, Zhejiang University, Hangzhou, Zhejiang, China (1987).

18. L.L. Schumaker, in: Mathematics of Finite Elements VII, (J. Whiteman, Ed.), Academic Press, London (1991) pp. 535–562.

19. D. Hong, Approx. Theory and Appl., 11, 67–75 (1995).

20. C.K. Chui and R.H. Wang, Math. Comp., 41, 131–142 (1983).

21. C.K. Chui and R.H. Wang, Scientia Sinica, 27, 1129–1142 (1984).

22. W. Dahmen and C.A. Micchelli, in: Approximation Theory IV, (C.K. Chui, L.L. Schumaker, and J.D. Ward, Eds.), Academic Press, New York (1983) pp. 27–121.

23. L. L. Schumaker, Computer Aided Geometric Design, 10, 329–345 (1993).

24. M.J. Lai and L.L. Schumaker, Quadrangulating scattered data points, manuscript (1996).

25. C. de Boor and R. DeVore, Trans. Amer. Math. Soc., 276, 775–788 (1983).

8th Int. Coll. on Differential Equations, pp. 227-236
D. Bainov (Ed)

Solitary-wave Solution of Turbulence with Application to Bēnard Convection

K. ISHIBASHI
Technical Research and Development Institute, JDA, 1-2-24, Ikejiri, Setagaya-ku, Tokyo 154, Japan

S. TSUGĒ
TARA Center, University of Tsukuba, Tsukuba 305, Japan

T.M.S. NAKAGAWA
Kanazawa Institute of Technology, Ishikawa 921, Japan

Abstract. A modified classical statistical theory of turbulence designed to include inhomogeneous turbulence is applied to high Rayleigh number regime ($Ra > 5\times10^5$) of Bēnard convection with infinite aspect ratio. The fluctuation correlations are solved using the methods of variable separation in the form of solitary waves in a transformed space. Measurable quantities, e.g.,turbulent intensities are obtained through a simple integration of the solitary-wave solution. The mean temperature distribution is calculated, and thereby the Nusselt number is determined as a function of the Rayleigh number. The power spectrum is also obtained through an inverse-Fourier transform of the solitary-wave solution. These quantities are compared with data from existing experiments. The agreement is satisfactory considering that the theory does not include any empirical parameters, as contrast to k-ϵ models or eddy viscosity concepts.

Keywords: Mathematical Analysis, Numerical Analysis, Turbulence, Bēnard Convection

1. INTRODUCTION

The statistical theory of turbulence emerged from the classical work of Kármán and Howarth [1]. This was originally limited to homogeneous, isotropic turbulence, but it has since been revised to include inhomogeneous turbulence [2, 3, 4]. Two key issues to make this approach pratical are:

i) Introduction of a method of separation of variables which converts equations for multi-point correlations into those for respective points [5].

ii) Devising a closure condition which allows for the separation with full nonlinear terms included.

These key findings have led to a set of equations that govern the mean flow and mean fluctuations respectively, coupled each other through turbulent transports.

This paper describes a method for predicting the turbulence characteristics of Bēnard convection with an infinite aspect ratio for Rayleigh numbers higher than 5×10^5. The

Bēnard problem is an example with enough experimental data available for comparison, and is also known as an example that is a colorful kaleidoscope [5] of flow regimes which changes with the controlling parameter. These regimes are, in order of increasing Rayleigh number, thermally strained quiescence, buoyancy driven convection, oscillation, chaos, transition to stochastic regime, and turbulence.

The area of interest in this paper is the last regime: The objective here is to solve the equations derived in ref.3 with a few simplifying assumptions based on physical insights which also are supported by flow visualizations (Sec.2.). The system of equations is solved by CFD which provides the solution of the solitary wave (Sec.3.). These solutions are integrated to give turbulent fluctuations and transfers. The quality of the solution is also discussed in the light of comparison with existing experiments (Sec.4.).

2. FORMULATION OF THE PROBLEM

The equations governing compressible turbulent flows are given in ref.3. Here we will adapt these equations to the Bēnard problem which can be characterized by the following physical/geometrical features:

i) Two infinite parallel plates aligned horizontally with a separation L is filled with an ideal gas. Turbulence will occur if the lower plate has a temperature sufficiently higher than the upper one, or in nondimensional terms, if the Rayleigh number

$$Ra = \alpha g L^3 \Delta T / \kappa \nu \tag{1}$$

is large enough ($Ra \geq 5\times 10^5$). The temperature difference, ΔT, is denoted by

$$\Delta T = T_0 - T_1 > 0 \tag{2}$$

where T_0 and T_1 are temperatures of the lower and upper plates, respectively, α is the isobaric thermal expansion, and κ, ν and g denote the thermal diffusivity, the kinematic viscosity and the gravitational constant, respectively.

ii) The average fluid convection becomes less important with an increase in the aspect ratio of the apparatus. In the limit of infinite aspect ratio there are no edge effects causing convection. Under this condition, only the energy equation needs to be solved.

iii) Visualization experiments [6] show that instantaneous velocity fluctuations near the wall where turbulence is most vigorous are essentially two-dimensional. Therefore we may assume that the smoothed fluctuation is also two-dimensional:

$$q = (q_1, 0, q_3) \tag{3}$$

where subscript 3 denotes the component normal to the wall, and 1 denotes a horizontal, but otherwise unspecified, component.

iv) The Boussinesq approximation is employed; namely, the density fluctuation only in the equation of continuity is ignored.

Before going on to the equations, it may be relevant to summarize the base-line of the formalism and the physical implications of variables to appear in the system. (For details see ref.4.). The proposed formalism deals not with instantaneous flow quantities $\underline{Z}_\alpha$, but with their average $Z_\alpha \equiv < \underline{Z}_\alpha >$ and average fluctuation-correlation $< \Delta Z_\alpha \Delta \hat{Z}_\beta >$,where $\Delta Z_\alpha \equiv \underline{Z}_\alpha - Z_\alpha$, and $\hat{Z}_\beta \equiv Z_\beta(\hat{x}, t)$. The latter quantities are decomposed by means of variable separation

$$< \Delta Z_\alpha \Delta \hat{Z}_\beta >= R.P. \int_{-\infty}^{\infty} q_\alpha(x, \Omega) q_\beta^*(\hat{x}, \Omega) d\Omega \tag{4}$$

where R.P. stands for the real part, asterisk (*) denotes the complex conjugate, and Ω is the nondimensional frequency. There are five q variables required to describe the turbulent medium under consideration, each corresponding to the stochastic fluctuation counterpart Δ as follows [4]:

$$q_0 \leftrightarrow \Delta\rho, \quad q_1 \leftrightarrow \rho\Delta u_1, \quad q_3 \leftrightarrow \rho\Delta u_3, \quad q_{40} \leftrightarrow \Delta p, \quad q_4 \leftrightarrow \rho R \Delta T \tag{5}$$

where ρ, p, T and R are the density, the pressure, the temperature and the gas constant per unit of mass.

With these preliminaries, the equations for the present problem are in order: The averaged energy equation, integrated once, is given by

$$-\lambda\frac{dT}{dx_3} + \frac{1}{\rho}\frac{\gamma}{\gamma - 1}\int_{-\infty}^{\infty} q_4 q_3^* d\Omega = -\lambda(\frac{dT}{dx_3})_w \tag{6}$$

where λ is the heat conductivity and γ is the adiabatic index. This equation claims that the total heat transport, namely, the molecular plus the turbulent transport, remain constant across the thermal layer. This constant value, namely the quantity on the r.h.s of Eq.(6) represents the total heat flux density through the wall

$$Q = -\lambda(dT/dx_3)_w = Nu\lambda\Delta T/L$$

The latter part of this formula is the definition of the Nusselt number; $Nu - 1$ gives a measure for the contribution due to the turbulent heat transfer. The turbulent heat flux density, namely, the second term on the l.h.s of Eq.(6) equals $c_p\rho < \Delta T\Delta u_3 >$ (c_p;the specific heat under constant pressure).

The equations for variables q have been derived in refs.2 and 4 for the most general case of compressible turbulence. A simplification in the analysis is achieved by the following variable separation,

$$q_\alpha(x_1, x_3, \Omega) = exp(i\Omega x_1/L)Q_\alpha(x_3, \Omega) \tag{7}$$

Furthermore, the Fourier transform,

$$Q_\alpha(x_3, \Omega) = \frac{1}{2\pi}\int_{-\infty}^{\infty} exp(-i\Omega s)F_\alpha(x_3, s)ds \tag{8}$$

The new variable s, introduced here is nondimensional time just as Ω is a nondimensional frequency. This allows us to write the equations subject to the assumptions ii) through iv) in the (z, s) space in nondimensional form

$$\frac{d\theta}{dz} = -Nu + \int_{-\infty}^{\infty} f_4 f_3 ds \tag{9}$$

$$\frac{\partial f_1}{\partial s} + \frac{\partial f_3}{\partial z} = 0 \tag{10}$$

$$\frac{\partial f_1}{\partial t} + \frac{\partial f_{40}}{\partial s} - \frac{1}{Re}\nabla^2 f_1 + \frac{1}{\sqrt{B}}(f_1\frac{\partial f_1}{\partial s} + f_3\frac{\partial f_1}{\partial z}) = 0 \tag{11}$$

$$\frac{\partial f_3}{\partial t} + \frac{\partial f_{40}}{\partial z} - \frac{1}{Re}\nabla^2 f_3 + \frac{1}{\sqrt{B}}(f_1\frac{\partial f_3}{\partial s} + f_3\frac{\partial f_3}{\partial z}) - \frac{1}{Pr}f_4 = 0 \tag{12}$$

$$f_3\frac{d\theta}{dz} - \frac{1}{PrRe}\nabla^2 f_4 + \frac{1}{\sqrt{B}}(f_1\frac{\partial f_4}{\partial s} + f_3\frac{\partial f_4}{\partial z}) = 0 \tag{13}$$

$$(\nabla^2 \equiv \partial^2/\partial s^2 + \partial^2/\partial z^2)$$

and the following nondimensional quantities have been introduced:

$$\theta = (T - T_1)/\Delta T, \ f_j = B^{1/2}F_j/\rho_0 u^* \ (j = 1, 3), \ f_4 = B^{1/2}F_4/\rho_0 R\Delta T,$$
$$f_{40} = B^{1/2}F_{40}/\rho_0 u^{*2}, \ z = x_3/L, \ Re = u^* L/\nu, \ Pr = \rho_0 c_p \nu/\lambda \quad (14)$$

u^* the characteristic velocity,and introducing the scaling of dependent variables:

$$Re = Ra^{1/2} \quad (15)$$

$$f_\alpha = B^{1/2}\bar{f}_\alpha (\alpha = 1, 3, 4, 40), \ B = PrRe/2\pi \quad (16)$$

Eq.(9) integrated from $z = 0$ to 1 gives the Nusselt number;

$$Nu = 1 + \int_0^1 dz \int_{-\infty}^{\infty} f_4 f_3 ds \quad (17)$$

The boundary conditions for the function f is that all fluctuations except the pressure fluctuation should vanish at a solid wall,

$$f_\alpha(0, s) = f_\alpha(1, s) = 0, \quad (\alpha = 1, 3, 4,) \quad (18)$$

On the other hand, by integrating Eq.(12) across the narrow sublayer, using Eq.(10), we obtain the boundary condition for the pressure fluctuation

$$\partial f_{40}/\partial z = 0, \ at \ z = 0 \ and \ 1 \quad (19)$$

3. THE SOLITARY WAVE SOLUTION

As is obvious from the integral expression of Eq.(9), the functions f must be square-integrable. Therefore, we have

$$f_\alpha(z, \pm\infty) = 0 \quad (20)$$

Thus, with conditions (18) and (19) we see that our task is to seek a solitary wave solution of Eqs.(9) through (13) in a transformed plane of (z, s).

The procedure for obtaining such a solitary wave solution is the following: Given a Prandtl number(Pr) and a Rayleigh number (Ra), the set of Eqs.(10) through (13), where $d\theta/dz$ is eliminated using Eq.(9), constitutes an initial value problem to determine f_1,f_3,f_4 and f_{40}. The actual steps are:

i) For easier formulation of the initial value problem an artificial unsteady term $\partial f_4/\partial t$ is added to the l.h.s. of the energy fluctuation equation (13). This does not appear to interfere with the steady state solution to be eventually reached.

ii) Assume initial values for each function f in the (z, s) plane subject to boundary conditions (18),(19) and (20), and start the time marching process using the marker-and-cell(MAC) method. This method consists of an alternating process of solving a Poisson equation of the form

$$\nabla^2 f_{40} = -\frac{\partial}{\partial t}(\frac{\partial f_1}{\partial s} + \frac{\partial f_3}{\partial z}) + \frac{1}{Re}\nabla^2(\frac{\partial f_1}{\partial s} + \frac{\partial f_3}{\partial z}) + \frac{1}{Pr}\frac{\partial f_4}{\partial z} - \frac{1}{\sqrt{B}}(f_1\frac{\partial}{\partial s}(\frac{\partial f_1}{\partial s} + \frac{\partial f_3}{\partial z})$$
$$+ f_3\frac{\partial}{\partial s}(\frac{\partial f_1}{\partial s} + \frac{\partial f_3}{\partial z}) + 2\frac{\partial f_1}{\partial z}\frac{\partial f_3}{\partial s} + (\frac{\partial f_1}{\partial s})^2 + (\frac{\partial f_3}{\partial z})^2) = 0 \quad (21)$$

and fulfilment of solenoidal condition (10), which is rechecked at each time step.

iii) The process is continued until a steady state ($\partial/\partial t \sim 0$) is reached.

The computational process was found to be relatively stable. Cautious treatment, however, such as the use of a staggered cell to suppress spurious errors, the use of up-wind finite difference of second order accuracy adopted for nonlinear terms, and the use of optimize value of Courant number (Cr=0.2).

4. OBSERVABLE QUANTITIES: COMPARISON WITH EXPERIMENTS

Figure 1 shows typical examples of the solitary wave solutions at Rayleigh numbers $Ra = 5\times10^5$. Unlike other solitary waves such as solitons, these solutions have multiple peaks and valleys which increase in number with increasing Rayleigh number.

Once the solitary wave solution has been obtained, various physical quantities, the mean and the turbulent fluctuations, can be calculated though a simple integration. In fact, the general expressions(4) for fluctuation correlations are rewritten in terms of the Fourier variables by the use of relationships (7) and (8),

$$< \Delta Z_\alpha \Delta \widehat{Z}_\beta >= \frac{1}{2\pi}\int_{-\infty}^{\infty} F_\alpha(x_3, s)F_\beta(\widehat{x}_3, s-(x_1-\widehat{x}_1)/L)ds \tag{22}$$

where the identity $\delta(x) = \frac{1}{2\pi}\int_{-\infty}^{\infty} exp(i\Omega x)d\Omega$ has been utilized. Turbulent intensities for physical quantitites Z_α are given by setting $\widehat{x} = x$ and $\alpha = \beta$ in this formula to yield

$$< (\Delta Z_\alpha)^2 >= \frac{1}{2\pi}\int_{-\infty}^{\infty} F_\alpha^2(x_3, s)ds \tag{23}$$

Shown in Figure 2 is calculated turbulent intensities (from above) of the longitudinal and vertical components of the velocity, and the temperature fluctuation, each compared with experimental data by Deardorff and Willis [7]. As predicted in the previous section, the distribution is <u>not</u> symmetric with respect to the centerline $z = 1/2$. Although the experimental data are available for only the upper half of the entire region, we believe that the agreement is reasonable, because the theory includes no adjustable parameters.

Figure 3 shows nondimensional temperature θ calculated from Eq.(9), and compared with experiment data reported by Garon and Goldstein [8]. The agreement is satisfactory, although the theory has a tendency to underestimates the boundary layer thickness. An interesting feature common to both the theory and the experiment is that the mean temperature distribution is <u>not</u> monotonous, but has under- and over- shoots at the high and low temperature sides, respectively, of the central core. This feature has been observed more clearly at lower Rayleigh numbers [9]. There is no reason, however, that the temperature gradient should have only one sign, because in this region the molecular heat transport is overwhelmed by the turbulent transport.

The inverse Fourier-transform back to Ω-space gives us the spectrum: The general expression for power-spectra is given by Eqs.(4) and (14) which give rise to

$$S_{\alpha\alpha}(\Omega) = \left\{\frac{1}{2\pi}\int_{-\infty}^{\infty} f_\alpha exp(-i\Omega s)ds\right\} \times \left\{\frac{1}{2\pi}\int_{-\infty}^{\infty} f_\alpha exp(-i\Omega s)ds\right\}^* \tag{24}$$

This formula, applied to the temperature fluctuation gives

$$S(\Omega) = \left\{\frac{1}{2\pi}\int_{-\infty}^{\infty} f_4 exp(-i\Omega s)ds\right\}^2 \tag{25}$$

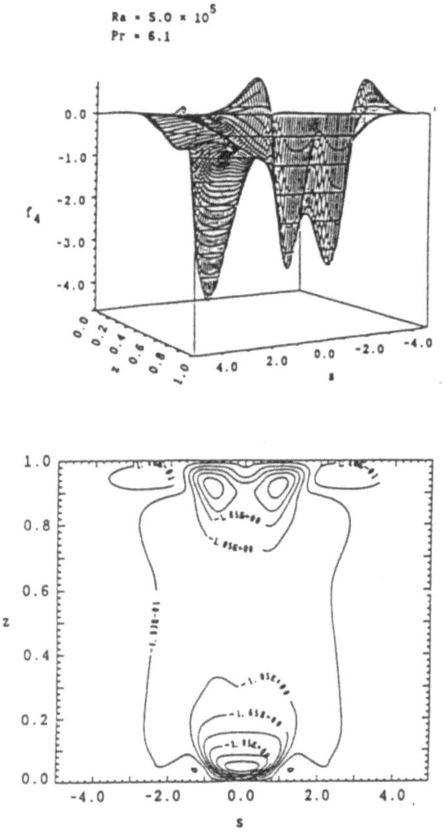

Figure 1: Solitary wave solutions for separated fluctuation variable f_α in the (z, s)-plane near the lower limit of the turbulent region; $Ra = 5.0 \times 10^5$, Pr=6.1, (f_4 ; temperature fluctuation)

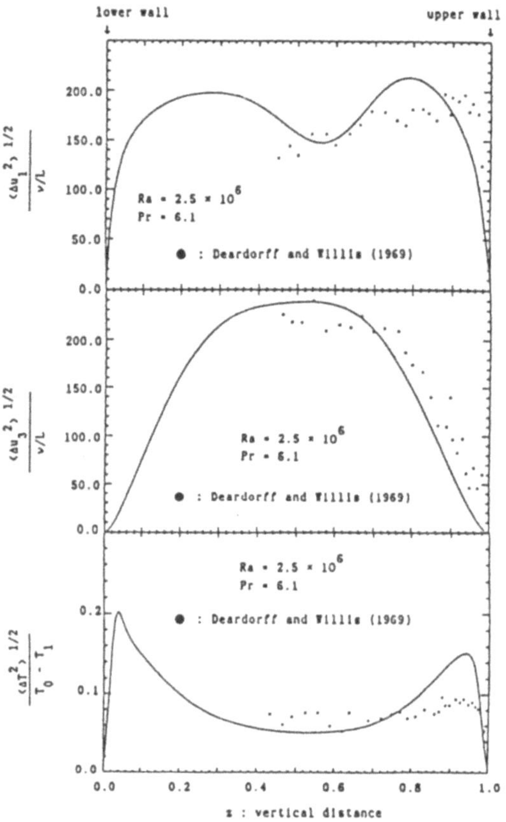

Figure 2: Turbulence intensity profiles of horizontal and vertical components of the velocity.Included are the temperature profiles at three different Rayleigh numbers and for the Prandtl number corresponding to the water. The experimental data used in the comparison are available only for half the medium despite the apparent asymmetry of the solution.

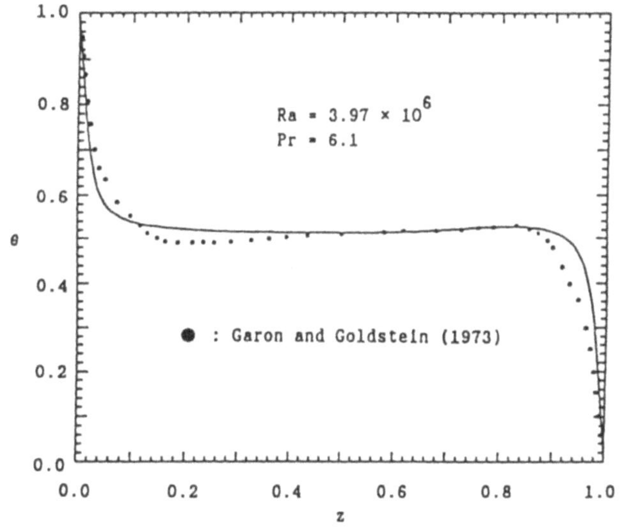

Figure 3: The mean temperature distribution at $Ra = 3.97 \times 10^6$ and Pr=6.1 (water) calculated from Eq.(18) plotted against vertical distance z, and compared with experiment for water (ref.8).

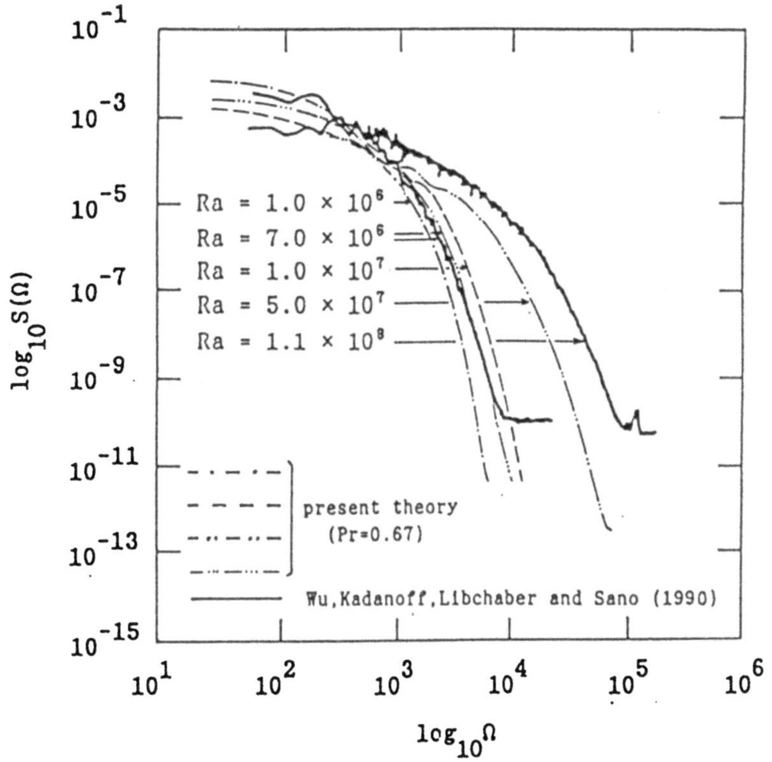

Figure 4: Power spectrum of the temperature fluctuation calculated using formula by expression (25), plotted against the nondimensional frequency, and compared with experiment (ref.10).

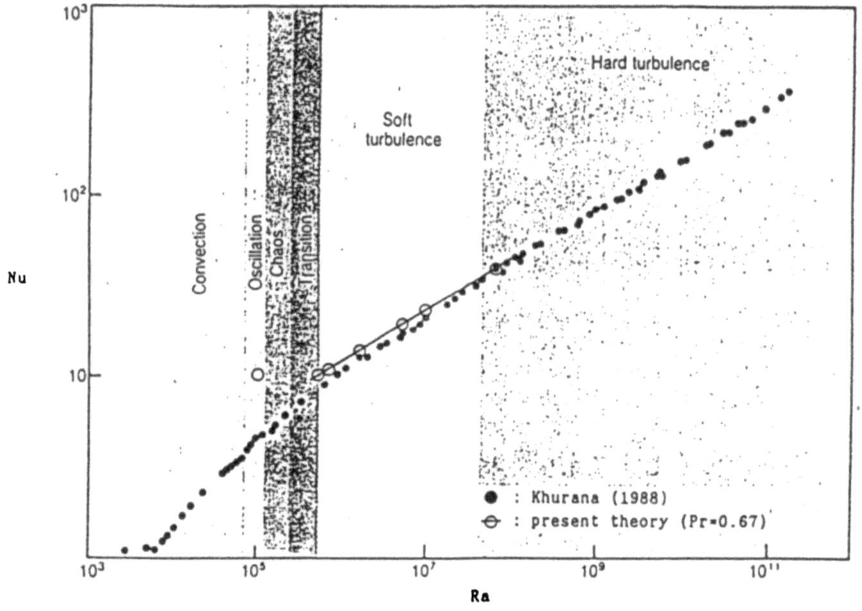

Figure 5: The Nusselt number (Nu; the nondimensional heat transfer) calculated using (17) and plotted against the Rayleigh number (Ra; the nondimensional temperature difference between walls) on the original figure from ref.5.

The calculated result is shown in Figure 4 and is compared with existing experimental data [10]. The two log-log plots are parallel-shifted to provide the best fit to each other. This is the only place throughout the paper where the use of 'best-fit' is needed. This is to make up for the indeterminancy of the constant $\omega^+ = \omega/\Omega$ which relates the frequency ω measured to the frequency Ω introduced in the theory. The low frequency hump observed in the experimental data reflects the organized convective motion characteristic of flows with a finite aspect ratio, which is not to be expected in the present theory.

Finally Figure 5 shows Nusselt number Nu calculated using Eq.(17) and plotted against the Rayleigh number plotted on the original figure by Khurana(ref.5). Agreement with experiment is again satisfactory over the calculated region. However, the calculated Nusselt number starts to deviate from the measured value for $Ra < 5\times10^5$. Below this Rayleigh number the obtained solitary wave solutions have a tendency to 'locked-in' to that at $Ra = 5\times10^5$, and thereby the predicted Nusselt numbers are frozen to a constant value. Since this Rayleigh number is known as the lower limit of the fully-turbulent region [5] below which deterministic flow regime starts, this phenomenon might be a sign of the onset of the organized convective motion. In high Rayleigh number region ($Ra > 10^8$), on the other hand, the computation is limited by the resolvablity of the boundary layer near the solid walls. However, this problem is surmountable in principle by employing finer unequal mesh in the thermal boundary layer.

As a whole, the theoretical curve of $Nu = f(Ra)$ fits the experimental data better in the high Rayleigh number region of so-called hard turbulence [5]. The slight deviation between the theory and the experiment is caused by the finite aspect ratio of the experimental apparatus that generates a organized convecting flow. With increase in the Reynolds number($\sim O(Ra^{1/2})$) the discrepancy tends to diminish since the edge effect causing the mean convective flow is overwhelmed by increasing heat transfer by turbulent eddies.

5. CONCLUSIONS

The classical statistical theory of turbulence has been revised by introducing the variable separation, and is applied to the problem of Bēnard turbulent convection, known as one of the simplest forms of inhomogeneous turbulence. In the Fourier-transformed space, the separated fluctuation variables are solved in the form of standing solitary waves. Turbulence intensities and the mean temperature distribution are then obtained through simple integrations of the solitary-wave solutions. The results compare favorably with experimental data. The power spectrum of turbulence is also calculated by inverse Fourier transform, which also agrees well with the measured values. Finally, the calculated Nusselt number is plotted against the Rayleigh number, and satisfactory agreement with experiment is observed. Together with the fact that the theory does not include any adjustable parameters, these coincidences indicate the soundness of the proposed methodology.

REFERENCES

1. Th. von Kármán and L. Howarth, Proc. Roy. Soc., A164,192(1938).

2. S. Tsugē, Phys. Fluids, 27, 1370 (1984).

3. S. Tsugē, K. Sagara and S. Kadowaki, in: Rarefied Gasdynamics Volume I, H.Oguchi (Edi.), Univ. of Tokyo Press, (1984) p. 19.

4. S. Tsugē and S. Ogawa, Combust. Sci. Tech. (Japanese Edition with English summary) in press.

5. A.Khurana, Physics Today, 41, 17 (June, 1988).

6. G. Zocchi, E. Moses and A. Libchaber, Physica, A166, 387 (1990).

7. J.W. Deardorff and G.E. Willis, J. Fluid Mech., 28, 675 (1969).

8. A.M. Garon and R.J. Goldstein,Phys. Fluids, 16, 1818 (1973).

9. T.Y. Chu and R.J. Goldstein, J. Fluid Mech., 60, 141 (1973).

10. X. Z. Wu, L. Kadanoff, A. Libchaber and M. Sano, Phys. Rev. Lett., 64, 2140 (1990).

8th Int. Coll. on Differential Equations, pp. 237–242
D. Bainov (Ed)

Hermite Rational Matrix Approximation of Riccati Differential Equations

LUCAS JÓDAR AND EMILIO DEFEZ
Departamento de Matemática Aplicada, Universidad Politécnica de Valencia, P.O.Box 22012, Valencia, Spain

Abstract. In this paper Hermite rational matrix approximations of Riccati equations are constructed using the Hermite matrix polynomial approximations of matrix exponentials.
Keywords: Hermite matrix polynomial, Riccati differential equation.

1. Introduction and preliminaries

Riccati matrix differential equations of the type

$$W'(t) = C - DW(t) - W(t)A - W(t)BW(t) \ , \quad W(0) = W_0 \ , \qquad (1)$$

where $W(t)$, C, D, B and W_0 are $r \times r$ complex matrices, elements of $C^{r\times r}$, arise in the imbedding context [1] and in the solution of linear boundary value problems [2], [3]. For the symmetric case, where A is the adjoint matrix of D, equation (1) is frequent in optimal control theory [4], [5] and filtering theory [6]. Most general methods to solve problem (1) are based on the transformation of (1) into the extended first order system

$$\begin{bmatrix} U'(t) \\ V'(t) \end{bmatrix} = S \begin{bmatrix} U(t) \\ V(t) \end{bmatrix} , \begin{bmatrix} U(0) \\ V(0) \end{bmatrix} = \begin{bmatrix} I \\ W_0 \end{bmatrix} ; S = \begin{bmatrix} A & B \\ C & -D \end{bmatrix} . \qquad (2)$$

Then from [7], the solution $W(t)$ of (1) is defined in the interval where $U(t)$ is invertible and in such domain $W(t) = V(t)\left[U(t)\right]^{-1}$. Since the computation of

$$U(t) = \begin{bmatrix} I & 0 \end{bmatrix} \exp St \begin{bmatrix} I \\ W_0 \end{bmatrix} , \quad V(t) = \begin{bmatrix} 0 & I \end{bmatrix} \exp St \begin{bmatrix} I \\ W_0 \end{bmatrix} , \qquad (3)$$

requires the computation of the matrix exponential $\exp St$, and it presents well known difficulties [8], this motivates the use of numerical integration techniques [9]. An excelent survey about the matrix exponential is [8]. Some of the drawbacks of the existing methods are (i) the computation of eigenvalues or eigenvectors, [8], (ii) the computation of the inverses of matrices, (see Padé methods, [10]), [8], (iii) storage problems and expensive computational time, [10], (iv) round-off accumulation errors,[10]. If D_0 is the complex plane cut along the negative real axis and $Log(z)$ denotes the principal logarithm of z, [11], then $z^{\frac{1}{2}}$ represents $\exp\left(\frac{1}{2}Log(z)\right)$. If A is a matrix in $C^{r\times r}$, its 2-norm denoted

$\|A\|_2$ is defined by $\|A\|_2 = \frac{\|Ax\|_2}{\|x\|_2}$, where for a vector y in C^r, $\|y\|_2$ denotes the usual Euclidean norm of y, $\|y\|_2 = \left(y^T y\right)^{\frac{1}{2}}$. The set of all the eigenvalues of A is denoted by $\sigma(A)$. If $f(z)$ and $g(z)$ are holomorphic functions of the complex variable z, which are defined in an open set Ω of the complex plane, and A is a matrix in $C^{r\times r}$ such that $\sigma(A) \subset \Omega$, then from the properties of the matrix functional calculus, [12], it follows that $f(A)g(A) = g(A)f(A)$. If A is a matrix with $\sigma(A) \subset D_0$, then $A^{\frac{1}{2}} = \sqrt{A}$ denotes the image by $z^{\frac{1}{2}}$ of the matrix functional calculus acting on the matrix A. We say that A is a positive stable matrix if $Re(z) > 0$ for all $z \in \sigma(A)$. The identity matrix in $C^{r\times r}$ will be denoted by I. For the sake of clarity in the presentation we recall some properties of the Hermite matrix polynomials which will be used below and that have been established in [13]. If A is a positive stable matrix in $C^{r\times r}$, the n−th Hermite matrix polynomial is defined by

$$H_n(x, A) = n! \sum_{k=0}^{[\frac{n}{2}]} \frac{(-1)^k \left(\sqrt{2A}\right)^{n-2k}}{k!(n-2k)!} x^{n-2k} , \tag{4}$$

and satisfy the three terms recurrence relationship

$$\left.\begin{aligned} H_n(x, A) &= Ix\sqrt{(2A)}H_{n-1}(x, A) - 2(n-1)H_{n-2}(x, A) \quad , \quad n \geq 1 \\ H_{-1}(x, A) &= 0 \quad , \quad H_0(x, A) = I. \end{aligned}\right] \tag{5}$$

If B is a matrix in $C^{r\times r}$ and n_0 is a positive integer we denote by $\mathcal{A}(B, n_0)$ the real number:

$$\mathcal{A}(B, n_0) = \sum_{n=0}^{n_0} \sum_{k=0}^{[\frac{n}{2}]} \frac{(\|B\|)^{n-2k}}{k!(n-2k)!} . \tag{6}$$

The proof of the next results may be found in [14]:

Lemma 1.1 *Let be A a positive stable matrix, $K > 2$ and $n \geq 0$ integer. Then*

$$\|H_n(x, A)\|_2 \leq \sqrt{n!}\,\sqrt{2^n}\,K^n e^{x^2} \quad , \quad |x| < \frac{K}{\|\sqrt{2}\,A\|_2} . \tag{7}$$

Theorem 1.1 *Let B be a matrix in $C^{r\times r}$ and let $\gamma \geq 0$ be defined by*

$$\gamma = \begin{cases} 0 \,,\; if\; |Re(z)| > |Im(z)| \quad , \quad \forall z \in \sigma(B) \\ \max\{|Re(z)| + |Im(z)|, z \in \sigma(B)\} \,,\; in\ other\ case \end{cases} .$$

Then

$$e^{Bx} = \sum_{n \geq 0} \frac{e^{-\gamma x+1}}{n!} H_n\left(x, \frac{1}{2}(B + \gamma I)^2\right) \quad , \quad -\infty < x < +\infty. \tag{8}$$

Let $E(B + \gamma I, x, n)$, be the n−th partial sum of series (8) and $\mathcal{A}$ is defined by (6). Then for $|x| < c$ and an integer $n \geq 0$ one gets

$$\left\|e^{Bx} - E(B + \gamma I, x, n)\right\|_2 \leq \exp(c\|B + \gamma I\|_2 + 1) - \mathcal{A}((B + \gamma I)c, n) . \tag{9}$$

The following result that may be regarded as a matrix version of the algorithm given in [15], provides an efficient procedure for computing finite matrix polynomial series expansions in terms of matrix polynomials satisfying a three terms formula. In particular, it is applicable to the Hermite matrix polynomials. Note the remarkable fact from a computational point of view that the evaluation of a matrix polynomial at $\bar{x}$ is only expressed in terms of $P_0(\bar{x})$.

Theorem 1.2 *Let $\{P_n(x)\}_{n\geq 0}$ be a sequence of matrix polynomials such that*

$$A_n P_n(x) = (xI - B_n)\, P_{n-1}(x) - C_n P_{n-2}(x) \quad , \quad n \geq 1 \ ,$$

where A_n is an invertible matrix in $C^{r\times r}$ and the degree of $P_n(x)$ is n. Let $Q(x)$ be a matrix polynomial defined by $Q(x) = \sum_{j=0}^{n} E_j P_j(x)$ where E_j is a matrix in $C^{r\times r}$. Let $\bar{x}$ be any real number and consider the sequence of matrices defined by

$$D_n = E_n \ , \ D_{n-1} = E_{n-1} + D_n A_n^{-1}(\bar{x}I - B_n) \ ,$$

and for $j = n-2, \ldots, 0$,

$$D_j = E_j + D_{j+1}A_{j+1}^{-1}(\bar{x}I - B_{j+1}) - D_{j+2}A_{j+2}^{-1}C_{j+2} \ .$$

Then $Q(\bar{x}) = D_0 P_0(\bar{x})$.

In this paper we propose a new method for computing rational matrix approximations of Riccati equations with a prefixed accuracy, by means of Hermite matrix polynomials approximations of the exponential of an appropriate matrix. This paper is organized as follows. In section 2 analytic-numerical approximations of problem (1) are contructed in terms of Hermite matrix polynomial series expansions. Given an admissible error $\epsilon > 0$ and a bounded interval $[\,0, t\,]$, an approximation in terms of Hermite matrix polynomials is contructed so that the error with respect to the exact solution is uniformly upper bounded by ϵ in $[\,0, t\,]$. In section 3 an example is included.

2. Construction of accurate rational matrix approximations

In this section we construct rational approximations of problem (1) expressed in terms of Hermite matrix polynomials. We start from the following result:

Theorem 2.1 **([16])** *Let $\omega_0 = \left\| \begin{bmatrix} I \\ W_0 \end{bmatrix} \right\|_2$, let $\delta_0 = \|S\|_2^{-1} \ln(1 + \omega_0)$. Then $U(t)$ is invertible in $|t| < \delta_0$. Let δ_1 a positive number verifying*

$$\ln \delta_1 + \|S\|_2 \delta_1 < -\ln(\omega_0 \ \|[A \ \ B]\|_2) \ , \ \delta_1 < \delta_0 \ , \tag{10}$$

and let M be defined by

$$M = \|[A \ \ B]\|_2 \exp(\delta_1 \ \|S\|_2)\, \omega_0 \ , \tag{11}$$

then for $|t| < \delta_1$ it follows that

$$\left\| [U(t)]^{-1} \right\|_2 < (1 - M\delta_1)^{-1} \ . \tag{12}$$

Definitions of $U(t)$ and $V(t)$, see (3), suggest the approximation:

$$W(t,n) = U(t,n)\,[U(t,n)]^{-1} , \tag{13}$$

where

$$U(t,n) = \begin{bmatrix} I & 0 \end{bmatrix} E\,(S+\gamma I, t, n) \begin{bmatrix} I \\ W_0 \end{bmatrix}, V(t,n) = \begin{bmatrix} 0 & I \end{bmatrix} E\,(S+\gamma I, t, n) \begin{bmatrix} I \\ W_0 \end{bmatrix}. \tag{14}$$

By (9) and (14) one gets

$$\|U(t) - U(t,n)\|_2 \leq \omega_0 \left(\exp\left(\delta_1 \|S+\gamma I\|_2 + 1\right) - \mathcal{A}\left((S+\gamma I)\,\delta_1, n\right)\right), \ |t| < \delta_1 . \tag{15}$$

Let n_0 be the first positive integer satisfying

$$\mathcal{A}\left((S+\gamma I)\,\delta_1, n_0\right) \geq \exp\left(\delta_1 \ \|S+\gamma I\|_2 \ + \ 1\right) \ - \ \frac{(1 - M\,\delta_1)}{2\,\omega_0} , \tag{16}$$

then by (15)-(16) one gets

$$\|U(t) \ - \ U(t,n)\|_2 \ \leq \ \frac{(1 - M\,\delta_1)}{2} \ < \ (1 - M\,\delta_1) \ < \ \frac{1}{\left\|[U(t)]^{-1}\right\|_2} , \tag{17}$$

and by the perturbation lemma [1] it follows

$$\left\|[U(t,n)]^{-1}\right\|_2 \leq \frac{(1 - M\,\delta_1)^{-1}}{1 - (1 - M\,\delta_1)^{-1}\omega_0 \left(\exp\left(\delta_1 \|S+\gamma I\|_2 + 1\right) - \mathcal{A}\left((S+\gamma I)\,\delta_1, n_0\right)\right)} . \tag{18}$$

By (17)-(18) one gets

$$\left\|[U(t,n)]^{-1}\right\|_2 \ \leq \ 2(1 - M\,\delta_1)^{-1} \ , \ \ n \geq n_0 \ , \ \ |t| < \delta_1 , \tag{19}$$

and by (9) and (14) it follows

$$\|V(t) - V(t,n)\|_2 \leq \omega_0 \left(\exp\left(\delta_1 \|S+\gamma I\|_2 + 1\right) - \mathcal{A}\left((S+\gamma I)\,\delta_1, n\right)\right), \ |t| < \delta_1 . \tag{20}$$

Let K be defined by

$$K \ = \ \max\left\{\delta_1 \|S+\gamma I\|_2, 2\right\} \ + \ 1 , \tag{21}$$

and note that by lemma 1.1 one gets

$$\|V(t,n)\|_2 \ \leq \ L \ , \ \ L \ = \ \omega_0 \ e^{\delta_1^2+1} \sum_{j \geq 0} \sqrt{\frac{(2K)^j}{j!}} \ , \ \ |t| < \delta_1 . \tag{22}$$

Taking into account the formula

$$[U(t)]^{-1} \ - \ [U(t,n)]^{-1} \ = \ [U(t,n)]^{-1} \ (U(t,n) - U(t)) \ [U(t)]^{-1} , \tag{23}$$

for $|t| < \delta_1$ and $n \geq n_0$ one gets:

$$\begin{aligned} &\|W(t) \ - \ W(t,n)\|_2 \\ &= \left\|U(t)\,[U(t)]^{-1} \ - \ U(t,n)\,[U(t,n)]^{-1}\right\|_2 \\ &= \left\|(V(t) - V(t,n))\,[U(t)]^{-1} + V(t,n)\left([U(t)]^{-1} - [U(t,n)]^{-1}\right)\right\|_2 \\ &\leq \|V(t) - V(t,n)\|_2 \left\|[U(t)]^{-1}\right\|_2 + \|V(t,n)\|_2 \left\|[U(t)]^{-1} - [U(t,n)]^{-1}\right\|_2 \\ &\leq \|V(t) - V(t,n)\|_2 \left\|[U(t)]^{-1}\right\|_2 + \|V(t,n)\|_2 \left\|[U(t,n)]^{-1}\right\|_2 \left\|[U(t)]^{-1}\right\|_2 \|U(t,n) - U(t)\|_2 \\ &\leq \left\|[U(t)]^{-1}\right\|_2 \left(\|V(t) - V(t,n)\|_2 + \|V(t,n)\|_2 \left\|[U(t,n)]^{-1}\right\|_2 \|U(t,n) - U(t)\|_2\right) . \end{aligned} \tag{24}$$

By (15),(19),(20),(22) and (24) it follows that

$$\|W(t) - W(t,n)\|_2 \leq P \left(\exp(\delta_1 \|S+\gamma I\|_2 + 1) - \mathcal{A}((S+\gamma I)\delta_1, n)\right) , \quad (25)$$

where

$$P = \omega_0 (1 - M\delta_1)^{-1} \left(1 + 2L(1 - M\delta_1)^{-1}\right) . \quad (26)$$

Given a prefixed $\epsilon > 0$, take the first positive integer n_1 such that

$$\mathcal{A}((S+\gamma I)\delta_1, n_1) \geq \exp(\delta_1 \|S+\gamma I\|_2 + 1) - \frac{\epsilon}{P} , \quad n_1 \geq n_0 , \quad (27)$$

then by (25) and (27) it follows that

$$\|W(t) - W(t,n)\|_2 \leq \epsilon , \quad |t| < \delta_1 , \quad n \geq n_1 . \quad (28)$$

Thus the following result has been established:

Theorem 2.2 *Let us consider the initial value problem (1), let $\epsilon > 0$ an admissible error, let M, ω_0 and δ_1 be the positive numbers defined in theorem 2.1, let K and L the positive constants defined by (21) and (22), let n_0 be the first positive integer satisfying (16) and let n_1 be the first positive integer satisfying (27). If n is a positive integer verifying $n \geq n_1$ and $W(t,n)$ is the rational matrix function defined by (13), then $W(t,n)$ is an approximate solution of problem (1), whose error is smaller than ϵ uniformly in the interval $|t| < \delta_1$.*

3. Example

Now we illustrate the previous result with an example in $C^{2\times 2}$, i. e.,with $r = 2$. Let us consider the problem (1) with data

$$A = D = \begin{bmatrix} 1 & -1 \\ 0 & 1 \end{bmatrix} , \quad B = \begin{bmatrix} 0 & 2 \\ 0 & 0 \end{bmatrix} , \quad C = \begin{bmatrix} -1 & 0 \\ 0 & -1 \end{bmatrix} , \quad W_0 = 0 \in C^{2\times 2} . \quad (29)$$

Straightforward computations using *Mathematica, v.2.2.1* show that in this case we have $\|S\|_2 = 2.592862368051453$. The number γ defined in Theorem 1.1 is $\gamma = 0$. The numbers ω_0 and δ_0 defined in Theorem 2.1 take the values $\omega_0 = 1$, $\delta_0 = 0.267328523966705$. This value of δ_0 does not satisfy the inequality (10) but, taking $\delta_1 = 0.22$, we have

$$\ln(0.22) + \|S\|_2 \times 0.22 = -0.943698011658456 ,$$

$$\left\| \begin{bmatrix} A & B \end{bmatrix} \right\|_2 = 2.488489984622652 , \ \ln(2.488489984622652) = 0.911676094643978 .$$

Hence we may take $\delta_1 = 0.22$ and the constant M defined in Theorem 2.1 takes the value $M = 4.4022061661541037$. The first positive integer n_0 verifying the inequality (16) is $n_0 = 11$. The constant K given by (21) is $K = 2$ and L given by (22) is $L = 65.21752882734062$. Given and admissible error $\epsilon = 10^{-3}$, the first positive integer n_1 satisfying (27) is $n_1 = 25$. Thus from Theorem 2.2, the matrix function

$$W(t,25) = U(t,25)\left[U(t,25)\right]^{-1} , \quad |t| < 0.22 , \quad (30)$$

is an approximate solution of (1) whose error is uniformly upper bounded by $\epsilon = 10^{-3}$ in all the interval $|t| < 0.22$.

Acknowledgements

This work has been partially supported by the *Generalitat Valenciana* grant GV-C-CN-1005796 and the Spanish D.G.I.C.Y.T grant PB96-1321-CO2-02.

References

1. J.M. Ortega, Numerical Analysis. A Second Course, McGraw-Hill, New York (1960) pp. 32.

2. L. Dieci, M.R. Osborne and R.D. Russel, SIAM J. Numer. Anal, 25(2), 1055-1073 (1988).

3. M.R. Osborne and R.D. Russel, SIAM J. Numer. Anal, 23, 1023-1033 (1986).

4. M. Athans and P. Falb, Optimal Control, McGraw-Hill, New York (1960).

5. W. Kratz, Quadratic Functional in Variatonal Analysis and Control Theory, Akademie Verlag, Munchen (1995).

6. R.S. Bucy and P.D. Joseph, Filtering for Stochastic Processes with Applications to Guindance, Interscience, New York (1968).

7. W.T. Reid, Riccati differential equations, Academic Press, New York (1971) pp. 28.

8. C.B. Moler and C.F. Van Loan, SIAM Rev, 20, 801-836 (1978).

9. C.R. Kenney and R.B. Leipnik, IEEE Trans. Automatic Control, AC-30, 962-970 (1985).

10. G. Golub y C.F. Van Loan, Matrix Computations, Johns Hopkins Univ. Press, Baltimore, M.A (1989) pp. 551-557.

11. S. Saks and A. Zygmund, Analytic Functions. Elsevier, Ansterdam (1971) pp. 72.

12. N. Dunford and J. Schwartz, Linear Operators, Part I, Interscience, New York (1957) pp. 558.

13. L. Jódar and R. Company, J. Approximation Theory Appl, 12(2), 20-30 (1996).

14. E. Defez and L. Jódar, Some applications of the Hermite matrix polynomials series expansions, Proceedings of the 8 Congreso de Polinomios Ortogonales y Aplicaciones (Sevilla, September 1997). (To appear).

15. A.G. Law, C.N. Zhang, A. Rezazadeh and L. Jódar, Numer. Algorit, 3, 265-272 (1992).

16. L. Jódar, Applied Mathematics and Computation, 50, 23-31 (1992).

8th Int. Coll. on Differential Equations, pp. 243-246
D. Bainov (Ed)

A Construction of Jordan Superalgebras from Triple Systems

NORIAKI KAMIYA
Shimane University, Japan

SUSUMU OKUBO
University of Rochester, U.S.A

Introduction

The concept of the standard embedding plays an important role in the theory of a construction of Lie algebras from triple systems, in particular, Frudenthal-Kantor triple systems and Lie triple systems are intimately related to Lie algebras as well as Lie superalgebras . The notation of a Jordan-Lie triple system is introduced by authors and is investigated for the quasi-classicality.

It is interesting for us to study how the standard embedding behaves for Jordan-Lie triple systems with familiar connection to Jordan superalgebras.

Our aim is to give an unified and explicit frame work for the construction of simple Jordan superalgebras and is to construct from our triple systems called a Jordan-Lie supertriple system.

Main results

Let V be a superspace. Suppose that a triple priduct $[\,.,\ .,\ .] : V \otimes V \otimes V \to V$ satisfies conditions

(1.0) $$\sigma([x, y, z]) = \{\sigma(x) + \sigma(y) + \sigma(z)\}\mathrm{mod}(2)$$

(1.1) $$[xyz] = -\delta(-1)^{xy}[yxz]$$

(1.2) $$(-1)^{xz}[xyz] + (-1)^{yx}[yzx] + (-1)^{zx}[zxy] = 0$$

(1.3) $$[uv[xyz]] = [[uvx]yz] + (-1)^{(u+v)x}[x, [uvy]z] + \delta(-1)^{(u+v)(x-y)}[xy[uvz]]$$

for $\delta = \pm 1$.

Remark. For $\delta = 1$, these equations mean the Lie triple system. For $\delta = -1$, the triple system is intimately related to Jordan-superalgebra , so that we call it a Jordan–Lie supertriple system.

It is often convenient to study both the case of $\delta = 1$ and $\delta = -1$, and we call the triple system satisfying above conditions $(1.0) \sim (1.3)$ a δ Jordan-Lie supertriple system.

If a triple product satisfies (1.1) and (1.3), but not necessarily (1.2), call the system a generalized δ-Jordan Lie triple system.

Def. A generalized δJLTS is called quasi-classical, if there exists a bilinear nondegenerate form $< .| . >$ satisfying conditions:

$$(i) \qquad < x|y >= 0 \; unless \; \sigma(x) = \sigma(y)$$

$$(ii) \qquad < y|x >= (-1)^{xy} < x|y >$$

$$(iii) \qquad < y|[xuv] >= (-1)^{xy+uv} < x|[yvu] > .$$

Def. Let V be a generalized δJLTS. We introduce the left multiplication operations: $V \otimes V \to \mathrm{End}\, V$ by

$$L(x,y)z = [xyz]$$

with product given by

$$(1.4) \qquad L(x,y) \bullet L(u,v) := L(x,y)L(u,v) - \delta(-1)^{(u+v)(x+y)} L(u,v)L(x,y).$$

Def. Let $L(x,y)$ be as in the above. We introduce a vector space M which a linear span of all $L(x,y)'s$, i,e,

$$(1.5) \qquad M = \{X | X = \sum_{j,k} C_{j,k} L(x_j, x_k), x_j, x_k \in V\}$$

for constants $C_{j,k}$ and write $M = L(V,V)$.

Proposition 1([1]). Let V be a quasi-classical generalized $\delta JLTS$. Then the vector space M defined by (1.5) is a quasi-classical Jordan superalgebra for $\delta = -1$.

Example 1.1. Let $x * y$ be an associative and commutative superalgebra, i.e.,

$$\begin{aligned} (x * y) * z &= x * (y * z), \\ y * x &= (-1)^{xy}(x * y). \end{aligned}$$

Then $xyz := (x * y) * z$ defines a generalized JLTS for $\delta = -1$.

Example 1.2. Let V be a Z_2-graded vector space equipped with $< .| . >$ and let $< .| . >$ be a nondegenerate bilinear form satisfying

$$\begin{aligned} < x|y > &= 0 \quad \text{unless } \sigma(x) = \sigma(y) && (1) \\ < x|y > &= (-1)^{xy} < y|x > . && (2) \end{aligned}$$

Then

$$xyz := < y|z > x - \delta(-1)^{xy} < x|z > y$$

defines a quasi-classical Lie supertriple system for $\delta = 1$, and this triple product defines a quasi-classical generalized Jordan-Lie supertriple system for $\delta = -1$.

Example 1.3([2]). Let $W = \sum_i \oplus W_i$ be a supervector space equipped with a supersymmetric bilinear form

$$< x_i, y_j >= (-1)^{ij} < y_j, x_i > \delta_{i,j}.$$

Then

$$\begin{aligned} < x_i y_j z_k >: &= < x_i, y_j > z_k + < y_j, z_k > x_i \\ &- (-1)^{(i+j)k} < z_k, x_i > y_j \end{aligned}$$

defines a $(-1,1)$-Jordan supertriple system on W, (for the details, to see [2]), that is,

$$< x_i y_j z_k > = (-1)^{(i+j)k+ij} < z_k y_j x_i > \tag{1.6a}$$

$$< x_i y_j < z_k v_l w_m >> = << x_i y_j z_k > v_l w_m > \tag{1.6b}$$

$$-(1)^{(i+j)k+ij} < z_k < y_j x_i v_l > w_m > + (-1)^{(k+l)(i+j)} < z_k v_l < x_i y_j w_m >> .$$

Then $(W, [\ ,\ ,\ ])$ is a Lie supertriple system with respect to the new triple product

$$[x_i y_j z_k] := < x_i y_j z_k > - (-1)^{ij} < y_j x_i z_k > .$$

When V is a JLTS,we can construct a large Jordan-superalgebra as follows. That is, we consider a direct sum

$$L_0 = V \oplus M = V \oplus L(V, V).$$

We introduce a bilinear product $X \bullet Y$ $(X, Y \in L_0)$ in L_0 by

$$\begin{aligned} x \bullet y &:= L(x, y) \in M \\ L(x, y) \bullet z &:= -\delta(-1)^{(x+y)z} z \bullet L(x, y) \\ &= [xyz] \end{aligned}$$

for $x, y, z \in V$ in addition to $L(x, y) \bullet L(u, v)$ given by (1.4).

Theorem 2. Let L_0 be given as in the above, then L_0 is a Jordan superalgebra for $\delta = -1$. Moreover, L_0 is quasi-classical, if V is quasi-classical.

Remark ([2], [3], [4]). For $\delta = 1, L_0$ is the standard embedding Lie superalgebra associated with Lie supertriple system V.

Conclusion Remark. Our articles are in mathematical Physics and nonassociative algebras.But it seems that our studying themes are connected with differential equations.

References

1. S. Okubo and N. Kamiya, Univ.of Rochester, Report UR–1467 (1996).

2. N. Kamiya, in: International Symposium on a Nonassociative Algebras and Related Topics, Hiroshima, Japan, World Scientific Pub. (1991) pp. 65–75.

3. S. Okubo and N. Kamiya, Univ.of Rochester, Report UR–1462 (1996).

4. S. Okubo, J. Math. Phys., 35, 2785–2803 (1994).

5. L. Hogben and V.G. Kac, Comm. Alg., 11, 1155–1156 (1983).

6. V.G. Kac, Comm. Alg., 5, 1375–1400 (1977).

7. N. Kamiya, Mem. Fac. Sci., Shimane Univ., 23, 33–51 (1989).

8. N. Kamiya and S. Okubo, in: Proc. 7th Int. Coll. on Differential Equations, D. Bainov (Ed), VSP (1997) pp. 189–196.

9. R.D. Schafer, An Introduction to Nonassociative Algebras, Academic Press (1966).

8th Int. Coll. on Differential Equations, pp. 247-251
D. Bainov (Ed)

Small Amplitude Fundamental Periodic Solutions on a Piecewise Linear Differential Equation

Kaoru KATORI
Faculty of Humanities, Hirosaki University, Bunkyo-cho 1, Hirosaki-shi, Aomori 036, Japan

Masamitu OTAKE
School of Science and Technology, Meiji University, Higasi-Mita 1-1-1, Tama-ku, Kawasaki, Kanagawa 214, Japan

Yoichi MANOME
Yokogawa Hewlett Packard, Takakura-cho 9-1, Hachiouji-shi, Tokyo 192, Japan

Taketoshi MISHIMA
Dept. of Infor. and Comp. Science, Saitama University, 255 Simo-okubo, Urawa-shi, Saitama, Japan

Abstract. In previous papers [1], [2], we had proposed an our new method for solving a piecewise linear ordinary differential equation and showed some examples.

That is, supposing that a resistor and a capacitor are linear but an inductor is piecewise linear, the equation is resulted from a series circuit consisting of the resistor, capacitor and inductor connected with an alternating electro-motive force. Since the nonlinearity is piecewise linear the solution can be given by the continuation of each component which is an analytic solution in the corresponding linear interval. But such troublesome continuation prevents us from grasping the whole aspect of the solutions. In this paper , we will show some results of the calculation with our method. It is important that results obtained by our method are not approximate results, but equivalent to the analytic ones.

Keywords: nonlinear, piecewise linear, exact solution

1. Introduction

We consider the following differential equation (1) that is obtained in the circuit as Figure 1.

$$\frac{d\Psi}{dt} + IR + \frac{Q}{C} = E\cos\omega t, \qquad I = \frac{dQ}{dt}. \tag{1}$$

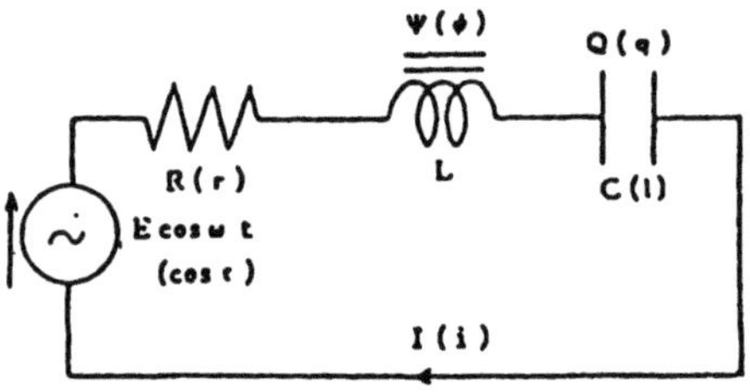

Figure 1: Nonlinear Circuit

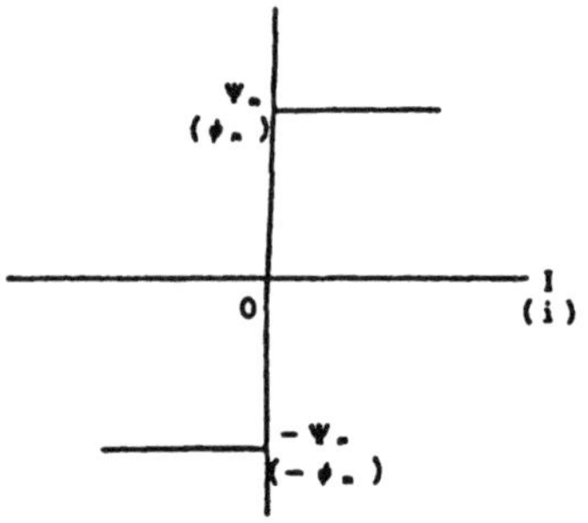

Figure 2: Nonlinear Characteristic

All terms in Equation (1) are normalized by the mentioned below transformations

$$\tau \equiv \omega t, \qquad q \equiv \frac{Q}{EC}, \qquad \psi \equiv \omega\frac{\Psi}{E},$$

$$i \equiv \frac{I}{\omega EC}, \qquad r \equiv \omega CR, \qquad \psi_m \equiv \omega\frac{\Psi_m}{E}. \tag{2}$$

All parametres in the round brackets in Figure 1 and Figure 2 show the terms resulted from above normalization. Then, we have the next equation (3).

$$\frac{d\psi}{d\tau} + ir + q = \cos\tau, \qquad i = \frac{dq}{d\tau} \tag{3}$$

The symbols Ψ, Q and I represent the number of magneticflux interlinkages in the inductor, the electric charge in the capacitor and the electric current running through the circuit, respectively. It is assumed that the resistance and the capacitance are independent of both the electric current and the voltage, but the inductance depends on the current as shown in Figure 2.

2. Notation and results

In order to solve equation (3), we consider the next two cases.

(A) When τ is between τ_1 and τ_2; In this case, q is equal to $q1$. So equation (3) is as follows:

$$\frac{d\psi}{d\tau} + ir + q1 = \cos\tau_1$$

and then,

$$\psi = \sin\tau - q1\tau + C \qquad (C \text{ is an integral constant}).$$

When $\tau = \tau_1$, $\psi = \psi_m$. So, $C = \psi_m - \sin\tau_1 + q1\tau_1$

$$-\psi_m \le \psi = \sin\tau - \sin\tau_1 - (\tau - \tau_1)q \pm \psi \le \psi_m. \tag{4}$$

(B) When τ is between τ_2 and τ_3; In this case, ψ is a constant. So equation (3) is as follows:

$$ir + q = \cos\tau, \qquad i = \frac{dq}{d\tau}.$$

Then we can get the next equation,

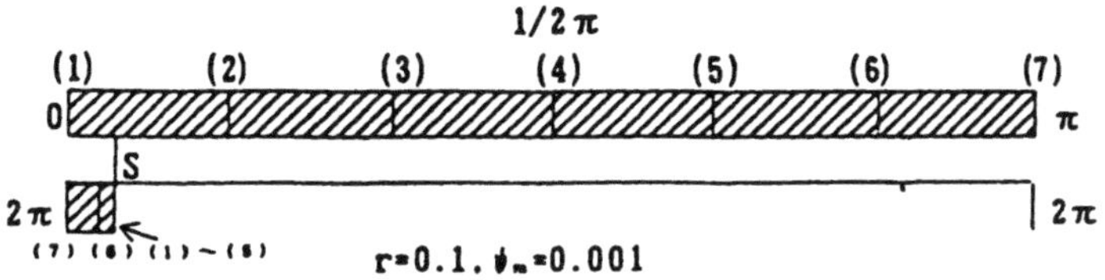

Figure 3-1: $r = 0.1$, $\psi_m = 0$

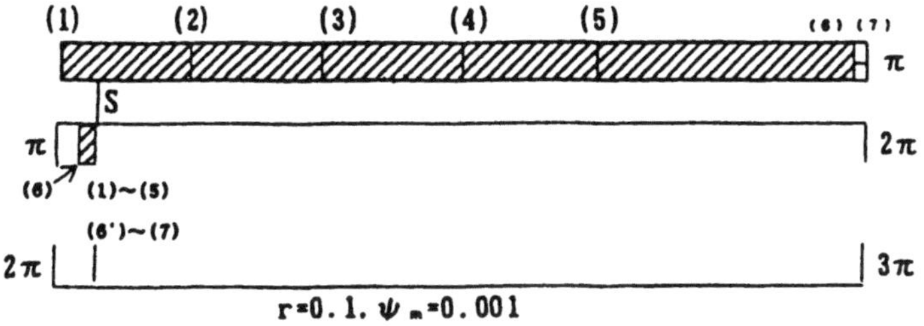

Figure 3-2: $r = 0.1$, $\psi_m = 0.001$

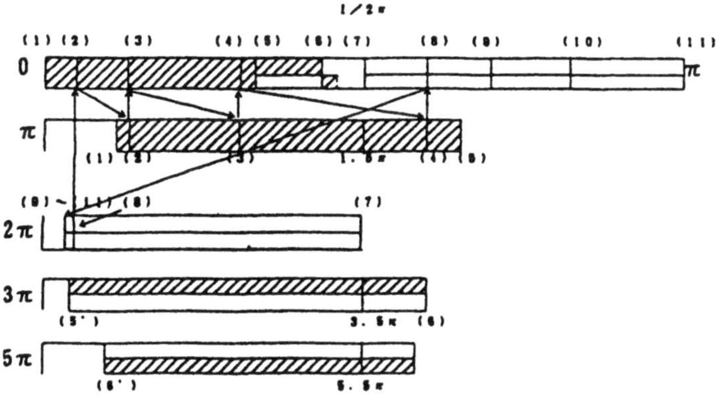

Figure 3-3: $r = 0.1$, $\psi_m = 1.890$

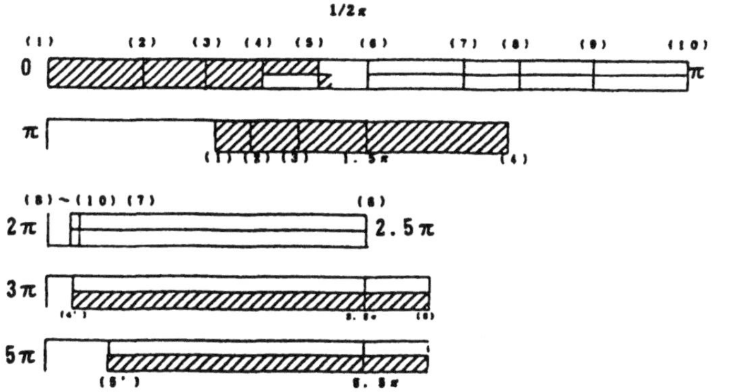

Figure 3-4: $r = 0.1$, $\psi_m = 2$

$$i = \frac{dq}{d\tau} + \frac{q}{r} = \frac{1}{r}\cos\tau,$$

(5) $$ir = \frac{r^2\cos\tau - r\sin\tau}{1+r^2} - \left(q - \frac{\cos\tau_2 + r\sin\tau_2}{1+r^2}\right)^{*}\exp\left(-\frac{\tau-\tau_2}{r}\right).$$

Double signs are used in the same order. It is difficult to understand the behaviour of the solution in Equation (3) according to the expressions (4) and (5) even if the analytic solution is obtained for each linear interval.

The schematic explanation for (A) and (B) of the expressions (4) and (5) is shown first for understanding the behaviour of the solution.

Figure 3 shows the phase of τ_3 (between π and 3π) obtained from the expressions (4) and (5) corresponding to an arbitrary phase τ_1.

The numbers (1), (2), ... , (9) between π and 3π show the phases of τ_3 corresponding to the phases (1), (2), ... , (9) of τ_1, respectively.

Figure 3 explains that any oscillation starting at an arbitrary given phase τ_1 gradually tends to the fixed oscillation S. The oscillation S is a fundamental periodic oscillation because of $\tau_3 = \tau_1 + \pi$.

3. Examples of analysis

We will show with some examples the schematic representation of solutions.

Table 1 and Table 2 show some calculatig values when ψ_m = 0, 0.0001, 1.894, 2, respectively.

ψ_m	0	(Figure 3-1)	0.001	(Figure 3-2)	1.894	(Figure 3-3)
(1)	0	$\pi + 0.0997$	0	$\pi + 0.0997$	0	$\pi + 0.6841$
(2)	0.5	$\pi + 0.0997$	0.5	$\pi + 0.0997$	0.1225	$\pi + 0.6933$
(3)	1	$\pi + 0.0997$	1	$\pi + 0.0997$	0.6933	$\pi + 1.0506$
(4)	$\pi/2$	$\pi + 0.0997$	$\pi/2$	$\pi + 0.0997$	1.0506	$\pi + 1.9717$
(5)	$\pi - 1$	$\pi + 0.0997$	$\pi/2 + 0.5$	$\pi + 0.0997$	$\pi/2 - 0.513$	$1.5\pi + 0.513$ $3\pi + 0.0998$
(6)	$\pi - 0.5$	$\pi + 0.0982$	$\pi - 0.0670$	$\pi + 0.0670$ $2\pi + 0.0097$	$\pi/2 - 0.187$	$3.5\pi + 0.187$ $5\pi + 0.469$
(7)	π	π	π	$2\pi + 0.0997$	$\pi/2$	2.5π
(8)					1.9717	$2\pi + 0.1125$
(9)					$\pi/2 + 0.6350$	$2\pi + 0.0997$
(10)					$\pi - 0.5$	$2\pi + 0.0997$
(11)					π	$2\pi + 0.0997$

Table 1: Calculating example (ψ_m = 0, 0.0001, 1.894)

ψ_m	2 (Figure 3-4)				
(1)	0	$\pi + 0.7795$	(6)	$\pi/2$	2.5π
(2)	0.5	$\pi + 0.9771$	(7)	$\pi/2 + 0.3670$	$2\pi + 0.1288$
(3)	0.7	$\pi + 1.1918$	(8)	$\pi/2 + 0.6350$	$2\pi + 0.0997$
(4)	$\pi/2 - 0.5716$	$1.5\pi + 0.5716$ $3\pi + 0.0998$	(9)	$\pi - 0.5$	$2\pi + 0.0997$
(5)	$\pi/2 - 0.2091$	$3.5\pi + 0.2091$ $5\pi + 0.4690$	(10)	π	$2\pi + 0.0997$

Table 2: Calculating example ($\psi_m = 2$)

4. Conclusion

In this paper, we showed some examples, by using of a piecewise linear approximation method, for solving nonlinear differential equations.

Our method is not an approximation method. So we can obtain all kinds of solutions which are to exist. In this meaning, our method is equivalent to the general analytic solution.

References

1. K. Katori, M. Otake, Y. Manome, T. Mishima, Proc. of the Sixth International Colloquium on Differential Equations, pp. 119-126 (1995).

2. K. Katori, M. Otake, Y. Manome, T. Mishima, Proc. of the Seventh International Colloquium on Differential Equations, pp. 205-212 (1996).

3. M. Goto, Res. of ETL, No. 281 (1930).

4. H.D. Block, IRE Trans. on Circuit Theory, Dec. (1960).

5. G.Takeuti, Res. of ETL, No. 480 (1946).

6. Y.Kane et al., The Jour. of IEE of Japan, Vol. 91, No. 11 (1971).

8. K.Hatta et al., IEICE Tech. Rept., NLP91-101 (1992).

9. K.Hatta et al., IEICE Tech. Rept., NLP92-110 (1993).

10. K.Okumura et al.,. IEICE, Vol. J62-A, No. 5 (1979).

11. M.Otake, K.Katori, Res.Rept. of the Fac. of ENG., Meiji Univ., No. 42 (1982).

12. K. Katori, M. Otake, M. Morisue, IEEE Trans. on MAG., MAG-23, No. 2 (1987).

8th Int. Coll. on Differential Equations, pp. 253-260
D. Bainov (Ed)

Monotonicity Methods for Stability Analysis of the Characteristic Polynomials Whose Coefficients are Polynomials

TAKESHI KAWAMURA
Dept. of Electrical and Electronic Eng., Kitami Inst. of Tech., 165 Koen-cho, Kitami, Hokkaido 090, Japan

MASASUKE SHIMA
Div. of Systems and Information Eng., Fac. of Eng., Hokkaido Univ., North 13, West 8, Sapporo 060, Japan

Abstract. In this paper, we study the stability analysis of the characteristic polynomials by means of monotonicity methods. We will modify the Kharitonov's theorem for the polynomials whose coefficients are polynomials of interval parameters.
Keywords: Monotonicity, Peak monotonicity, Stability analysis.

1. INTRODUCTION

In the field of the control engineering, there are many studies concerned with the stability of the characteristic polynomials whose coefficients contain the interval parameters. Most of them are studied the stability problem in which interval parameters appear linearly or multi-linearly in the coefficients of polynomials. But even if the system to be controlled is described by the linear state space form, its parameters appear nonlinearly in the coefficients of characteristic polynomials.

We take this point of view in this paper, so we set our goal to derive the stability conditions of polynomials whose coefficients are interval polynomials. In our previous works with notion of monotonicity, we have derived the following two results, one is an extension of the mapping theorem and another is a modification of the Kharitonov's theorem. In these studies, if a polynomial is monotone with respect to parameters, the polynomial has the relative maximum and minimum values on the endpoints of parameters. These relative maximum and minimum values have played an important role in the stability analysis.

In this paper, we study the case where coefficients are not monotone with respect to interval parameters. We will define the notion of "peak monotonicity" and estimate the intervals which contain the maximum and minimum value of polynomials and derive the stability conditions.

2. DESCRIPTION OF THE SYSTEM

We study the characteristic polynomial

$$F(s) = s^n + c_1(\boldsymbol{p})s^{n-1} + c_2(\boldsymbol{p})s^{n-2} + \cdots + c_{n-1}(\boldsymbol{p})s + c_n(\boldsymbol{p}) \quad (1)$$

where $c_j(\boldsymbol{p})$'s are polynomials of $\boldsymbol{p} = (p_1, \ldots, p_m)$,
$p_i \in [\, \underline{p}_i,\ \overline{p}_i \,]$, $p_i \in \Re$, $i = 1, 2, \ldots, m$.

For each $c_j(\boldsymbol{p})$, there exists an interval $[\, \underline{c}_j,\ \overline{c}_j \,]$ to which the value of $c_j(\boldsymbol{p})$ belongs under the restrictions $p_i \in [\, \underline{p}_i,\ \overline{p}_i \,]$ $(i = 1, \ldots, m;\ j = 1, \ldots, n)$.

The relations between the interval $[\, \underline{c}_j,\ \overline{c}_j \,]$ and the interval $[\, \underline{p}_i,\ \overline{p}_i \,]$ are complicated in general. For example, the system with structural uncertainties which expressed in the state space form

$$\begin{aligned} \dot{\boldsymbol{x}}(t) &= \boldsymbol{A}\boldsymbol{x}(t); \quad \boldsymbol{A} = \boldsymbol{A}_0 + p_1\boldsymbol{A}_1 + \cdots + p_m\boldsymbol{A}_m \\ \text{where} &\quad p_i \in [\, \underline{p}_i, \overline{p}_i \,],\ p_i \in \Re;\ \boldsymbol{A}, \boldsymbol{A}_i \in \Re^{n \times n},\ i = 1, 2, \ldots, m. \end{aligned} \tag{2}$$

This system has the characteristic polynomial in the form of equation (1).

3. MONOTONICITY AND MONOTONICITY TESTS

We consider the monotonicity in two cases, one is a polynomial with one interval parameter and another is a polynomial with m interval parameters $(m \geq 2)$.

Definition 1.[Monotone function of p_i][1] If a function $c(\boldsymbol{p}) = c(p_1, \ldots, p_m)$ is a polynomial of p_i, $i = 1, \ldots, m$, and satisfies the following condition for some i,

$$\frac{\partial c(\boldsymbol{p})}{\partial p_i} \leq 0 \quad \text{or} \quad \frac{\partial c(\boldsymbol{p})}{\partial p_i} \geq 0 \tag{3}$$
$$\text{for each } p_i \in (\, \underline{p}_i,\ \overline{p}_i \,), p_k \in [\, \underline{p}_k,\ \overline{p}_k \,], k \neq i;\ k = 1, 2, \cdots, m$$

it is called a monotone function of p_i. □

3.1. Monotonicity test for one-parameter polynomial

In our previous works, we gave the monotonicity test for polynomial (1). This monotonicity test is based on the Sturm's theorem which can be used to count the number of roots on given interval by the Euclidean algorithm. According to the Euclidean algorithm, we have

$$r_1 = \frac{dc(p)}{dp}, \quad r_2 = \frac{d^2c(p)}{dp^2} \tag{4}$$

$$r_1 = q_1r_2 - r_3,\ r_2 = q_2r_3 - r_4,\ \ldots,\ r_{w-2} = q_{w-2}r_{w-1} - r_w \tag{5}$$

where r_w is the greatest common factor of r_1 and r_2.

Evaluating the values $r_1(\underline{p})$, $r_2(\underline{p})$, …, $r_w(\underline{p})$ and counting how many times $r_1(\underline{p})$, $r_2(\underline{p})$, …, $r_w(\underline{p})$ change its signs, we denote its number by $V(\underline{p})$. The same procedure is applied for $V(\overline{p})$. Then we have the theorem with $V(p)$.

Theorem 1.[Monotonicity test(1)][1] In the above system, if the following condition 1) or 2) holds, $c(p)$ is a monotone function of p on the interval $[\, \underline{p},\ \overline{p} \,]$.

$$1) \quad V(\underline{p}) - V(\overline{p}) = 0 \tag{6}$$

$$2) \quad V(\underline{p}) - V(\overline{p}) = 1 \text{ and } \operatorname{sgn}(r_1(\underline{p})r_1(\overline{p})) > 0 \tag{7}$$

□

Remark 1.[1] If r_1 has α-tuple roots at $p = \underline{p}$ or/and β-tuple roots at $p = \overline{p}$, then all functions (4) and (5) become 0. In this case, $V(p)$ can not be calculated in the neighborhood of $\underline{p}$ or/and $\overline{p}$.

So we must eliminate the factor $(p-\underline{p})^{\alpha-1}$ or/and $(p-\overline{p})^{\beta-1}$ in r_1 and r_2. □

3.2. Monotonicity test for multi-parameter polynomial

For multi-parameter polynomials, we will examine the monotonicity in this section. Firstly, we give the definition for the convenience of the expression, because in the multi-parameter case there are many subscripts.

Definition 2.[1] For parameters $p_1, p_2, \ldots, p_m$, we define the set of parameters p_i $(i = 1, 2, \ldots, m)$ as $p(1_{j_1}2_{j_2}3_{j_3}\cdots m_{j_m})$ where each $j_1, j_2, \cdots, j_m$ takes 0 or 1 respectively. In this form, we denote $j_i = 0$ if $p_i = \underline{p}_i$ and also $j_i = 1$ if $p_i = \overline{p}_i$.
For example, $p(i_0 j_1)$ means $p_i = \underline{p}_i$ and $p_j = \overline{p}_j$. □

In this section, we suppose a function $c(\boldsymbol{p}) = c(p_1, \ldots, p_m)$ be a polynomial of p_i, $i = 1, \ldots, m$, where $p_i \in [\,\underline{p}_i,\ \overline{p}_i\,]$ and $c(\boldsymbol{p}) \in [\,\underline{c},\ \overline{c}\,]$. In this case, if $\frac{\partial c(\boldsymbol{p})}{\partial p_i} \neq 0$ on $(\,\underline{p}_i,\ \overline{p}_i\,]$ for some i, then the function $c(\boldsymbol{p})$ is monotone in p_i by Definition 1.
In the same way as one parameter case, we examine the existence of roots of $\frac{\partial c(\boldsymbol{p})}{\partial p_i} = 0$ on $(\,\underline{p}_i,\ \overline{p}_i\,]$. We have the sequence

$$r_{i\,1} = \frac{\partial c(\boldsymbol{p})}{\partial p_i}, \quad r_{i\,2} = \frac{\partial^2 c(\boldsymbol{p})}{\partial p_i^2} \tag{8}$$

$$r_{i\,1} = q_{i\,1}r_{i\,2} - r_{i\,3},\ r_{i\,2} = q_{i\,2}r_{i\,3} - r_{i\,4},\ \ldots,\ r_{i\,w_i-2} = q_{i\,w_i-2}r_{i\,w_i-1} - r_{i\,w_i} \tag{9}$$

where $r_{i\,1}, r_{i\,2}, \ldots, r_{i\,w_i}$ are polynomials of $p_1, p_2, \ldots, p_m$.

And $r_{i\,w_i}$ is the greatest common factor of $r_{i\,1}$ and $r_{i\,2}$.
As in the case of one-parameter system, $V(p(1_0 2_0 \cdots (i-1)_0 (i+1)_0 \cdots m_0))(\underline{p}_i)$ is calculated by counting the sign changes of the sequence

$$r_{i\,1}(p(1_0 2_0 \cdots (i-1)_0 (i+1)_0 \cdots m_0))(\underline{p}_i),\ r_{i\,2}(p(1_0 2_0 \cdots (i-1)_0 (i+1)_0 \cdots m_0))(\underline{p}_i),\ \ldots,\ r_{i\,w_i}(p(1_0 2_0 \cdots (i-1)_0 (i+1)_0 \cdots m_0))(\underline{p}_i).$$

And $V(p(1_0 2_0 \cdots (i-1)_0 (i+1)_0 \cdots m_0))(\overline{p}_i)$ is calculated in the same way. Using these expressions, we have the following theorem.

Theorem 2.[Monotonicity test(2)][1] If the following conditions 1), and 2) or 3) hold, $c(\boldsymbol{p})$ is a monotone function of p_i on $[\,\underline{p}_1,\ \overline{p}_1\,] \otimes [\,\underline{p}_2,\ \overline{p}_2\,] \otimes \cdots \otimes [\,\underline{p}_m,\ \overline{p}_m\,]$.
1) Let $r_{i\,1}$ be regarded as a polynomial of p_i.
All coefficients of p_i in $r_{i\,1}$ are monotone of p_k. ($k \neq i, k = 1, 2, \ldots, m$).
2) For all pairs $\underline{p}_k$ and $\overline{p}_k$ $(k \neq i)$,

$$\begin{aligned} & V(p(1_{j_1}2_{j_2}\cdots(i-1)_{j_{i-1}}(i+1)_{j_{i+1}}\cdots m_{j_m}))(\underline{p}_i) \\ - \; & V(p(1_{j_1}2_{j_2}\cdots(i-1)_{j_{i-1}}(i+1)_{j_{i+1}}\cdots m_{j_m}))(\overline{p}_i) = 0 \end{aligned} \tag{10}$$

where $j_1, j_2, \ldots, j_{i-1}, j_{i+1}, \ldots, j_m = 0, 1$.

3) For all pairs $\underline{p}_k$ and $\overline{p}_k$ $(k \neq i)$,

$$\begin{aligned} & V(p(1_{j_1}2_{j_2}\cdots(i-1)_{j_{i-1}}(i+1)_{j_{i+1}}\cdots m_{j_m}))(\underline{p}_i) \\ - \; & V(p(1_{j_1}2_{j_2}\cdots(i-1)_{j_{i-1}}(i+1)_{j_{i+1}}\cdots m_{j_m}))(\overline{p}_i) = 1 \end{aligned} \tag{11}$$

and

$$\begin{aligned} & \mathrm{sgn}(r_{i\,1}(p(1_{j_1}2_{j_2}\cdots(i-1)_{j_{i-1}}(i+1)_{j_{i+1}}\cdots m_{j_m})(\underline{p}_i)) \\ & r_{i\,1}(p(1_{j_1}2_{j_2}\cdots(i-1)_{j_{i-1}}(i+1)_{j_{i+1}}\cdots m_{j_m})(\overline{p}_i))) > 0 \end{aligned} \tag{12}$$

where $j_1, j_2, \ldots, j_{i-1}, j_{i+1}, \ldots, j_m = 0, 1.$ □

Remark 2.[1] As in Remark 1, if $r_{i\,1}$ has α-tuple roots at $p_i = \underline{p}_i$ or/and β-tuple roots at $p_i = \overline{p}_i$, we must eliminate $(p_i - \underline{p}_i)^{\alpha-1}$ or/and $(p_i - \overline{p}_i)^{\beta-1}$ in $r_{i\,1}$ and $r_{i\,2}$.
After the elimination, we can calculate the term

$$\begin{aligned} & V(p(1_{j_1}2_{j_2}\cdots(i-1)_{j_{i-1}}(i+1)_{j_{i+1}}\cdots m_{j_m}))(\underline{p}_i) \\ - & V(p(1_{j_1}2_{j_2}\cdots(i-1)_{j_{i-1}}(i+1)_{j_{i+1}}\cdots m_{j_m}))(\overline{p}_i) \end{aligned}$$

as in the case of one-parameter function. □

Remark 3.[1] In the Theorem 2, we have to check whether $r_{i\,1}$ is monotone in p_k or not. If $r_{i\,1}$ can be expressed as a polynomial of p_i, the coefficients of the powers of p_i are expressed by polynomials of p_k $(k \neq i)$. Applying this procedure repeatedly to the coefficients, we finally obtain coefficient functions with only one parameter. Then we can check its monotonicity as in the case of one-parameter system and calculate the maximum values and the minimum values of these coefficient functions.

Using these maximum and minimum values, we obtain interval polynomials whose monotonicity can be checked. Calculating its maximum and minimum values, values yield another interval polynomials and so on. This backward procedure terminates with finite steps and we can check the monotonicity of $r_{i1}(\boldsymbol{p})$. □

4. PEAK MONOTONICITY

Firstly, we give the notion of peak monotonicity for non-monotone polynomials.

Definition 3.[Peak monotonicity of one parameter polynomial] Let $c(p)$ be a polynomial of $p \in [\underline{p},\ \overline{p}]$. Dividing $c(p)$ by $r_1(p) = dc(p)/dp$, we have

$$\begin{aligned} c(p) &= q_1(p)r_1(p) + r_2(p) \qquad (13)\\ \text{where} \quad & q_1(p) \text{ and } r_2(p) \text{ are polynomial of } p,\ p \in [\underline{p},\ \overline{p}]. \end{aligned}$$

If the residual polynomial $r_2(p)$ is a monotone function of p, $c(p)$ is called a peak-wise monotone polynomial. □

For the multi-parameter case, we consider a polynomial $c(\boldsymbol{p})$ of $\boldsymbol{p} = (p_1, p_2, \ldots, p_m)$ with $p_i \in [\underline{p}_i,\ \overline{p}_i]$, $i = 1, 2, \ldots, m$ and $c(\boldsymbol{p}) \in [\underline{c},\ \overline{c}]$. It is natural that the ridges and the valleys of $c(\boldsymbol{p})$ are distributed over the whole regions $[\underline{p}_1,\ \overline{p}_1] \otimes [\underline{p}_2,\ \overline{p}_2] \otimes \cdots \otimes [\underline{p}_m,\ \overline{p}_m]$.

Definition 4.[Peak monotonicity of multi-parameter polynomial] Let $c(\boldsymbol{p})$ be a polynomial of $\boldsymbol{p} = (p_1, p_2, \ldots, p_m)$, $p_i \in [\underline{p}_i,\ \overline{p}_i]$. Dividing $c(\boldsymbol{p})$ by $r_{i1}(\boldsymbol{p}) = \partial c(\boldsymbol{p})/\partial p_i$, we have

$$\begin{aligned} c(\boldsymbol{p}) &= q_{i1}(\boldsymbol{p})r_{i1}(\boldsymbol{p}) + r_{i2}(\boldsymbol{p}) \qquad (14)\\ \text{where} \quad & q_{i1}(\boldsymbol{p}) \text{ and } r_{i2}(\boldsymbol{p}) \text{ are polynomial of } \boldsymbol{p} = (p_1, p_2, \ldots, p_m),\\ & p_i \in [\underline{p}_i,\ \overline{p}_i],\ i = 1, 2, \ldots, m. \end{aligned}$$

If the residual polynomial $r_{i2}(\boldsymbol{p})$ is a monotone function of p_i, $c(\boldsymbol{p})$ is called a peak-wise monotone polynomial of p_i. □

Using the criteria of the peak monotonicity, we can check the stability of the characteristic polynomials. In our previous works, our monotonicity conditions are applied to the extension of the mapping theorem and the modification of Kharitonov's theorem. In these studies, the maximum values and the minimum values of coefficients are necessary in the stability analysis. In this paper, we also estimate the maximum and the minimum values of $c(p)$ and $c(\boldsymbol{p})$, using the notion of peak monotonicity.

Theorem 3.[One parameter polynomial] Let $c(p)$ be a polynomial of the only one parameter $p \in [\underline{p},\ \overline{p}]$.

If $c(p)$ is a peak-wise monotone function of p, then the values of $c(p)$ on $[\underline{p},\ \overline{p}]$ lie on the interval $[\underline{c}',\ \overline{c}']$; where $\underline{c}' = \min(c(\underline{p}), r_2(\underline{p}), c(\overline{p}), r_2(\overline{p}))$, $\overline{c}' = \max(c(\underline{p}), r_2(\underline{p}), c(\overline{p}), r_2(\overline{p}))$. □

Proof of Theorem 3. If $r_2(p)$ is a monotone function of p, $r_2(p)$ takes its maximum and minimum values at the endpoints of parameter interval. At the peaks and the bottoms of $c(p)$, $r_1(p) = 0$ and $c(p) = r_2(p)$. Then the values of $c(p)$ at the peaks and the bottoms lie on the interval composed from the values of $r_2(p)$ at the boundary of the interval $[\underline{p},\ \overline{p}]$.

Thus, the maximum value and the minimum value of $c(p)$ on the interval $[\underline{p},\ \overline{p}]$ lie necessarily on the interval $[\underline{c}',\ \overline{c}']$, where $\underline{c}' = \min(c(\underline{p}), r_2(\underline{p}), c(\overline{p}), r_2(\overline{p}))$, $\overline{c}' = \max(c(\underline{p}), r_2(\underline{p}), c(\overline{p}), r_2(\overline{p}))$. □

As in the case of the monotonicity test, we give the notations for the sake of expressions for multi-parameter case.

Definition 5.[Notations] We define $\underline{c}' = \min(\underline{c}(\boldsymbol{p}_e), r_{12}(\boldsymbol{p}_e), r_{22}(\boldsymbol{p}_e), \ldots, r_{m2}(\boldsymbol{p}_e))$, and $\overline{c}' = \max(\overline{c}(\boldsymbol{p}_e), r_{12}(\boldsymbol{p}_e), r_{22}(\boldsymbol{p}_e), \ldots, r_{m2}(\boldsymbol{p}_e))$, $i = 1, 2, \ldots, m$ where $\underline{c}(\boldsymbol{p}_e)$ denotes the minimum value of polynomial $c(\boldsymbol{p})$ at all the combinations of endpoints of p_i's and $\overline{c}(\boldsymbol{p}_e)$ the maximum value of polynomial $c(\boldsymbol{p})$ at all the combinations of endpoints of p_i's. And $r_{i2}(\boldsymbol{p}_e)$ denotes values of polynomial $r_{i2}(\boldsymbol{p})$ at all the combinations of endpoints of p_i's. □

Using these notations, we have the following theorem.

Theorem 4.[Multi-parameter polynomial] Let $c(\boldsymbol{p})$ be a polynomial of $\boldsymbol{p} = (p_1, p_2, \ldots, p_m)$ with $p_i \in [\underline{p}_i,\ \overline{p}_i]$, $i = 1, 2, \ldots, m$.

If the polynomial $c(\boldsymbol{p})$ is a peak-wise monotone function of p_i, $i = 1, 2, \cdots, m$, then every $r_{i2}(\boldsymbol{p})$ is monotone and the values of $c(\boldsymbol{p})$ on the interval $[\underline{p}_i,\ \overline{p}_i]$ lie on the interval $[\underline{c}',\ \overline{c}']$. □

Proof of Theorem 4. If polynomial $c(\boldsymbol{p})$ is peak-wise monotone in p_i, every $r_{i2}(\boldsymbol{p})$ takes the maximum value and the minimum value at the endpoints of parameter intervals. At the ridges and the valleys of $c(\boldsymbol{p})$, $r_{i1}(\boldsymbol{p}) = 0$ and $c(\boldsymbol{p}) = r_{i2}(\boldsymbol{p})$. The values of $c(\boldsymbol{p})$ at the peaks and bottoms lie on the interval composed from $r_{i2}(\boldsymbol{p})$ at the boundary of the interval $[\underline{p}_1,\ \overline{p}_1] \otimes [\underline{p}_2,\ \overline{p}_2] \otimes \cdots \otimes [\underline{p}_m,\ \overline{p}_m]$. The maximum and the minimum values of $c(\boldsymbol{p})$ on the interval $[\underline{p}_1,\ \overline{p}_1] \otimes [\underline{p}_2,\ \overline{p}_2] \otimes \cdots \otimes [\underline{p}_m,\ \overline{p}_m]$ lie necessarily on the interval $[\underline{c}',\ \overline{c}']$. □

5. MODIFICATION OF KHARITONOV'S THEOREM FOR NONLINEAR COEFFICIENTS

For all coefficients $c_j(\boldsymbol{p})$ of $F(s)$, we can check the monotonicity and the peak monotonicity with respect to p_i by our tests. If all coefficients $c_j(\boldsymbol{p})$ are monotone or peak-wise monotone in p_i, the maximum and minimum value of $c_j(\boldsymbol{p})$ are derived or estimated at either endpoints of interval $[\ \underline{p}_i,\ \overline{p}_i\ ]$. Using the maximum and minimum values of coefficients, we can check the stability by the Kharitonov's Theorem.

Theorem 5.[Modified Kharitonov's theorem] If the coefficients $c_j(\boldsymbol{p})$'s of $F(s)$ defined by (1) are monotone or peak-wise monotone in p_i, $i = 1, 2, \ldots, m$, then the maximum value and the minimum value of c_j are given by $\underline{c}_j$, $\overline{c}_j$, $\underline{c}'_j$, or $\overline{c}'_j$.

Moreover, $F(s)$ belongs to the convex hull F_h whose 2^n vertices are constructed with the maximum values and the minimum values of c_j, $j = 1, 2, \ldots, n$ and correspond to the 2^m vertices of $[\,\underline{p}_1,\ \overline{p}_1\,] \otimes [\,\underline{p}_2,\ \overline{p}_2\,] \otimes \ldots \otimes [\,\underline{p}_m,\ \overline{p}_m\,]$.

If all vertices polynomials of $F_h(s)$ are stable, then $F(s)$ is stable. □

Proof of Theorem 5. If only a parameter p_i is varied on $[\,\underline{p}_i,\ \overline{p}_i\,]$ with the other parameters fixed, in the case of the monotonicity, it is shown that $c_j(\boldsymbol{p})$ has the relative maximum value at $\underline{p}_i$ or $\overline{p}_i$ due to the monotonicity. And the relative minimum value of $c_j(\boldsymbol{p})$ is determined in the same way.

In the case of the peak monotonicity, the coefficient $c_j(\boldsymbol{p})$ lies on $[\underline{c}'_j,\ \overline{c}'_j]$ ($\underline{c}'_j$, $\overline{c}'_j$ are given in Definition 5).

Another parameter p_k can be varied and so on. And the number of combination the maximum value of c_i and the minimum value of c_j is 2^n.

According to the continuity of polynomials' roots with respect to parameters, $F(s)$'s roots belong to the convex hull F_h as in the proof of the Kharitonov's theorem 1. □

6. EXAMPLE

Let us consider the characteristic polynomial with interval parameters

$$\begin{aligned} F(s) &= s^3 + (3 - p_1 - p_2)s^2 + (p_1^2 + p_1 + p_1 p_2 - 2p_2^2 + 2p_2 + 1)s \\ &+ \frac{1}{4}p_1^3 + \frac{1}{3}p_1^2 p_2 + \frac{1}{8} \qquad (15) \\ &= s^3 + c_1(\boldsymbol{p})s^2 + c_2(\boldsymbol{p})s + c_3(\boldsymbol{p}) \\ \text{where} \quad & p_1 \in [0, 1],\ p_2 \in [0, 1]. \end{aligned}$$

In order to study the stability of the system, we have to check the monotonicity and peak monotonicity of c_j, $j = 1, 2, 3$.

6.1. Monotonicity and peak monotonicity of coefficients

6.1.1. $c_1(\boldsymbol{p})$. $c_1(\boldsymbol{p})$ is monotone in p_1 and p_2. (c_1 is an affine function of p_1 and p_2)

6.1.2. $c_2(\boldsymbol{p})$. Firstly, we consider $c_2(\boldsymbol{p})$ as a polynomial of p_1,

$$\frac{\partial c_2(\boldsymbol{p})}{\partial p_1} = 2p_1 + (1 + p_2) > 0. \qquad (16)$$

All coefficients of p_1 are monotone on the interval $p_2 \in [0, 1]$ and $c_2(\boldsymbol{p})$ is monotone in p_1.

Next, we consider $c_2(\boldsymbol{p})$ as a polynomial of p_2.

$$\frac{\partial c_2(\boldsymbol{p})}{\partial p_2} = -4p_2 + (2 + p_1) \qquad (17)$$

All coefficients of p_2 are monotone on the interval $p_1 \in [0, 1]$, but

$$\left.\frac{\partial c_2(\boldsymbol{p})}{\partial p_2}\right|_{p_1=0,\, p_2=1} \leq 0. \text{ and } \left.\frac{\partial c_2(\boldsymbol{p})}{\partial p_2}\right|_{p_1=1,\, p_2=0} \geq 0 \qquad (18)$$

This means that $c_2(\boldsymbol{p})$ is **NOT** monotone in p_2. Therefore, we examine the peak monotonicity of c_2. According to Definition 4, we have

$$c_2(\boldsymbol{p}) = q_{21}\frac{\partial c_2}{\partial p_2} + r_{22} \qquad (19)$$

where $\quad q_{21}(\boldsymbol{p}) = \frac{1}{2}p_2 - \frac{1}{8}(p_1+2),\ r_{22}(\boldsymbol{p}) = \frac{9}{8}p_1^2 + \frac{3}{2}p_1 + \frac{3}{2}.$

Using r_{22}, now we check the peak monotonicity.

$$\frac{\partial r_{22}}{\partial p_1} = \frac{9}{4}p_1 + \frac{3}{2} \geq 0 \tag{20}$$

This means directly that c_2 is peak-wise monotone in p_2. Then the estimated interval of c_2 is $[1,\ \frac{14}{4}]$.

$$\begin{aligned} r_{22}\Big|_{p_1=0} &= \frac{9}{8},\ r_{22}\Big|_{p_1=1} = \frac{14}{4},\ c_2\Big|_{p_1=0,\ p_2=0} = 1,\ c_2\Big|_{p_1=0,\ p_2=1} = 1, \\ c_2\Big|_{p_1=1,\ p_2=0} &= 3,\ c_2\Big|_{p_1=1,\ p_2=1} = 4. \end{aligned}$$

6.1.3. $c_3(\boldsymbol{p})$. For this coefficients, we consider $c_3(\boldsymbol{p})$ as a polynomial of p_1.

$$\frac{\partial c_3(\boldsymbol{p})}{\partial p_1} = \frac{3}{4}p_1^2 + \frac{2}{3}p_1p_2 \geq 0. \tag{21}$$

All coefficient of p_1 are monotone on the interval $p_2 \in [0, 1]$ and $c_3(\boldsymbol{p})$ is monotone in p_1.
Next, we consider $c_3(\boldsymbol{p})$ as a polynomial of p_2.

$$\frac{\partial c_3(\boldsymbol{p})}{\partial p_2} = \frac{1}{3}p_1^2 \geq 0 \tag{22}$$

In this case $c_3(\boldsymbol{p})$ is monotone in p_2. Thus, $c_3(\boldsymbol{p})$ is monotone in p_1, p_2.

6.2. Making the Kharitonov's 4 polynomials

Calculating the maximum values and the minimum values of c_1, c_3, then we have

$$\begin{aligned} F(s) &= s^3 + c_1s^2 + c_2s + c_3 \\ \text{where} \quad & c_1 \in [1,\ 3],\ c_2 \in \left[1,\ \frac{14}{4}\right],\ c_3 \in \left[\frac{1}{8},\ \frac{17}{24}\right]. \end{aligned} \tag{23}$$

Then we have the Kharitonov's 4 polynomials

$$F_1(s) = s^3 + \underline{c}_1s^2 + \overline{c}_2s + \overline{c}_3 = s^3 + 1s^2 + \frac{14}{4}s + \frac{17}{24} \tag{24}$$

$$F_2(s) = s^3 + \overline{c}_1s^2 + \underline{c}_2s + \underline{c}_3 = s^3 + 3s^2 + 1s + \frac{1}{8} \tag{25}$$

$$F_3(s) = s^3 + \overline{c}_1s^2 + \overline{c}_2s + \underline{c}_3 = s^3 + 3s^2 + \frac{14}{4}s + \frac{1}{8} \tag{26}$$

$$F_4(s) = s^3 + \underline{c}_1s^2 + \underline{c}_2s + \overline{c}_3 = s^3 + 1s^2 + 1s + \frac{17}{24}. \tag{27}$$

According to the Hurwitz stability criterion, all F_1, F_2, F_3, F_4 are stable. Thus, this system is stable.

7. CONCLUDING REMARKS

We present stability analysis methods of interval polynomials, using the concepts of the monotonicity and the peak monotonicity. Introducing the concepts of "peak monotonicity", we present sufficient conditions to estimate the maximum and the minimum value of non-monotone function on the intervals of parameters. Applying our conditions to the stability analysis, we obtain a criterion of the stability of a characteristic polynomial $F(s)$ whose coefficients are non-monotone polynomials of interval parameters.

Using the notions of monotonicity and peak monotonicity, we modify the Kharitonov's Theorem to the case of characteristic polynomials whose coefficients are interval polynomials.

In the example, both the criteria of monotonicity and peak monotonicity are made use of together.

References

1. T. Kawamura and M. Shima, Journal of Mathematical Systems, Estimation, and Control, 6, 461–464 (1996).

2. T. Kawamura and M. Shima, Nonlinear Analysis, (in press).

3. T. Takagi, Lecture in Algebra, Kyouritu-syuppan (1965) (in Japanese).

4. F.R. Gantmacher, The Theory of Matrices, Vol. I, II, Chelsea (1959).

5. J. Ackermann, Robust Control System with Uncertain Physical Prameters, Springer-Verlag (1993).

6. T. Kawamura and M. Shima, in: System and Networks: Mathematical Theory and Applications, Mathematical Research 79, Akademie Verlag (1994) pp.279-282.

8th Int. Coll. on Differential Equations, pp. 261-273
D. Bainov (Ed)

Employing Transform Orthogonality in the Determination of Integral Identities Involving Arbitrary Functions

VICTOR KOWALENKO
School of Physics, University of Melbourne, Parkville, Victoria, 3052, Australia

Abstract. Integral transforms involving an arbitrary function $f(t)$ can be a very powerful means of evaluating sophisticated integrals. In general, however, the same type of integral transform is involved when determining such identities. Here, new integral transform identities are developed not necessarily involving the same type of transform on both sides. To derive these identities, we study the general integral $I = < f, g > = \int_0^\infty dt f(t) g(t)$. It is shown via a suitable choice of kernel function $g(t)$, appropriate changes of variable and the exploitation of transform orthogonality how to develop powerful identities involving different types of integral transforms on both sides. From this study some general theorems will be presented.

Keywords: *Integral transform identities, evaluation of integrals, transform orthogonality*

Mathematics Subject Classification: 26A33, 33C40, 44A10, 45P05.

I. Introduction

Recently, Glasser and Kowalenko (GK) [1] developed a new method for evaluating Laplace transforms (LTs) and other integrals by studying the following LT involving an arbitrary function $f(t)$:

$$s^{-1} g(s) = \int_0^\infty dy \, e^{-py} \int_0^y dt \; J_0\Big(a\sqrt{y^2 - t^2}\Big) f(t) \; , \tag{1}$$

where $s = (p^2 + a^2)^{1/2}$ and $g(s)$ is the LT of $f(t)$ with respect to (w.r.t.) s, denoted here by $\mathcal{L}_s[f(t)]$. Eq. (1) is a standard result in LT theory, which appears as No. 1.1.5.33 in Ref. [2]. Of particular interest to GK was the $p = 0$ case, which yields

$$\int_0^\infty dy \int_0^y dt \, J_0\Big(a\sqrt{y^2 - t^2}\Big) f(t) = a^{-1} \mathcal{L}_a[f(t)] \, . \tag{2}$$

Thus, if the LT of the function $f(t)$ is known and the inner integral can be evaluated, then one can often determine new integrals as demonstrated in Ref. [1]. In some instances, the double integral on the lhs of Eq. (2) can be evaluated, thereby yielding a new LT pair. However, its greatest power can be seen when one first makes the substitution $s = yt$ and then the substitution $u = \sqrt{1 - y^2}/y$ to obtain

$$\int_0^\infty du \sqrt{\frac{u}{u^2+1}} \int_0^\infty ds \; (aus)^{1/2} J_0(aus) s^{1/2} f(s) = a^{-1/2} \mathcal{L}_a[f(s)] \, . \tag{3}$$

In Eq. (3) the inner integral can be regarded as a Hankel transform (HT) according to Oberhettinger's notation [3]. Therefore, Eq. (3) can be written as

$$\int_0^\infty du\, \left(\frac{u}{u^2+1}\right)^{1/2} \mathcal{H}^0_{au}[\sqrt{s}\, f(s)] = a^{-1/2}\, \mathcal{L}_a[f(s)]\,, \tag{4}$$

where the HT is defined as $\mathcal{H}^\nu_u[f(t)] = \int_0^\infty dt\, \sqrt{ut}\, f(t)\, J_\nu(ut)$. Interestingly, Eq. (4) can be generalised by using No. 1.1.5.34 instead of No. 1.1.5.33 in Ref. [2], whereupon one obtains

$$\int_0^\infty du\, \frac{u^{\nu+1/2}}{\sqrt{u^2+1}} \left(\sqrt{u^2+1}+1\right)^{-\nu} \mathcal{H}^\nu_{au}[\sqrt{s}\, f(s)] = a^{-1/2} \mathcal{L}_a[f(s)]\,. \tag{5}$$

Eq. (4) or rather its generalisation given by Eq. (5) is thus a very powerful identity in determining new integrals. Basically, if one knows the LT of a function $f(t)$ and the HT of $\sqrt{s}\, f(s)$, then one can invariably obtain a new integral as exemplified by the large sample of results given in Ref. [1]. As a consequence, the question to be addressed here is whether there is a systematic approach to determining other identities with arbitrary functions, especially those requiring other types of transforms. If a multitude of identities involving arbitrary functions can be derived, then it follows that an enormous number of heretofore unknown integrals can be evaluated very easily. Imagine, for example, if Eq. (5) involved Fourier transforms (FTs) and LTs rather than HTs and LTs. Since we have seen that it is in our interest to obtain further identities involving arbitrary functions, we now need to prove Eq. (1) in the hope that modification of this proof may yield more fruitful identities than Eq. (5).

II. The Proof

Rather than prove Eq. (1) here, we shall consider its generalisation given by No. 1.1.5.34 in Ref. [2]. This is

$$\int_0^\infty dy \int_0^y dt\, \left(\frac{y-t}{y+t}\right)^{\nu/2} J_\nu\!\left(a\sqrt{y^2-t^2}\right) f(t) = a^{-1} \mathcal{L}_a[f(t)]\,, \tag{6}$$

where $\Re\, \nu > -1$ and $\Re\, p > |\Im\, a|$.

Proof. In order to proceed with the proof, let us make the following definition:

$$I = \langle f, g \rangle = \int_0^\infty dt\, f(t) g(\beta, t)\,, \tag{7}$$

where $f(t)$ is an arbitrary function and $g(\beta, t)$ will be referred to as the kernel. All integrals of this form will be assumed to exist in the Lebesgue sense. It should also be noted that the definition incorporates integrals defined over a sub-interval of the integration range since for $I = \int_a^b dt\, f(t) g(\beta, t)$, the kernel can be put equal to $g(t)H(t-a)H(b-t)$, where $H(x)$ is the Heaviside step-function.

Consider the kernel given by

$$g(\beta, t) = \frac{\beta^\nu}{\sqrt{\beta^2+t^2}} \left(t + \sqrt{\beta^2+t^2}\right)^{-\nu}\,. \tag{8}$$

From No. 3.12.1.1 of Ref. [2] we have

$$\int_0^\infty ds\, e^{-ts} J_\nu(\beta s) = g(\beta, t)\,, \tag{9}$$

where $\Re\,\nu > -1$ and $\Re\,t > |\mathrm{Im}\,\beta|$. Under these conditions and assuming that the LT of $f(t)$ exists, which enables Fubini's theorem [4] to be employed, we have

$$I = \int_0^\infty ds\, J_\nu(\beta s)\, \mathcal{L}_s[f(t)]\,. \tag{10}$$

On the other hand, if the substitution $t = \beta y/\sqrt{1-y^2}$ is made, then the integral I becomes

$$I = \int_0^1 dy\, \left(1-y^2\right)^{-1} \left(\frac{1-y}{1+y}\right)^{\nu/2} f\left(\frac{\beta y}{\sqrt{1-y^2}}\right). \tag{11}$$

By using the properties of the Dirac delta function [5], we can write Eq. (11) as

$$I = \int_0^1 dy\, (1-y^2)^{-1/2} \left(\frac{1-y}{1+y}\right)^{\nu/2} \int_0^\infty dt\, f(yt)\, \delta\left(t\sqrt{1-y^2} - \beta\right). \tag{12}$$

We now replace the delta function in Eq. (12) by the orthogonality identity for Bessel functions (see p. 943 of Ref. [6]), i.e.,

$$\delta\left(\sqrt{1-y^2}\,t - \beta\right) = \sqrt{1-y^2}\,t \int_0^\infty du\, u\, J_\nu(\beta u)\, J_\nu\left(\sqrt{1-y^2}\,tu\right). \tag{13}$$

Introducing Eq. (13) into Eq. (12) and using Eq. (10), we obtain

$$I = \int_0^\infty ds\, J_\nu(\beta s)\mathcal{L}_s[f(t)] = \int_0^\infty ds\, s J_\nu(\beta s) \int_0^1 dy \left(\frac{1-y}{1+y}\right)^{\nu/2} \int_0^\infty dt\, t f(yt) J_\nu\left(s\sqrt{1-y^2}\,t\right), \tag{14}$$

where, again, in accordance with Fubini's theorem [4], the order of the integration on the rhs has been interchanged. Whenever two integrals are equal to another, it does not necessarily mean that their integrands are equal to one another. However, the integrals in Eq. (14) are HTs and from the uniqueness theorem for HTs we know that if two HTs are equal to each other, then their integrands must be equal to one another. Since both integrands are equal to one another, we have thus derived Eq. (6).

III. Further Results

In the previous section we presented the proof of the generalised version of Eq. (2) given by Eq. (6). Our aim, however, is to obtain results such as Eq. (5), which involve different types of transforms. We have seen that to obtain such results further changes of variable are required to results like Eq. (6). As a result, not only modifications to the proof of the previous section are made in this section, but also alternative changes of variable are employed in arriving at identities with different transforms of an arbitrary function on both sides. By carrying out this study we shall be able to achieve our ultimate aim of developing general theorems incorporating all the results in this paper.

The first notable feature about the proof in Sec. 2 is that the parameter β must remain arbitrary throughout the analysis. That is, it must only appear as the transformed variable in the final equation, e.g. in the same Bessel functions in Eq. (14) or else the uniqueness theorems for transforms cannot be invoked, which means that the integrands cannot be equated to one another. To see this more clearly, consider the following theorem.

Theorem 1. Given that $\mathcal{H}^{\nu}_{au}[t^{1/2-\nu}f(t)]$ and $\mathcal{L}_a[f(t)]$ are analytic along the positive real a-axis and that $\mathcal{H}^{\nu}_{au}[t^{1/2-\nu}f(t)]=O(u^{\gamma})$, as $u\to 0$, where $\Re\,\gamma>-\Re\,\nu-3/2$ and also is $o(u^{\beta})$ as $u\to\infty$, where $\Re\,\beta<\Re\,\nu-1/2$, then for $\Re\,\nu>-1/2$

$$\int_0^{\infty} du\, \frac{u^{\nu+1/2}}{(u^2+1)^{\nu+1/2}}\, \mathcal{H}^{\nu}_{au}[t^{1/2-\nu}f(t)] = \frac{\sqrt{\pi}\, a^{\nu-1/2}}{2^{\nu}\Gamma(\nu+1/2)}\, \mathcal{L}_a[f(t)]\,. \tag{15}$$

Remark. Note that the above theorem reduces to Eq. (4) for $\nu=0$.

Proof. Consider the integral I with the following kernel:

$$I=< f,g >= \int_0^{\infty} dt\, f(t)\,(t^2+\beta^2)^{-\nu-1/2}\,. \tag{16}$$

As before, we take $\beta>0$. Then according to No. 3.12.1.3 from Ref. [2], for $\Re\,\nu>-1/2$ we can write I as

$$I=\frac{\sqrt{\pi}}{\Gamma(\nu+1/2)}\int_0^{\infty} da\,(a/2\beta)^{\nu}J_{\nu}(\beta a)\,\mathcal{L}_a[f(t)]\,, \tag{17}$$

where the t and a integrations have been interchanged since $\mathcal{L}_a[f(t)]$ is analytic along the positive real a-axis. Making the change of variable $y=t/(t^2+\beta^2)^{1/2}$ transforms I to

$$I=\beta^{-2\nu}\int_0^1 dy\,(1-y^2)^{\nu-1}f\Big(\frac{\beta y}{\sqrt{1-y^2}}\Big)\,. \tag{18}$$

Thus, equating Eq. (18) with Eq. (17) yields

$$\int_0^{\infty} da\, a^{\nu}J_{\nu}(\beta a)\,\mathcal{L}_a[f(t)] = \frac{\Gamma(\nu+1/2)}{\sqrt{\pi}}\Big(\frac{2}{\beta}\Big)^{\nu}\int_0^1 dy\,(1-y^2)^{\nu-1}f\Big(\frac{\beta y}{\sqrt{1-y^2}}\Big)\,. \tag{19}$$

Now we are in a position to introduce the delta function as in the earlier proof. However, here we need to ensure that the final form of Eq. (19) has the arbitrary variable β only in its transform kernel, which in this case will be $J_{\nu}(\beta s)$. If this is not the case, then the uniqueness theorem for HTs cannot be employed. Thus, while we intend to introduce an integral over $\delta(\sqrt{1-y^2}\,t-\beta)$ in Eq. (19), we also need to introduce powers of t to remove the factor of $\beta^{-\nu}$. The previous proof did not require this step.

As a consequence, Eq. (19) can be written as

$$\int_0^1 dy(1-y^2)^{(\nu-1)/2}\int_0^{\infty} dt\,t^{-\nu}f(yt)\,\delta\Big(\sqrt{1-y^2}\,t-\beta\Big)=\frac{\sqrt{\pi}}{2^{\nu}\Gamma(\nu+1/2)}\int_0^{\infty} da\,a^{\nu}J_{\nu}(\beta a)\,\mathcal{L}_a[f(t)]. \tag{20}$$

Introducing Eq. (13) into the above equation gives

$$\begin{aligned}\int_0^{\infty} da\, a J_{\nu}(\beta a)\int_0^1 dy\,(1-y^2)^{\nu/2}\int_0^{\infty} dt\, t^{1-\nu}f(yt)\,J_{\nu}\Big(a\sqrt{1-y^2}\,t\Big) &= \\ =\frac{\sqrt{\pi}}{\Gamma(\nu+1/2)}\int_0^{\infty} da\,(a/2)^{\nu}J_{\nu}(\beta a)\mathcal{L}_a[f(t)]. &\end{aligned} \tag{21}$$

Now β appears as an arbitrary HT transform variable on both sides and hence, employing the uniqueness theorem for HTs gives

$$\int_0^1 dy\,(1-y^2)^{\nu/2}\int_0^{\infty} dt\, t^{1-\nu}f(yt)\,J_{\nu}\Big(a\sqrt{1-y^2}\,t\Big) = \frac{\sqrt{\pi}\,a^{\nu-1}}{2^{\nu}\Gamma(\nu+1/2)}\mathcal{L}_a[f(t)]\,. \tag{22}$$

By replacing yt by s and putting $t = \sqrt{1-y^2}/y$, we obtain Eq. (16). The conditions on γ and β are needed to ensure that the integral on the lhs of Eq. (15) is convergent.

As another example of ensuring that the arbitrariness of β is maintained, let us prove the following theorem.

Theorem 2. Given that $\mathcal{H}_s[t^{3/2}f(t)]$ and $s^{-1}\mathcal{L}_s[f(t)]J_\nu(\beta s)$ are analytic for positive real values of s, and that $\mathcal{H}^\nu_{sx}[t^{3/2}f(t)] = O(x^\alpha)$ as $x \to 0$, where $\Re\alpha > -\Re\nu - 3/2$, and is $o(x^\beta)$ as $x \to \infty$, where $\Re\beta < -3/2$, then for $\Re\nu > !0$

$$\int_0^\infty dx\,\sqrt{sx}\left(\frac{\sqrt{x^2+1}-1}{\sqrt{x^2+1}+1}\right)^{\nu/2} \mathcal{H}^\nu_{sx}[t^{3/2}f(t)] = \nu s^{-1}\mathcal{L}_s[f(t)]\,. \tag{23}$$

Proof. To prove this result using the preceding material the kernel in $< f, g >$ must be replaced by $g(\beta, t) = (t + \sqrt{t^2+\beta^2})^{-\nu}$ where $\Re\nu > 0$. Then we can use No. 3.12.1.7 from Ref. [2] to write

$$I = \beta^\nu \int_0^\infty dt\, f(t)\left(t + \sqrt{t^2+\beta^2}\right)^{-\nu} = \nu \int_0^\infty ds\, s^{-1} J_\nu(\beta s)\, \mathcal{L}_s[f(t)]\,. \tag{24}$$

Making the change of variable $t = \beta y/\sqrt{1-y^2}$ and introducing the δ-function, one obtains

$$I = \int_0^1 dy\,\frac{1}{\sqrt{1-y^2}}\left(\frac{1-y}{1+y}\right)^{\nu/2} \int_0^\infty dt\, t\, f(yt)\, \delta\Big(\sqrt{1-y^2}\,t - \beta\Big)\,. \tag{25}$$

Again, replacing the δ-function in Eq. (25) by the orthogonality identity for Bessel functions, i.e., Eq. (13), one obtains

$$s^2 \int_0^1 dy \left(\frac{1-y}{1+y}\right)^{\nu/2} \int_0^\infty dt\, t^2 f(yt)\, J_\nu\Big(\sqrt{1-y^2}\,st\Big) = \nu\, \mathcal{L}_s[f(t)]\,. \tag{26}$$

Making the two changes of variable are made, first $u = yt$ and second $x = \sqrt{1-y^2}/y$, tranforms Eq. (26) into Eq. (23). The conditions on α and β follow from ensuring that the lhs of Eq. (23) remains convergent.

We now consider a different type of situation where an operator facilitates the evaluation of $< f, g >$. Let us prove the following theorem, which involves LTs on both sides.

Theorem 3. Given that (i) $\mathcal{L}_b\Big[tf(yt)J_0\Big(\sqrt{1-y^2}\,st\Big)\Big]$ is analytic in y over [0,1], (ii) $\mathcal{L}_{\sqrt{s^2+b^2}}[f(t)]$ is analytic for positive real values of s and (iii) $\mathcal{L}_b\Big[tf(yt)J_0\Big(\sqrt{1-y^2}\,st\Big)\Big] = O(y^\alpha)$ as $y \to 0$, where $\Re\alpha > -1$, then

$$\int_0^1 dy\, \mathcal{L}_b\Big[t f(yt) J_0\Big(\sqrt{1-y^2}\,st\Big)\Big] = (s^2+b^2)^{-1/2}\mathcal{L}_{\sqrt{s^2+b^2}}[f(t)]\,. \tag{27}$$

Proof. To prove this result let $g(\beta, t) = exp(-b\sqrt{t^2+\beta^2})$ in $< f, g >$. Then $< f, g >$ can be written as

$$I = -\frac{\partial}{\partial b}\int_0^\infty dt\, f(t)\, \frac{e^{-b\sqrt{t^2+\beta^2}}}{\sqrt{t^2+\beta^2}}\,. \tag{28}$$

The introduction of the partial derivative enables us to employ No. 3.12.6.4 from Ref. [2], whereupon Eq. (28) becomes

$$I = -\frac{\partial}{\partial b}\int_0^\infty ds\, \frac{s\, J_0(\beta s)}{\sqrt{s^2+b^2}}\, \mathcal{L}_{\sqrt{s^2+b^2}}[f(t)]\,. \tag{29}$$

In Eq. (29) the order of the t and s integrations has been interchanged. Putting $t=\beta y/\sqrt{1-y^2}$ and introducing the δ-function, one obtains

$$I=-\frac{\partial}{\partial b}\int_0^1 dy\,\frac{1}{\sqrt{1-y^2}}\int_0^\infty dt\,f(yt)\,e^{-bt}\,\delta\Big(\sqrt{1-y^2}\,t-\beta\Big)\,. \tag{30}$$

Introducing the orthogonality identity for Bessel functions in Eq. (30) and equating the ensuing result with Eq. (29), one obtains Eq. (27) after applying the uniqueness theorem for HTs. The condition on α follows from ensuring that the lhs of Eq. (27) remains convergent at the lower limit. From here on, we shall only state where necessary the conditions for ensuring the convergence of the integrals in ensuing theorems. These are left for the reader to evaluate.

Eq. (27) can be verified by setting $f(t)=1$, which yields $(s^2+b^2)^{-1}$ on the rhs. The LT on the lhs can be evaluated by using No. 2.12.8.4 from Ref. [7] and thus, we are left with

$$b^{-2}\int_0^1 dy\,{}_1F_0\Big(\frac{3}{2};-\frac{(1-y^2)s^2}{b^2}\Big)=\frac{b}{2}\int_0^1 dy\,\frac{(1-y)^{-1/2}}{(s^2y+b^2)^{3/2}}\,. \tag{31}$$

The integral in Eq. (31) can be evaluated by using No. 2.2.5.4 in Ref. [8] and as a consequence, we find that the entire lhs yields the same result of $(s^2+b^2)^{-1}$ as the rhs. Furthermore, if we replace $\exp(-b\sqrt{t^2+\beta^2})$ in Eq. (28) by $(t^2+\beta^2)^{-\nu/2-1/2}K_{\nu+1/2}(b\sqrt{t^2+\beta^2})$, employ No. 3.12.6.7 from Ref. [2] and follow the same procedure used to derive Eq. (27), then we eventually obtain

$$\int_0^1 dy\,(1-y^2)^{\nu/2}\int_0^\infty dt\,t^{3/2}f(yt)\,K_{\nu+1/2}(bt)\,J_\nu\Big(\sqrt{1-y^2}\,st\Big)=$$
$$=\sqrt{\frac{\pi}{2(s^2+b^2)}}\,\frac{s^{\nu/2}}{b^{\nu+1/2}}\,\mathcal{L}_{\sqrt{s^2+b^2}}[f(t)]. \tag{32}$$

This result reduces to Eq. (27) when $\nu=0$, thereby showing the former equation is a generalisation of the latter. In addition, making the changes of variable, $t'=yt$ and $x=\sqrt{1-y^2}/y$ transforms Eq. (32) to

$$\int_0^\infty dx\,\frac{x^{\nu+1/2}}{(x^2+1)^{\nu/2+1/4}}\mathcal{H}^\nu_{sx}\Big[tf(t)\,K_{\nu+1/2}\Big(bt\sqrt{x^2+1}\Big)\Big]=$$
$$=\sqrt{\frac{\pi}{2(s^2+b^2)}}\,\frac{s^{(\nu+1)/2}}{b^{\nu+1/2}}\,\mathcal{L}_{\sqrt{s^2+b^2}}[f(t)]\,. \tag{33}$$

The previous example involved an LT on the rhs, whose transform variable was not linear, but dependent on $\sqrt{s^2+b^2}$. Let us consider another example not involving a non-linear transform variable.

Theorem 4. Given that $\mathcal{L}_{s^2}[f(t)]$ is analytic for positive real values of s and that $\mathcal{L}_{t^2/4}[xe^{-xt^2/4}f(x)J_\nu(sxt)]$ is analytic for positive real values of t, then

$$\int_0^\infty dt\,t^{\nu+1}\mathcal{L}_{t^2/4}[xf(x)J_\nu(stx)]=2^{\nu+1}s^\nu\mathcal{L}_{s^2}[f(t)]\,. \tag{34}$$

Proof. To prove this theorem consider the kernel $g(t)=(\beta/2)^\nu t^{-\nu-1}\exp(-\beta^2/4t)$. For $\Re\,\nu>-1$ we can use No. 3.12.3.8 from Ref. [2] to obtain

$$I=(\beta/2)^\nu\int_0^\infty dt\,t^{-\nu-1}f(t)\,\exp(-\beta^2/4t)=2\int_0^\infty ds\,s^{\nu+1}J_\nu(\beta s)\,\mathcal{L}_{s^2}[f(t)]\,. \tag{35}$$

With the change of variable $t = \beta y/(1-y)$, I becomes

$$I = \int_0^1 dy \, \frac{(1-y)^{\nu-1}}{2^\nu y^{\nu+1}} \, f\left(\frac{\beta y}{(1-y)}\right) \exp\left(-\frac{\beta(1-y)}{4y}\right) . \tag{36}$$

Introducing the δ-function yields

$$I = \int_0^1 dy \, \frac{(1-y)^\nu}{2^\nu y^{\nu+1}} \int_0^\infty dt \, f(yt) \exp\left(-\frac{(1-y)^2 t}{4y}\right) \delta\big((1-y)t - \beta\big) . \tag{37}$$

Using the orthogonality identity for Bessel functions again, one obtains

$$\int_0^1 dy \, (1/y - 1)^{\nu+1} \int_0^\infty dt \, t \exp\left(-\frac{(1-y)^2 t}{4y}\right) f(yt) \, J_\nu((1-y)st) = 2^{\nu+1} s^\nu \mathcal{L}_{s^2}[f(t)] \, . \tag{38}$$

The inner t-integral in Eq. (38) can be regarded as an LT and one can verify this result by putting $f(t) = 1$. Then, making a change of variable and using No. 2.12.8.4 from Ref. [7], one obtains

$$\int_0^\infty dt \, {}_2F_1\left(\frac{\nu}{2} + 1, \frac{\nu+3}{2}; \nu+1; -at\right) = \frac{4}{a\,(\nu+1)} . \tag{39}$$

Eq. (39) is just a special case of No. 2.21.1.1 in Ref. [9]. Finally, to obtain Eq. (34), we must make the substitution $s = yt$ in Eq. (38) and then put $t = (1-y)/y$. Note that a different change of variable has been employed compared with the preceding examples, which represents another modification to the proof in Sec. 2.

The previous examples have dealt with expressing the rhs in terms of the LTs of arbitrary functions, but one need not only consider LTs. As an example, let us prove the following:

Theorem 5. Given that (i) $\mathcal{H}^\nu_{sx}[t^{\nu+1/2} f(t)]$ is analytic in x over $[1, \infty)$ where $\Re\,\nu > -1/2$, (ii) $\mathcal{H}^\nu_{sx}[t^{\nu+1/2} f(t)] = o(x^\gamma)$ as $x \to \infty$ where $\Re\,\gamma < -\Re\,\nu - 1/2$ and (iii) the Fourier cosine transform of $f(t)$ is analytic for positive real values of its transform variable s, then

$$\int_1^\infty dx \, x^{1/2-\nu} (x^2-1)^{\nu-1/2} \mathcal{H}^\nu_{sx}\left[t^{\nu+1/2} f(t)\right] = \frac{2^\nu}{\sqrt{\pi}} \frac{\Gamma(\nu+1/2)}{s^{\nu+1/2}} \mathcal{F}^C_s[f(t)] \, . \tag{40}$$

Proof. Consider $< f, g >$ with $g(\beta, t) = (\beta^2 - t^2)^{\nu-1/2} H(\beta - t)$. Therefore,

$$I = \int_0^\beta dt \, f(t)(\beta^2 - t^2)^{\nu-1/2} . \tag{41}$$

Using No. 13.11 on p. 67 of Ref. [10], we find for $\Re\,\nu > -1/2$ that

$$\int_0^\infty dy \, \frac{\beta^\nu}{(y^2+1)^{\nu+1}} f\left(\frac{\beta y}{\sqrt{y^2+1}}\right) = \frac{2^\nu}{\sqrt{\pi}} \Gamma(\nu+1/2) \int_0^\infty ds \, s^{-\nu} J_\nu(\beta s) \, \mathcal{F}^C_s[f(t)] \, . \tag{42}$$

In Eq. (42) $\mathcal{F}^C_s[f(t)]$ denotes the Fourier cosine transform w.r.t. to y, which is defined as

$$\mathcal{F}^C_s[f(t)] = \int_0^\infty dt \, f(t) \cos(st) \, . \tag{43}$$

In obtaining Eq. (42) we have also made the substitution $y = t/\sqrt{\beta^2 - t^2}$. Then using the properties of the delta function, we can write the lhs of Eq. (42) as

$$\int_0^\infty dy \, (y^2+1)^{-\nu/2-1} \int_0^\infty dt \, t^\nu f(yt) \, \delta\left(t - \frac{\beta}{\sqrt{y^2+1}}\right) = \int_0^\infty dy \, (y^2+1)^{-\nu/2} \int_0^\infty dt \, t^{\nu+1}$$
$$\times \, f(yt) \int_0^\infty ds \, s \, J_\nu(\beta s) \, J_\nu\left(\sqrt{y^2+1}\, ts\right) . \tag{44}$$

Thus by equating the rhs of Eq. (42) with the rhs of Eq. (44), we obtain

$$\int_0^\infty dy\,(y^2+1)^{-\nu/2}\int_0^\infty dt\,t^{\nu+1}f(yt)\,J_\nu\Big(\sqrt{y^2+1}\,st\Big)=\frac{2^\nu}{\sqrt{\pi}}\,\frac{\Gamma(\nu+1/2)}{s^{\nu+1}}\,\mathcal{F}_s^C[f(t)]\,. \qquad (45)$$

Eq. (45) is a powerful identity much like Eq. (6) and can be verified by putting $f(t)=e^{-t}$ and then using Nos. 3.1 on p. 11 of Ref. [10] and 2.12.8.1 from Ref. [7]. By making the change of variable $s=yt$ and then putting $x=\sqrt{y^2+1}/y$, we eventually obtain Eq. (41). The condition on γ is necessary to ensure that the integral on the lhs of Eq. (40) is convergent.

It should also be pointed out that Y-transforms can be used to develop identities involving arbitrary functions. Before considering an example, let us introduce **H**-transform as defined in Ref. [3]. This is defined as

$$S_s^\nu[f(t)]=\int_0^\infty dt\,\sqrt{st}\,f(t)\mathbf{H}_\nu(st)\,,$$

where $\mathbf{H}_\nu(x)$ represents a Struve function. Now we can turn our attention to the following theorem:

Theorem 6. Given that (i) the LT of $f(t)$ is analytic for positive real values of its transform variable and (ii) $S_{sx}^\nu[\sqrt{t}\,f(t)]$ is analytic for positive real values of x, then

$$\int_0^\infty dx\,\sqrt{\frac{x}{x^2+1}}\left(\cos(\nu\pi)\left(\frac{\sqrt{x^2+1}-1}{\sqrt{x^2+1}+1}\right)^{\nu/2}-\left(\frac{\sqrt{x^2+1}+1}{\sqrt{x^2+1}-1}\right)^{\nu/2}\right)S_{sx}^\nu[\sqrt{t}\,f(t)]=$$
$$=\frac{\sin(\nu\pi)}{\sqrt{s}}\,\mathcal{L}_s[f(t)]\,, \qquad (46)$$

provided $|\Re\,\nu|<1$, but $\nu\neq 0$.

Consider $I=<f,g>$ with $g(t)$ given by

$$g(t)=(t^2+\beta^2)^{-1/2}\left[\frac{\beta^\nu\cot(\nu\pi)}{\left(t+\sqrt{t^2+\beta^2}\right)^\nu}-\frac{\left(t+\sqrt{t^2+\beta^2}\right)^\nu}{\beta^\nu\sin(\nu\pi)}\right]. \qquad (47)$$

Making the substitution $t=\beta y/\sqrt{1-y^2}$, we find that $<f,g>$ can be written as

$$\frac{1}{\sin(\nu\pi)}\int_0^1 dy\,(1-y^2)^{-1/2}\left[\cos(\nu\pi)\left(\frac{1-y}{1+y}\right)^{\nu/2}-\left(\frac{1+y}{1-y}\right)^{\nu/2}\right]\int_0^\infty dt\,f(yt)$$
$$\times\,\delta\Big(\sqrt{1-y^2}\,t-\beta\Big)=\int_0^\infty ds\,Y_\nu(\beta s)\,\mathcal{L}_s[f(t)]\,, \qquad (48)$$

where we have used No. 3.13.1.1.from Ref. [2]. Noting that the orthogonality condition for the Y-transform is given by

$$\delta\Big(\sqrt{1-y^2}\,t-\beta\Big)=\sqrt{1-y^2}\,t\int_0^\infty ds\,s\,\mathbf{H}_\nu\Big(\sqrt{1-y^2}\,ts\Big)\,Y_\nu(\beta s)\,, \qquad (49)$$

we arrive at

$$\int_0^1 dy\,\left(\cos(\nu\pi)\left(\frac{1-y}{1+y}\right)^{\nu/2}-\left(\frac{1+y}{1-y}\right)^{\nu/2}\right)\int_0^\infty dt\,t\,f(yt)\,\mathbf{H}_\nu\Big(\sqrt{1-y^2}\,st\Big)=$$
$$=\frac{\sin(\nu\pi)}{s}\,\mathcal{L}_s[f(t)]\,, \qquad (50)$$

where $|\Re\,\nu|<1$, but $\nu\neq 0$. Making the changes of variable, first $s=yt$ and second $x=\sqrt{1-y^2}/y$ yields Eq. (46).

IV. General Theorems

In the previous section we presented additional examples of theorems based on the proof of Eq. (6) given in Sec. 2. Here we consider the essential features of these results in order to develop more general theorems incorporating all the preceding results.

The first feature that we note is that the kernel in our examples can be written in one of two forms. In many of the examples we have been able to make the subsitution $t = \beta p(y)$, so that $g(\beta,t)$ can be written as $\beta^\alpha g_1(y)$, where $g_1(y)$ does not depend upon β. Obviously, Theorem 4 represents an exception when $t! = \beta y/(1-y)$ is introduced into $\exp(-\beta^2/4t)$. We shall consider this case later.

We have seen too that the kernel must be expressible in the form of

$$g(\beta,t) = \int_0^\infty ds\, \beta^\gamma u(s) h(w(s)t)\, Z(\beta s) , \tag{51}$$

where $u(s)$ is a function of s, but not dependent upon β, and γ can be any real number. The function $h(at)$ represents the kernel of a transform. For example, for an LT $h(at)$ would $\exp(-at)$ while for an FT it would be $\exp(iat)$. In fact, $h(at)$ could be any one of $\left\{e^{-at}, e^{iat}, \cos(at), \sin(at), J_\nu(at), t^\alpha, (t+a)^{-1}, \sqrt{t}\, K_\nu(at)\right\}$, the last three indicating it could be the kernel of a Mellin transform (MT), a Hilbert transform or a Kontorovich- Lebedev transform. The function $w(s)$ represents the transform variable for the t-integration and from Theorem 4 we have seen that it need not be a linear function of s. In Eq. (51), however, the most important function is $Z(\beta s)$, which we shall refer to as an orthogonal kernel. This means that it must satisfy

$$\delta(t-\beta) = \int_0^\infty du\, ut Z(\beta u) Z_2(tu) , \tag{52}$$

where $Z_2(tu)$ also becomes the kernel of a transform. In this paper, we have considered

$$\delta(t-\beta) = \int_0^\infty du\, ut\, J_\nu(\beta u) J_\nu(ut) = \int_0^\infty du\, ut\, \mathbf{H}_\nu(ut) Y_\nu(\beta t) . \tag{53}$$

Note that putting $\nu = \pm 1/2$ in Eq. (53) yields cosine and sine functions for $Z_2(tu)$. It is $Z_2(ut)$ that has determined the type of transform appearing on the lhs of the preceding theorems. On the other hand, those transform kernels not possessing the orthogonality of Eq. (52) such as the LT and MT cannot be considered.

Now we are in a position to present the first general theorem, which is:

Theorem 7. Given that the kernel $g(\beta,t)$ of $I = \langle f, g \rangle$ can be written as $\beta^\alpha g_1(y)$ when the substitution $t = \beta y$ is made and that it can also be written as an integral involving an orthogonal kernel, $Z(\beta, s)$, such that

$$g(\beta,t) = \int_0^\infty ds\, \beta^\gamma u(s) h(w(s)t)\, Z(\beta s) ,$$

where $h(at)$ is a transform kernel, then

$$\int_0^\infty dx\, g_1(1/x) x^{\alpha-\gamma} \int_0^\infty dt\, f(t) Z_2(stx) = s^{-1} u(s) \mathcal{A}_{w(s)}[f(t)] , \tag{54}$$

where $\mathcal{A}_{w(s)}[f(t)] = \int_0^\infty dt\, h(w(s)t) f(t)$.

Proof. Consider the $I = \langle f, g \rangle = \int_0^\infty dt\, f(t) g(\beta,t)$. Then making the substitution $t = \beta y$,

$$I = \beta^{\alpha+1} \int_0^\infty dy\, f(\beta y) g_1(y) , \tag{55}$$

where $g_1(y)$ is independent of β. Since I can also be written in terms of an integral with the orthogonal kernel $Z(\beta s)$ as given above, equating the two forms for I yields

$$\int_0^\infty dy\,\beta^{\alpha+1-\gamma} f(\beta y) g_1(y) = \int_0^\infty ds\, u(s) Z(\beta s) \int_0^\infty dt\, h(w(s)t) f(t)\,, \tag{56}$$

where in accordance with Fubini's theorem, we have interchanged the s and t integrals on the rhs of Eq. (56). Introducing the δ-function, we find that the lhs of Eq. (56) can be written as

$$\int_0^\infty dy\, g_1(y) \int_0^\infty dt\, t^{\alpha+1-\gamma} f(yt)\delta(t-\beta) = \int_0^\infty dt\, t^{\alpha+1-\gamma} f(yt) \int_0^\infty ds\, st Z(\beta s) Z_2(ts)\,, \tag{57}$$

where we have used the fact that $Z(\beta s)$ is an orthogonal kernel. If we equate Eq. (57) with the rhs of Eq. (56), then we obtain integrals involving the same transform kernel $Z(\beta s)$. Since such transforms obey a uniqueness theorem, their integrands must be equal to another, whereupon we obtain

$$\int_0^\infty dy\, g_1(y) \int_0^\infty dt\, t^{\alpha+1-\gamma} f(yt) Z_2(st) = s^{-1} u(s) \mathcal{A}_{w(s)}[f(t)]\,, \tag{58}$$

where $\mathcal{A}_{w(s)}[f(t)]$ is given above. Making the change of variable, $z=yt$, yields

$$\int_0^\infty dy\, y^{\gamma-\alpha-2} g_1(y) \int_0^\infty dz\, f(z) Z_2(sz/y) = s^{-1} u(s) \mathcal{A}_{w(s)}[f(t)]\,. \tag{59}$$

Finally, making the change of variable $x=y^{-1}$ yields the desired result.

In the previous section we made changes of variable of the form:

$$t=\beta y/(b-dy^\nu)^\mu\,, \tag{60}$$

so that β could be eliminated by employing orthogonality. In actual fact, b and d were set equal to unity. In many of the examples, $\nu=2$ and $\mu=1/2$, but an example where $\nu=1$ and $\mu=1$ was also considered. Thus, for all the theorems in Sec. 3 $\mu\nu=1$. As a consequence, the change of variable given by Eq. (60) led to the appearance of transform kernels containing $(b-dy^\nu)^\mu$ on the lhs of the various results in the previous section. We now adapt the previous theorem to accommodate this important change of variable.

Theorem 8. Given that the kernel $g(\beta,t)$ of $I=<f,g>$ can be written as $\beta^\alpha g_1(y)$ when the substitution $t=\beta y/(b-dy^\nu)^\mu$ with $\nu\mu=1$ is made and that it can also be written as an integral involving an orthogonal kernel, $Z(\beta s)$ as given in Thm. 7, then

$$\int_0^\infty dx\, x^{\alpha-\gamma+1} g_1\left(\left(\frac{b}{x^{1/\mu}+d}\right)^\mu\right) \int_0^\infty du\, u^{\alpha+2-\gamma} f(u) Z_2(sxu) = s^{-1} u(s) \mathcal{A}_{w(s)}[f(t)]\,, \tag{61}$$

where $\mathcal{A}_{w(s)}[f(t)]$ is defined in Theorem 7.

Proof. The proof is identical to Thm. 7 except that the lhs must now accommodate the change of variable given by Eq. (60). With this change of variable, I becomes

$$I = b\int_0^{(b/d)^{1/\nu}} dy\, \frac{\beta^{\alpha+1} g_1(y)}{(b-dy^\nu)^{\mu+1}} f\left(\frac{\beta y}{(b-dy^\nu)^\mu}\right)\,. \tag{62}$$

Equating Eq. (62) to the form involving the transform kernel $h(w(s)t)$ yields

$$b\int_0^{(b/d)^{1/\nu}} dy\, \frac{\beta^{\alpha+1-\gamma} g_1(y)}{(b-dy^\nu)^{\mu+1}} f\left(\frac{\beta y}{(b-dy^\nu)^\mu}\right) = \int_0^\infty ds\, u(s) Z(\beta s) \mathcal{A}_{w(s)}[f(t)]\,, \tag{63}$$

where in accordance with Fubini's theorem we have interchanged the order of the s and t integrations. We now introduce the δ-function into the lhs of Eq. (63), which yields

$$b\int_0^{(b/d)^{1/\nu}} dy\, g_1(y)(b-dy^\nu)^{\mu(\alpha-\gamma+1)-1}\int_0^\infty dt\, t^{\alpha+1-\gamma} f(yt)\delta((b-dy^\nu)^\mu t-\beta)$$
$$= \int_0^\infty ds\, u(s)Z(\beta s)\mathcal{A}_{w(s)}[f(t)]\,. \tag{64}$$

Since $Z(\beta s)$ is an orthogonal kernel, the δ-function can be replaced by the following result:

$$\delta((b-dy^\nu)^\mu t-\beta) = \int_0^\infty ds\, Z(\beta s)Z_2((b-dy^\nu)^\mu ts)\,(b-dy^\nu)^\mu ts\,. \tag{65}$$

Introducing Eq. (65) into Eq. (64) yields

$$b\int_0^{(b/d)^{1/\nu}} dy\,(b-dy^\nu)^{\mu(\alpha-\gamma+2)-1} g_1(y)\int_0^\infty dt\, t^{\alpha-\gamma+2} f(yt)\int_0^\infty ds\, sZ(\beta s)Z_2((b-dy^\nu)^\mu ts)$$
$$= \int_0^\infty ds\, u(s)Z(\beta s)\mathcal{A}_{w(s)}[f(t)]\,. \tag{66}$$

Now interchange the order of the integrals on the lhs of Eq. (66) in accordance with Fubini's theorem. Then the first integral becomes an orthogonal integral transform with $Z(\beta s)$ as the kernel. Since $Z(\beta s)$ appears on both sides for arbitrary β, we now have an identity involving the same orthogonal transform. Because such transforms are unique, the integrands must be equal. Hence, we obtain

$$\int_0^{(b/d)^{1/\nu}} dy\,(b-dy^\nu)^{\mu(\alpha-\gamma+2)-1} g_1(y)\int_0^\infty dt\, t^{\alpha+2-\gamma} f(yt)Z_2((b-dy^\nu)^\mu ts)$$
$$= (bs)^{-1}\, u(s)\, \mathcal{A}_{w(s)}[f(t)]\,. \tag{67}$$

In this form we already have a powerful identity, but as noted after Eq. (2), making further changes of variable will yield an even greater identity. Thus, putting $u=yt$ first and then $x=y^{-1}(b-dy^\nu)^\mu$ whilst simultaneously taking into account that $\nu\mu=1$, one eventually arrives at Eq. (61).

In one of the examples in the previous section it was necessary to introduce a differential operator in order to derive the integral identity with an arbitrary function. In the same example, the kernel $g(\beta,t)$ did not turn into the form of $\beta^\alpha g_1(y)$ when $t=\beta y/\sqrt{1-y^2}$, which leads to the following theorem.

Theorem 9. If the integral $I=\langle f,g\rangle$ can be expressed via a differential operator L in terms of two transform kernels, one of which is orthogonal and linear in the arbitrary variable β and if $g(\beta,\beta y)=\beta^\gamma g_1(\beta p(y))$, then

$$\int_0^\infty dy\, p(y)^{\alpha-\gamma-3}\int_0^\infty dt\, tf\left(\frac{yt}{p(y)}\right) g_1(t)Z_2\left(\frac{st}{p(y)}\right) = s^{-1}L\,\mathcal{A}_{w(s)}[f(t)]\,. \tag{68}$$

Proof. Since I can be expressed via a differential operator L in terms of two transform kernels, one of which is $Z(\beta s)$, it can be written as

$$I=\beta^\alpha L\int_0^\infty dt\, f(t)g_2(\beta,t) = \beta^\alpha L\int_0^\infty dt f(t)\int_0^\infty ds\, h(w(s)t)Z(\beta s)\,, \tag{69}$$

where $h(w(s))$ represents the other transform kernel with transform variable $w(s)$ and the factor β^α is assumed to arise by the action of the differential operator L. Of course, α

can also equal zero. Using the fact that $g(\beta, y) = \beta^\gamma g_1(\beta p(y))$ when $t = \beta y$ and equating the ensuing result to Eq. (69), we obtain

$$\int_0^\infty dy\, \beta^{\gamma+1-\alpha} f(\beta y)\, g_1(\beta p(y)) = L \int_0^\infty dt\, f(t) \int_0^\infty ds\, h(w(s)t) Z(\beta s) \,. \tag{70}$$

Introducing the δ-function yields

$$\int_0^\infty dy p(y)^{\alpha-\gamma-2} \int_0^\infty dt\, f\left(\frac{yt}{p(y)}\right) g_1(t)\, \delta(t - \beta p(y)) = L \int_0^\infty dt f(t) \int_0^\infty ds h(w(s)t) Z(\beta s). \tag{71}$$

Since $Z(\beta s)$ is an orthogonal transform, we can replace the δ-function by an orthogonality identity as has been done previously. Then by the uniqueness theorem for transforms with kernel $Z(\beta s)$, we obtain Eq. (68), which is different from the previous theorems since the arbitrary function is dependent upon two variables. This has arisen because of the substitution of t by βy in the kernel $g(\beta, t)$.

V. Conclusion

In this paper various integral identities involving arbitrary functions have been derived culminating in some very general theorems presented in Sec. 4. Whilst the material presented here is complete, it should be pointed out that the work is very much in its infancy and much remains to be done. In the first instance, it is quite clear that tables of integral identities involving arbitrary functions need to be produced before the task of producing new integrals from these identities can be accomplished. Another question is whether there exists a master theorem capable of incorporating all the material presented here. These issues are to be addressed in the future.

VI. Acknowledgements

V.K. is grateful to the Australian Research Council for its support via an Australian Research Fellowship and acknowledges the support provided by the School of Physics, University of Melbourne. He also thanks Professor D. Bainov of Plovdiv Technical University, Bulgaria, for his hospitality, which enabled the completion of this work.

References

1. M.L. Glasser and V. Kowalenko, Integral Transforms and Special Functions, 5, 161-184 (1997).

2. A.P. Prudnikov, Yu.A. Brychkov and O.I. Marichev, Integrals and Series, Vol. 4: Direct Laplace Transforms, Gordon and Breach, New York (1992).

3. F. Oberhettinger, Tables of Bessel Transforms, Springer-Verlag, Berlin (1972).

4. H. Kestelman, Modern Theories of Integration, Dover, New York (1960), p. 206.

5. M.J. Lighthill, Introduction to Fourier Analysis and Generalised Functions, Cambridge University Press, Cambridge (1975).

6. P.M. Morse and H. Feshbach, Methods of Theoretical Physics, McGraw-Hill, New York (1953).

7. A.P. Prudnikov, Yu. A. Brychkov and O.I. Marichev, Integrals and Series, Vol. 2: Special Functions, Gordon and Breach, New York (1986).

8. A. P. Prudnikov, Yu.A. Brychkov and O.I. Marichev, Integrals and Series, Vol. 1: Elementary Functions, Gordon and Breach, New York (1986).

9. A.P. Prudnikov, Yu.A. Brychkov and O.I. Marichev, Integrals and Series, Vol. 3: More Special Functions, Gordon and Breach, New York (1990).

10. F. Oberhettinger, Tables of Fourier Transforms and Fourier Transforms of Distributions, Springer-Verlag, Berlin (1990).

8th Int. Coll. on Differential Equations, pp. 275–284
D. Bainov (Ed)

Parallel Algorithm for Construction of Singular Surfaces in Linear Differential Games

S. S. KUMKOV AND V. S. PATSKO
Institute of Mathematics & Mechanics, S.Kovalevskaya str., 16, Ekaterinburg, 620219, Russia
E-mail: kumkov.jr@ods.imm.intec.ru, patsko@ods.imm.intec.ru

Abstract. The paper deals with an algorithm of construction and classification of singular surfaces in linear differential games with fixed terminal moment and convex payoff function which depends on two components of the phase vector. The algorithm is imbedded into the backward procedures for building level sets of the value function. In this backward procedure, the level set section at the next moment is calculated using the section for the previous one. The data obtained during these calculations are used for detection and classification of the singular points in the new constructed section. Unrolled through the time, these singular points give the singular lines which go along over the surface of the level set. Singular lines are joined in the singular surfaces in the space of the game. The main idea of the parallel computations is briefly described.

Keywords: Differential games, value function, singular surfaces, numerical methods, parallel computation algorithms.

INTRODUCTION

As singular surfaces in the theory of differential games, such sets in the game space are named on which the optimal motions have some peculiarities (dispersion, refraction, junction, etc.). The classification of the singular surfaces was suggested in the R.Isaacs' book [1]. Necessary conditions which characterize this or that type of singularity were studied by P.Bernhard [2] and A.A.Melikyan [3]. In many papers (see, for example, [4, 5]), the analysis of the singular surfaces appearing in concrete differential games carried out.

In this paper, a computer algorithm for construction, analysis and classification of singular surfaces is shortly described.

We consider a linear antagonistic differential game [6]

$$\dot{x} = A(t)x + B(t)u + C(t)v, \tag{1}$$
$$x \in R^n, \quad u \in P, \quad v \in Q, \quad \varphi(x_i(T), x_j(T))$$

with fixed terminal time T and convex payoff function φ which depends on two coordinates x_i, x_j of the phase vector. The first (second) player governs the control u (v) choosing that from the convex compact P (Q) and minimizes (maximizes) the value of the function φ at the moment T.

It is known that the substitution $y(t) = X_{i,j}(T,t)x(t)$, where $X_{i,j}(T,t)$ is a matrix combined of two rows of the fundamental Cauchy matrix, provides the transformation to the equivalent differential game of the second order.

For the linear differential game (1) at the beginning of the 80's, in the Institute of Mathematics and Mechanics, Ekaterinburg, there were elaborated [7–11] the backward constructions for building level sets of the value function.

Described in this paper algorithm for constructing the singular surfaces is imbedded into the algorithm [8] for building level sets of the value function. The singular surfaces in the three-dimensional space t, y_1, y_2 of the equivalent game are combined of the singular lines. In its turn, the singular lines are built of the singular points which are selected on the sections of the concrete level set during the process of constructing such sections. If the original game is of the second order on its phase variable, then the reverse transition to the original phase coordinates t, x_1, x_2 is possible.

Till now, the algorithm has been elaborated only for the case of scalar controls of the first and second players. Hence, calculating the singular surfaces, it is supposed that the sets P and Q are segments.

The complex of programs was elaborated which permits to build the collection of level sets of the value function onto the given grid of its values. In the case of scalar players' controls, it additionally permits to construct and classify the singular surfaces. There exists the special version of these programs intended for the parallel multiprocessor computers.

1. CONSTRUCTING LEVEL SETS OF THE VALUE FUNCTION

Below, the algorithm from the paper [8] for constructing level sets of the value function is considered.

1.1. Backward procedure

Assume that the transfer from the game (1) with the payoff function φ depending on 2 coordinates of the phase vector to the equivalent game

$$\dot{y} = D(t)u + E(t)v, \qquad y \in R^2, \quad u \in P, \quad v \in Q, \quad \varphi(y(T)) \tag{2}$$
$$D(t) = X_{i,j}(T,t)B(t), \quad E(t) = X_{i,j}(T,t)C(t),$$

is already done. Let on the interval $[0,T]$ the sequence of moments $t_i: \ t_N = T, \dots, t_i = t_{i+1} - \Delta, \dots, t_0 = 0$ dividing the interval with the step Δ is given. The interest is in finding the level sets $W_c(t_i) = \{y \in R^2 : V(t_i, y) \le c\}$ of the value function V for the given value of parameter c.

Replace the dynamics (2) by the piecewise-constant dynamics

$$\dot{y} = \mathbf{D}(t)u + \mathbf{E}(t)v, \quad \mathbf{D}(t) = D(t_i), \quad \mathbf{E}(t) = E(t_i), \quad t \in [t_i, t_{i+1}). \tag{3}$$

Instead of the sets P and Q, let us consider their polyhedral approximations $\mathbf{P}$, $\mathbf{Q}$. Let $\hat{\varphi}$ be the approximating payoff function. For any c, its level set $\mathbf{M}_c = \{y : \hat{\varphi}(y) \le c\}$ is a convex polygon.

The approximating game (3) is taken so that, for each step $[t_i, t_{i+1}]$ of the backward procedure, we deal with the game with simple motions, polyhedral convex control constraints and the convex polyhedral target set. Put $\mathbf{W}_c(t_N) = \mathbf{M}_c$, next, find the game solvability set $\mathbf{W}_c(t_{N-1})$, then $\mathbf{W}_c(t_{N-2})$, and so on. As a result, the collection of convex sets is obtained which approximate in the Hausdorf metrics the level sets $W_c(t_i)$ of the value function in the game (2).

Let $\mathcal{P}(t_i) = -\Delta D(t_i)\mathbf{P}$, $\mathcal{Q}(t_i) = \Delta E(t_i)\mathbf{Q}$. The support function $l \to \rho(l, \mathbf{W}_c(t_i))$ of the set $\mathbf{W}_c(t_i)$ is the convex hull [12] of the function

$$\gamma(l, t_i) = \rho(l, \mathbf{W}_c(t_i)) + \rho(l, \mathcal{P}(t_i)) - \rho(l, \mathcal{Q}(t_i)).$$

The function $\gamma(\cdot, t_i)$ is positively-homogeneous and piecewise-linear. The property of local convexity of this function can be violated only at the frontier of the linearity cones of the function $\rho(\cdot, \mathcal{Q}(t_i))$, i.e. at the frontier of the cones generated by the normals to the edges of the polygon $\mathcal{Q}(t_i)$ which have the common vertex.

1.2. Algorithm of convex hull construction

Let us agree to omit the argument t_i in the notation of the function γ. The linearity cones of γ are determined by the normals to the convex polygons $\mathbf{W}_c(t_i)$, $\mathcal{P}(t_i)$, $\mathcal{Q}(t_i)$. Gathering the outer normals of these sets and ordering them clockwise, the collection L of the vectors is obtained. The collection of values $\gamma(l)$ of the function γ on the vectors $l \in L$ is denoted by Φ. The collections L, Φ describe completely the function γ.

The collection of the normals to $\mathcal{Q}(t_i)$ ordered clockwise is denoted by S. The collection S is called the collection of "suspicious" vectors. This name is connected with the fact that the function γ is locally convex on the cones which interior does not contain the normals of the set $\mathcal{Q}(t_i)$. The violation of the local convexity can appear only on the cones which interior contains at least one normal of the polygon $\mathcal{Q}(t_i)$.

Let $L^{(1)} = L$, $\Phi^{(1)} = \Phi$, $S^{(1)} = S$. The $k+1$ step of the multistep convexing process consists in replacing the collections $L^{(k)}$, $\Phi^{(k)}$ by the collections $L^{(k+1)} \subset L^{(k)}$, $\Phi^{(k)} \subset \Phi^{(k+1)}$. The collection $S^{(k)}$ is also replaced by the new one $S^{(k+1)}$.

Describe now one step of the convexing process. Suppose that the angle between two neighboring vectors from the collection $L^{(k)}$ counted clockwise is less than π. Let $l \to \gamma^{(k)}(l)$ be the piecewise-linear function determined by the collections $L^{(k)}$, $\Phi^{(k)}$.

Since $L^{(k)} \subset L^{(k-1)} \subset \ldots \subset L^{(1)}$, $\Phi^{(k)} \subset \Phi^{(k-1)} \subset \ldots \subset \Phi^{(1)}$, then for any vector $\bar{l} \in L^{(k)}$ the value $\gamma^{(k)}(\bar{l})$ is equal to $\gamma(\bar{l})$.

Take some vector $l_* \in S^{(k)}$ and check the local convexity of the function $\gamma^{(k)}$ on the cone generated by the vector l_* and two its neighboring vectors l_- and l_+ selected counterclockwise and clockwise from the collection $L^{(k)}$. In other words, check whether the inequality $l_*'y \le \gamma(l_*)$ is active in the triple of the inequalities $l_-'y \le \gamma(l_-)$, $l_*'y \le \gamma(l_*)$, $l_+'y \le \gamma(l_+)$. If the system of three inequalities is compatible, then (by virtue of the ordering the vectors l_-, l_*, l_+) only the middle one can be inactive.

The algorithm of verification: find the intersection point y_* of the straight lines $l_-'y = \gamma(l_-)$ and $l_*'y = \gamma(l_*)$, and then check the inequality $l_+'y_* < \gamma(l_+)$. If it holds, the local convexity takes place (the middle inequality is active). In the opposite case, the local convexity is absent (the middle inequality is inactive).

In the first case, the vector l_* is taken away from the collection $S^{(k)}$, and the remained set is denoted by $S^{(k+1)}$. Let $L^{(k+1)} = L^{(k)}$, $\Phi^{(k+1)} = \Phi^{(k)}$.

In the second case, two situations are distinguished. Let α be the angle counted clockwise from l_- to l_+:

- $\alpha < \pi$. The vector l_* is taken away from the collection $S^{(k)}$, and, simultaneously, the vectors l_- and l_+ are included into the collection (one of them or even both could be already presented in the collection $S^{(k)}$). Denote the new collection of the "suspicious" vectors by $S^{(k+1)}$. The difference of the new collection $L^{(k+1)}$ from the collection $L^{(k)}$ is that the vector l_* is absent in $L^{(k+1)}$. When processing $\Phi^{(k)}$ to $\Phi^{(k+1)}$, the values $\gamma^{(k)}(l_*) = \gamma(l_*)$ are taken away;
- $\alpha \ge \pi$. It means that the discussed triple of inequalities is incompatible. Thus, the convex hull of the function γ does not exist, i.e. $\mathbf{W}_c(t_i) = \emptyset$. The constructing is ceased.

One step of the convexing algorithm has been described. The algorithm finishes at the step with the number j, when for the first time $S^{(j)} = \emptyset$, i.e. when the collection of the "suspicious" vectors is empty. It means that the function $\gamma^{(j)}$, which corresponds to the collections $L^{(j)}$ and $\Phi^{(j)}$, is locally convex everywhere. Thus, the function $\gamma^{(j)}$ is the convex hull of the function γ. The second variant of termination is the following: the angle α between the vectors l_- and l_+ becomes greater or equal to π after rejecting the checked vector l_* from the collection of the "suspicious" vectors at some step. It means that $\mathbf{W}_c(t_i) = \emptyset$.

In Figure 1, three level sets of the value function are shown for the differential game

$$\begin{aligned} \dot{x}_1 &= x_2 + v \\ \dot{x}_2 &= u \end{aligned} \qquad |u| \le 1, \ |v| \le 1, \quad \varphi(x_1, x_2) = |x_1| + |x_2|. \tag{4}$$

The picture is represented in the equivalent coordinates. The symbol $\tau = T - t$ denotes the backward time. All three sets have terminated.

In Figure 2, the level set of the value function is shown for the differential game

$$\begin{aligned} \dot{x}_1 &= x_2 + v \\ \dot{x}_2 &= -x_1 + u \end{aligned} \qquad |u| \le 1, \ |v| \le 0.9, \quad \varphi(x_1, x_2) = x_1^2 + x_2^2. \tag{5}$$

The picture is given in the original coordinates.

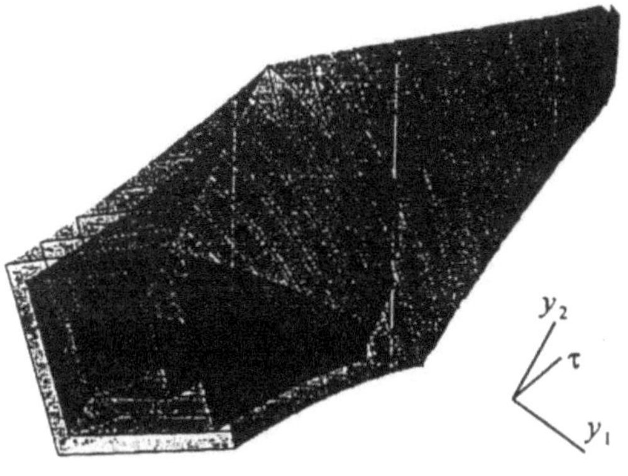

Figure 1: Three level sets of value function for game (4).

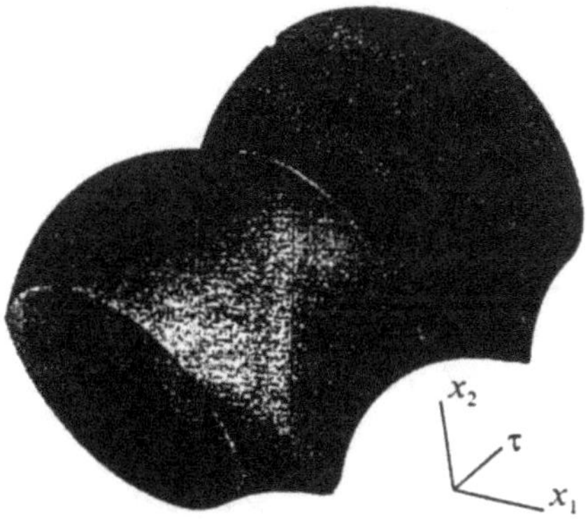

Figure 2: One level set of value function for game (5).

2. CONSTRUCTION OF SINGULAR SURFACES

In this section, it is supposed that the controls of both players are scalar. So, the sets $\mathcal{P}(t_i)$ $\mathcal{Q}(t_i)$ are segments in the space y_1, y_2.

On each time interval $[t_i, t_{i+1})$, we have the game with simple motions [1] as the approximating game. The first player tries to pass the system from the polygon $\mathbf{W}_c(t_i)$ onto the polygon $\mathbf{W}_c(t_{i+1})$, the second player hinders to prevent it. Entering the discrimination of the second player, determine the optimal motions.

Let us fix some arbitrary point y_0 on the boundary of the polygon $\mathbf{W}_c(t_i)$. The second player's control constant on the interval $[t_i, t_{i+1})$ is called the optimal one if the first player can not direct the motion from the point y_0 inside the polygon $\mathbf{W}_c(t_{i+1})$. For fixed optimal control of the second player, the first player's control (also constant on the same interval) is called the optimal one if the motion gets onto the boundary of the set $\mathbf{W}_c(t_{i+1})$. The system motion generated by the optimal players' controls is called the optimal.

The point y_0 is called the regular if the optimal motion getting out from this point is unique and generated by the extreme players' controls. A point, which is not regular, is called the singular one.

Below, the classification of the singular points of the polygon $\mathbf{W}_c(t_i)$ is described. It is based on the analysis of the character of the optimal motions getting out from these points. For the classification, we shall use two marks. These marks attach to normals participated in the process of convex hull construction of the function $\gamma(\cdot, t_i)$. The mark FS ("former suspicious") is added to such normals which, during the convexing process, were denoted as "suspicious" ones, but, after the process, were remained in the collection which determines the polygon $\mathbf{W}_c(t_i)$. The mark $N\mathcal{P}$ is added to the normals which were

Table of singularities classification

Number of normals	Flags	Additional condition	Object for marking	Secondary condition	Type of singularity
2	*FS*, *FS*	$\mathcal{Q}$-normal is strictly between	Vertex	$\mathcal{P}$-normal is not between	Dispersal for 2nd
				$\mathcal{P}$-normal is between	Dispersal for both
2	*FS*, $N\mathcal{P}$+*FS*	$\mathcal{Q}$-normal is strictly between	$\mathcal{P}$-normal edge		Equivocal
1	$N\mathcal{P}$		Edge	Normal is not equal to W-normal	Switching for 1st
				Normal is equal to W-normal	"Fuzzy" switching for 1st
1	*FS*	Normal is equal to $\mathcal{Q}$-normal	Edge		"Fuzzy" dispersal for 2nd
1	$N\mathcal{P}$+*FS*	Normal is equal to $\mathcal{Q}$-normal	Edge	Normal is not equal to W-normal	Switching for both
				Normal is equal to W-normal	"Fuzzy" switching for both

taken from the set $\mathcal{P}(t_i)$. The classification is represented in Table.

Table deals with the normals from the collection obtained after the process of convexing. Individual normals or pairs of neighbor ones are analyzed. The number in the first column describes the number of normals considered in this or that combination of normals. In the second column, the marks are represented whose presence is checked out on the considered normals. With that, if only one mark is shown for some concrete vector, then the second mark is considered to be absent. The third column contains the additional conditions. Marks in the second column, together with the satisfaction of the additional condition, determine in whole the singularity type of the object from the forth column. Such object can be the polygon $\mathbf{W}_c(t_i)$ vertex which is incident for edges determined by normals (row 1), either the edge correspondent to the considered normal (rows 3–5) or one of two shown normals (row 2). Inside some cases, there exists subdivision which is determined by the condition from the fifth column. In the last column, the type of singularity is indicated which is finally prescribed to the mentioned object.

The term "$\mathcal{P}$-normal" ("$\mathcal{Q}$-normal", "W-normal") means a normal taken from the set $\mathcal{P}(t_i)$ ($\mathcal{Q}(t_i)$, $\mathbf{W}_c(t_{i+1})$).

Note, that length of the singular edge is in the linear proportion to the time step Δ if the normal to such edge was absent in the normals collection which determined the polygon $\mathbf{W}_c(t_{i+1})$.

Denominations of the singular surfaces are determined by a character of the optimal motions getting out from the marked point. The names are coordinated with ones used in the R.Isaacs' book.

The term "dispersal" corresponds to the situation when the bundle of the optimal motions getting out from the given point contains two motions generated by different extreme controls of the second player and one or two extreme controls of the first player.

The term "equivocal" corresponds to the situation when, from the vertex determined by the normals with marks $\mathcal{NP}+FS$ and FS, two optimal motions go out and those are generated by different extreme controls of the second player. Herewith, the first player's control for one of these motion is of extreme value, but is not such for the other motion. The latter motion comes to such vertex of the polygon $\mathbf{W}_c(t_{i+1})$ in which the cone of normals contains the considered normal with the mark $\mathcal{NP}+FS$. All other points of the marked edge are routine, namely, each point generates only one motion which comes to the mentioned vertex.

It is a very complicated case, formally included in Table into the "equivocal" class, when the normal marked as $\mathcal{NP}+FS$ is equal to a W-normal. Maybe, this case have to be considered and classified separately. However, it appears quite rarely and therefore the "equivocal" row in Table is not divided.

The term "switching" corresponds to the situation when the bundle of the optimal motions from each internal point of the singular edge contains a motion generated by one extreme control of the second player and non-extreme control of the first player.

On different sides of the equivocal edge, both players' controls change. In dispersal and switching situations, only one of players can change its control or they can change their controls together. In these two latter situations, a special subdivision appears. With that, it is indicated which players change their controls: the first, the second or both.

The term "fuzzy" means the special situation when two or three normals coincide. For example, the normals from the sets $\mathcal{Q}(t_i)$ and $\mathbf{W}_c(t_{i+1})$ are equal. Herewith, the infinite non-uniqueness of the optimal controls appears, and the bundle of the optimal motions diffuses and becomes "fuzzy". Such case is not one of the general position.

It can be shown that any point which is neither marked vertex nor included to any marked edge is the regular one. And vice versa, any point which is marked vertex or included to a marked edge is the singular, except, maybe, the case when it is an endpoint of a marked edge.

The data for the algorithm of singularities classification are accumulated during the process of the convex hull constructing, and after, the algorithm of classification begins to work. Really, its work is reduced to the verification of the presence of situations described by rows of Table. As the result, the collection of the points and intervals with description of type of their singularity is obtained for the next backward time section of the level set.

For graphical presentation of the singular surfaces, a collection of the level sets is calculated on the given grid of magnitudes of the value function, and, simultaneously, the singular points and edges are detected on each section of each level set. Constructing the singular surfaces on the base of the singular points and intervals is carried out not in this program, but later, in the program of visualization.

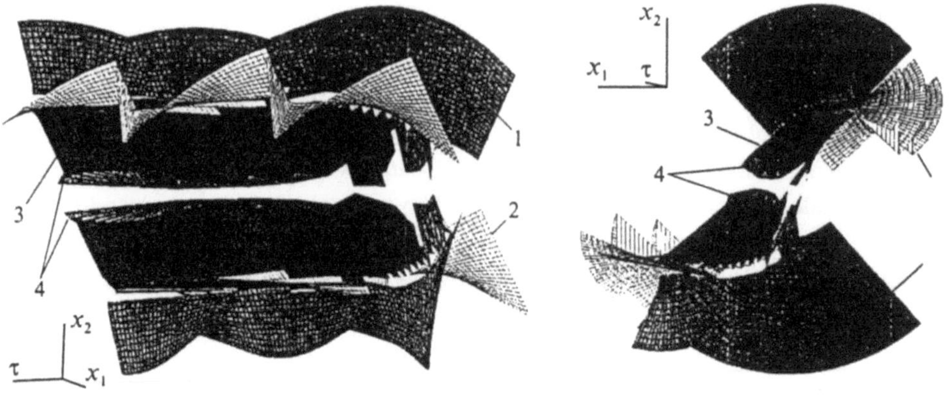

Figure 3: Singular surfaces for game (6). *Left*: View from the axis x_1. *Right*: View from the axis τ. *Notations*: 1 – dispersal surface for the second player, 2 – switching surface for the first player, 3 – equivocal surface, 4 – dispersal surface for both players.

The validity of the elaborated algorithm was verified on the test example with the dynamics from the game (4) but with different payoff function $\varphi(x_1,x_2) = |x_1|$, $x_2 = 0$; $\varphi(x_1,x_2) = +\infty$, $x_2 \neq 0$. The singular surfaces appeared in this example had been investigated in details in [13]. Results of our calculations coincide well with the results of the mentioned paper.

Figure 3 shows the picture of the singular surfaces for the differential game

$$\begin{aligned} \dot{x}_1 &= x_2 + v \\ \dot{x}_2 &= -x_1 + u \end{aligned} \qquad |u| \leq 1, \ |v| \leq 0.9, \quad \varphi(x_1,x_2) = \max\{|x_1|, |x_2|\}. \tag{6}$$

The surface close to the axis τ is dispersal for both players. On some distance from the axis τ, the equivocal surface situated. It comes into the switching surface for the first player and dispersal one for the second. The "empty" parts round the time axis visible in Figure 3 can be filled by calculations with smaller time step Δ of the backward procedure.

Calculations of many examples show that the presence of the dispersal for both players surface passing through the time axis and the presence of the equivocal surface on some distance from the axis are typical for the differential games of the second order with the dynamics and payoff function symmetrical relatively zero.

3. ORGANIZATION OF PARALLEL COMPUTATIONS

Parallelizing of computations was fulfilled for the massively-parallel computer of the *Paragon* type. It includes the collection of the i860 Intel processors with distributed memory. Connection between separate processors is hold by INMOS transputers. Information is transferred to a customer through the host IBM-compatible computer included into the local area network. The access is also possible via the Internet.

In the discussed problem, the level sets of the value function and singular lines on them are built independently for different values of the function. So, the parallelizing of computations is natural, and computation and building each separate level set are also independent processes and are carried out on a separate processors. In the case of many sets calculations, some processors can calculate several sets sequentially. Different level sets can demand different computer time. So, the effective procedure of distribution of calculation tasks for each separate processor is the central point in the parallel algorithm.

The parallel algorithm for constructing level sets of the value function is described in [14]. Now, it has been modified for implementing the additional calculations on detection and classification of the singular surfaces. The algorithm of the singular surfaces building imbeds well into the algorithm of the level sets calculations. So, the modification is joined only with including the block of the singular points classification into the main calculation procedure. With that, all governing procedures including the task manager are hold.

ACKNOWLEDGEMENT

This research was supported by the Russian Foundation of Basic Researches under Grant No. 97-01-00458.

References

1. R.Isaacs, Differential Games, John Wiley and Sons, New York (1965).
2. P.Bernhard, in: Lecture Notes in Control and Information Sciences, 3, P.Hagedorn, H.W.Knobloch and G.J.Olsder, (Eds.), Springer-Verlag, Berlin (1977) pp. 1–33.
3. A.A.Melikyan, in: Lecture Notes in Control and Information Sciences, 156, R.P.Hämäläinen and H.K.Ehtamo, (Eds.), Springer-Verlag, Berlin (1991) pp. 81–90.
4. A.W.Merz, The Homicidal Chauffeur - a Differential Game, PhD dissertation, Stanford University (1971).
5. J.Shinar, M.Medinah and M.Biton, JOTA, 43(3), 431–456 (1984).
6. N.N.Krasovskii and A.I.Subbotin, Game-theoretical Control Problems, Springer-Verlag, New York (1988).
7. A.I.Subbotin and V.S.Patsko (Eds.), Algorithms and Programs for Solving Linear Differential Games, Inst. of Math. and Mech., Sverdlovsk (1984) (in Russian).
8. E.A.Isakova, G.V.Logunova, and V.S.Patsko, in: ref. [7] (1984) pp. 125–158 (in Russian).
9. N.D.Botkin, in: ref. [7] (1984) pp. 5–38 (in Russian).
10. A.M.Taras'ev and V.N.Ushakov, in: ref. [7] (1984) pp. 159–190 (in Russian).
11. A.M.Taras'ev and V.N.Ushakov, in: Investigations of Minimax Control Problems, A.I.Subbotin and V.S.Patsko, (Eds.), Inst. of Math. and Mech., Sverdlovsk (1985), pp. 82–90 (in Russian).

12. B.N.Pschenichnyi and M.I.Sagaidak, Cybernetics, 6(2), 72–83 (1970).

13. V.S.Patsko and S.I.Tarasova, Engrg. Cybernetics, 22(5), 59–67 (1985).

14. S.S.Kumkov, in: Developing and Application of High Performance Computational Facilities and Programs on the Base of Parallel Technologies in IMM UrO RAS, Inst. Math. and Mech., Report, Ekaterinburg (1996) pp. 114–121 (in Russian).

8th Int. Coll. on Differential Equations, pp. 285–292
D. Bainov (Ed)

Algorithms and Software for Selfadjoint ODE Eigenproblems

CARSTEN R. MAPLE AND MARCO MARLETTA
Department of Mathematics and Computer Science, University of Leicester, University Road, Leicester LE1 7RH, United Kingdom

Abstract. We describe recent developments in theory and software for selfadjoint eigenproblems involving linear Hamiltonian systems.

AMS (MOS) Subject Classification. 34B, 34L, 65L15.

1 Introduction

In 1923, Prüfer [1] published an elegant paper in which he showed how the use of polar coordinates (r, θ) in the (py', y) phase plane can be used to derive all the classical results on eigenvalues for regular second-order Sturm-Liouville problems,

$$-(p(x)y')' + q(x)y = \lambda w(x)y, \quad x \in [a, b], \tag{1.1}$$

$$A_1 y(a) + A_2 py'(a) = 0, \quad B_1 y(b) + B_2 py'(b) = 0. \tag{1.2}$$

Here $[a, b]$ is a finite interval and the coefficient functions are such that $1/p$, q and w are all in $L^1[a, b]$; the functions p and w are assumed to be strictly positive in $[a, b]$, while $(A_1, A_2) \neq (0, 0)$ and $(B_1, B_2) \neq (0, 0)$ for nontrivial boundary conditions. Over the ensuing twenty years a lot of work was done to extend the Prüfer theory to equations of higher order and equations with matrix coefficients $p(x)$, $q(x)$ and $w(x)$ and vector dependent variables y. This work is summarized in the books of Atkinson [2] and Reid [3].

The development of the theoretical side of the subject has been reflected in software for solution of these problems.

First consider second order scalar problems. The now famous SLEIGN code of Bailey, Gordon and Shampine [4] appeared in 1978, and used the Prüfer theory extensively. Pryce's D02KEF in the NAG library appeared shortly after and allowed the treatment of problems with nonlinear λ-dependence. In 1991, Marletta and Pryce published the code SL02F [5] to compete with the existing codes: it was significantly faster than SLEIGN or D02KEF, but within a short period of time it was surpassed in speed by the SLEDGE code of Fulton and Pruess [6]. SLEIGN2 [7] is another recently published code: it is not as fast as SL02F or SLEDGE (nor indeed as fast as SLEIGN on many problems), but it is the first to be able to handle limit circle oscillatory problems, which have eigenvalues extending to $\pm\infty$. Pryce has produced a special package for running all the different codes. See

`ftp://ftp.rmcs.cranfield.ac.uk/pub/departments/seas/cised/numanal`

Now let us consider more general problems. In 1987, Atkinson, Krall, Leaf and Zettl [8] proposed to use the the oscillation theory of Atkinson [2] for matrix-vector Sturm-Liouville problems. These still have the form (1.1), but the coefficients $p(x)$, $q(x)$ and $w(x)$ are now symmetric real $n \times n$ matrices, with $p(x)$ and $w(x)$ positive definite, and y is an n-vector. In the boundary conditions (1.2), A_1, A_2, B_1 and B_2 are $n \times n$ matrices satisfying certain rank and symmetry conditions. Although there was a problem with the treatment of the boundary conditions in [8], this paper pointed the way to go in the theory by developing a 'spectral function'

$$N(\lambda) = \text{number of eigenvalues} < \lambda,$$

and showing that it could be computed, for each fixed λ, by solving at most two initial value problems. In [9] Marletta proposed an algorithm which got round the problem in [8] and which allowed eigenvalues λ_k to be found at a cost which did not increase unboundedly with k. At the same time, Greenberg [10] considered an eigenvalue problem for a selfadjoint ODE of arbitrary even order and with matrix coefficients. Completely independently, Abramov published very similar work [11][1]. In fact the 1989 paper of Baur and Kratz [12] seems to have pre-empted some of the work of Abramov and Greenberg.

The work of Greenberg was exploited by Marletta, who published two codes (SL11F and SL12F, see [13]) to solve general eigenvalue problems for Hamiltonian systems. These codes are available from `netlib` in the `aicm` section. SL11F can be quite expensive to run due to its generality, so Greenberg and Marletta [14] developed a special code called SLEUTH [15] for fourth order Sturm-Liouville equations

$$(p(x)y'')'' - (s(x)y')' + q(x)y = \lambda w(x)y. \tag{1.3}$$

SLEUTH has been used by pure mathematicians (see, e.g. [16], [17], [18]) and has revealed an interesting theoretical result which we mention at the end of this paper.

Abramov and Aslanyan have also used the theory of Abramov to develop a number of algorithms for various types of regular and singular Hamiltonian system (see, e.g., [19], [20]).

2 Hamiltonian Systems

We start our more detailed discussion with eigenvalue problems which can be cast in the form of Hamiltonian systems. Much of the theoretical work here can be found in Abramov [11] and Greenberg [10] while the more algorithmic material, in particular the use of the Baker-Campbell-Hausdorff formula, is based on [13].

2.1 Theory for Hamiltonian Systems

Many equations can be cast in the form of Hamiltonian systems,

$$Jz' = S(x, \lambda)z, \tag{2.1}$$

[1]The English version of this paper was actually not available until 1993

where z is a $2n$-vector and J and S are $2n \times 2n$ matrices of the form

$$J = \begin{pmatrix} \mathbf{0} & -I \\ I & \mathbf{0} \end{pmatrix}, \quad S(x,\lambda) = \begin{pmatrix} S_{11}(x,\lambda) & S_{12}(x,\lambda) \\ S_{12}^T(x,\lambda) & S_{22}(x,\lambda) \end{pmatrix}.$$

Here $\mathbf{0}$ is the $n \times n$ zero matrix, I is the $n \times n$ identity, and S_{11}, S_{12}, S_{22} are real matrices with S_{11} and S_{22} symmetric. For the boundary conditions we partition the vector z as $z = (u^T, v^T)^T$ where u and v are n-vectors. The boundary conditions we wish to consider are those which have the form

$$A_1 u(a) + A_2 v(a) = 0 \in \mathbf{R}^n, \quad B_1 u(b) + B_2 v(b) = 0 \in \mathbf{R}^n, \tag{2.2}$$

where (A_1, A_2) and (B_1, B_2) are $n \times 2n$ matrices of full rank, and

$$A_1 A_2^T = A_2 A_1^T, \quad B_1 B_2^T = B_2 B_1^T. \tag{2.3}$$

Definition 2.1 *A number μ is called an* **eigenvalue** *if, when $\lambda = \mu$, the differential equation (2.1) has a nonzero solution z which satisfies both the boundary conditions (2.2).*

We denote by $Z_L = (U_L^T, V_L^T)^T$ the $2n \times n$ solution of the initial value problem $JZ' = S(x,\lambda)Z$, $Z(a) = (-A_2, A_1)^T$, and by $Z_R = (U_R^T, V_R^T)^T$ the $2n \times n$ solution of the initial value problem $JZ' = S(x,\lambda)Z$, $Z(b) = (-B_2, B_1)^T$. We now define the Θ-matrices of Atkinson [2].

Definition 2.2 *Let $Z = (U^T, V^T)^T$ be a full-rank $2n \times n$ solution of the ODE $JZ' = S(x,\lambda)Z$ satisfying $Z^T J Z \equiv 0$. The associated $n \times n$ Θ-matrix is $\Theta = (V + iU)(V - iU)^{-1}$.*

We shall denote by Θ_L and Θ_R the Θ-matrices associated with Z_L and Z_R respectively. The properties $Z_L^T J Z_L \equiv 0$ and $Z_R^T J Z_R \equiv 0$ imply the following.

Lemma 2.1 *The matrices $\Theta_L(x;\lambda)$ and $\Theta_R(x;\lambda)$ are unitary.*

We shall denote the eigenvalues of Θ_L by $\exp(i\theta_j^L(x;\lambda))$, $j = 1,\dots,n$, and the eigenvalues of Θ_R by $\exp(i\theta_j^R(x;\lambda))$, $j = 1,\dots,n$. Of course, the so-called **phase angles** θ_j^L and θ_j^R are not uniquely determined by these conditions. To ensure uniqueness we impose the following conditions:

- $\theta_j^L(a;\lambda) \in [0, 2\pi)$ for $j = 1,\dots,n$;
- $\theta_j^R(b;\lambda) \in (0, 2\pi]$ for $j = 1,\dots,n$;
- $\theta_1^L(x;\lambda) \le \theta_2^L(x;\lambda) \le \dots \le \theta_n^L(x;\lambda) \le \theta_1^L(x;\lambda) + 2\pi$, for $x \in [a,b]$;
- $\theta_1^R(x;\lambda) \ge \theta_2^R(x;\lambda) \ge \dots \ge \theta_n^R(x;\lambda) \ge \theta_1^R(x;\lambda) - 2\pi$, for $x \in [a,b]$;
- all the phase angles are continuous functions of x and λ.

Note that the first two requirements in this list are compatible with the last, because the $\theta_j^L(a;\lambda)$ and $\theta_j^R(b;\lambda)$ are actually independent of λ. This is because $Z_L(a)$ and $Z_R(b)$ are independent of λ.

In addition to Θ_L and Θ_R it is useful to fix a point $c \in [a,b]$ and study the matrix

$$\Theta_{RL}(x;\lambda) := \Theta_R^*(x;\lambda)\Theta_L(x;\lambda).$$

Note that $\Theta_{RL}(c;\lambda)$ is a unitary matrix; we shall denote its eigenvalues by $\exp(i\omega_j(c;\lambda))$, $j=1,\dots,n$, and we shall normalise them so that $\omega_j \in [0,2\pi)$, together with $\omega_1 \le \omega_2 \le \dots \le \omega_n$.

Suppose the Hamiltonian system (2.1) has the property that for any nontrivial solution z and any real numbers α, β in $[a,b]$ with $\alpha<\beta$,

$$\int_\alpha^\beta z^T \frac{\partial S}{\partial \lambda}(x,\lambda) z\,dx > 0. \tag{2.4}$$

[For example, equations arising from reduction of $2n$-th order problems to Hamiltonian form have this property.] We have the following result.

Theorem 2.1 *Suppose that the Hamiltonian system (2.1) is normal and has the property (2.4). Let $c\in[a,b]$ be fixed and let*

$$N(\lambda) = \frac{1}{2\pi}\left\{\sum_{j=1}^n \theta_j^L(c;\lambda) - \sum_{j=1}^n \theta_j^R(c;\lambda) - \sum_{j=1}^n \omega_j(c;\lambda)\right\} - n. \tag{2.5}$$

Then $N(\lambda)$ is a monotone increasing integer-valued function of λ whose discontinuities are precisely the eigenvalues of the problem given by (2.1) subject to the boundary conditions (2.2). At an eigenvalue λ^* of (2.1), (2.2) whose eigenspace has dimension d,

$$N(\lambda) - N(\lambda -) = d.$$

Notice that this theorem does not claim that $N(\lambda)$ is the number of eigenvalues less than λ. To make such a claim we would have to know about the value of $\lim_{\lambda\to-\infty} N(\lambda)$, if this exists. If our Hamiltonian system arises from a selfadjoint differential equation of order $2n$, then Greenberg [10] has shown that $N(\lambda)$ is indeed the number of eigenvalues of (2.2), (2.2) which are less than λ, counted according to multiplicity. Similar results are available in the work of Abramov [11] and Baur and Kratz [12].

Proof See [9]. □

2.2 Algorithms

Theorem 2.1 provides the basis for all our numerical algorithms for Hamiltonian systems and fourth (and higher) order problems. The codes SL11F, SL12F and SLEUTH all take different approaches to computing the quantities in (2.5), but the philosophy is the same: compute $N(\lambda)$ numerically for a number of different values of λ in order to locate the discontinuities, which are the eigenvalues.

The code SL11F, for example, uses the fact that the Θ-matrices satisfy a system of ODEs of the form $\frac{d\Theta}{dx} = i\Theta\Omega(x,\Theta)$, $\frac{d}{dx}\text{Argdet}\Theta = \text{Trace}(\Omega)$, where Ω is Hermitian. These are integrated by representing Θ as the exponential of an Hermitian matrix and using the Baker-Campbell-Hausdorff formula. SLEUTH, on the other hand, approximates fourth order Sturm-Liouville equations by equations with piecewise constant coefficients, which are then integrated in terms of elementary functions.

3 Numerical experiments

All the numerical experiments reported in this section were performed on a Silicon Graphics workstation with an R3000 processor.

3.1 Experiments on Hamiltonian systems

In [13] we applied SL11F both to a periodic second-order Sturm-Liouville problem (the Mathieu equation) and to a problem derived by a separation-of-variables technique from a PDE eigenvalue problem on an annular region. In this paper we apply SL11F to some sixth order problems.

Problem 1 Let D denote d/dx. The differential equation

$$\left(D^2 - a^2\right)\left(D^2 - a^2 - \nu\right)\left(D^2 - a^2 - \nu/P\right)y = a^2Ry, \quad x \in (0,1),$$

arises in the analysis of a Bénard problem [21, p. 43]. The parameters a, ν, R and P are constant, and ν is regarded as the eigenparameter. The problem arises with both non-selfadjoint and selfadjoint boundary conditions. We consider it with the selfadjoint conditions

$$y = y'' = y^{(iv)} = 0 \text{ at } x = 0 \text{ and at } x = 1.$$

The dependence on ν is obviously nonlinear, and one possibility is to convert the problem to a linear eigenproblem

$$\text{(3.1)} \qquad \left(D^2 - a^2\right)\left(D^2 - a^2 - \nu\right)\left(D^2 - a^2 - \nu/P\right)y - a^2Ry = \lambda y \quad x \in (0,1);$$

the linear eigenvalues λ_0, λ_1, $\lambda_2, \ldots$ are functions of ν and the nonlinear eigenvalues are those values of ν for which some λ_k is zero. In fact the ν-eigenvalues are the numbers $\nu_j^{(\pm)}$ $(j = 1, 2, \ldots)$ given by

$$\text{(3.2)} \qquad -(1+P)(j^2\pi^2 + a^2)/2 \pm \sqrt{(1-P)^2(j^2\pi^2+a^2)^2/4 + a^2RP(j^2\pi^2+a^2)^{-1}}.$$

The linear eigenvalues turn out to depend non-monotonically on the parameter ν, so that if ν is regarded as the eigenparameter then (2.4) will not be satisfied. As an illustration of the effect this has on SL11F if the problem is solved directly, consider the results in Table 1, computed for $a = P = 1$ at a tolerance of 10^{-7}. For example, SL11F finds that there is an eigenvalue at $\nu = -11.172934$, which is a reasonable approximation to the true value of $\nu = -11.172919$; however it cannot find the eigenvalue at $\nu = -10.566290$ because the spectral function $N(\nu)$ is actually *decreasing* at this ν. The poor accuracy of the approximations which the code does manage to find is due to the insensitivity of the phase-angles of Θ_{RL} to changes in ν near these eigenvalues: rather like trying to find a zero of a function in a region where its derivative is small.

k	ν found	ν exact	CPU (secs)
0	-11.172934	-11.172919	28
0	not found	-10.566290	**
0	-40.636378	-40.635594	65
0	not found	-40.321241	**

Table 1: Results for Problem 1: TOL = 10^{-7}.

Problem 2 We wanted to test SL11F on a problem whose differential equation exhibits stiffness in at least part of the range: we chose the problem

$$-y^{(vi)} + (3x^2y'')'' + ((8 - 3x^4)y')' + (x^6 - 14x^2)y = \lambda y, \quad x \in (-10, 10),$$

with boundary conditions $y = y' = y'' = 0$ at $x = \pm 10$. The differential operator here is the cube of the 'harmonic oscillator' operator $-d^2/dx^2 + x^2$. Some results are shown

k	λ_k(approx.)	λ_k('exact')	CPU (secs)
0	1.0000010	1	142
1	27.000012	27	235
2	125.00002	125	259
3	343.00005	343	329
9	6859.0142	6859	680
19	59319.212	59319	1818

Table 2: Results for Problem 2: TOL = 10^{-7}.

in Table 2. For information, the exact eigenvalues for the corresponding problem on $(-\infty, \infty)$ are $\lambda_k = (2k+1)^3$. In [22] we were able to solve this problem over a much larger interval using a coefficient approximation technique and some oscillation theory, but SL11F finds the large coefficients in the differential equation on $(-100, 100)$ (they are $O(10^{12})$ near the endpoints) too difficult, and the run times become unrealistic. On $(-10, 10)$, however, the resuls in Table 2 show that mild stiffness is not too problematic. Notice that on this problem the accuracy is consistently better than on Problem 1: here the phase angles of Θ_{RL} are much more sensitive to the value of λ, making accurate rootfinding much easier. We also remark that for this problem, it is not too difficult to show that interval truncation is not the dominant contribution to the error for an interval of $[-10, 10]$, for modest values of k. SL11F was quite a bit slower than the new sixth order code [22], particularly for large k. The sixth order code took 113 seconds for $k = 0$ and 475 seconds for $k = 19$.

3.2 Experiments with SLEUTH

One of the ways we tested the fourth order code SLEUTH during its development was to apply it to problems of the form

$$\ell^2 y = \lambda y,$$

where ℓ was a second order operator. Consider, for example, the case where ℓ is the 'harmonic oscillator' operator, $\ell z = -\frac{d^2 z}{dx^2} + x^2 z$. This gives

$$\ell^2 y = y^{(iv)} - 2(x^2 y')' + (x^4 - 2)y.$$

We therefore considered two problems. These both used the same differential equation,

$$y^{(iv)} - 2(x^2 y')' + (x^4 - 2)y = \lambda y, \quad x \in [0, \infty), \tag{3.3}$$

with no boundary conditions required at infinity, but different boundary conditions at $x = 0$: we compared Dirichlet conditions $y(0) = 0 = y'(0)$ with Neumann conditions $y''(0) = 0 = y'''(0)$. The results are shown in Table 3, where the interval $[0, \infty)$ has been truncated to $[0, 100]$ and the problem solved with a tolerance of 10^{-6}. This particular problem is very insensitive to the choice of artificial boundary conditions at the point $x = 100$.

It seemed clear from these results that for this problem, λ_{k+1}(Neumann BCs) = λ_k(Dirichlet BCs). This result runs counter to the intuition one gains from working with second order problems. However it transpires that a result of this nature is true for a wide class of problems, as Brown and Marletta proved in [17].

Theorem 3.2 *Suppose that $\mathcal{L}$ is a fourth order differential operator on the domain of functions f in $L^2[0, \infty)$ which are four times continuously differentiable and are such that*

Table 3: Tests on Problem (3.3).

k	λ_k(Dirichlet)	CPU	λ_k(Neumann)	CPU
0	16.000000	9 secs	2E-06	5 secs
1	63.999838	3 secs	15.999999	12 secs
2	144.00002	5 secs	64.000003	4 secs
99	159999.99	251 secs	156815.5	42 secs
100	163216.01	261 secs	159999.5	39 secs

$\mathcal{L}f \in L^2[0,\infty)$. *Suppose also that* $\mathcal{L}$ *has the form* $\mathcal{L} = \ell^2$ *where* ℓ *is a second order operator*

$$\ell(y)(x) = -y''(x) + q(x)y(x)$$

and q *is twice continuously differentiable with* $q(0) = 0$, $q'(0) = 0$. *Suppose also that* $\mathcal{L}$ *is strong limit-point at infinity. Then any eigenfunction* y *of* $\mathcal{L}$ *subject to Neumann boundary conditions* $y'''(0) = 0 = y''(0)$ *which is not in the null-space of* ℓ *generates an eigenfunction* $z = \ell y$ *of* $\mathcal{L}$ *subject to Dirichlet conditions* $z(0) = 0 = z'(0)$.

Proof Suppose y is as described, so that $\mathcal{L}y = \lambda y$ for some real λ, and let $z = \ell y$. Because y is not in the null-space of ℓ, z is non-trivial. Also $z(0) = -y''(0) + q(0)y(0) = 0$ because $y''(0) = 0$ and $q(0) = 0$, and $z'(0) = -y'''(0) + q(0)y'(0) + q'(0)y(0) = 0$ since $y'''(0) = 0$ and $q(0) = q'(0) = 0$. Thus z satisfies the Dirichlet boundary conditions. Clearly $\mathcal{L}z = \ell^2\ell y = \ell\mathcal{L}y = \ell\lambda y = \lambda\ell y = \lambda z$, so z satisfies the differential equation $\mathcal{L}z = \lambda z$: as the coefficient q is twice continuously differentiable, this makes it easy to see that z is four times continuously differentiable. Finally, z is square integrable. This follows because

$$\langle z, z\rangle = \langle \ell y, \ell y\rangle = \langle \ell^2 y, y\rangle = \lambda\langle y, y\rangle,$$

the penultimate equality using the fact that $\ell^2 = \mathcal{L}$ is strong limit-point at infinity together with the fact that $\ell y(0) = (\ell y)'(0) = 0$. □

ACKNOWLEDGEMENT This work has been supported by the EPSRC, by the Royal Society of London and the University of Leicester Research Committee.

References

1. H. Prüfer, Math. Ann., 95, 499 (1923).
2. F.V. Atkinson, Discrete and Continuous Boundary Problems, Acadamic Press, New York (1964).
3. W.T. Reid, Sturmian Theory for Ordinary Differential Equations, Springer, New York (1980).
4. P.B. Bailey, M.K. Gordon and L.F. Shampine, ACM Trans. Math. Software, 4, 193 (1978).
5. M. Marletta and J.D. Pryce, J. Comp. Appl. Math., 39, 57 (1992).
6. C.T. Fulton and S. Pruess, ACM Trans. Math. Software, 19, 360 (1993).
7. P.B. Bailey, B.S. Garbow, H.G. Kaper and A. Zettl, ACM Trans. Math. Software, 17, 500 (1991); also by same authors, in same volume, pp. 491 ff.

8. F.V. Atkinson, A.M. Krall, G.K. Leaf and A. Zettl, Argonne Laboratories Technical Report (1987).

9. M. Marletta, Numerical Algorithms, 4, 65 (1994).

10. L. Greenberg, University of Maryland Technical Report, TR91-24 (1991).

11. A.A. Abramov, Computational Mathematics and Mathematical Physics, 31, 27 (1991).

12. G. Baur and W. Kratz, Rend. Circ. Mat. Palermo (2), 38, 329 (1989).

13. M. Marletta, Advances in Computational Mathematics, 2, 155 (1994).

14. L. Greenberg and M. Marletta, IMA J. Numer. Analysis, 15, 319 (1995).

15. L. Greenberg and M. Marletta, *The code SLEUTH for solving fourth order Sturm-Liouville problems*, ACM Trans. Math. Software, in press (1997).

16. C. Bennewitz, B.M. Brown, W.D. Evans, D.K.R. McCormack and M. Marletta, Proc. Royal Soc. Lond., A452, 1765 (1996).

17. B.M. Brown and M. Marletta, *On a conjecture of Bennewitz, and the behaviour of the Titchmarsh-Weyl matrix near a pole*, submitted to Proc. Royal Soc. Lond., (1997).

18. N.G.J. Dias, On an Integral Inequality Associated with a $2n$th Order Quasidifferential Expression, PhD thesis, Department of Computing Mathematics, University of Wales, Cardiff (1994).

19. A.A. Abramov and A.A. Aslanyan, Computational Mathematics and Mathematical Physics, 34, 1629 (1994).

20. A.A. Aslanyan, Computational Mathematics and Mathematical Physics, 35, 1331 (1995).

21. P.G. Drazin and W.H. Reid, Hydrodynamic Stability, CUP, Cambridge (1981).

22. L. Greenberg and M. Marletta, *Oscillation theory and numerical solution of sixth and higher order Sturm-Liouville problems*, SIAM J. Numer. Analysis, in press (1998).

23. S. Pruess, SIAM J. Numer. Analysis, 10, 55 (1973).

8th Int. Coll. on Differential Equations, pp. 293-300
D. Bainov (Ed)

$\mathcal{L}^{2,\lambda}$-Theory for Nonvariational Basic Parabolic Systems

MARIO MARINO AND ANTONINO MAUGERI
Dipartimento di Matematica, Università di Catania, Viale "A. Doria" 6, 95125 Catania, Italy

Abstract. The Hölder continuity of the solutions to nonvariational nonlinear basic parabolic systems is proved for small values of n.
Keywords: *Non-linear parabolic systems, Caccioppoli-type inequality, $\mathcal{L}^{2,\lambda}$-theory.*

1. Introduction

Let Ω be a bounded open set in $\mathbb{R}^n$, $n \geq 2$, of class C^2, with generic point $x = (x_1, x_2, \dots, x_n)$. If T is a positive real number, we denote by Q the cylinder $\Omega \times (-T, 0)$ and by X the point (x,t) of $\mathbb{R}^n_x \times \mathbb{R}_t$. If $u(X)$ is a vector $Q \to \mathbb{R}^N$, N integer ≥ 1, we set:

$$D_i u = \frac{\partial u}{\partial x_i}, \qquad Du = (D_1 u, D_2 u, \dots, D_n u),$$

$$H(u) = \{D_{ij} u\} = \{D_i D_j u\}, \qquad i,j = 1, \dots, n;$$

obviously, Du and $H(u)$ are elements of $\mathbb{R}^{nN}$ and $\mathbb{R}^{n^2 N}$ respectively.

Setting

$$W^p\left(Q, \mathbb{R}^N\right) = \left\{v : v \in L^p\Big(-T, 0, H^{2,p}(\Omega, \mathbb{R}^N)\Big), \frac{\partial v}{\partial t} \in L^p\left(Q, \mathbb{R}^N\right)\right\},$$

$$W^p_0\left(Q, \mathbb{R}^N\right) = \left\{v \in W^p\left(Q, \mathbb{R}^N\right) : v \in L^p\Big(-T, 0, H^{1,p}_0(\Omega, \mathbb{R}^N)\Big), v(x, -T) = 0\right\},$$

where $p \in [1, +\infty[$, let $u \in W^2\left(Q, \mathbb{R}^N\right)$ be a solution of the parabolic system

$$a\Big(H(u)\Big) - \frac{\partial u}{\partial t} = f(X) \text{ in } Q. \tag{1.1}$$

We are concerned with regularity properties of the solutions to system (1.1); this study is preliminary to the regularity of solutions $u \in W^2\left(Q, \mathbb{R}^N\right)$ of the complete system

$$a\left(X, u, Du, H(u)\right) - \frac{\partial u}{\partial t} = f(X, u, Du).$$

It is well known that the Cauchy-Dirichlet problem

$$\begin{cases} u \in W^2_0\left(Q, \mathbb{R}^N\right) \\ a(H(u)) - \frac{\partial u}{\partial t} = f(X) \in L^2(Q, \mathbb{R}^N) \end{cases}$$

is in general not solvable; it is necessary to require some conditions on the operator

$$a(\xi) : \mathbb{R}^{n^2N} \to \mathbb{R}^N.$$

We assume that $a(\xi)$ is continuous in $\mathbb{R}^{n^2N}$ and verifies the conditions:
(A) *there exist three positive constants α, γ and δ with $\gamma + \delta < 1$, such that*

$$\left\| \sum_{i=1}^{n} \tau_{ii} - \alpha\Big[a(\tau+\eta) - a(\eta)\Big] \right\|^2 \le \gamma\|\tau\|^2 + \delta\left\|\sum_{i=1}^{n}\tau_{ii}\right\|^2, \forall \tau, \eta \in \mathbb{R}^{n^2N};$$

$$a(0) = 0. \tag{1.2}$$

Condition (A), introduced by S. Campanato (see [1]), is equivalent to the following "pseudo monotonicity condition"

$$\|a(\tau+\eta) - a(\eta)\| \le \frac{c(n)}{\alpha}\|\tau\|, \ \forall \tau, \eta \in \mathbb{R}^{n^2N}, \tag{1.3}$$

$$\left(a(\tau+\eta) - a(\eta) | \sum_{i=1}^{n}\tau_{ii}\right) \ge \frac{1-\delta}{2\alpha}\left\|\sum_{i=1}^{n}\tau_{ii}\right\|^2 - \frac{\gamma}{2\alpha}\|\tau\|^2, \ \forall \tau, \eta \in \mathbb{R}^{n^2N}. \tag{1.4}$$

In fact, let us consider

$$\alpha^2\|a(\tau+\eta) - a(\eta)\|^2 = \left\| -\sum_{i=1}^{n}\tau_{ii} + \alpha\Big[a(\tau+\eta) - a(\eta)\Big] + \sum_{i=1}^{n}\tau_{ii}\right\|^2 \le$$

$$\le 2\left\|\sum_{i=1}^{n}\tau_{ii}\right\|^2 + 2\left(\delta\left\|\sum_{i=1}^{n}\tau_{ii}\right\|^2 + \gamma\|\tau\|^2\right) \le c(n)\|\tau\|^2;$$

then we obtain

$$\|a(\tau+\eta) - a(\eta)\|^2 \le \frac{c(n)}{\alpha^2}\|\tau\|^2.$$

On the other hand one has

$$\left\|\sum_{i=1}^{n}\tau_{ii} - \alpha\Big[a(\tau+\eta) - a(\eta)\Big]\right\|^2 = \left\|\sum_{i=1}^{n}\tau_{ii}\right\|^2 +$$

$$+\left\|\alpha\Big[a(\tau+\eta) - a(\eta)\Big]\right\|^2 - 2\alpha\left(a(\tau+\eta) - a(\eta) | \sum_{i=1}^{n}\tau_{ii}\right) \le \gamma\|\tau\|^2 + \delta\left\|\sum_{i=1}^{n}\tau_{ii}\right\|^2$$

and hence

$$\left(a(\tau+\eta) - a(\eta) | \sum_{i=1}^{n}\tau_{ii}\right) \ge \frac{1-\delta}{2\alpha}\left\|\sum_{i=1}^{n}\tau_{ii}\right\|^2 - \frac{\gamma}{2\alpha}\|\tau\|^2.$$

From (1.4), by choosing $\tau = \Big\{\lambda_i\lambda_j\zeta\Big\}$, $\forall \lambda \in \mathbb{R}^n$, $\forall \zeta \in \mathbb{R}^N$, $\forall \eta \in \mathbb{R}^{n^2N}$, we get

$$\left(a(\Big\{\lambda_i\lambda_j\zeta\Big\} + \eta) - a(\eta) | \ \|\lambda\|^2\zeta\right) \ge \frac{1-\delta}{2\alpha}\|\lambda\|^4 \ \|\zeta\|^2 - \frac{\gamma}{2\alpha}\|\lambda\|^4 \ \|\zeta\|^2$$

and hence we achieve the ellipticity condition

$$\left(a(\Big\{\lambda_i\lambda_j\zeta\Big\} + \eta) - a(\eta)|\zeta\right) \ge \frac{1-(\gamma+\delta)}{2\alpha}\|\lambda\|^2 \ \|\zeta\|^2. \tag{1.5}$$

On the other hand condition (1.3) ensures that there exist the derivatives $\dfrac{\partial a^h(\eta)}{\partial \eta_{ij}^k}, h, k = 1,2,\dots,N,\ i,j = 1,2,\dots,n$, a.e. in $\mathbb{R}^{n^2N}$, which belong to $L^\infty(\mathbb{R}^{n^2N})$. Then from (1.5), substituting ζ with $t\zeta, t \in \mathbb{R}$, and using the formula for the differential to evaluate $a(t\{\lambda_i\lambda_j\zeta\} + \eta) - a(\eta)$, we obtain

$$\left(t \sum_{ij=1}^{n} \lambda_i\lambda_j \frac{\partial a(\eta)}{\partial \eta_{ij}}\zeta + 0(t)|t\zeta\right) \geq \frac{1-(\gamma+\delta)}{2\alpha}\, t^2\|\lambda\|^2\ \|\zeta\|^2,$$

where

$$\frac{\partial a(\eta)}{\partial \eta_{ij}} = \left\{\frac{\partial a^h(\eta)}{\partial \eta_{ij}^k}\right\}_{h,k=1,2,\dots,N} \quad \text{and} \quad \lim_{t\to 0}\frac{0(t)}{t} = 0.$$

Finally, dividing by t^2 and passing to the limit as $t \to 0$, yields

$$\sum_{ij=1}^{n} \lambda_i\lambda_j \left(\frac{\partial a(\eta)}{\partial \eta_{ij}}\zeta|\zeta\right) \geq \frac{1-(\gamma+\delta)}{2\alpha}\|\lambda\|^2\ \|\zeta\|^2. \tag{1.6}$$

We have thus obtained the usual ellipticity condition. Let us note that, if $N > 1$, then condition (A) does not imply that $a(\eta)$ be strongly elliptic. In fact, the system of linear elasticity satisfies condition (A) and nevertheless it is not strongly elliptic.

The ellipticity of the vector $a(\eta)$, in turn, does not imply condition (A), as the following simple counterexample in the case $N = 1$ shows

$$a\left(H(u)\right) = \Delta u + \frac{2(n-1)}{n-2}D_{nn}\, u, \quad n \geq 3.$$

In particular, from (1.3), with $\eta = 0$, and by virtue of condition (1.2), we get

$$\|a(\tau)\| \leq \frac{c(n)}{\alpha}\|\tau\|, \quad \forall \tau \in \mathbb{R}^{n^2N}.$$

The aim of this paper is to study the global Hölder-continuity of the solutions to system (1.1), under assumptions (A) and (1.2). To this end it is enough to study the $\mathcal{L}^{2,\lambda}(Q,\mathbb{R}^N)$ regularity for the solutions to system (1.1).

Let us recall that $\mathcal{L}^{2,\lambda}(Q,\mathbb{R}^N)$, $0 \leq \lambda \leq n+4$, is the space of all functions $u \in L^2(Q,\mathbb{R}^N)$ such that

$$[u]^2_{\mathcal{L}^{2,\lambda}(Q,\mathbb{R}^N)} = \sup_{\substack{X^0\in\overline{Q}\\ 0<\sigma\leq\delta(Q)}} \frac{1}{\sigma^\lambda} \fint_{Q(X^0,\sigma)\cap Q} \|u(X) - u_{Q(X^0,\sigma)\cap Q}\|^2 dX < +\infty,$$

where $Q(X^0,\sigma) = B(x^0,\sigma) \times (t^0-\sigma^2, t^0), \delta(Q) = \max\left\{\text{diam } \Omega, \sqrt{T}\right\}, u_{Q(X^0,\sigma)\cap Q} = \fint_{Q(X^0,\sigma)\cap Q} u dX = \dfrac{1}{\text{meas } [Q(X^0,\sigma)\cap Q]}\displaystyle\int_{Q(X^0,\sigma)\cap Q} u(X)dX.$

If Q is suitably regular, the space $\mathcal{L}^{2,\lambda}(Q,\mathbb{R}^N)$ has the following properties:

i) if $\lambda > n+2$, $\mathcal{L}^{2,\lambda}(Q,\mathbb{R}^N) = C^{0,\alpha}(\overline{Q},\mathbb{R}^N)$, $\alpha = \frac{\lambda-(n+2)}{2}$;

ii) if $0 \leq \lambda < n+2$, $\mathcal{L}^{2,\lambda}(Q,\mathbb{R}^N) = L^{2,\lambda}(Q,\mathbb{R}^N)$, where $L^{2,\lambda}(Q,\mathbb{R}^N)$ is the space of functions $u \in L^2(Q,\mathbb{R}^N)$ such that

$$\|u\|^2_{L^{2,\lambda}(Q,\mathbb{R}^N)} = \sup_{\substack{X^0\in\overline{Q}\\ 0<\sigma\leq\delta(Q)}} \frac{1}{\sigma^\lambda} \int_{Q(X^0,\sigma)\cap Q} \|u\|^2 dX < +\infty;$$

iii) for $\lambda = n+2$, $\mathcal{L}^{2,n+2} =$ BMO $= \mathcal{L}^{p,n+2}$, $\forall p \geq 1$.

The main result we establish is the following:

Theorem 1.1. *There exists a number $\tilde{\lambda} \in (2, n+2]$ such that if $u \in W^2(Q, \mathbb{R}^N)$ is a solution to the system*

$$a(H(u)) - \frac{\partial u}{\partial t} = f(X) \quad \text{in } Q,$$

with $f(X) \in \mathcal{L}^{2,\mu}(Q, \mathbb{R}^N)$, $0 < \mu < \tilde{\lambda}$, then

$$Du \in \mathcal{L}^{2,\mu+2}_{\text{loc}}(Q, \mathbb{R}^{nN}), u \in \mathcal{L}^{2,\mu+4}_{\text{loc}}(Q, \mathbb{R}^N).$$

(see [2], Theorem 4.1).

From this theorem it follows that

"*if $n = 2$, Du is Hölder-continuous in Q*";

"*if $n \leq 4$, u is Hölder-continuous in Q*".

These results say that a solution of the nonlinear system (1.1) is regular in Q for small values of n, whereas for large values of n it is partially regular (see [3]); however, it is an open problem whether the solution $u \in W^2(Q, \mathbb{R}^N)$ to system (1.1) is Hölder-continuous in Q also for $n \geq 5$.

To prove Theorem 1.1 we need some preliminary results, which are of independent interest.

Let $u \in W^2(Q, \mathbb{R}^N)$ be a solution of the "basic" parabolic system

$$a(H(u)) - \frac{\partial u}{\partial t} = 0 \quad \text{in } Q, \tag{1.7}$$

and suppose that $a(\xi)$ satisfies conditions (A) and (1.2). The following theorems hold.

Theorem 1.2. (Caccioppoli-type inequality) *$Du \in W^2_{\text{loc}}(Q, \mathbb{R}^{nN})$ and, $\forall\ Q(2\sigma) = Q(X^0, 2\sigma) \subset Q$, the following inequality holds:*

$$\int_{Q(\sigma)} \left(\|H(Du)\|^2 + \|\frac{\partial(Du)}{\partial t}\|^2 \right) dX \leq$$

$$\leq c\sigma^{-2} \left\{ \sigma^{-2} \int_{Q(2\sigma)} \|D(u - P_{Q(2\sigma)})\|^2 dX + \int_{Q(2\sigma)} \|H(u - P_{Q(2\sigma)})\|^2 dX \right\}, \tag{1.8}$$

where $P_{Q(2\sigma)}$ is the vector-polynomial in x, of degree ≤ 2, such that

$$\int_{Q(2\sigma)} D^\alpha(u - P_{Q(2\sigma)}) dX = 0, \quad \forall \alpha: \ |\alpha| \leq 2. \tag{1.9}$$

(see [4], [2]).

Theorem 1.3. (L^q_{loc}-regularity). *There exists a number $\tilde{q} > 2$ such that, $\forall q \in (2, \tilde{q})$, it results*

$$Du \in W^q_{\text{loc}}(Q, \mathbb{R}^{nN})$$

and, $\forall Q(2\sigma) \subset\subset Q$, with $\sigma \in (0,1)$, the following estimate holds:

$$\left[\fint_{Q(\sigma)} \left(\|H(Du)\|^q + \left\| \frac{\partial(Du)}{\partial t} \right\|^q \right) dX \right]^{\frac{1}{q}} \leq \tag{1.10}$$

$$\leq c \left[\fint_{Q(2\sigma)} \left(\|H(Du)\|^2 + \left\| \frac{\partial(Du)}{\partial t} \right\|^2 \right) dX \right]^{\frac{1}{2}}.$$

(see [2], Section 2).

Theorem 1.4. (Fundamental interior estimate for $H(Du)$ and $\frac{\partial(Du)}{\partial t}$). *For every $Q(\sigma)\subset\subset Q$, with $\sigma < 2$, $\forall\tau \in (0,1)$ and $\forall q \in (2,\tilde{q})$ $\tilde{q}$ is the constant which appears in Theorem 1.3., we have*

$$\int_{Q(\tau\sigma)} \left(\|H(Du)\|^2 + \|\frac{\partial(Du)}{\partial t}\|^2\right)dX \le \tag{1.11}$$

$$\le c\tau^{(n+2)(1-\frac{2}{q})} \int_{Q(\sigma)} \left(\|H(Du)\|^2 + \|\frac{\partial(Du)}{\partial t}\|^2\right)dX.$$

(see [2], Theorem 3.1).

Theorem 1.5. (Fundamental interior estimate for $H(u)$ and $\frac{\partial u}{\partial t}$). *For every $Q(\sigma) \subset Q$, with $\sigma < 2, \forall\tau \in (0,1)$ and $\forall q \in (2, \min(\tilde{q}, n+2))$ (1), we have*

$$\int_{Q(\tau\sigma)} \left(\|H(u)\|^2 + \left\|\frac{\partial u}{\partial t}\right\|^2\right)dX \le \tag{1.12}$$

$$\le c\tau^{2+(n+2)(1-\frac{2}{q})} \int_{Q(\sigma)} \left(\|H(u)\|^2 + \left\|\frac{\partial u}{\partial t}\right\|^2\right)dX.$$

(see [2], Theorem 3.2).

We also need the following existence result.

Theorem 1.6. *If Ω is of class C^2 and convex while the vector $a(\xi)$ satisfies conditions (1.2) and (A), then, $\forall f \in L^2(Q, \mathbb{R}^N)$ and $\forall u \in W^2(Q, \mathbb{R}^N)$, the Cauchy-Dirichlet problem*

$$\begin{cases} w \in W_0^2(Q, \mathbb{R}^N) \\ a\,(H(w) + H(u)) - \frac{\partial w}{\partial t} = f(X) \text{ in } Q \end{cases}$$

has a unique solution and it results

$$\int_Q \left(\|H(w)\|^2 + \left\|\frac{\partial w}{\partial t}\right\|^2\right)dX \le c\int_Q \|f - a(H(u))\|^2 dX. \tag{1.13}$$

(see [2], Theorem 1.2).

The following Poincaré-type inequalities play an important role in proving Theorems 1.3–1.5 and then in the $\mathcal{L}^{2,\lambda}$ theory.

Lemma 1.1. *If $u \in W^p(Q, \mathbb{R}^N)$, then*

$$\int_Q \|u - u_Q\|^p dX \le c\,\delta^p(Q) \int_Q \|Du\|^p dX + \tag{1.14}$$

$$+ c\,\delta^{2p}(Q) \int_Q \left\|\frac{\partial u}{\partial t}\right\|^p dX$$

and

$$\int_Q \|Du - (Du)_Q\|^p dX \le c\delta^p(Q) \int_Q \left(\|H(u)\|^p + \left\|\frac{\partial u}{\partial t}\right\|^p\right)dX, \tag{1.15}$$

where

$$\delta(Q) = \max\left\{\text{diam } \Omega, \sqrt{T}\right\}.$$

(see [5], Lemma 2.1 and 2.II).

Lemma 1.2. *If $u \in W^2(Q, \mathbb{R}^N)$ is a solution of system (1.7), if $Q(X^0, \sigma) \subset\subset Q$ with $\sigma \in (0,2)$ and if $P_{Q(X^0,\sigma)}$ is the vector-polynomial in x, of degree ≤ 2, satisfying the condition*

$$\int_{Q(X^0,\sigma)} D^\alpha(u - P_{Q(X^0,\sigma)})dX = 0, \quad \forall\alpha : |\alpha| \leq 2,$$

then

$$\sigma^{-2}\int_{Q(X^0,\sigma)} \|D(u - P_{Q(X^0,\sigma)})\|^2 dX + \int_{Q(X^0,\sigma)} \|H\left(u - P_{Q(X^0,\sigma)}\right)\|^2 dX \leq \tag{1.16}$$

$$\leq c\left[\int_{Q(X^0,\sigma)} \left(\|H(Du)\|^{\frac{2(n+2)}{n+4}} + \left\|\frac{\partial(Du)}{\partial t}\right\|^{\frac{2(n+2)}{n+4}}\right)dX\right]^{\frac{n+4}{n+2}}.$$

(see [2], Section 2).

2. Proof of Theorem 1.1

Let $f: Q \to \mathbb{R}^N$ be a vector of class $\mathcal{L}^{2,\mu}(Q, \mathbb{R}^N)$, $0 < \mu < \tilde{\lambda} = \min\left\{2 + (n+2)(1 - \frac{2}{q}), n+2\right\}$, and let $u \in W^2(Q, \mathbb{R}^N)$ be a solution of the parabolic system

$$a(H(u)) - \frac{\partial u}{\partial t} = f(X) \text{ in } Q. \tag{2.1}$$

Fixed $Q(\sigma) = Q(X^0, \sigma) \subset Q$, with $\sigma < 2$, let w be the solution of the Cauchy-Dirichlet problem

$$\begin{cases} w \in W_0^2(Q(\sigma), \mathbb{R}^N) \\ a(H(w) + H(u)) - \frac{\partial w}{\partial t} = \frac{\partial u}{\partial t} \quad \text{in } Q(\sigma). \end{cases} \tag{2.2}$$

Theorem 1.6 ensures the existence of a unique solution to problem (2.2) satisfying inequality (1.13) with $f = \frac{\partial u}{\partial t}$ and $Q = Q(\sigma)$.

Setting in $Q(\sigma)$, $v = w + u$, the function v belongs to $W^2(Q(\sigma), \mathbb{R}^N)$ and satisfies the system

$$a(H(v)) - \frac{\partial v}{\partial t} = 0 \quad \text{in } Q(\sigma). \tag{2.3}$$

For v, by means of Theorem 1.5, we have

$$\int_{Q(\tau\sigma)} \left(\|H(v)\|^2 + \left\|\frac{\partial v}{\partial t}\right\|^2\right)dX \leq \tag{2.4}$$

$$\leq c\tau^{2+(n+2)(1-\frac{2}{q})}\int_{Q(\sigma)} \left(\|H(v)\|^2 + \left\|\frac{\partial v}{\partial t}\right\|^2\right)dX,$$

$\forall\tau \in (0,1)$ and $\forall q \in (2, \min(\tilde{q}, n+2))$.

On the other hand, from (1.13) we get

$$\int_{Q(\sigma)} \left(\|H(w)\|^2 + \left\|\frac{\partial w}{\partial t}\right\|^2\right)dX \leq c\int_{Q(\sigma)} \left\|\frac{\partial u}{\partial t} - a(H(u))\right\|^2 dX \tag{2.5}$$

and also, by virtue of (2.1),

$$\int_{Q(\sigma)} \Big(\|H(w)\|^2 + \|\frac{\partial w}{\partial t}\|^2\Big) dX \le c \int_{Q(\sigma)} \|f\|^2 dX. \tag{2.6}$$

Using (2.4) and taking into account that $u = v - w$, produces, $\forall \tau \in (0,1)$,

$$\int_{Q(\tau\sigma)} \Big(\|H(u)\|^2 + \|\frac{\partial u}{\partial t}\|^2\Big) dX \le c\tau^{2+(n+2)(1-\frac{2}{q})} \int_{Q(\sigma)} \Big(\|H(v)\|^2 + \|\frac{\partial v}{\partial t}\|^2\Big) dX +$$

$$+2 \int_{Q(\sigma)} \Big(\|H(w)\|^2 + \|\frac{\partial w}{\partial t}\|^2\Big) dX \le$$

$$\le c\tau^{2+(n+2)(1-\frac{2}{q})} \int_{Q(\sigma)} \Big(\|H(u)\|^2 + \|\frac{\partial u}{\partial t}\|^2\Big) dX +$$

$$+c \int_{Q(\sigma)} \Big(\|H(w)\|^2 + \|\frac{\partial w}{\partial t}\|^2\Big) dX,$$

from which, due to (2.6), we deduce

$$\int_{Q(\tau\sigma)} \Big(\|H(u)\|^2 + \|\frac{\partial u}{\partial t}\|^2\Big) dX \le c\tau^{2+(n+2)(1-\frac{2}{q})} \int_{Q(\sigma)} \Big(\|H(u)\|^2 + \|\frac{\partial u}{\partial t}\|^2\Big) dX +$$

$$+c \int_{Q(\sigma)} \|f\|^2 dX.$$

By the assumption $f \in \mathcal{L}^{2,\mu}(Q, \mathbb{R}^N)$, we get

$$\int_{Q(\tau\sigma)} \Big(\|H(u)\|^2 + \|\frac{\partial u}{\partial t}\|^2\Big) dX \le \tag{2.7}$$

$$\le c\tau^{2+(n+2)(1-\frac{2}{q})} \int_{Q(\sigma)} \Big(\|H(u)\|^2 + \|\frac{\partial u}{\partial t}\|^2\Big) dX + c\sigma^\mu \|f\|^2_{\mathcal{L}^{2,\mu}(Q,\mathbb{R}^N)}.$$

Now, choosing $q \in (2, \min(\tilde{q}, n+2))$ in such a way that $2 + (n+2)(1-\frac{2}{q}) > \mu$ (recall that $\mu < \tilde{\lambda} = \min\big\{2 + (n+2)(1-\frac{2}{\tilde{q}}), n+2\big\}$), bearing in mind (2.7) (written for this value of q), and making use of an algebraic lemma, we obtain

$$\int_{Q(\tau\sigma)} \Big(\|H(u)\|^2 + \|\frac{\partial u}{\partial t}\|^2\Big) dX \le c\tau^\mu \Big\{ \int_{Q(\sigma)} \Big(\|H(u)\|^2 + \|\frac{\partial u}{\partial t}\|^2\Big) dX + \tag{2.8}$$

$$+\sigma^\mu \|f\|^2_{\mathcal{L}^{2,\mu}(Q,\mathbb{R}^N)} \Big\}.$$

Estimate (2.8) ensures that

$$H(u) \in L^{2,\mu}_{loc}(Q, \mathbb{R}^{n^2N}), \frac{\partial u}{\partial t} \in L^{2,\mu}_{loc}(Q, \mathbb{R}^N). \tag{2.9}$$

Hence, by Lemma 1.1, we achieve

$$Du \in \mathcal{L}^{2,\mu+2}_{loc}(Q, \mathbb{R}^{nN}). \tag{2.10}$$

Making use of (2.9), (2.10) as well as of Lemma 1.1, we obtain

$$u \in \mathcal{L}^{2,\mu+4}_{loc}(Q, \mathbb{R}^N). \tag{2.11}$$

From (2.10), if $n < 2\tilde{q} - 2$ (and in particular if $n = 2$), it results $n < \tilde{\lambda}$ and hence, for $f \in \mathcal{L}^{2,\mu}(Q, \mathbb{R}^N)$, with $\mu \in (n, \tilde{\lambda})$, we get

$$Du \in \mathcal{L}^{2,\mu+2}_{loc}(Q, \mathbb{R}^{nN}), \quad \text{with } \mu + 2 > n + 2,$$

and consequently

Du is Hölder-continuous in Q.

Similarly, if $n < 3\,\tilde{q} - 2$ (and in particular if $n \leq 4$), it results $n - 2 < \tilde{\lambda}$ and, for $f \in \mathcal{L}^{2,\mu}(Q, \mathbb{R}^N)$, with $\mu \in (n-2, \tilde{\lambda})$, by (2.11) we get

$$u \in \mathcal{L}^{2,\mu+4}_{loc}(Q, \mathbb{R}^{N}), \quad \text{with } \mu + 4 > n + 2,$$

and hence

u is Hölder-continuous in Q.

References

1. S. Campanato, Rend. Sem. Mat. Fis. Milano, 60, 113–131 (1990).

2. M. Marino and A. Maugeri, Le Matematiche, 51, 349–368 (1996).

3. M. Marino and A. Maugeri, Rend. Mat., (7), 13, 499–527 (1993).

4. S. Campanato, Atti Accad. Naz. Lincei Rend. Cl. Sc. Fis. Mat. Natur., (9), 2, 129–136 (1991).

5. P. Cannarsa, Boll. U.M.I., (5), 18-C, 291–315 (1981).

8th Int. Coll. on Differential Equations, pp. 301–310
D. Bainov (Ed)

A New Approach to the Description of Deep Water Waves Using Fractional Derivatives

KAZUO MATSUUCHI
Institute of Engineering Mechanics, University of Tsukuba, Tennodai 1-1-1, Tsukuba, Ibaraki, 305 Japan

Abstract. In fluid mechanics fractional derivatives often play an important role in describing the motion. We discuss the role focusing on the behavior of water waves. A single equation governing the weakly nonlinear gravity waves travelling on deep water has been derived from Laplace's equation combined with the corresponding boundary conditions. Introducing two kind of fractional derivatives, Riemann-Liouville and Weyl ones, of order 1/2 enables us to derive it. In spite of the singleness, the equation can describe any deformation of waves under the influence of second order nonlinearity. Deriving dynamic wave hahavior from the equation is much easier than it is with using the original set of equations. In order to know the wave-wave interaction, several solutions were examined.
Keywords: *Fractional Derivative, Potential Flow, Stokes Wave train, Nonlinear Interaction.*
AMS Subject Classification: 26A33

1. Introduction

A thorough understanding of deep water waves will be achieved when we have come to understand the behavior of waves of arbitrary shape. To determine the exact time evolution of the initial wave forms, we derive a single equation governing the free surface taking the nonlinearity of the second order into account. This derivation can be made only by virtue of using the derivatives of fractional order. This type of derivatives has not been yet utilized in describing the physical systems except in extremely limited cases (Lighthill [1], Sugimoto and Kakutani [2], Sugimoto [3]). However, the adaption of the derivatives to the description of the present wave system is shown to produce a powerful means of handling the system and interpreting its physics.

2. Basic equations

When flow is irrotational, the velocity potential ϕ is governed by two dimensional Laplace's equation in the Cartesian coordinates (x, z):

$$\frac{\partial^2 \phi}{\partial x^2} + \frac{\partial^2 \phi}{\partial y^2} = 0, \tag{1}$$

where x is measured horizontally to the right and z vertically upwards. At the free surface, given by $z = \zeta(x,t)$, the velocity potential must satisfy the boundary conditions:

$$\frac{\partial \phi}{\partial z} = \frac{\partial \zeta}{\partial t} + \frac{\partial \phi}{\partial x}\frac{\partial \zeta}{\partial x}, \tag{2}$$

$$\frac{\partial \phi}{\partial t} + \frac{1}{2}\left[\left(\frac{\partial \phi}{\partial x}\right)^2 + \left(\frac{\partial \phi}{\partial z}\right)^2\right] + \zeta = 0, \tag{3}$$

where t is time; all the quantities have been normalized in terms of a characteristic length L, which will be specified later, and a characteristic velocity $\sqrt{gL}$, g being the acceleration dues to gravity. For infinitely deep water, the remaining boundary condition is that the velocity decreases to zero as $z \to -\infty$.

3. Linear theory

Let us first consider a monochromatic wave of infinitesimal amplitude. For the perturbed quantities ϕ_1 and ζ_1 to ϕ and ζ, respectively, equations (1)–(3) yield

$$\zeta_1(x,t) = a\mathrm{e}^{\mathrm{i}(kx-\omega t)}, \tag{4}$$

$$\phi_1(x,z,t) = -\mathrm{i}a\frac{\omega}{k}\mathrm{e}^{kz}\mathrm{e}^{\mathrm{i}(kx-\omega t)}. \tag{5}$$

In the above expression a is the complex amplitude, k the wave number, and ω the frequency. The latter two must satisfy the dispersion relation,

$$\omega^2 = k, \tag{6}$$

which is satisfied for waves travelling in the positive and negative directions. Any waves of arbitrary shape can be expressed, by superposition of the above monochromatic ones, as

$$\zeta_1(x,t) = \int_{-\infty}^{\infty} a(k)\mathrm{e}^{\mathrm{i}(kx-\omega t)}\mathrm{d}k. \tag{7}$$

Thus far we have assumed that k is positive. For the purpose of using equation (7), the value of k should be extended to a negative sign. Confining ourselves to the waves travelling only in the positive x-direction, instead of the dispersion relation (6) it should be written for both signs of k,

$$\omega = \mid k \mid^{1/2} \operatorname{sgn} k. \tag{8}$$

Similarly, for the perturbed potential we obtain,

$$\phi_1(x,z,t) = -\int_{-\infty}^{\infty} a(k)\mathrm{i} \mid k \mid^{-1/2} \operatorname{sgn} k\mathrm{e}^{|k|z}\mathrm{e}^{\mathrm{i}(kx-\omega t)}\mathrm{d}k. \tag{9}$$

Differentiating equation (7) and using relation (8), we have

$$\frac{\partial \zeta_1}{\partial t} + \int_{-\infty}^{\infty} \mathrm{i} \mid k \mid^{1/2} \operatorname{sgn} k a(k)\mathrm{e}^{\mathrm{i}(kx-\omega t)}\mathrm{d}k = 0. \tag{10}$$

To transform expression (10) into a more tractable form, we utilize the relations known from the theory of Fourier transform. The formal adaption of the convolution theorem (Bracewell [4], p. 108) gives

$$(11)\qquad \int_{-\infty}^{\infty} \hat{f}(k) \mid k \mid^{-\beta-1} e^{ikx} dk = \frac{1}{2\cos\pi(\beta+1)/2}(\mathrm{I}^{\beta+1} + \mathrm{K}^{\beta+1})f(x),$$

$$(12)\qquad \int_{-\infty}^{\infty} \hat{f}(k) \mid k \mid^{-\beta-1} \operatorname{sgn} k e^{ikx} dk = \frac{-1}{2i\sin\pi(\beta+1)/2}(\mathrm{I}^{\beta+1} - \mathrm{K}^{\beta+1})f(x),$$

where the function $f(x)$ and its Fourier transform $\hat{f}(k)$ satisfy

$$(13)\qquad f(x) = \int_{-\infty}^{\infty} \hat{f}(k) e^{ikx} dk.$$

In deriving identities (11) and (12), the functions such as $\mid k \mid^{\beta}$ and $\mid k \mid^{\beta} \operatorname{sgn} k$ have been regarded as generalized functions (Lighthill [5]). Identities (11) and (12) include the fractional integrals $\mathrm{I}^{\lambda} f(x)$ and $\mathrm{K}^{\lambda} f(x)$ defined by

$$(14)\qquad \mathrm{I}^{\lambda} f(x) = \frac{1}{\Gamma(\lambda)} \int_{-\infty}^{x} f(x')(x-x')^{\lambda-1} dx',$$

$$(15)\qquad \mathrm{K}^{\lambda} f(x) = \frac{1}{\Gamma(\lambda)} \int_{x}^{\infty} f(x')(x'-x)^{\lambda-1} dx',$$

which are called, respectively, the Riemann-Liouville integral and the Weyl integral of order λ (see, for example, Magnus et al [6]). By use of equation (12), (10) can be rewritten as

$$(16)\qquad \frac{\partial \zeta_1}{\partial t} + \mathrm{J}_{-}^{-1/2} \zeta_1 = 0,$$

where, for simplicity, we have set

$$(17)\qquad \mathrm{J}_{\pm}^{-1/2} = \frac{1}{\sqrt{2}}(\mathrm{I}^{-1/2} \pm \mathrm{K}^{-1/2}).$$

Equation (16) including the fractional derivatives of order 1/2, or the fractional integrals of order −1/2, can describe any waves of infinitesimal amplitude travelling in the positive x-direction.

4. Weakly nonlinear theory and derivation of a single equation

In the following, we consider the weakly nonlinear wave whose amplitude is assumed to be small compared with the length L. To derive a single equation for waves of small but finite amplitude we expand the dependent variables $\phi(x,z,t)$ and $\zeta(x,t)$ as

$$(18)\qquad \phi(x,z,t) = \sum_{n=1}^{\infty} \varepsilon^n \phi_n(x,z,t),$$

(19) $$\zeta(x,t) = \sum_{n=1}^{\infty} \varepsilon^n \zeta_n(x,t),$$

where ε is a measure of smallness of amplitude. The explicit form for $\phi_1(x,z,t)$ and $\zeta_1(x,t)$ has already been given in equations (4) and (5), respectively. The problem for order ε^2 is reduced to solving the following equations:

(20) $$\frac{\partial^2\phi_2}{\partial x^2} + \frac{\partial^2\phi_2}{\partial z^2} = 0 \quad \text{for} \quad -\infty < z < 0,$$

(21) $$\frac{\partial\phi_2}{\partial z} - \frac{\partial\zeta_2}{\partial t} = \frac{\partial\phi_1}{\partial x}\frac{\partial\zeta_1}{\partial z} - \frac{\partial^2\phi_1}{\partial z^2}\zeta_1 \quad \text{at} \quad z = 0,$$

(22) $$\frac{\partial\phi_2}{\partial t} + \zeta_2 = -\frac{1}{2}\left[\left(\frac{\partial\phi_1}{\partial x}\right)^2 + \left(\frac{\partial\phi_1}{\partial z}\right)^2\right] - \frac{\partial^2\phi_1}{\partial t\partial z}\zeta_1 \quad \text{at} \quad z = 0.$$

Inserting ϕ_1 and ζ_1 into equations (21) and (22), we obtain

(23) $$\frac{\partial\phi_2}{\partial z} - \frac{\partial\zeta_2}{\partial t} = \mathrm{J}_+^{-1/2}\zeta_1\frac{\partial\zeta_1}{\partial x} + \mathrm{J}_+^{-1/2}\frac{\partial\zeta_1}{\partial x}\cdot\zeta_1,$$

(24) $$\frac{\partial\phi_2}{\partial t} + \zeta_2 = -\frac{1}{2}[(\mathrm{I}^{-1/2}\zeta_1)^2 + (\mathrm{K}^{-1/2}\zeta_1)^2] + \mathrm{I}^{-1/2}\mathrm{K}^{-1/2}\zeta_1\cdot\zeta_1,$$

where the operator $\mathrm{J}_+^{-1/2}$ has been defined by equation (17).

The velocity potential of the second order $\phi_2(x,z,t)$ can also be expressed by a superposition of the monochromatic waves in a form similar to equation (9), such that

(25) $$\phi_2(x,z,t) = \int_{-\infty}^{\infty} \hat{g}(k,t)\mathrm{e}^{|k|z}\mathrm{e}^{ikx}\mathrm{d}k,$$

where $\hat{g}(k,t)$ is an arbitrary function of k and t. Inserting this expression into equation (23), we have

(26) $$\mathrm{I}^{-1/2}\mathrm{K}^{-1/2}g(x,t) - \frac{\partial\zeta_2}{\partial t} = \mathrm{J}_+^{-1/2}\zeta_1\cdot\frac{\partial\zeta_1}{\partial x} + \mathrm{J}_+^{-1/2}\frac{\partial\zeta_1}{\partial z}\cdot\zeta_1,$$

where $g(x,t)$ is the inverse Fourier transform of $\hat{g}(k,t)$. On the other hand, the insertion of equation (25) into equation (24) gives

(27) $$\frac{\partial g}{\partial t} + \zeta_2 = -\frac{1}{2}[(\mathrm{I}^{-1/2}\zeta_1)^2 + (\mathrm{K}^{-1/2}\zeta_1)^2] + \mathrm{I}^{-1/2}\mathrm{K}^{-1/2}\zeta_1\cdot\zeta_1.$$

Elimination of $g(x,t)$ by the cross-differentiation of equations (26) and (27) yields for $\zeta_2(x,t)$,

(28) $$\begin{aligned}\frac{\partial^2\zeta_2}{\partial t^2} + \mathrm{I}^{-1/2}\mathrm{K}^{-1/2}\zeta_2 &= -\frac{1}{2}\mathrm{I}^{-1/2}\mathrm{K}^{-1/2}[(\mathrm{I}^{-1/2}\zeta_1)^2 + (\mathrm{K}^{-1/2}\zeta_1)^2 - 2\mathrm{I}^{-1/2}\mathrm{K}^{-1/2}\zeta_1\cdot\zeta_1] \\ &\quad + \frac{1}{2}\frac{\partial^2\zeta_1^2}{\partial x^2} + \frac{1}{2}\frac{\partial}{\partial x}[(\mathrm{I}^{-1/2}\zeta_1)^2 - (\mathrm{K}^{-1/2}\zeta_1)^2].\end{aligned}$$

In the above equation, we have assumed that the first order waves travel only in the positive x-direction. In other words, they must satisfy equation (16). It should, however, be noted that the operator appearing on the left-hand side denotes the waves propagating in both directions. This will easily be shown as follows. The expression (16) describing the waves propagating in the positive x-direction can be replaced, for the waves in both directions, by

$$\left(\frac{\partial}{\partial t}+D_+\right)\left(\frac{\partial}{\partial t}+D_-\right)\zeta_1=0, \tag{29}$$

where $D_\pm = \pm \mathrm{J}_-^{-1/2}$. It is easily seen that the operators in (29) coincide with those in the left-hand side of equation (28). From (29) and (28), the actual displacement $\zeta(x,t)(=\varepsilon\zeta_1(x,t)+\varepsilon^2\zeta_2(x,t)+\mathrm{O}(\varepsilon^3))$ satisfies

$$\left(\frac{\partial}{\partial t}+D_+\right)\left(\frac{\partial}{\partial t}+D_-\right)\zeta=N\zeta+\mathrm{O}(\varepsilon^3), \tag{30}$$

where N denotes the nonlinear operator given in the right-hand side of equation (28).

5. Several solutions

Except for a few simple cases, it is difficult to seek analytical solutions for equation (30), since it is expressed in terms of fractional derivatives. We first attempt to obtain periodic wave solutions with period 2π. This periodic condition means that the characteristic length L, which has not been specified till now, is determined to be $\lambda_{max}/2\pi$, λ_{max} being the maximum wavelength. Solutions satisfying the condition can be written as

$$\zeta(x,t)=\varepsilon\sum_{n=-\infty}^{\infty}X_n(t)\mathrm{e}^{\mathrm{i}nx} \qquad (X_{-n}=X_n^*), \tag{31}$$

where the asterisk stands for the complex conjugate. Inserting this into equation (30), we have for positive n,

$$\frac{\mathrm{d}^2X_n}{\mathrm{d}t^2}+nX_n=\varepsilon B_n, \tag{32}$$

where the nonlinear term B_n has the following form,

$$B_n=-n\sum_{l=-\infty}^{\infty}\sqrt{|\,l\,||\,n-l\,|-|\,l\,|})X_lX_{n-l}-\frac{n^2}{2}\sum_{l=-\infty}^{\infty}X_lX_{n-l}. \tag{33}$$

It should be noted that the expression (33) can also be derived simply from

$$N\zeta(x,t)=-\mathrm{I}^{-1/2}\mathrm{K}^{-1/2}[(\mathrm{J}_+^{-1/2}\zeta)^2-\mathrm{I}^{-1/2}\mathrm{K}^{-1/2}\zeta\cdot\zeta]+\frac{1}{2}\frac{\partial^2\zeta^2}{\partial x^2}. \tag{34}$$

For a large number n, $|\,X_n\,|$ is generally assumed to be small. Truncating the number of waves by a properly large number, say N, the problem is then reduced to solving a finite set of ordinary differential equation (32). Numerical manipulations for solving the set of equations are simple and established. Actually, we can perform the procedure easily using, for example, the well-known Runge-Kutta method.

The interaction terms expressed in equation (33) are also written as

$$B_n = B_n^h + B_n^l, \tag{35}$$

where the first term B_n^h represents the effect due to the overall wave numbers and B_n^l, the effect due only to wave numbers lower than n. These two terms are

$$B_n^h = -2n\sum_{l=1}^{\infty}\left(\sqrt{l(n+l)} - l\right) X_l^* X_{n+l}, \tag{36}$$

$$B_n^l = -n\sum_{l=1}^{n-1}\sqrt{l(n-l)}X_l X_{n-l}. \tag{37}$$

The former expression shows that the wave of wave number n can be influenced simultaneously by two waves of wave number l and $n+l$. In other words, this means that if no waves of the wave number larger than n exist, this type of interaction has no effect on the wave of wave number n. Therefore, this interaction is different from the well-known nonlinearity generating higher harmonics. On the other hand, the latter is close to the ordinal interaction of the second order where the wave of wave number n is influenced only by two waves of smaller wave number. The coefficient becomes maximum when $l = n/2$. This means that the second harmonic plays an important role in the latter interaction.

5.1. Two-wave interaction

As an example of wave-wave interactions, we take up two-wave interaction in which only two waves interact. Truncating the series on the right-hand side of equation (32) by two waves, we have

$$\frac{d^2X_1}{dt^2} + X_1 = -2\varepsilon\left(\sqrt{2} - 1\right)X_1^* X_2, \tag{38}$$

$$\frac{d^2X_2}{dt^2} + 2X_2 = -2\varepsilon X_1^2. \tag{39}$$

These equations coincide with those derived in another way within the present order of accuracy (Matsuuchi [7]).

One of the interesting solutions to the above equations is the steadily-travelling wave, which was first investigated by Stokes (1847). It is easy to derive the solution for the surface displacement $\zeta(x,t)$ as

$$\zeta(x,t) = \varepsilon A_1 e^{i(x-t)} + \varepsilon^2 A_1^2 e^{2i(x-t)} + \text{c.c.}, \tag{40}$$

where c.c. stands for the complex conjugate to the preceding expression. This steady wave solution coincides with that obtained by Stokes within the present order of approximation, and shows a steep crest and shallow trough as a characteristic inherent in nonlinear gravity waves.

In relation to the realization of the steady wave solution (40) we will take up an approximate solution to equations (38) and (39). This solution is also related to the waves developing from a sinusoidal wave. Consider X_1 to be of order unity and X_2 to be zero in the initial state. It is thus plausible that the second harmonic generated remains small while the fundamental remains nearly constant during a whole interacting process. By setting

$$X_1 = r_1 \mathrm{e}^{-\mathrm{i}t}, \qquad X_2 = r(t)\mathrm{e}^{\mathrm{i}\theta(t)} \tag{41}$$

the imaginary and real parts of equation (39), respectively, give

$$2\frac{\mathrm{d}\theta}{\mathrm{d}t}\frac{\mathrm{d}r}{\mathrm{d}t} + \frac{\mathrm{d}^2\theta}{\mathrm{d}t^2}r = 2\varepsilon r_1^2 \sin(2t+\theta), \tag{42}$$

$$\frac{\mathrm{d}^2 r}{\mathrm{d}t^2} + \left[2 - \left(\frac{\mathrm{d}\theta}{\mathrm{d}t}\right)^2\right] r = -2\varepsilon r_1^2 \cos(2t+\theta). \tag{43}$$

In deriving these, we have assumed r_1 to remain fixed. The above set of equations permits an oscillatory solution in the following form,

$$r(t) = a\cos(2t - \alpha t + \beta), \quad \frac{\mathrm{d}\theta}{\mathrm{d}t} = -\alpha, \tag{44}$$

α, β, and a being real constants. The constant α is given by solving a quadratic equation. The phase velocity $c_2(=\alpha/2)$ of this generated wave is determined, leading to

$$c_2 = \frac{2 \pm \sqrt{2}}{4}. \tag{45}$$

Therefore, as an approximate solution we have

$$\zeta(x,t) = 2\varepsilon r_1 \cos(x-t) + 2\varepsilon a\cos(2t - \alpha t + \beta)\cos(2x - \alpha t + \beta), \tag{46}$$

where the amplitude a has been determined to be $2\varepsilon r_1^2$.

It is well known that the steady wave train found by Stokes becomes unstable to small perturbations of long wave length (Benjamin and Feir [8]). This instability is brought about owing to the nonlinearity of the third order. This phenomenon is beyond our present consideration, because our approximation treats the lowest order, i.e., the second order, nonlinearity. Any instabilities of the second order are not known to exist. Hence, it is interesting to know whether the initial wave train will asymptotically approach the Stokes wave train (40) which is a unique steady solution, or in other words, whether or not the Stokes wave train can be observed.

5.2. Boundary value problem

Next, we consider a case in which the displacement at a point $x = 0$ varies as $\zeta(0,t) = \mathrm{e}^{\mathrm{i}\Omega t}$ for $t > 0$. Such a situation is especially interesting when nonlinearity plays an important role. However, to show the availability of our present approach, we will solve the linear problem.

Even in the case including the fractional power of wave number in the dispersion relation, the solution can be derived as well. According to Thau [9], we can write

$$\zeta(x,t) = \frac{1}{2\pi \mathrm{i}}\int_L \mathrm{e}^{h(s)x}\frac{\mathrm{d}s}{s - \mathrm{i}\Omega}, \tag{47}$$

where $h(s)$ is determined, by taking the dispersion relation into account, as

$$h(s) = is^2 + \frac{s}{m}, \quad m = x/t. \tag{48}$$

The integral path L is a vertical line lying to the right of all singularities. In our present case only one singularity exists at $s = i\Omega$.

Changing the integral path L to the imaginary axis except at the singular point, we can obtain

$$\zeta(x,t) = \frac{1}{2\pi i} \int_{-\infty}^{\infty} e^{h(\lambda)x} \frac{d\lambda}{\lambda - \Omega} + \frac{1}{2} e^{i(\Omega t - \Omega^2 x)}, \tag{49}$$

where $h(\lambda) = i(-\lambda^2 + \lambda/m)$. The contribution consists of the integral along the axis of the complex s–plane and a half of the residue at $s = i\Omega$. Setting $\xi = \lambda - \Omega$, finally we have

$$\zeta(x,t) = \frac{1}{2} e^{i(-\Omega^2 x + \Omega t)} \left[\frac{2}{\pi} g(x,t) + 1\right]. \tag{50}$$

The function $g(x,t)$ is expressed in an integral form as

$$g(x,t) = \int_0^{\infty} \sin \xi (u - \xi x) \frac{d\xi}{\xi} \tag{51}$$

where $u = t - 2\Omega x$. In Figure 1 the wave form at $t = 100$ is depicted together with $g(x,t)$ obtained numerically. It is easy to see that the wave front travels at the group velocity $c_g = 1/2\Omega$, which is a half of the phase velocity.

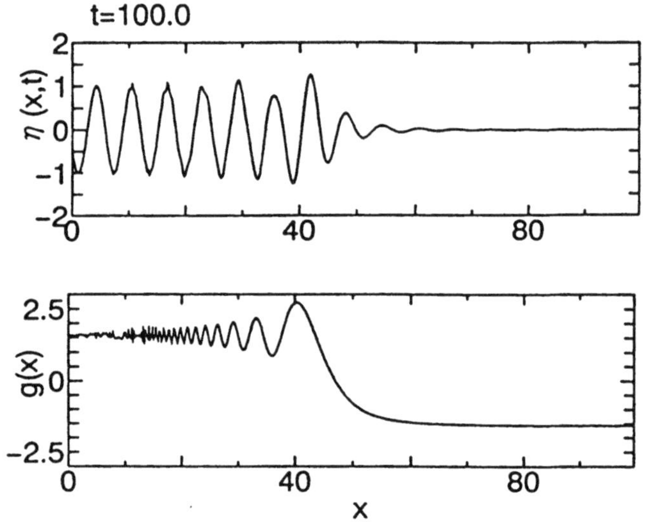

Figure 1: Wave profile at $t = 100$ and the corresponding $g(x,t)$

6. Finite difference algorithm

In nonlinear problems, we cannot avoid numerical way for finding the solution. The present new approach with fractional derivatives might be easy to handle because it is a single equation. However, it is not true unless we might estimate numerically the derivatives in a simple manner. For example, if we calculate the integrals expressed as equations (14) and (15) in each x using Simpson's formula, the procedure would be complex and hence the present approach would not be useful.

We will propose here a new algorithm for the estimation using a finite difference scheme. The ordinary derivative $\mathrm{I}^{-1}f(x)$ is approximated as

$$\mathrm{I}^{-1}f(x) = \frac{1}{\delta}[f(x) - f(x-\delta)]. \tag{52}$$

Noting that the following identity,

$$f(x-\delta) = \mathrm{e}^{-\delta\frac{d}{dx}}, \tag{53}$$

is established formally, it is easy to see that

$$\mathrm{I}^{-1}f(x) = \frac{1}{\delta}(1 - \mathrm{e}^{-\delta\frac{d}{dx}})f(x). \tag{54}$$

Therefore we have for the derivative of order λ,[1]

$$\begin{aligned} \mathrm{I}^{-\lambda}f(x) &= \frac{1}{\delta^\lambda}(1 - \mathrm{e}^{-\delta\frac{d}{dx}})^\lambda f(x) \\ &= \frac{1}{\delta^\lambda}\sum_{j=0}^{\infty}\frac{\Gamma(\lambda+1)}{\Gamma(j+1)\Gamma(\lambda-j+1)}(-1)^j f(x-j\delta), \end{aligned} \tag{55}$$

where the binomial theorem has been adopted in deriving the last expression. In the above equation $\Gamma(x)$ is the gamma function of x. Similarly, we can get for Weyl integral of order $-\lambda$,

$$\begin{aligned} \mathrm{K}^{-\lambda}f(x) &= \frac{1}{\delta^\lambda}(1 - \mathrm{e}^{-\delta\frac{d}{dx}})^\lambda f(x) \\ &= \frac{1}{\delta^\lambda}\sum_{j=0}^{\infty}\frac{\Gamma(\lambda+1)}{\Gamma(j+1)\Gamma(\lambda-j+1)}(-1)^j f(x+j\delta), \end{aligned} \tag{56}$$

since the first order derivative for Weyl integral is approximated as

$$\mathrm{K}^{-1}f(x) = \frac{1}{\delta}[f(x) - f(x+\delta)]. \tag{57}$$

It is much easier to code the summation in equations (55) and (56) than it is with the original definition of the Riemann-Liouville and Weyl integrals (14) and (15).

Acknowledgements

The author would like to express his hearty thanks to Prof. H. Konno of University of Tsukuba for discussing the way of estimating the fractional derivatives based upon Hirota's method.

[1]The present scheme based upon Hirota's method was suggested by H. Konno.

References

1. M. J. Lighthill, Waves in Fluids, Cambridge University Press (1978).

2. N. Sugimoto and T. Kakutani, Wave Motion, 7, 447-458 (1985).

3. N. Sugimoto, Nonlinear Wave Motion, ed. A. Jeffrey, Longman, 162-179 (1989).

4. R. N. Bracewell, The Fourier Transform and Its Applications, McGraw-Hill, 2nd ed. (1978).

5. M. J. Lighthill, An Introduction to Fourier Analysis and Generalized Functions, Cambridge University Press (1964).

6. W. Magnus, F. Oberhettinger and F.G. Tricomi, Tables of Integral Transforms, Vol.2, ed. A. Erderyi, McGraw-Hill (1954).

7. K. Matsuuchi, Fluid Dyn. Res., 12, 281-291 (1993).

8. T. B. Benjamin and J.E. Feir, J. Fluid Mech., 27, 417-430 (1967).

9. S. A. Thau, Nonlinear Waves, ed. S. Leibovich and A.R.Seebass, Conell University Press, 45-78 (1974).

8th Int. Coll. on Differential Equations, pp. 311-318
D. Bainov (Ed)

Application of a Spectral Finite Difference Scheme to a Complex Configuration

Yoshihiro MOCHIMARU
Dept. of International Development Engineering, Tokyo Institute of Technology, Tokyo 152, Japan

Abstract. A newly developed spectral finite difference scheme is applied under a conformal mapping to a circular region. Concrete examples are given to a natural convection problem to show its effectiveness.

1. INTRODUCTION

A spectral finite difference scheme[1][2][3] has been developed for numerically solving partial differential equations linked with boundary and / or initial condition(s). Great benefit will be achieved for problems under a conformal boundary-fitted coordinate system. For a two-dimensional Navier-Stokes equation (with or without a bouyancy term, substantially incompressible) supplemented with a continuity equation and / or an energy equation. a coordinate transformation to a circular region of a unit radius does work for an internal heat / fluid flow problem occupied in a simply-connected region.

2. ANALYSIS

2.1. Coordinate System in General

The following form of coordinate transformation is considered:

(1) $$\exp(\alpha + i\beta) = \frac{1+i\eta}{1-i\eta} \ , \ \eta \equiv f(x+iy) \ ,$$

where (x, y) is a Cartesian coordinate system in a physical plane, (α, β) is a mapped one, and f is an analytic function such that along a closed boundary f is real and varies monotonously from $-\infty$ to $+\infty$ with increasing β. Also it is assumed f^{-1} (the inverse function of f) can be assigned to a single value for the said domain. Under this condition, the coordinate system (α, β) can be a boudary-fitted coordinate, so that the boundary is expressed by $\alpha = 0$. Such a function is possibly given by a complex velocity potential function corresponding to an ideal fluid flow or one multiplied by a constant. Throughout these expressions the coordinates x and y are made dimensionless with respect to a suitable reference length a .

2.2. Basic Equations

Two-dimensional laminar transient natural convection enclosed in the cavity ($-\infty < \alpha \leq 0$) is considered. The acceleration of gravity is assumed to be in the negative y-direction. A dimensionless vorticity transport equation, the relation between vorticity and a stream function, and a dimensionless energy equation neglecting dissipation terms are given by

(2) $$\frac{K}{r^2}\frac{\partial \zeta}{\partial t} + \frac{1}{r}\frac{\partial(\zeta, \psi)}{\partial(r, \beta)} = \frac{1}{\sqrt{Gr}}\left(\frac{\partial^2}{\partial r^2} + \frac{1}{r}\frac{\partial}{\partial r} + \frac{1}{r^2}\frac{\partial^2}{\partial \beta^2}\right)\zeta + \frac{1}{r}\frac{\partial(T, y)}{\partial(r, \beta)} ,$$

(3) $$\frac{K}{r^2}\zeta + \left(\frac{\partial^2}{\partial r^2} + \frac{1}{r}\frac{\partial}{\partial r} + \frac{1}{r^2}\frac{\partial^2}{\partial \beta^2}\right)\psi = 0 ,$$

(4) $$\frac{K}{r^2}\frac{\partial T}{\partial t} + \frac{1}{r}\frac{\partial(T, \psi)}{\partial(r, \beta)} = \frac{1}{Pr\sqrt{Gr}}\left(\frac{\partial^2}{\partial r^2} + \frac{1}{r}\frac{\partial}{\partial r} + \frac{1}{r^2}\frac{\partial^2}{\partial \beta^2}\right)T$$

respectively, where

(5) $$K \equiv \frac{\partial(x, y)}{\partial(\alpha, \beta)} = \left|\frac{dz}{d(\alpha + i\beta)}\right|^2 ,$$

(6) $$z \equiv x + iy \ , \quad r \equiv \exp\alpha \ .$$

ζ : dimensionless vorticity based on U/a, ψ : dimensionless stream function based on Ua, Gr : Grashof number based on the length a and a reference temperature difference ΔT, [in case of Dirichlet condition ΔT is specified by the temperature difference between specified two points; in case of Neumann condition $\Delta T \equiv \overline{q}a/\kappa$, $\overline{q}$: mean heat flux, κ : thermal conductivity of fluid], T : dimensionless temperature given by $(T' - T_r)/\Delta T$, T' : local temperature, T_r : reference temperature, U : reference velocity defined by $(\nu/a)\sqrt{Gr}$, ν : kinematic viscosity of the fluid, Pr : Prandtl number of the fluid, t : dimensionless time based on a/U . For the mapping of the type (1),

(7) $$\frac{\sqrt{K}}{r} = \frac{2}{|f'|\,|1 + \exp(\alpha + i\beta)|^2} \ .$$

The ranges of r and β are specified as

(8) $$0 \leq r \leq 1 \ ,$$

(9) $$-\pi < \beta \leq \pi \ .$$

2.3. Boundary Conditions at $r = 1$

No slip flow on the solid boundary (dynamic conditions) reduces to

(10) $$\psi(r = 1, \beta, t) = 0 \ ,$$

$$\frac{\partial}{\partial r}\psi(r=1,\beta,t)=0 \ . \tag{11}$$

In case a part of the boundary consists of free surface or symmetrical plane, a relevant condition will apply. Thermal boundary conditions may depend on the problem of application and will be mentioned later.

2.4. Auxiliary Conditions at $r = 0$

The condition that scalar quantities and its gradients are continuous at $r = 0$ will apply.

2.5. Mean Nusselt Number

For a full Dirichlet type problem, mean Nusselt number Nu_m based on the temperature difference ΔT and reference length a is given by

$$Nu_m = \int \frac{\partial T}{\partial r} d\beta \Big/ \int \sqrt{K} d\beta \ , \tag{12}$$

where integration is carried out over a designated part. For a (mostly) Neumann type problem, mean Nusselt number Nu_m is given by

$$1/Nu_m = \int T\sqrt{K} d\beta \Big/ \int \sqrt{K} d\beta \ . \tag{13}$$

3. SPECTRAL FINITE DIFFERENCE FORMULATION

3.1. General

Spectral decomposition for any quantity ϕ (scalar or vector component) is specified by a form

$$\phi = \sum_{n=1}^{\infty} \phi_{sn}(r,t)\sin n\beta + \sum_{n=0}^{\infty} \phi_{cn}(r,t)\cos n\beta \ . \tag{14}$$

Especialy for the current case ϕ may be ζ, ψ , or T.

3.2. Discretization in the Interior Region

Equations(2),(3), and (4) can be decomposed, using addition formulae of trigonometric functions etc., into Fourier components of β [1]. The resulting set of decomposed equations can be discretized in time and space by a finite difference approximation, using a non-uniform grid spacing in r, where the n-th coordinate r_n (numbered from the boundary) is specified, using a suitable positive constant γ, by

$$r_n = 1 - h\left\{\frac{\sinh(n-1)\gamma}{\sinh\gamma} + 1\right\} \ , \tag{15}$$

where $1 - h$ is the coordinate of the grid point adjacent to the boundary.
Bouyancy terms can be evaluated as follows:

$$\int_{-\pi}^{\pi} \frac{\partial y}{\partial \beta}\begin{bmatrix}\sin n\beta \\ \cos n\beta\end{bmatrix} d\beta = n \ \mathrm{Im} \int_{-\pi}^{\pi} z \begin{bmatrix}-\cos n\beta \\ \sin n\beta\end{bmatrix} d\beta \ , \tag{16}$$

(17) $$\int_{-\pi}^{\pi} \frac{\partial y}{\partial \alpha} \begin{bmatrix} \sin n\beta \\ \cos n\beta \end{bmatrix} d\beta = n \ \text{Real} \int_{-\pi}^{\pi} z \begin{bmatrix} \cos n\beta \\ -\sin n\beta \end{bmatrix} d\beta \ .$$

If necessary, the following identity may be used:

(18) $$\int_{-\pi}^{\pi} z \cos n\beta d\beta = \oint \frac{z}{2i} \left(e^{-n\alpha} \xi^{n-1} + e^{n\alpha} \xi^{-n-1} \right) d\xi \ ,$$

(19) $$\int_{-\pi}^{\pi} z \sin n\beta d\beta = - \oint \frac{z}{2} \left(e^{-n\alpha} \xi^{n-1} - e^{n\alpha} \xi^{-n-1} \right) d\xi \ ,$$

where

$$\xi \equiv \exp(\alpha + i\beta) \ .$$

3.3. Discretization at $r = 1$

Not only Dirichlet / Neumann conditions but also their mixed ones along the boundary will apply[2]. For example, Dirichlet condition of the type (10) $\psi(1,t) = 0$ along whole boundary reduces to

(20) $$\psi_{sn}(1,t) = 0 \ (n \geq 1) \ ,$$

(21) $$\psi_{cn}(1,t) = 0 \ (n \geq 0) \ .$$

Likewise Eq.(11) reduces to

(22) $$\frac{\partial}{\partial r} \psi_{sn}(1,t) = 0 \ , (n \geq 1) \ ,$$

(23) $$\frac{\partial}{\partial r} \psi_{cn}(1,t) = 0 \ , (n \geq 0) \ ,$$

which can be replaced (for getting higher numerical stability) by

(24) $$\frac{1}{h^2} \psi_{sn}(1-h,t) + \frac{1}{2\pi} \int_{-\pi}^{\pi} (K\zeta)_{r=1-h} \sin n\beta d\beta = 0 \ , (n \geq 1) \ ,$$

(25) $$\frac{1}{h^2} \psi_{cn}(1-h,t) + \frac{1}{2\pi} \int_{-\pi}^{\pi} (K\zeta)_{r=1-h} \cos n\beta d\beta = 0 \ , (n \geq 1) \ ,$$

(26) $$\frac{1}{h^2} \psi_{c0}(1-h,t) + \frac{1}{4\pi} \int_{-\pi}^{\pi} (K\zeta)_{r=1-h} d\beta = 0 \ .$$

3.4. Treatment of Auxiliary Condition at $r = 0$

(27) $$\frac{\partial}{\partial r} T_{c0}(0,t) = 0 \ ,$$

(28) $$T_{cn}(0,t) = T_{sn}(0,t) = 0 \ , \ (n \geq 1) \ ,$$

(29) $$\frac{\partial}{\partial r}\psi_{c0}(0,t) = \frac{\partial}{\partial r}\zeta_{c0}(0,t) = 0 \ ,$$

(30) $$\psi_{sn}(0,t) = \psi_{cn}(0,t) = \zeta_{sn}(0,t) = \zeta_{cn}(0,t) = 0 \ , \ (n \geq 1) \ .$$

4. EXAMPLES OF MAPPING

4.1. Case (I)

(31) $$f = \frac{1}{2}\left(z + \frac{1}{z}\right) \ ; \ \mathrm{Im}(z) \leq 0 \qquad \text{(Semi-circular cavity).}$$

or equivalently

(32) $$z = \frac{(\xi - 1) + \sqrt{2\,(1+\xi^2)}}{i\,(1+\xi)} \ , \ \xi \equiv \exp\left(\alpha + i\beta\right) \ ,$$

where $\sqrt{\ }$ means a principal value.

4.2. Case (II)

(33) $$f = \frac{1}{2}\left(z + \frac{1}{z}\right) - c\,i \ ; \ \mathrm{Im}(z) \leq 0 \ , \ c > 0 \ \text{(Like semi-circular cavity)}$$

Or equivalently

(34) $$z = \frac{\xi - 1 - c\,(\xi + 1) + \sqrt{2 + 2c + c^2}\sqrt{1 + A\,(\xi + B\xi^2)}}{i\,(\xi + 1)} \ ,$$

where

(35) $$A \equiv \frac{2c^2}{2 + 2c + c^2} \ , \ B \equiv \frac{2 - 2c + c^2}{2c^2} \ .$$

4.3. Case (III)

(36) $$f = -i - \frac{i}{z^2} \ ; \ \mathrm{Im}(z) \leq 0 \ \text{(Cassini cavity)}$$

Or equivalently

(37) $$z = -\sqrt{-\left(\frac{1+\xi}{2}\right)} \ .$$

4.4. Case(IV)

(38) $$f = \cos \pi z \ ; -1 \leq \mathrm{Real}(z) \leq 0 \, , 0 \leq \mathrm{Im}(z)$$

(Semi-infinite rectangular cavity / cavity cell unit)

Or equivalently

$$z = \frac{1}{\pi i} \ln \left[i \left\{ \frac{1 - \xi + \sqrt{2(1+\xi^2)}}{1+\xi} \right\} \right] . \tag{39}$$

4.5. Case (V)

$$f = -\{\exp(iz) + \exp(-2iz)\} \quad ; \quad 0 \le \mathrm{Real}(z) \le \frac{\pi}{2}, \mathrm{Im}(z) > -1 \tag{40}$$

(Deformed semi-infinite cavity)

Or equivalently

$$\begin{aligned} \exp(-iz) &= -\omega \sqrt[3]{\frac{1}{2} + \sqrt{\frac{1}{4} - \frac{i}{27}\left(\frac{1-\xi}{1+\xi}\right)^3}} \\ &\quad + \frac{i}{3}\frac{1-\xi}{1+\xi} \Big/ \left\{ \omega \sqrt[3]{\frac{1}{2} + \sqrt{\frac{1}{4} - \frac{i}{27}\left(\frac{1-\xi}{1+\xi}\right)^3}} \right\} , \end{aligned} \tag{41}$$

where $\omega = (-1 + \sqrt{3}i)/2$, and $\sqrt[3]{\ }$ stands for a principal value.

4.6. Case (VI)

$$f = iz - \frac{1}{z^2} , \quad \mathrm{Im}(z) \le 0 , \mathrm{Real}(z) \le 0 \quad \text{(Half of a deformed cavity)} \tag{42}$$

Or equivalently

$$\begin{aligned} \frac{1}{z} &= \omega \sqrt[3]{\frac{i}{2} + \sqrt{-\frac{1}{4} - \frac{i}{27}\left(\frac{1-\xi}{1+\xi}\right)^3}} \\ &\quad - \frac{i}{3}\frac{1-\xi}{1+\xi} \Big/ \left\{ \omega \sqrt[3]{\frac{i}{2} + \sqrt{-\frac{1}{4} - \frac{i}{27}\left(\frac{1-\xi}{1+\xi}\right)^3}} \right\} , \end{aligned} \tag{43}$$

$x = 0$ may be a physical symmetrical plane instead of a solid boundary. In this case along $x = 0 (\alpha = 0, 2\tan^{-1}(3/\sqrt[3]{4}) < \beta \le \pi)$ the identity $\psi = \zeta = \partial T/\partial\alpha = 0$ holds. Treatment of such combined boundary conditions (e.g. on a solid boundary and on a symmetrical plane) is numerically possible[2].

5. NUMERICAL SOLUTION PROCEDURE

By truncating the series (14) up to a suitable order, the system of differential equations can be integrated semi-implicitly with respect to time to get a steady-state solution.

6. NUMERICAL RESULTS

Concrete numerical simulations were carried out for cases (I), (III), and (IV) to get a steady-state solution at $Pr = 0.7$ or $Pr = 3$. Thermal boundary conditions for case (I) of Dirichlet type (heated from below) is

(44) $$T(r=1,|\beta|>\pi/2,t>0)=0\ ,\ \ T(r=1,|\beta|\le\pi/2,t>0)=1\ .$$

Initial conditions are

(45) $$\psi=\zeta=T=0\ .$$

In this case due to symmetry of the field $\psi_{cn}(r,t) = \zeta_{cn}(r,t) = 0, (n \geq 0); T_{sn}(r,t) = 0, (n \geq 1)$. Roughly if $Gr \lesssim 1500$, circulation is relatively small, i.e. $||\psi||$ is small and nearly proportinal to $\sqrt{Gr}$, and mean Nusselt number Nu_m along the heated boundary is nearly the same as that for a purely heat conduction case ($Gr = 0$) , whereas if Gr gets greater than about 1500, Nu_m increases with increasing Gr . At least up to $Gr \doteq 4000$, steady-state solutions can be obtained for $Pr = 0.7$. Under the thermal boundary condition (44), although $|T| < +\infty$, mathematicall $Nu_m = +\infty$ for infinite series. However, for the first N-terms $Nu_m \sim C_1 + C_2 \ln N$, $C_2 > 0$, which is consistent with a spectral finite difference scheme of finite number of components.

Thermal boundary conditions for case (III) of Dirichlet type (heated from the lower part) is

(46) $$T(r=1,|\beta|>2\pi/3,t>0)=0\ ,\ \ T(r=1,|\beta|\le 2\pi/3,t>0)=1\ .$$

Initial conditions are

(47) $$\psi=\zeta=T=0\ .$$

In this case due to symmetry of the field $\psi_{cn}(r,t) = \zeta_{cn}(r,t) = 0, (n \geq 0); T_{sn}(r,t) = 0, (n \geq 1)$. For these conditions at $Pr = 0.7$ (air) , steady-state solutions can be obtained at least $Gr \lesssim 3 \times 10^4$. If $Gr \lesssim 7000$, mean Nusselt number Nu_m at a steady-state along the heated boundary is nearly independent of Gr, i.e. nearly equal to the one corresponding to a purely heat conduction case, and $||\psi||$ is nearly proportional to $\sqrt{Gr}$. If $Gr \gtrsim 7000$, Nu_m at a steady-state increases with Gr . If $Gr > 7000$, circulation is so strong that Max($|\psi|$ at $Gr = 20000$) / Max($|\psi|$ at $Gr = 7000$) $\gg \sqrt{20000/7000}$.

Thermal and dynamic boudary conditions for case (IV) considered (with fins of height 1.5) [4] are

(48) $$\psi=0(-1<x<0,y=0 \text{ or } x=-1 \text{ or } x=0)\ ,$$

(49) $$\partial\psi/\partial y=0(-1<x<0,y=0)\ ,$$

(50) $$\partial\psi/\partial x=0(x=0,0\le y\le 1.5 \text{ or } x=-1,0\le y\le 1.5)\ ,$$

(51) $$\zeta=0(x=0,y>1.5 \text{ or } x=-1,y>1.5)\ ,$$

(52) $$T=1(-1<x<0,y=0 \text{ or } x=-1,0\le y\le 1.5 \text{ or } x=0,0\le y\le 1.5)\ ,$$

(53) $$\partial T/\partial x = 0 (x = -1, 1.5 < y < 1.7 \text{ or } x = 0, 1.5 < y < 1.7) \ ,$$

(54) $$T = 0 (x = -1, y \geq 1.7 \text{ or } x = 0, y \geq 1.7).$$

Under these conditions, steady-state thermal and flow fields are obtained at $Gr = 5000$, $Pr = 3$ (water at a suitable temperature) (in a nearly pure heat conduction state). Steady-state mean Nusselt number Nu_m along the fin surfaces facing the cavity and the base is 0.361 .

7. CONCLUSIONS

It was demonstrated that with proper transformation of governing equations and boundary conditions to a circular region it is possible to solve natural convection heat transfer, using a spectral finite difference method.

REFERENCES

1. Y. Mochimaru, Proc. 26th Ann. Iranian Mathematics Conference, 2, 245–254 (1995).

2. Y. Mochimaru, Int. Sym. Transport Phenomena in Thermal-Fluids Engineering, 1, 315–320 (1996).

3. Y. Mochimaru, 3rd Int. Sym. Experimental and Computational Aerothermodynamics of Internal Flows, 909–916 (1996).

4. Y. Mochimaru and C.Y. Liu, Book of Abstracts, International Sym. Advances in Computational Heat Transfer, 109–111 (1997).

8th Int. Coll. on Differential Equations, pp. 319-326
D. Bainov (Ed)

A New Minimum Curvature Multi-Step Method for Unconstrained Optimization

I.A.R. MOGHRABI AND S.E. OBEID
Natural Science Division, Lebanese American University, P.O. Box 13 — 5053, Beirut, Lebanon, e-mail: imoghrbi@lau.edu.lb

Abstract. Multi-step quasi-Newton methods for optimization were derived in [1]-[3], where it was shown how an interpolation in the variable-space could be used to generate "better" Hessian approximations. The work presented in [4] concentrated a choice of the curve parameters that ensure a "smooth" interpolation. In this paper, we carry on with a similar idea and define a rational model with a free parameter. Our derivation of the new algorithm is based on determining some value of the parameter that minimizes the curvature in some chosen metric. It is shown how such value can be "cheaply" calculated at each iteration. Numerical comparison between the new algorithm and other multi-step algorithms reveal the merits of the new approach.
Keywords: Quasi-Newton methods, multi-step methods, unconstrained optimization

1. INTRODUCTION

One of the most serious deficiencies of Newton's method for unconstrained optimization is the high cost involved in evaluating the real Hessian of the objective function. While retaining the basic structure of the Newton's method, quasi-Newton methods are based on constructing approximations to the Hessian that satisfy the so-called Secant Equation, namely:

$$B_{i+1}\underline{s}_i = \underline{y}_i,$$

where $\underline{s}_i \equiv \underline{x}_{i+1} - \underline{x}_i$ and $\underline{y}_i \equiv \underline{g}(\underline{x}_{i+1}) - \underline{g}(\underline{x}_i)$.

Multi-step methods are based on satisfying a variant of the Secant Equation of the form:

$$G(\underline{x}_{i+1})\underline{x}'(\tau_m) = \hat{\underline{g}}'(\underline{x}(\tau_m)),$$

of which the secant equation is a special case for $\underline{x}(\tau)$ and $\hat{\underline{g}}(\tau)$ taken as half lines (where $\hat{\underline{g}}(\underline{x}_i) \cong \underline{g}(\underline{x}_i)$). The quantity $\underline{x}(\tau)$ is not necessarily linear in form, as in the standard one-step methods. In fact, it is a polynomial of degree m which interpolates the last $m+1$ iterates such that

$$\underline{x}(\tau_k) = \underline{x}_{i-m+k+1} \qquad \text{for} \quad k = 0, 1, \ldots, m.$$

Similarly, the required derivative $d\underline{g}/d\tau$ is estimated by higher order differencing techniques that interpolate the $m+1$ most recent available gradient points.

It has been found convenient to represent the interpolating curve in its Lagrangian form as:

$$\underline{x}(\tau) \equiv \sum_{k=0}^{m}(\underline{x}_{i-m+k+1}L_k(\tau_k)),$$

where $\{L_k(\tau)\}_{k=0}^{m}$ are the standard Lagrange polynomials, with similar expression for the gradient.

In general, the new Hessian approximation will satisfy a relation of the form

$$B_{i+1}\underline{r}_i = \underline{w}_i$$

for $\underline{r}_i \equiv \underline{x}'(\tau_m)$ and $\underline{w}_i \equiv \hat{\underline{g}}'(\tau_m)$, or, more conveniently, as:

$$\underline{r}_i = \sum_{j=0}^{m-1} \underline{s}_{i-j} \left\{ \sum_{k=m-j}^{m} L'_k(\tau_m) \right\} \quad \text{and} \quad \underline{w}_i = \sum_{j=0}^{m-1} \underline{y}_{i-j} \left\{ \sum_{k=m-j}^{m} L'_k(\tau_m) \right\}.$$

Ford and Moghrabi [3] addressed the curve parameterization issue where it was shown that it would be better to choose *τ-values* such that $\tau_2 - \tau_1$ reflects the size of $\underline{s}_i$ and $\tau_1 - \tau_0$ to reflect the size of $\underline{s}_{i-1}$. In this respect, the authors in [3] distinguish between two important approaches for defining distances, for some metric $\phi_M(z_1, z_2) = \{(z_1, z_2)^T M(z_1, z_2)\}^{1/2}$ defined on $\mathbb{R}^n$.

The two approaches are: 1) The ***Accumulative Approach*** for which one of the *τ-points* is chosen to be a focal point (say the j-th for which $\tau_j = 0$). The remaining *τ-values* are determined by accumulating the intermediate distances measured in ϕ_M [3]. For this approach

$$\forall\, k \quad \tau_k < \tau_{k+1} \qquad \tau_k = \begin{cases} -\sum_{p=k+1}^{j} \phi_M(\underline{x}_{i-m+p+1}, \underline{x}_{i-m+p}) & k < j, \\ 0 & k = j, \\ \sum_{p=j+1}^{k} \phi_M(\underline{x}_{i-m+p+1}, \underline{x}_{i-m+p}) & k > j. \end{cases}$$

2) The ***Fixed-Point Approach*** is also based on choosing an origin, namely τ_j. Then the values $\{\tau_k\}$ (with $k \neq j$), are found by measuring the distance from the origin or focus point to any other point on the curve directly, thus, neglecting intermediate distances. Different choices of the origin will result in different *τ-values*. In [3], the parameter corresponding to the most recent iterate $\underline{x}_{i+1}$ was taken as the focal point. Therefore, $\tau_m = 0$ and $\tau_k = -\phi_M(\underline{x}_{i+1}, \underline{x}_{i+k-m+1})$ for $k = 0, 1, \ldots, m-1$.

As a special instance of the more general multi-step methods, we shall consider throughout this paper the two-step quasi-Newton methods [1], [3] for which the Hessian should satisfy

$$B_{i+1}(\kappa_i \underline{s}_i + \pi_i \underline{s}_{i-1}) = \kappa_i \underline{y}_i + \pi_i \underline{y}_{i-1}.$$

Here, $\underline{r}_i \equiv L'_2(\tau_2)\left[\underline{s}_i - \frac{L'_0(\tau_2)}{L'_2(\tau_2)}\underline{s}_{i-1}\right]$ and $\underline{w}_i \equiv L'_2(\tau_2)\left[\underline{y}_i - \frac{L'_0(\tau_2)}{L'_2(\tau_2)}\underline{y}_{i-1}\right]$.

It can be shown ([5]) for the quadratic interpolating curve $\underline{x}(\tau)$ that:

$$\underline{x}'(\tau) \equiv (1-\tau_1)^{-1}(2\tau - \tau_1)\underline{s}_i - \tau_1^{-1}(2\tau - \tau_1 - 1)\underline{s}_{i-1}$$

and

$$\underline{x}''(\tau) \equiv 2(1-\tau_1)^{-1}\underline{s}_i - 2\tau_1^{-1}\underline{s}_{i-1}.$$

If $\tau_0 = 0$ and $\tau_2 = 1$, then τ_1 is a parameter at our disposal and can thus be chosen so as to minimize $\underline{x}''(\tau)$ in some norm to ensure "smooth" interpolation [5]. Upon defining

$\delta = \dfrac{\tau_2 - \tau_1}{\tau_1 - \tau_0}$, then the parameter τ_1 can be expressed in terms of δ as $\tau_1 = (1+\delta)^{-1}$, where the aim now is:

$$\min_{\delta \in \mathbf{R}} \phi(\delta) \equiv \frac{\|\underline{x}''(\tau)\|_M^2}{4} = (1+\delta^{-1})^2\sigma_i - 2(1+\delta^{-1})(1+\delta)\mu_i + (1+\delta)^2\sigma_{i-1},$$

for $\sigma_j \equiv \|\underline{s}_j\|_M^2 \geq 0$ and $\mu_j \equiv \underline{s}_{j-1}^T M \underline{s}_j$.

This is equivalent to finding roots of

$$\Psi(\delta) \equiv \delta^3 - \mu\delta^2 + \mu\delta - \sigma, \qquad \text{for} \quad \mu = \frac{\mu_i}{\sigma_{i-1}} \quad \text{and} \quad \sigma = \frac{\sigma_i}{\sigma_{i-1}}.$$

2. THE NEW APPROACH

2.1. The rational model

We begin our derivation of the new algorithm by selecting a rational form for the interpolating curve in the space of variables as follows

$$\underline{x}(\tau, \theta) = \frac{\underline{q}(\tau)}{1 + \theta\tau} \tag{1}$$

where (the vector) $\underline{q}(\tau)$ is of a quadratic form in τ expressed as (for $m = 2$):

$$\underline{q}(\tau) = \sum_{k=0}^{m} L_k(\tau)(1 + \theta\tau_k)\underline{x}_{i-m+k+1} \qquad \text{for} \quad k = 0, 1, \dots, m, \tag{2}$$

$L_j(\tau)$ being the j-th Lagrange polynomial of degree m associated with the abscissae $\{\tau_k\}_{k=0}^m$. In the two-step method, the curve $\underline{x}(\tau, \theta)$ for a specified set of distinct τ values $\{\tau_k\}_{k=0}^2$ is hoped to satisfy:

$$\underline{x}(\tau_k, \theta) = \underline{x}_{i+k-1} \qquad \text{for} \quad j = 0, 1, 2. \tag{3}$$

The set of abscissae $\{\tau_k\}_{k=0}^2$ has been chosen as:

$$\tau_2 = 0, \quad \tau_1 = -\rho_i \qquad \text{and} \quad \tau_0 = -\rho_i - \rho_{i-1} \tag{4}$$

where, for instance, ρ_k could be chosen to be $\|\underline{s}_k\|_M$, for some matrix M defining the metric. The choices (4) of the τ-values conform to the *Fixed-Point* Approach discussed above. It also follows that:

$$\delta = \frac{\tau_2 - \tau_1}{\tau_1 - \tau_0} = \frac{\rho_i}{\rho_{i-1}} \qquad \text{(always positive).} \tag{5}$$

2.2. The minimum curvature criteria

The problem here is that of ensuring that the curve at τ_2, the update coordinate, is minimal. We, therefore, wish to minimize $\|\underline{x}''(\tau_2, \theta)\|_M^2$ in order to derive expressions for $\underline{x}'(\tau_2, \theta)$ to be used in the update of the matrix.

From (1) and (2) it follows that:

$$\underline{x}'(\tau, \theta) \equiv \frac{(1+\theta\tau)\underline{q}' - \theta\underline{q}}{(1+\theta\tau)^2} \tag{6}$$

and

(7) $$\underline{q}(\tau) = L_0(\tau)(1+\theta\tau_0)x_{i-1} + L_1(\tau)(1+\theta\tau_1)x_i + L_2(\tau)x_{i+1}.$$

At τ_2, equation (6) becomes:

$$\underline{x}'(\tau_2,\theta) = \underline{q}' - \theta\underline{x} = L_0'(\tau_2)(1+\theta\tau_0)x_{i-1} + L_1'(\tau_2)(1+\theta\tau_1)x_i + L_2'(\tau_2)x_{i+1}$$
$$- \theta\Big(L_0(\tau_2)(1+\theta\tau_0)x_{i-1} + L_1(\tau_2)(1+\theta\tau_1)x_i + L_2(\tau_2)x_{i+1}\Big),$$

or equivalently:

(8) $$\underline{x}'(\tau_2,\theta) = -L_0'(\tau_2)\underline{s}_{i-1} + L_2'(\tau_2)\underline{s}_i - \theta\underline{s}_i + \theta\left(\frac{\rho_i}{\rho_{i-1}}\right)\underline{s}_{i-1}.$$

By substituting in (8) for the prespecified τ-values we obtain:

$$\underline{x}'(\tau_2,\theta) = -\frac{\rho_i\underline{s}_{i-1}}{\rho_{i-1}(\rho_i+\rho_{i-1})} + \frac{(2\rho_i+\rho_{i-1})\underline{s}_i}{\rho_i(\rho_i+\rho_{i-1})} - \theta\underline{s}_i + \theta\left(\frac{\rho_i}{\rho_{i-1}}\right)\underline{s}_{i-1}.$$

Upon defining α as:

(9) $$\alpha \equiv (\rho_i + \rho_{i-1})^{-1}$$

and for δ as in (5) we get:

(10) $$\underline{x}'(\tau_2,\theta) = \underline{s}_i\left(\alpha - \theta + \frac{1}{\rho_i}\right) + \underline{s}_{i-1}\delta(\theta-\alpha).$$

Then, we proceed to derive an expression for the second derivative $\underline{x}''(\tau,\theta)$ as follows:

(11) $$\underline{x}''(\tau,\theta) = \frac{(\underline{q}'' - \theta\underline{x}')(1+\theta\tau) - \theta(\underline{q}' - \theta\underline{x})}{(1+\theta\tau)^2}$$

or, equivalently,

(12) $$\underline{x}''(\tau_2,\theta) = \underline{q}'' - 2\theta(\underline{q}' - \theta\underline{x}(0)) = \underline{q}'' - 2\theta\underline{q}'(\tau_2) + 2\theta^2 x_{i+1},$$

where

$$\underline{q}''(\tau) = -L_0''(\tau)\underline{s}_{i-1} + L_2''(\tau)\underline{s}_i + \theta\Big[L_0''(\tau)\tau_0 x_{i-1} - (L_0''(\tau) + L_2''(\tau)\tau_1)x_i\Big];$$
$$\underline{q}'(\tau_2) = L_0'(\tau_2)(1+\theta\tau_0)x_{i-1} - (L_0'(\tau_2) + L_2'(\tau_2))(1+\theta\tau_1)x_i + L_2'(\tau_2)x_{i+1}.$$

Using $\hat{\theta} = \theta/\alpha$, equation (12) can be expressed as:

(13) $$\underline{x}''(\tau_2,\hat{\theta}) = 2\alpha^2\Big[(1+\delta^{-1}) - (2+\delta^{-1})\hat{\theta} + \hat{\theta}^2\Big]\Big[\underline{s}_i - \delta\underline{s}_{i-1}\Big].$$

Thus, our aim now is to minimize for $\hat{\theta}$ the function

(14) $$\Phi(\hat{\theta}) = \|\underline{x}''(\tau_2,\hat{\theta})\|_M^2 = \underline{x}''(\tau_2,\hat{\theta})^T M \underline{x}''(\tau_2,\hat{\theta})$$
$$= 4\alpha^4\Big[(1+\delta^{-1}) - (2+\delta^{-1})\hat{\theta} + \hat{\theta}^2\Big]^2\Big(\|\underline{s}_i\|_M^2 + \delta^2\|\underline{s}_{i-1}\|_M^2 - 2\delta\mu_i\Big)$$
$$= 4\alpha^4\Big[(1+\delta^{-1}) - (2+\delta^{-1})\hat{\theta} + \hat{\theta}^2\Big]^2\Big(\sigma_i + \delta^2\sigma_{i-1} - 2\delta\mu_i\Big),$$

where $\sigma_k = \|\underline{s}_k\|_M^2 \geq 0$ and $\mu_k = \underline{s}_{k-1}^T M \underline{s}_k$, for any $\hat{\theta}$ and for

$$\Gamma \equiv 8\alpha^4(\sigma_i + \delta^2\sigma_{i-1} - 2\delta\mu_i).$$

Now our problem is essentially that of minimizing $\left[(1+\delta^{-1}) - (2+\delta^{-1})\hat{\theta} + \hat{\theta}^2\right]^2$, where if we define λ, ν and ω as:

$$\lambda \equiv \frac{-3\delta^{-1} - 6}{2}, \qquad \nu \equiv \frac{\delta^{-2} + 6\delta^{-1} + 6}{2} \quad \text{and} \quad \omega \equiv \frac{-\delta^{-2} - 3\delta^{-1} - 2}{2},$$

we obtain the following cubic equation:

$$\Phi'(\hat{\theta}) = 2\Gamma\left[\hat{\theta}^3 + \hat{\theta}^2\lambda + \hat{\theta}\nu + \omega\right],$$

which has 3 real roots, given by

$$\hat{\theta}_1 = 1, \qquad \hat{\theta}_2 = \frac{\delta^{-1}}{2} + 1 \quad \text{and} \quad \hat{\theta}_3 = \delta^{-1} + 1.$$

The second root corresponds to a maximum curvature and will hence be disregarded.

The two values of $\hat{\theta}$ corresponding to a minimum thus giving rise to two possible expressions for $\underline{x}'(\tau_2, \theta)$ as follows

1) for $\hat{\theta} = 1$, $\underline{x}'(\tau_2, \theta) = \underline{s}_i/\rho_i$;

2) for $\hat{\theta} = \delta^{-1} + 1$, $\underline{x}'(\tau_2, \theta) = \alpha(\underline{s}_i + \underline{s}_{i-1})$.

It is obvious that if the rational form (1) is used for modeling the gradient behaviour as well as $x(\tau)$ and $\hat{\theta} = 1$, then the Secant Equation is naturally a minimum curvature form.

3. NUMERICAL RESULTS AND CONCLUSIONS

The numerical experiments were carried out using a set of sixty test functions (each with four different starting-points), giving a total of 240 test problems with dimensions ranging from two to eighty. This set (with some modifications to starting-points and convergence criteria) is described in [4], where more details concerning the general mode of implementation of such methods may be found. As in [4], the sixty functions were classified into subsets of "low" ($2 \leq n \leq 15$), "median" ($6 \leq n \leq 45$), and "high" ($46 \leq n \leq 80$) dimension. Space does not permit a full description of the results.

Summaries (total number of function/gradient evaluations, total iterations, total time) of the performance of the methods on each of these subsets and on the complete set. For each problem, the method yielding the best performance (judged by the number of iterations required) is awarded one point. The total number of points gained by each method over the test set under consideration is indicated in each table of results by the entry against the heading "Scores".

The comparisons between the BFGS method and the new method are given in Tables 1 to 4. These results show clearly that the new method exhibits a superior numerical performance, by comparison with the BFGS method, and that this superiority becomes apparent as the problem dimension increases. For problems of low dimension, the gain in evaluations and iterations is not tangible.

Table 1. 40 problems of dimension 2 to 15.

	BFGS	New method
Total evaluations	5122	5227
Total iterations	3928	4025
Scores	25	15
Total time (sec.)	7.44	8.21

Table 2. 100 problems of dimension 16 to 45.

	BFGS	New method
Total evaluations	20871	18131
Total iterations	18910	16012
Scores	18	85
Total time (sec.)	199.7	139.0

Table 3. 100 problems of dimension 46 to 80.

	BFGS	New method
Total evaluations	18575	15133
Total iterations	17694	14007
Scores	17	83
Total time (sec.)	717.0	575.3

Table 4. Overall results for 240 problems.

	BFGS	New method
Total evaluations	44568	38491
Total iterations	40532	34044
Scores	60	183
Total time (sec.)	923.9	722.51

Table 5. 40 problems of dimension 2 to 15.

	MC	New method
Total evaluations	5060	5360
Total iterations	3818	3840
Scores	23	18
Total time (sec.)	7.64	8.39

Table 6. 100 problems of dimension 16 to 45.

	MC	New method
Total evaluations	17016	17313
Total iterations	14633	15116
Scores	61	43
Total time (sec.)	155.4	168.0

Table 7. 100 problems of dimension 46 to 80.

	MC	New method
Total evaluations	13376	14527
Total iterations	12245	13447
Scores	73	31
Total time (sec.)	514.2	568.4

Table 8. Overall results for 240 problems.

	MC	New method
Total evaluations	35452	37200
Total iterations	30696	32403
Scores	157	102
Total time (sec.)	677.0	744.8

The results reported here are those for $\hat{\theta}_3$. It is worth noting that the performance of the derived algorithm is sensitive to the choice of the matrix M, used in the definition of the metric in

$$\rho_i = \|\underline{s}_i\|_M \qquad \text{and} \qquad \rho_{i-1} = \|\underline{s}_{i-1}\|_M.$$

Work currently in progress is focused upon alternatives (and, possibly, more appropriate) choices for the matrix M which defines the norm in which the curvature minimization is carried out. We anticipate that this work will be reported in a future paper.

A new criterion for determining the parameterization of the interpolating curves in two-step quasi-Newton methods has been proposed and it has been shown how the minimization problem which thus arises has cheaply computable roots at each iteration. Numerical experiments have shown that a method based on this criterion (using the standard 2-norm) yields a substantial improvement in numerical performance over the BFGS method, but that it is not quite competitive with the method NC recently introduced by the authors.

Acknowledgements

The authors gratefully acknowledge the support of the Lebanese National Council for Scientific Research, in the person of its Secretary General, in this work and the support received by the *Council for International Exchange of Scholars,* Washington D.C. Special thanks go also to professor John Ford of Essex University for the valuable comments on an earlier version of this paper.

References

1. J.A. Ford and I.A. Moghrabi, Scientia Iranica, 1, 327–334 (1995).
2. J.A. Ford and I.A. Moghrabi, J. Comput. Appl. Math., 50, 305–323 (1994).
3. J.A. Ford and I.A. Moghrabi, Optim. Meth. Software, 2, 357–370 (1993).
4. J.A. Ford and I.A. Moghrabi, Computers Math. Applic., 31, 179–186 (1996).
5. J.A. Ford and A.F. Saadallah, Colloq. Math. Soc. Jan. Bolyai, 50, 539–563 (1986).
6. J.A. Ford and I.A. Moghrabi, Department of Computer Science (University of Essex) Technical Report CSM-270 (1996).
7. J.A. Ford and I.A. Moghrabi, Department of Computer Science (University of Essex) Technical Report CSM-271 (1996).
8. J.E. Dennis and R.B. Schnabel, SIAM Review, 21, 443–459 (1979).
9. C.G. Broyden, J. Inst. Math. Applic., 6, 76–90 and 222–231 (1970).
10. R. Fletcher, Comput. J., 13, 317–322 (1970).
11. D. Goldfarb, Maths. Comp., 24, 23–26 (1970).
12. D.F. Shanno, Maths. Comp., 24, 647–656 (1970).

8th Int. Coll. on Differential Equations, pp. 327-334
D. Bainov (Ed)

Towards a Real RealRAM: a Prototype Using C++

NORBERT TH. MÜLLER
LG Mathematische Informatik / Modellierung, University of Trier - D-54286 Trier, Germany
e-mail: mueller@uni-trier.de

Abstract. We discuss an implementation of a package for error-free real arithmetic emerging from the concept of a RealRAM. Unique features are nondeterministic functions as well as operators computing limits of user-defined sequences.
Keywords: Computer Arithmetic, C++, Error Analysis, Computable Real Analysis, Random Access Machines
AMS Classification: 26E40, 65D99, 65G99, 65Y25

1 Introduction

The theory of computability and complexity on the field of real numbers is already well developed, see e.g. [1], [2]. On the other hand, the reality of computing with reals is still tied to the old concept of single and double precision approximations. Improvements like [3] did not find widespread usage, whereas newer approaches like [4] are not satisfying at least from the viewpoint of type 2 computability.

In 1995, Brattka and Hertling [5] presented a theoretical paper on random access machines working with reals as entities and their simulation by Turing machines, that could be a starting point for joining theory and practice. We chose to use C++ as a base for our prototype aiming to realize their approach, especially as there is a freely available GNU compiler offering high level concepts like overloading of operators as well as low level concepts like long jumps (being essential for the implementation).

A preliminary version of this paper has been presented at the Second Workshop on Computability and Complexity in Analysis, August, 24/25th 1996, Trier, Germany. The C++ sources can be found at `http://www.informatik.uni-trier.de/~mueller/`

2 Multiple Precision Arithmetic

There is a long tradition for implementations of multiple precision arithmetic. Best known surely is Brent's package [3] written in FORTRAN. An overview can be found e.g. in the file `BIGNUMS.TXT` [6] available at many FTP servers. One of the most recent packages is `GNU Multiple Precision (GMP)`.

One part of our prototype for a RealRAM is an implementation of an interface to such multiple precision packages that is easy to use and fits to the theory of type 2 computability. It is quite easy to adopt the interface to different multiple precision packages. In fact, at the moment one is able to choose between GMP and a package named LR (written by the author) during compile time.

The interface to the MP packages is a class named DYADIC: Its values can be interpreted as a subset of $\mathbb{R}$ that ideally should be dense (in the usual mathematical sense) in $\mathbb{R}$. In fact, the implementations of the MP packages only allow finite sets of values, but they are so rich that their finiteness is almost impossible to note. For example, the LR package uses a traditional half logarithmic representation, where each value consists of a 'mantissa' and an 'exponent'. The mantissa may have a length up to 100,000 words (3.1 million bits), which can easily be extended by changing Dyadic_Base_MAXSIZE. The exponents are of type long, i.e. they have 32 bits or 64 bits (on a DEC AXP).

The class additionally implements a small set of basic operations: (1) Methods to automatically create, initialise, and delete values of type DYADIC, (2) the basic arithmetic operations +, -, *, and /, (3) the usual tests <, = ... comparing values, and (4) a few additional functions like conversions from integer or for printing values. Where possible, overloaded operators are used.

The behaviour of the arithmetic operators is determined by a function setprec. Using this funcion, the precision of subsequent operations on DYADIC numbers is set in the following way: After a call to DYADIC::setprec(p), the result of any basic operation a op b differs at most by 2^p from the exact result performed in $\mathbb{R}$. This concept of 'precision' of *operations* is different from the concept used in most MP packages, where the precision of a *variable* denotes the length of the mantissa. In consequence, precisions will usually be of the form 2^p for negative integers p! Please note that this notion of precision only applies to single operations. In any sequence of operations, the result may of course differ more than 2^p from the exact result due to error propagation. Accordingly, the term 'error' (of a value) will denote the absolute difference between the value and the real number we want to approximate.

3 User interface to real numbers

In the previous chapter we outlined a simple interface to multiple precision numbers. A similar concept can be used for an interface to real numbers. Although the syntax almost remains the same, the semantics is totally different.

As we are aiming at a Real RAM, we implement a class REAL, that behaves like real numbers without any error. Again a small set of operations is allowed to be used directly with variables of this type: usual arithmetic operations, tests (with special semantics explained below), output, and conversion to/from integer and DYADIC numbers. In addition there are several routines with a certain nondeterministic behaviour, and routines to compute limits of certain user definable sequences. Overloading of operators is used again wherever possible.

The user is allowed to use normal programming methods like defining own data types (e.g. real matrices) and functions (even returning real values). He should *not* use the normal IO operations, as this could lead to surprising and annoying results. malloc should be treated with care, as the implementation in not yet protected against memory leaks. The use of alloca should be possible without problems. External libraries have to be handled with care, as they will normally not be suited for implementation of the special semantics of the Real RAM. With these restrictions, the language looses a lot of its universal character, but enough is left to do numerical work.

Because of implementational issues, the usual procedure main is not accessible to the user, it has a predefined implementation necessary to get the intended semantics. A procedure called compute takes its place as the starting procedure of any user defined program.

The basic arithmetic operations +, -, *, and /, have their intended meanings, i.e. they implement error-free arithmetic; the special case of division by zero is discussed in the next chapter.

The semantics of the comparison operators are partially defined tests: the arguments must be different, otherwise infinite loops will occur. In consequence, < and <= are *identical*. Furthermore, there is no test on equality or inequality. This implies that comparing of reals should only be done, if inequality is known in advance. So even 'trivial' programs like computing the maximum of two numbers simply by comparing and choosing the larger number will not behave in the expected way. In consequence, use of these tests is strongly discouraged. Very often techniques explained in chapters 6 and 7 can be used instead.

As an example consider the computation of the following 'logistic function: Compute the iteration $x_{i+1} = 3.75 \cdot x \cdot (1-x)$ starting with $x_0 = 0.5$:

```
void compute(){
  REAL x = REAL(1) / REAL(2);    REAL c = REAL(375) / REAL(100);
  int count; rscanf("%d",&count);
  for ( int i=0; i<=count; i++ )
    { x= c*x*(1-x); rwrite(x,-50,18); rprintf(" %d\n",i); }
}
```

The output of the program consists of a list of the first 10 decimal places of the values x_i together with the index i, where the statement rwrite(x,-50,18) ensures that any shown value differs at most by $2^{-50} \approx 10^{-15}$ from the exact result. For example, for x_{1000} we get +.7917467409E+0000 as result. To compute this value the Real RAM on a PPro 200 running Linux needs about one third of a second, about 4 hours were necessary to compute x_{50000}. Using standard double precision floats instead of REAL, starting from about x_{100} all significant digits turn out to be wrong!

4 Basic implementation issues

The basic idea of the implementation is similar to [5]: Essentially, all computations are done with approximations to the real numbers, each single operation is computed with a certain precision p_i. If the error of an approximation is to big e.g. to make a decision, *the whole computation* is repeated with a significantly better precision p_{i+1}. This may lead to many iterations of the computation with a sequence of precisions $p_0 >> p_1 >> p_2 \ldots$

It is well known that the asymptotic complexity of all these iterations is the same as the complexity of the last iteration, if we have $p_i = p - f^i$ for a fixed (e.g. rational) $f > 1$ and if all used operations and functions have a well behaved ('regular', see e.g. [7]) complexity. So theoretically, the choice of p and f is not essential.

Obviously, the starting precision as well as the increment in the iterated computations are critical points in practice. We chose $p_i = p + q \cdot \frac{f^i - 1}{f - 1}$, with $p = q = -50$ and $f = 1.25$ as default values, leading to the following sequence of precisions; p, q, and f can be changed at the start of any program to optimise its time complexity:

p_0	p_1	p_2	p_3	...	p_{10}	p_{15}	p_{20}	p_{25}	...
$p = -50$	$p + q = -100$	-162	-240	...	-1703	-5502	-17096	-52477	...

Internally, each variable x of type **REAL** is implemented as a multiple precision number d together with information on the absolute error $e \geq |d - x|$. This information on the error is not necessarily connected to the length of d. Essentially, in any stage of the computation, we work with intervals $(d - e, d + e)$ knowing that intended value of x must be included in the interval.

The bound e for the error could be stored in different resolutions: One might choose $\{2^{-n} | n \in \mathbb{N}\}$ or $\{z \cdot 2^p | z \in \mathbb{N}, p \in \mathbb{Z}\}$ as sets of error informations. Both extremes have disadvantages: With the first approach, we loose at least one bit of information on any operation, demanding a very high initial precision. The second approach implies storing e as an own multiple precision number (or storing both values $d - e$ and $d + e$, i.e. doing interval arithmetic), so the storage is doubled and operations are slower.

We chose an intermediate solution: The error is stored as a value from $\{z \cdot 2^p | z \in \mathbb{N}, p \in \mathbb{Z}\}$, but restricted to $z < 2^{gb}$ and $|p| < 2^{maxexp}$. Here gb and $maxexp$ depend on the CPU and the MP package: For usual 32 bit CPUs and LR we have $gb = 30$ and $maxexp = 29$, while the implementation uses $gb = 62$ and $maxexp = 61$ for a DEC AXP.

In consequence, there is a significant reduction in the growth of errors during single operations by only a small overhead, especially in computing scalar products of vectors where all components are of similar error.

As an example, we examine the resulting error propagation for the division of reals: Let $d_1, e_1 = z_1 \cdot 2^{p_1}$ and $d_2, e_2 = z_2 \cdot 2^{p_2}$ be given. In order to exclude division by zero, suppose $|d_2| > 2 \cdot e_2 > 0$. Suppose we compute d with $|d - d_1/d_2| \leq 2^p$ for a certain p.

Let $s_i \in \mathbb{Z}$ be as small as possible with $|d_i| < 2^{s_i}$, i.e. usually $2^{s_i - 1} \leq |d_i| < 2^{s_i}$. In case of $d_1 = 0$, s_i will be the above minimal value $2^{-maxexp}$ due to the limitations of the implementation of the multiple precision numbers.

For any x_1, x_2 with $|d_i - x_i| \leq e_i$ we have

$$\begin{aligned}
|\frac{x_1}{x_2} - d| &\leq |\frac{x_1}{x_2} - \frac{d_1}{d_2}| + 2^p = |\frac{x_1 \cdot d_2 - d_1 \cdot x_2}{x_2 \cdot d_2}| + 2^p \\
&= |\frac{(x_1 - d_1) \cdot d_2 + d_1 \cdot (d_2 - x_2)}{x_2 \cdot d_2}| + 2^p \\
&\leq \frac{e_1 \cdot |d_2| + e_2 \cdot |d_1|}{|d_2|/2 \cdot |d_2|}| + 2^p \leq \frac{z_1 2^{p_1 + s_2} + z_2 2^{p_2 + s_1}}{2^{2s_2 - 1}} + 2^p
\end{aligned}$$

Choosing $p \geq \max\{p_1 - s_2 + 1, p_2 + s_1 - 2s_2 + 1\}$ we have $|x_1/x_2 - d| \leq 2^p \cdot (z_1 \cdot 2^{p_1 - s_2 + 1 - p} + z_2 \cdot 2^{p_2 + s_1 - 2s_2 + 1 - p} + 1)$, where the value in the brackets consists of non-positive exponents of 2 and the integer factors z_1 and z_2 and so can easily be bounded by an integer z using appropriate right shifts of z_1 and z_2. leading to $|x_1/x_2 - d| \leq 2^p \cdot z$.

One of the assumptions above was $|d_2| > 2 \cdot e_2 > 0$ to exclude a division by zero. If this inequality does not hold, we may either have attempted a true division by zero or (usually) the error in the approximation d_2 is too large to exclude such a division. So this is a point where a reiteration with higher initial precision should be initiated.

The reiteration is implemented by a `longjmp`, i.e. an immediate return to an outer level of control. This jump (ususally) leaves the procedure `compute` and returns to the hidden implementation of `main`. Here the precision is incremented as above, then `compute` is restarted from the beginning.

This also implies that a division by zero initiates an infinite loop of iterations instead of an exception as in usual arithmetic. As division can not be extended to a total continuous function, this effect is unavoidable.

The same holds with the tests on the real numbers: They are undefined if applied to two identical operands, as this is the only possibility to obtain continuous (but partial) tests. In consequence, the tests may also imply reiterations of the computation if their arguments are similar and even lead to an infinite loop if they are identical.

On the other hand, if applied to different arguments, the tests will eventually lead to the correct result, maybe on the cost of a high precision.

5 Input and Output

The iterative character of the computations described above poses problems for interactive usage of the Real RAM: In each iteration each input from `stdin` and each output to `stdout` is repeated. Although this can be used to identify the iterations, any illusion of working with entities is destroyed.

To avoid this effect, the user should use the special IO-functions `rscanf` for reading one input and `rprintf` for writing (a variable number of) outputs.

In addition, there is a function `rwrite(x,p,w)`, which prints a decimal approximation to a real number `x` with an error of at most 2^p and with a width of w characters. If the error of x is bigger than 2^p in the actual iteration of the computational process, a reiteration in initiated. This ensures that the printed approximations will be correct within the given error bound.

The implementation of `rscanf` simply uses buffered input: On each iteration a pointer in the buffer is reset to the beginning of the buffer, which initially is empty. Then inputs are taken from the buffer moving this pointer. When the pointer reaches the end of the buffered input, new characters are appended to the buffer by reading from the standard input.

`rprintf` and `rwrite(x,p,w)` simply use a counter for the number of outputs from previous iterations to determine whether a call should actually lead to new output.

6 Nondeterministic Functions

In traditional numerical algorithms, there are many functions that cannot be adopted directly to the Real RAM: In its core, the semantics of the Real RAM is continuous, whereas these functions are not. Like in [5], there are operations with uncertainty that usually give the correct result but also are allowed to give a 'slightly' wrong answer near points where the original operations are not continuous in their arguments.

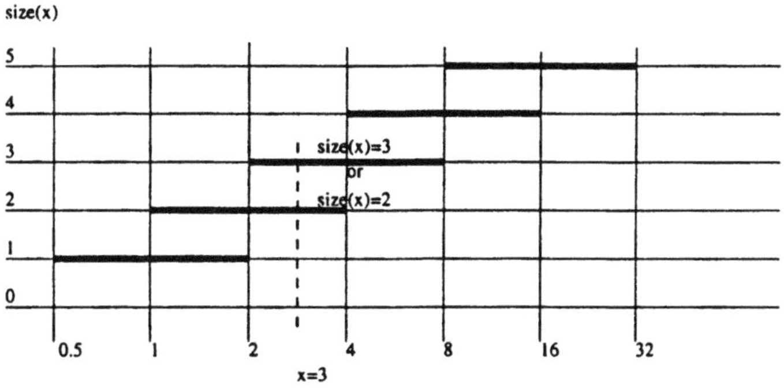

Figure 1: `size(x)`

Often this can be expressed as the computation of a multi-valued function (i.e. a relation) instead of a function. As an example consider `size(x)`: This function computes an integer value k such that $|x|$ lies between 2^{k-2} and 2^k. So the value of `size(x)` is not exactly fixed, but usually there are two possible values, as can be seen in figure 1.

The Real RAM is allowed to return any of the possible values. In consequent calls with the same real value x, we might even get different values for `size(x)`. From the view of the programmer, this is a kind of nondeterminism in the program. On the other hand, the program will *never* return a value k such that $x > 2^k$ or $x < 2^{k-2}$. To ensure this, calling `size(x)` may invoke reiteration. In addition, usage with argument zero leads to an infinite loop.

This concept can obviously be generalised. The procedure `rwrite` has a similar behaviour, as it is defined on the whole real line, but its values are discrete. Additional examples are `bound(x,k)` and `positive(x,k)`:

$$\texttt{bound}(x,k) = \begin{cases} \text{true}, & |x| < 2^{k-2} \\ \text{true or false}, & 2^{k-2} \le |x| \le 2^k \\ \text{false}, & 2^k < |x| \end{cases}$$

$$\texttt{positive}(x,k) = \begin{cases} \text{true}, & 2^k < x \\ \text{true or false}, & -2^k \le x \le 2^k \\ \text{false}, & x < -2^k \end{cases}$$

Calling `bound(x,k)` or `positive(x,k)` may obviously invoke reiteration.

These functions can be used in the construction of loops. E.g., for any p and $x \ge 0$ the following function returns a real $y \ge 0$ such that $|y - \sqrt{x}| \le 2^p$ using Heron's iteration algorithm for the square root.

```
REAL sqrt1(long p,REAL x) {
  REAL y=1; REAL error;
  do { y=(y+x/y)/2; error=y-x/y; } while ( ! bound(error,p) );
  return y;  }
```

For these multi-valued functions, reiterations with increased precision may lead to different results. This leads to problems with the interactive usage of the Real RAM, as simple tests like `if ( bound(x,p) ) rscanf(a) else rscanf(b)` could imply different behaviour in different iterations. We solved this problem by storing the results of each of those operations and recalling the previous results in reiterations. In consequence, any flow of information from the continuous world of reals to the discrete world stays unchanged through the reiterations. This again implies that the flow of control in a computation will be repeated in reiterations up to the point where the previous iteration ended. As a side effect, we only need to store a sequence of the results and do not need any information on the context of the operations to reproduce their results.

In general, these operations should be applied as rarely as possible, as their results permanently occupy memory. In the following chapters we discuss environments where storing the results is not necessary (or even causes errors), so there the operations can be used freely.

7 Limits of sequences

Almost any important object in analysis is defined as some kind of a limit. Our prototype has several corresponding operators: The most simple form is `limit(a)` computing the limit of a sequence $(a_p)_{p\in\mathbb{Z}}$, if $|a_p - \lim_{n\to-\infty} a_n| \le 2^p$ holds.

In the form `limit(a,x)`, this operator can also be applied to sequences of (uniformly continuous) functions $(a_p(x))_{p\in\mathbb{Z}}$, where now $|a_p(x) - \lim_{n\to-\infty} a_n(x)| \leq 2^p$ must hold. As an application, the following simple function computes square roots of nonnegative reals, where `sqrt1` is as above: `REAL sqrt(REAL x) {return limit(sqrt1,x);}`

The idea of the implementation is as follows: To compute $\lim_{n\to-\infty} a_n(x)$, we simply compute $a_m(x)$ for some m, and increase the error of the result by 2^m. In addition, we ensure that m tends to $-\infty$ with growing precision of the iterations.

The argument `x` in `limit(a,x)` will not be exact during any stage of the computation. For any fixed m, it might even be impossible to compute $a_m(x)$ with the current precision p_i. To solve this problem, we try to compute $a_m(x)$ for different values $m = p_0, p_1, \ldots, p_i$ (as defined above) and then take that result giving the smallest error (as the sum of the error from the computation of a_m and the bound 2^m for the difference between a_m and the limit of the sequence). If we are unable to compute even a_{p_0}, we reiterate in order to get a better starting approximation x.

Usually this way of computing the limit will not be optimal, e.g. in the case where the a_n are Lipschitz-bounded. If x is known as a pair (d, e) when computing the limit and if we know an upper bound 2^ℓ for Lipschitz constants in the interval $(d-e, d+e)$, we may simply choose m such that 2^m is slightly bigger than $2^\ell \cdot e$ and compute $a_m(d)$, where. Then $|a_m(d) - a_m(x)| \leq 2^\ell \cdot e \leq 2^m$, i.e. $a_m(d)$ differs from the limit by at most 2^{1+m}. d can be treated as an exact value, so we are able to compute $a_m(d)$ with an arbitrary small error. The corresponding implementation is of the form `limit(a,l,x)`, where `a` and `x` are as above and `l` is such a Lipschitz bound ℓ.

In any case, we need local reiterations of the computations, i.e. the `longjmp` from within the computation of the limit must no longer lead to `main`, but to the procedure implementing `limit`. In addition, in `limit` there is a `longjmp` to an outer level, if even a_{p_0} can not be computed due to the error in `x`.

Using concepts like `limit(a,l,x)`, the flow of control in different iterations will surely not be the same, as m depends on p_i in the iteration with precision p_i. To solve this problem, we simply stop recording the results of the tests and conversions introduced in the previous chapter as long as we compute such a limit. This will have no influence on the outer flow of control as we compute a real value, unless there are side effects. In consequence, neither input, output nor any other side effects are allowed in the algorithm computing the sequence.

8 Continuous functions

Although e.g. the function `bound` is not continuous, there are applications where we finally get to continuous results. One example is solving systems of linear equations, where the classical algorithms internally rearrange the matrices due to the size of pivot elements. Here the result (i.e. the solution of the system) is independent from the actual rearrangement. These algorithms can easily be rewritten for the Real RAM.

Here again, we do not need to record the results of tests or conversions, as their actual result does not influence the result of the whole computation. Algorithms like `sqrt1` are usually used as arguments in a `limit`, so the recording is usually already stopped. On the other hand, solving linear equations is likely to be found on the outer level of control with extensive usage (and extensive permanent use of memory). This suggests the use of additional attributes specifying ‘continuity’ for procedures or defining ‘continuous’ sections in programs, where recording is turned off.

This is implemented by two functions `continuous_begin()` and `continuous_end()`, that must be called at the begin resp. the end of such a continuous section. The programmer must take care himself that it is impossible to enter or leave the section without appropriate calls to the functions. As an example, we show how to compute the exponential: For $|x| < 1$ we use a limit operator computing the Taylor expansion, otherwise we scale (with continuous results) using $e^x = (e^{x/2})^2$.

```
REAL exp_approx (long prec, const REAL& x){   REAL z=1,y=1; long i=1;
  if ( prec >= 2 ) return REAL(long(0));
  while ( !bound(y,prec) ) { y=y*x/i; z=z+y; i+=1; }
  return z;  };

REAL exp (const REAL& x) { continous_begin (); REAL y;
  if ( bound(x,0) ) { y = limit(exp_approx,0,x); }
       else { y = exp(x/2); y = y*y; }
  continous_end (); return y;  };
```

9 Conclusion

Our prototype shows that a Real RAM is no pure theoretical concept. With its growing library, especially of matrix arithmetic, it might soon be a tool e.g. for verifying traditional numerical algorithms, as its users may work with reals like entities (as far as theory allows), while retaining most of their working methods.

References

1. K. Ko, Complexity Theory of Real Functions, Birkhäuser, Boston (1991).
2. K. Weihrauch, Informatik Berichte 171 - 2/1995, FernUniversität Hagen.
3. R.P. Brent, ACM Trans. Math. Software, 4, 57–70 (1978).
4. R. Klatte, U. Kulisch et al., C-XSC, a C++ Class Library for Extended Scientific Computing, Springer, Berlin (1993).
5. V. Brattka and P. Hertling, Informatik Berichte 193 - 12/1995, FernUniversität Hagen.
6. M. Riordan, `ftp://ripem.msu.edu /pub/bignum/BIGNUMS.TXT`
7. N.Th. Müller, Proc. 22 JAIIO - PANEL '93, Part 2, Buenos Aires, (1993), pp. 259-281, (also available at `http://www.informatik.uni-trier.de/~mueller`).

8th Int. Coll. on Differential Equations, pp. 335-340
D. Bainov (Ed)

Asymptotic Behaviour of Solutions of an Impulsive Semilinear Parabolic Cauchy Problem

KIYOKAZU NAKAGAWA

Department of Mathematics, Tohoku Gakuin University, Sendai 981-31, Japan

Abstract. In this paper we consider the Cauchy problem for the semilinear parabolic equation $\frac{\partial u}{\partial t} = \Delta u + u^p$ with an impulsive condition. We would like to make a sufficient impulsive condition to construct a global solution and investigate the behaviour as t tends to ∞.

Key words. Asymptotic behaviour, impulsive equation, semilinear parabolic equations, and Cauchy problem.

1 INTRODUCTION

Bainov and Minchev [1] formerly presented the paper "Dynamics of impulsive populations and their asymptotic behaviour" in 1993 and considered the equation with an impulsive condition

$$\begin{aligned} u_t(x,t) &= \alpha\Delta u(x,t) + f(u(x,t))u(x,t) && in \bigcup_{i=0}^{\infty}\{\Omega\times(t_k,t_{k+1})\} \\ u(x,0) &= \phi(x) && x\in\overline{\Omega}, \\ u|_{\partial\Omega} &= 0, \\ u(x,t_k) &= g_k(x,t_k,u(x,t_k^-)) && x\in\overline{\Omega}, k=1,2,\cdots. \end{aligned}$$

Here $\Omega \in R^n$ is a bounded domain with smooth boundary $\partial\Omega$. They discussed the estimates of the populations growth, uniqueness critetia and nonnegative solutions.

In the last year, they and I [2] have considered the more general type of an impulsive semilinear parabolic equation. We considered this equation from the point of view of semi-group theory. Suppose that $0 = t_0 < t_1 < t_2 < \cdots < t_p < \cdots$ are given numbers and $\lim_{n\to\infty} t_n = \infty$. Let X be a Banach space.

$$\begin{aligned} \frac{du(t)}{dt} + A(t)u(t) &= f(u(t)), \\ u(0) &= u_0, \\ u(t_k) &= g_k(t_k, u(t_k^-)) \qquad \text{for } k = 1,2,\cdots, \end{aligned}$$

where $g_k(t_k,\cdot): X \to X,\ k = 1,2,\cdots$ are given maps and $u(t_k^-) = \lim_{s\to t_k-0} u(s)$.

We made the two assumptions.

H1.

$$\| g_k(t_k, u(t_k^-)) \|_X \leq M_k \| u(t_k^-) \|_X,$$

where $M_k,\ k = 1, 2, \cdots$ are constants and

$$C \prod_{s<t_k<t} (CM_k) \leq L_0 e^{\gamma(t-s)},$$

where $L_0 \geq 0$, γ and C are constants .

H2. $f : X \to X$ is Lipschitz continuous and satisfies that $\| f(u(t)) \|_X \leq M \| u(t) \|_X$ where $M > 0$ is a constant.

Then we have obtained the theorem.

Theorem 1.1 *Let $A(t)A(s)^{-1}$ be uniformly bounded in t and s. We suppose H1 and H2. Then the following estimate holds*

$$\| u(t) \|_X \leq L_0 \| u_0 \|_X \ e^{(\gamma-\delta_0+CM)t},$$

where δ_0 is defined by the operator A.

1. *If we suppose that $\gamma + CM = \delta_0$ then $\| u(t) \|_X \leq L_0 \| u_0 \|_X$ is a bounded function.*
2. *If we suppose that $\gamma + CM < \delta_0$ then $\lim_{t\to\infty} \| u(t) \|_X = 0$.*

Unfortunately, the hypothesis H2 excludes the case that $f(u) = |u|^{p-1}u,\ p > 1$. So, we would like to discuss the following impulsive semilinear parabolic Cauchy problem

$$\frac{\partial u}{\partial t} = \Delta u + u^p \qquad in\ \bigcup_{i=0}^{\infty}\{R^n \times (t_k, t_{k+1})\}, \tag{1}$$

$$u(x,0) = \varphi(x) \qquad in\ R^n, \tag{2}$$

$$u(x,t_k) = g_k(u(x,t_k^-)) \qquad for\ x \in R^n,\ k = 1, 2, \cdots, \tag{3}$$

where Δ is the Laplace operator, $p > 1$ is a constant, φ is a nonnegative bounded continuous function in R^n. We suppose that $0 = t_0 < t_1 < t_2 < \cdots < t_p < \cdots$ are given numbers and $\lim_{n\to\infty} t_n = \infty$ and $g_k : R \to R,\ \ k = 1, 2, \cdots$ are continuous and each $g_k(v)$ is increasing for $v \in R$.

Due to the possible nonuniqueness of solutions of the Cauchy problem from (1) to (3), we shall restrict our attention to a certain class of solutions $u(x,t)$; namely, those with the following properties.

(i). $u(x,t) \geq 0\ in\ \bigcup_{i=0}^{\infty}\{R^n \times (t_k, t_{k+1})\}$,

(ii). u satisfies the integral equation for $(x,t) \in R^n \times [t_k, t_{k+1}),\ k = 0, 1, 2, \cdots$,

$$u(x,t) = \frac{1}{(4\pi(t-t_k))^{\frac{n}{2}}} \int_{R^n} e^{\frac{-|x-y|^2}{4(t-t_k)}} u(y,t_k)dy + \int_{t_k}^{t}\int_{R^n} \frac{1}{(4\pi(t-s))^{\frac{n}{2}}} e^{\frac{-|x-y|^2}{4(t-s)}} u^p(y,s)dyds,$$

(iii). $u(x,t_k) = g_k(u(x,t_k^-)),\ \ k = 1, 2, \cdots$.

On the other hand, it is proved that if u satisfies the integral equation and is bounded in $R^n \times [t_k, t_{k+1})$, $k = 0, 1, 2, \cdots$ then u is unique and is a classical solution of the differential equation; i.e. $u \in C^{2,1}(R^n \times (t_k, t_{k+1})) \bigcap C(R^n \times [t_k, t_{k+1}))$, $k = 0, 1, 2, \cdots$ and u satisfies the differential equation (1)-(3).

These types of the Cauchy problems of semilinear parabolic equations without an impulsive condition were considered by various authors.

In 1966, Fujita [3] proved that if $p < \frac{n+2}{n}$ then there is no global solution for all φ which are nonnegative and are not equivalent to zero in R^n, and in case $p > \frac{n+2}{n}$ there is a global solution if φ is bounded by $\varepsilon exp(-|x|^2)$ where ε is a small positive number.

In 1992, Lee and Ni [4] attempted to understand the behaviour of the solution $u(x,t)$ while the initial value φ is not so small near $x = \infty$. They obtained the following theorem. Let $C_b(R^n)$ be the space of all bounded continuous functions in R^n.

Theorem 1.2 *Suppose that $p > \frac{n+2}{n}$ and $a \geq \frac{2}{p-1}$. Put $\varphi = \lambda\psi$. Then the following conclusions hold.*

1. *For every $\psi \in \{\phi \in C_b(R^n) | \limsup_{|x|\to\infty} |x|^a \phi(x) < \infty$ and $\phi \geq 0\}$, there exists $\Lambda_0 > 0$, depending on p, n and ψ, such that there is a global solution $u(x, t; \lambda\psi)$for every $\lambda < \Lambda_0$. Moreover, for $\lambda < \Lambda_0$ we have*

$$\| u(\cdot, t; \lambda\psi) \|_{L^\infty(R^n)} = O(q(t; , a, n)) \quad as \quad t \to \infty.$$

Where $$q(t; a, n) = \begin{cases} t^{-\frac{1}{2}\min\{a,n\}} & if\ a \neq n \\ t^{-\frac{n}{2}} \log t & if\ a = n. \end{cases}$$

2. *Set $\lambda = 1$. Let μ_1 be the first eigenvalue of $-\Delta$ in B_1, the ball of radius 1 in R^n, with zero Dirichlet boundary value. If $\liminf_{x\to\infty} |x|^{\frac{2}{p-1}} \psi(x) > \mu_1^{\frac{1}{p-1}}$, then there is no global solution.*

2 Main result and its proof

The behaviour of the initial value near $x = \infty$ is essential to existence and non-existence of a global solution of the Cauchy problem. Instead of such a condition, we would like to make a sufficient impulsive condition to construct a global solution and investigate the behaviour as t tends to ∞. We present the main theorem now.

Theorem 2.1 *Put $M_0 = \| \psi \|_{L^\infty(R^n)}$. We assume that each $g_k : [0, \infty) \to [0, \infty)$ is a continuous function such that $g_k(v)$ is increasing and*

$$\| g_k(v) \|_{L^\infty(R^n)} \leq M_k \| v \|_{L^\infty(R^n)} \quad k = 1, 2, \cdots.$$

Let $M_1, M_2, \cdots, M_n, \cdots$ be constants and satisfy

$$M_0 M_1 \cdots M_k < [2(p-1)(t_{k+1} - t_k)]^{\frac{1}{1-p}} 2^{k(1-p)}, \qquad k = 0, 1, 2, \cdots.$$

Then there exists a unique solution $u(x, t)$ of (1) - (3) and we have

$$\| u(\cdot, t) \|_{L^\infty(R^n)} \leq 2^{p-1} [2(p-1)(t_{k+1} - t_k)]^{\frac{1}{1-p}} \tag{4}$$

$$for\ t \in [t_k, t_{k+1}), \quad k = 0, 1, 2, \cdots.$$

We will need the following lemma to prove the theorem.

Lemma 2.1 *Suppose that* $f \in C^1(R)$ *and* $\overline{u}(x,t), \underline{u}(x,t) \in C^{2,1}(R^n \times (T_0,T_1)) \bigcap C(R^n \times [T_0,T_1))$ *are bounded in* $R^n \times [T_0, T_1']$ *for all* $T_1' < T_1$. *If* $\overline{u}(x,T_0) \geq \underline{u}(x,T_0)$ *for all* $x \in R^n$ *and*

$$\overline{u}_t - \Delta\overline{u} \geq f(\overline{u}) \qquad \text{in } R^n \times (T_0,T_1), \tag{5}$$

$$\underline{u}_t - \Delta\underline{u} \leq f(\underline{u}) \qquad \text{in } R^n \times (T_0,T_1),$$

then $\overline{u}(x,t) \geq \underline{u}(x,t)$ *for all* $(x,t) \in R^n \times [T_0,T_1)$. *Furthermore, for any* $\varphi \in C_b(R^n)$ *with* $\overline{u}(x,T_0) \geq \varphi(x) \geq \underline{u}(x,T_0)$ *in* R^n, *there exists a unique solution* $u(x,t) \in C^{2,1}(R^n \times (T_0,T_1)) \bigcap C(R^n \times [T_0,T_1))$ *of the problem*

$$\begin{aligned} u_t &= \Delta u + f(u) && \text{in } R^n \times (T_0,T_1), \\ u(x,T_0) &= \varphi(x) && \text{in } R^n, \end{aligned}$$

with the property that $\overline{u} \geq u \geq \underline{u}$ *in* $R^n \times [T_0,T_1)$.

This lemma is well known. We refer the interested readers to the paper of T. Lee and W. Ni [4]. Now we would like to prove the main theorem with the aid of this lemma.

Proof of the main theorem.

1) We first assume that $0 = t_0 \leq t < t_1$. We shall constract $\underline{u} \equiv 0$ and $\overline{u}$ with the following property

$$\begin{aligned} \overline{u}_t &\geq \Delta\overline{u} + \overline{u}^p && \text{in } R^n \times (t_0,t_1), \\ \overline{u}(x,t_0) &\geq \psi(x) && \text{in } R^n. \end{aligned}$$

Then lemma 2.1 will guarantee that $\overline{u}(x,t) \geq u(x,t) \geq 0$ for all $(x,t) \in R^n \times [t_0,t_1)$.

We shall use the usual notation $e^{t\Delta}\varphi$ to represent the solution of the heat equation with initial value φ ; i.e.

$$(e^{t\Delta}\varphi)(x) = \frac{1}{(4\pi t)^{\frac{n}{2}}} \int_{R^n} e^{-\frac{|x-y|^2}{4t}} \varphi(y)dy.$$

$\overline{u}(x,t)$ takes the form $\overline{u}(x,t) = h_0(t)e^{t\Delta}\psi$ where

$$h_0(t) = [1-(p-1)\int_0^t \| e^{s\Delta}\psi \|^{p-1}_{L^\infty(R^n)} ds]^{\frac{1}{1-p}}.$$

Since $\| e^{s\Delta}\psi \|_{L^\infty(R^n)} \leq \| \psi \|_{L^\infty(R^n)}$ for all $s \geq 0$ and $\| \psi \|_{L^\infty(R^n)} \leq [2(p-1)(t_1-t_0)]^{\frac{1}{1-p}}$, we have

$$\begin{aligned} 2(p-1)\int_0^t \| e^{s\Delta}\psi \|^{p-1}_{L^\infty(R^n)} ds &\leq 2(p-1)(t_1-t_0) \| \psi \|^{p-1}_{L^\infty(R^n)} ds \\ &< 2(p-1)(t_1-t_0)\{[2(p-1)(t_1-t_0)]^{\frac{1}{1-p}}\}^{p-1} = 1. \end{aligned}$$

Then $h_0(t)$ is well defined in $[t_0,t_1)$ and we immediately have

$$2^{\frac{1}{1-p}} \leq [1-(p-1)\int_0^t \| e^{s\Delta}\psi \|^{p-1}_{L^\infty(R^n)} ds]^{\frac{1}{p-1}} \leq 1.$$

Namely, $1 \leq h_0(t) \leq 2^{p-1}$. So $\overline{u}(t)$ is also defined on $R^n \times [t_0,t_1)$ and

$$\| \overline{u}(\cdot,t) \|_{L^\infty(R^n)} \leq 2^{p-1} \| e^{t\Delta}\psi \|_{L^\infty(R^n)} \leq 2^{p-1}[2(p-1)(t_1-t_0)]^{\frac{1}{1-p}}.$$

On the other hand, the conclusion that $\overline{u}$ satisfies the differential inequality (5) follows from that fact that

$$\frac{dh_0}{dt} = \| e^{t\Delta}\psi \|^{p-1}_{L^\infty(R^n)} h_0^p \quad in\ t_0 < t < t_1,$$

which in turn may be verified by straightforward computation.

By virtue of lemma 2.1, there exists a unique solution $u(x,t)$ of (1)-(3) and it satisfies the inequality (4) in $[t_0, t_1)$.

We shoud notice that from the above argument the solution $u(x,t)$ can be uniquely continued beyond t_1 to some interval.

2) We now assume that $t_1 \le t < t_2$. In this case $\underline{u}(x,t) \equiv 0$ and $\overline{u}(x,t) = h_1(t)e^{(t-t_1)\Delta}\overline{u}(x,t_1)$ where

$$h_1(t) = [1-(p-1)\int_{t_1}^t \| e^{(s-t_1)\Delta}\overline{u}(\cdot,t_1) \|^{p-1}_{L^\infty(R^n)} ds]^{\frac{1}{1-p}}.$$

Since $\overline{u}(x,t_1) = g_1(\overline{u}(x,t_1^-))$ and $\| g_k(\overline{u}(\cdot,t_1^-)) \|_{L^\infty(R^n)} \le M_1 \| \overline{u}(\cdot,t_1^-)) \|_{L^\infty(R^n)}$, we have

$$\| \overline{u}(\cdot,t_1) \|_{L^\infty(R^n)} \le M_1 \| \overline{u}(\cdot,t_1^-) \|_{L^\infty(R^n)} .$$

From the first case, we obtained $\| \overline{u}(\cdot,t_1^-) \|_{L^\infty(R^n)} \le 2^{p-1} \| \psi \|_{L^\infty(R^n)}$. Thus we have

$$\| \overline{u}(\cdot,t_1) \|_{L^\infty(R^n)} \le M_1 2^{p-1} \| \psi \|_{L^\infty(R^n)} .$$

So we obtain that

$$\begin{aligned} 2(p-1)\int_{t_1}^t \| e^{(s-t_1)\Delta}\overline{u}(\cdot,t_1) \|^{p-1}_{L^\infty(R^n)} ds &\le 2(p-1)\int_{t_1}^{t_2} \| \overline{u}(\cdot,t_1) \|^{p-1}_{L^\infty(R^n)} ds \\ &\le 2(p-1)(t_2-t_1)M_1^{p-1}2^{(p-1)^2} \| \psi \|^{p-1}_{L^\infty(R^n)} \\ &\le 2(p-1)(t_2-t_1)2^{(p-1)^2} \times \{[2(p-1)(t_2-t_1)]^{\frac{1}{1-p}}2^{(1-p)}\}^{p-1} = 1. \end{aligned}$$

Therefore

$$1 \le [1-(p-1)\int_{t_1}^t \| e^{(s-t_1)\Delta}\overline{u}(\cdot,t) \|^{p-1}_{L^\infty(R^n)} ds]^{\frac{1}{1-p}} \le 2^{p-1}.$$

In other words, $h_1(t)$ is also well defined in $[t_1,t_2)$ and satisfies $1 \le h_1(t) \le 2^{p-1}$. This implies that $\overline{u}(x,t)$ is defined on $R^n \times [t_1,t_2)$. Then we have

$$\begin{aligned} \| \overline{u}(\cdot,t) \|_{L^\infty(R^n)} &\le 2^{p-1} \| e^{(t-t_1)\Delta}\overline{u}(\cdot,t_1) \|_{L^\infty(R^n)} \le 2^{p-1} \| \overline{u}(\cdot,t_1) \|_{L^\infty(R^n)} \\ &\le 2^{2(p-1)}M_1 \| \psi \|_{L^\infty(R^n)} \le 2^{p-1}[2(p-1)(t_2-t_1)]^{\frac{1}{1-p}}. \end{aligned}$$

As in the first case, the solution $u(x,t)$ can be uniquely continued from t_1 to t_2. Moreover $u(x,t)$ satisfies the inequality (4).

3) We assume that $t_k \le t < t_{k+1}$, $k = 1, 2, \cdots$. In this case $\underline{u}(x,t) \equiv 0$ and $\overline{u}(x,t) = h_k(t)e^{(t-t_k)\Delta}\overline{u}(x,t_k)$ where

$$h_k(t) = [1-(p-1)\int_{t_k}^t \| e^{(s-t_k)\Delta}\overline{u}(\cdot,t_k) \|^{p-1}_{L^\infty(R^n)} ds]^{\frac{1}{1-p}}.$$

Since we had the estimate of $h_{k-1}(t)$, we obtain the following inequality.

$$\begin{aligned} \| \overline{u}(\cdot,t_k) \|_{L^\infty(R^n)} &= \| g_k(\overline{u}(\cdot,t_k^-)) \|_{L^\infty(R^n)} \\ &\le M_k \| \overline{u}(\cdot,t_k^-) \|_{L^\infty(R^n)} \\ &\le M_k 2^{p-1} \| \overline{u}(\cdot,t_{k-1}) \|_{L^\infty(R^n)} \\ &\le M_k 2^{k(p-1)} M_1 M_2 \cdots M_{k-1} \| \psi \|_{L^\infty(R^n)} \end{aligned}$$

Because $M_1 M_2 \cdots M_k \| \psi \|_{L^\infty(R^n)} < 2^{k(1-p)}[2(p-1)(t_{k+1}-t_k)]^{\frac{1}{1-p}}$, we obtain that

$$2(p-1)\int_{t_k}^{t} \| e^{(s-t_k)\Delta}\overline{u}(\cdot,t_k) \|_{L^\infty(R^n)}^{p-1} ds \le 2(p-1)(t_{k+1}-t_k)\prod_{j=0}^{k} M_j^{p-1} 2^{k(p-1)^2}$$
$$\le 2(p-1)(t_{k+1}-t_k)2^{k(p-1)^2} \times \{[2(p-1)(t_{k+1}-t_k)]^{\frac{1}{1-p}} 2^{k(1-p)}\}^{p-1} = 1.$$

Then we have

$$1 \le [1-(p-1)\int_{t_k}^{t} \| e^{(s-t_k)\Delta}\overline{u}(\cdot,t) \|_{L^\infty(R^n)}^{p-1} ds]^{\frac{1}{1-p}} \le 2^{p-1}.$$

$h_k(t)$ is also well defined in $[t_k, t_{k+1})$ and satisfies $1 \le h_k(t) \le 2^{p-1}$. Moreover $\overline{u}(x,t)$ is defined on $R^n \times [t_k, t_{k+1})$ and we obtain

$$\begin{aligned} \| \overline{u}(\cdot,t) \|_{L^\infty(R^n)} &\le 2^{p-1} \| e^{(t-t_k)\Delta}\overline{u}(\cdot,t_k) \|_{L^\infty(R^n)} \\ &\le 2^{p-1} M_1 M_2 \cdots M_k 2^{k(p-1)} \| \psi \|_{L^\infty(R^n)} \\ &= 2^{p-1}[2(p-1)(t_{k+1}-t_k)]^{\frac{1}{1-p}}. \end{aligned}$$

As in the first case, the solution $u(x,t)$ can be uniquely continued from t_k to t_{k+1} and it also satisfies the inequality(4).

Thus the proof of theorem is complete. Finally, we give here the Corollary of the theorem.

Corollary 2.1 *Let s_k be the length of the time interval (t_k, t_{k-1}), $k = 0, 1, 2, \cdots$. We assume the same hypotheses of theorem 2.1. Then we have the following.*

1. *If $\lim_{k\to\infty} s_k = s \neq 0$, then $u(x,t)$ is a bounded function.*
2. *If $\lim_{k\to\infty} s_k = 0$, then $\lim_{k\to\infty} u(x,t) = 0$.*

References

1. D. Bainov and E. Minchev, Dynamics of impulsive populations and their asymptotic behaviour, (to appear).
2. D. Bainov, E. Minchev and K. Nakagawa, Asymptotic behaviour of solutions of impulsive semilinear parabolic equations, (to appear).
3. H. Fujita, J. Fac. Sci. Univ. Tokyo (I), 13, 109–124 (1966).
4. T. Lee and W. Ni, Trans. Amer. Math. Soc., 333, 365–378 (1992).
5. L. Erbe, H. Freedman, X. Liu and J. Wu, J. Austral. Math. Soc. Ser. B, 32, 382–400 (1991).

8th Int. Coll. on Differential Equations, pp. 341–346
D. Bainov (Ed)

Error Estimates of Optimal Orders for One Dimensional Stefan Problems

MI - RAY OHM
Department of Mathematics, Dong-Seo University, 617-716 Pusan, Korea

Abstract. Finite element approximations for the one-dimensional Stefan problem are discussed. The aim of this paper is to have the optimal convergence with respect to the order of h by Galerkin method.
AMS(MOS) subject classification. 65N30, 65G99

1. INTRODUCTION

The mathematical formulation of many problems arising in practice leads to boundary value problems for partial differential equations -especially of parabolic type- with the feature that the boundary is not prescribed in advance but depends on certain properties of the solution itself. Probably the oldest such free boundary problem is due to Stefan (1889). In one space dimension, it may be described as follows: In a domain

$$\Omega := \{(y, \tau) |\ \tau > 0, 0 < y < s(\tau)\}$$

a function u is sought as the solution of the heat equatioin

$$u_\tau - u_{yy} = 0 \qquad \text{in } \Omega.$$

The initial temparature $u(y, 0)$ with compatibility conditions as well as the water temparature at $y = 0$:

$$u(0, \tau) = f(\tau) \qquad 0 < \tau \leq T_0$$

are prescribed. The free boundary $y = s(\tau)$ is defined by the condition

$$u(s(\tau), \tau) = 0$$

and the additional condition

$$s_\tau + u_y(s(\tau), \tau) = 0.$$

The melting (or freezing) of an ice block is one of the physical interpretations.

The approach presented in this paper is based on the transformation of the Stefan problem in one space dimension to an initial-boundary value problem for the heat equation in a fixed domain. Of course, the problem is non-linear. The finite element approximation adopted here is the standard continuous Galerkin method in time. In this paper, only the regular case is discussed. This means the error analysis is based on the assumption that the solution is sufficiently smooth.

The aim of this paper is to have the optimal convergence with respect to the order of power of h by Galerkin method.

2. WEAK FORMULATION OF THE STEFAN PROBLEM, FINITE ELEMENT METHOD

The Stefan problem is Problem P_U: Given $T_0 > 0$, $g(y) \in C^1(0,1)$ with $g(1) = 0$ and $f(\tau) \in C^2[0, T_0]$. Find $\{(s(\tau), U(y,\tau))\}$ such that

$$\begin{aligned} s(\tau) &> 0 && \text{for } 0 < \tau \le T_0 \\ s(0) &= 1, \\ U_{yy} - U_\tau &= 0 && \text{in } \Omega = \{(y,\tau) | 0 < \tau \le T_0, 0 < y < s(\tau)\} \\ U(0,\tau) &= f(\tau) \\ U(s(\tau),\tau) &= 0 && \text{for } 0 < \tau \le T_0 \\ U(y,0) &= g(y) \end{aligned}$$

and in addition

$$\frac{ds}{d\tau} + U_y(s(\tau),\tau) = 0 \qquad \text{for } 0 < \tau \le T_0.$$

In order to reduce this problem to one with fixed boundaries, we introduce the new space variable

$$x = s^{-1}(\tau) y. \tag{2.1}$$

The corresponding transformation of τ defined by

$$\frac{d\tau}{dt} = s^2(\tau), \quad \tau(0) = 0 \tag{2.2}$$

leads for the formulation $u(x,t) = U(y,\tau)$ to Problem $P_u 1$: Find u such that

$$u_{xx} - u_t = x u_x(1,t) u_x(x,t) \quad \text{in } Q = \{(x,t) |\ x \in I, 0 < t \le T\} \tag{2.3}$$

$$\frac{ds}{dt} = -u_x(1,t)s, \quad s(0) = 1 \qquad \text{for } 0 < t \le T \tag{2.4}$$

$$\frac{d\tau}{dt} = s^2, \quad \tau(0) = 0 \tag{2.5}$$

with the boundary conditions

$$u(0,t) = f(\tau(t)), \quad u(1,t) = 0 \qquad \text{for } 0 < t \le T \tag{2.6}$$

and the initial condition

$$u(x,0) = g(x) \qquad \text{for } 0 < x < 1.$$

Here $t = T$ corresponds to $\tau = T_0$. The original Stefan problem is now split into a nonlinear parabolic initial boundary value problem for a fixed domain and two ordinary differential equations (2.4) and (2.5). If the boundary condition at $x = 0$ is time-dependent, the parabolic problem and the ordinary differential equations are coupled.

Let us introduce the space

$$\dot{H}_2 = \{w |\ w \in H_2, w(1) = 0\}.$$

Then u belongs to $\dot{H}_2$. Multiplication of (2.3) by w_{xx}, with $w \in \dot{H}_2$ and integration give

$$(u_{xx}, w_{xx}) - (u_t, w_{xx}) = u_x(1,t)(x u_x, w_{xx})$$

here $(\cdot,\cdot)$ denotes the $L_2(0,1)$-scalar product. For simplicity u' and $\dot{u}$ denote differentiation with respect to x and t. Differentiation the equation (2.6) with respect to time t gives

$$\begin{aligned}\dot{u}(0,t) &= \frac{df}{d\tau}(\tau(t))s^2(t)\\ \dot{u}(1,t) &= 0 \qquad \text{for } 0 < t \le T.\end{aligned}$$

In this way we come to Problem P_u2: Find a function u such that $u(\cdot,t) \in \dot{H}_2$ and

$$(\dot{u}',w') + (u'',w'') = u'(1,t)(xu',w'') - \frac{df}{d\tau}s^2w'(0) \tag{2.7}$$

$$\text{for } w \in \dot{H}_2 \text{ and } 0 < t \le T$$

with the initial condition

$$u(x,0) = g(x) \qquad \text{for } 0 < x < 1.$$

(2.7) is the weak formulation of (2.3) and (2.6).

There are many methods for getting finite element solutions. We will study only the standard one: Let $S_h = S_h^{2,r}, r \ge 4$ denote the continuous splines of order r with $r \ge 4$, i.e.,S_h consists of continuous piecewise polynomial functions of degree less than r for some regular subdivision Γ_h of (0,1). Up to fixed factors, h is a lower and an upper bound of the length of the subintervals of Γ_h. Furthermore

$$\dot{S}_h = \{w \mid w \in S_h, w(1) = 0\}.$$

In this way, $\dot{S}_h \subset \dot{H}_2$. The finite element method is Problem P_{u_h}: Find u_h such that $u_h(\cdot,t) \in \dot{S}_h$ and

$$(\dot{u}_h',\chi') + (u_h'',\chi'') = u_h'(1,t)(xu_h',\chi'') - \frac{df}{d\tau}s_h^2\chi'(0) \tag{2.8}$$

$$\text{for } \chi \in \dot{S}_h \text{ and } 0 < t \le T$$

with the initial condition

$$u_h(\cdot,0) = P_h g.$$

Here P_h is a projection onto $\dot{S}_h$.

3. ERROR ESTIMATES

We assume that the solution of problem P_u2 is sufficiently smooth. This is meant for the regular case. Then the error estimates are optimal with respect to the order of power h. The analysis given below could be defined to deal with the reduced smoothness of the solution and the corresponding reduced order of convergence.

Since S_h is finite dimensional, problem P_{u_h} results in a finite system of ordinary differential equations. Therefore u_h always exists in a certain interval $(0,\bar{t})$. Here, $\bar{t}$ may depend on g but not on $\dot{S}_h$.

Theorem 1. (Ohm [1,2]) The problem P_{u_h} has a unique solution $u_h \in \dot{H}_2$.

In order to simplify the presentation, we introduce the trilinear form

$$b(\psi,\eta,\phi) = \frac{1}{2}\psi'(1)(x\eta',\phi'') + \frac{1}{2}\eta'(1)(x\psi',\phi'') \tag{3.1}$$

here $\psi'(1) = \psi'(1,t)$. Most of time, we will supress the dependence on t,e.g., we write $v(1) = v(1,t)$ etc. From (2.7), u and u_h satisfy

$$(\dot{u}', \chi') + (u'', \chi'') = b(u, u, \chi) - \frac{df}{d\tau}(\tau)s^2\chi'(0) \tag{3.2}$$

$$(\dot{u}_h', \chi') + (u_h'', \chi'') = b(u_h, u_h, \chi) - \frac{df}{d\tau}(\tau_h)s_h^2\chi'(0). \tag{3.3}$$

For the error $e = u - u_h$, we get by substration the above relations

$$(\dot{e}', \chi') + (e'', \chi'') - 2b(u, e, \chi) = -b(e, e, \chi) + (\dot{f}(\tau_h)s_h^2 - \dot{f}(\tau)s^2)\chi'(0). \tag{3.4}$$

Since u is fixed,

$$a(\psi, \eta) = (\psi'', \eta'') - 2b(u, \psi, \eta) \tag{3.5}$$

is a bilinear form in ψ, η.
In the following lemma we can know this bilinear form $a(\cdot,\cdot)$ in (3.5) is bounded and coersive.

Lemma 1.(Ohm [3])

(i) $a(\cdot,\cdot)$ is bounded in $\cdot H_2$, i.e.,

$$|a(\psi, \eta)| \leq M\|\psi\|_{H_2}\|\eta\|_{H_2} \tag{3.6}$$

(ii)

$$a(\psi, \psi) \geq m\|\psi''\|^2 - \Lambda\|\psi'\|^2, \tag{3.7}$$

here $m > 0$, M and Λ depend only on $\|u'\|_{L_\infty(L_\infty)}$.
With (3.5),

$$a_\Lambda(\psi, \eta) = a(\psi, \eta) + \Lambda(\psi', \eta') \tag{3.8}$$

is positive in $\dot{H}_2$ and $\dot{H}_2 = \{w|\ w \in H_2 \cap w(1) = 0\}$. (3.4) can be rewritten

$$(\dot{e}', \chi') + a_\Lambda(e, \chi) = \Lambda(e', \chi') - b(e, e, \chi) + (\dot{f}(\tau_h)s_h^2 - \dot{f}(\tau)s^2)\chi'(0). \tag{3.9}$$

Let $\hat{u_h} = P_h u \in \dot{S}_h$ be the Galerkin approximation to u with respect to the form a_Λ:

$$a_\Lambda(u - \hat{u}_h, \chi) = 0 \quad \text{for } \chi \in \dot{S}_h. \tag{3.10}$$

In order to derive error estimates, we need a splitting of $e = u - u_h$. We put

$$e = (u - \hat{u}_h) - (u_h - \hat{u}_h) = \epsilon - \phi. \tag{3.11}$$

We can see $\phi \in \dot{S}_h$. From (3.9), we can have with (3.10)

$$\begin{aligned}(\dot{\phi}', \chi') + a_\Lambda(\phi, \chi) = \Lambda(\phi', \chi') + (\dot{\epsilon}', \chi') - \Lambda(\epsilon', \chi') + \\ + e'(1)(xe', \chi'') + (\dot{f}(\tau_h)s_h^2 - \dot{f}(\tau)s^2)\chi'(0).\end{aligned} \tag{3.12}$$

We replace in the quadratic term $e(1)$ by $E(1)$, for some function E such that E' is a L_∞-function and bounded by 1. Then

$$(\dot\phi', \chi') + a_\Lambda(\phi, \chi) = \Lambda(\phi', \chi') - E'(1)(x\phi', \chi'') + (\dot\epsilon', \chi') - \Lambda(\epsilon', \chi') + \\ + E'(1)(x\epsilon', \chi'') + (\dot f(\tau_h)s_h^2 - \dot f(\tau)s^2)\chi'(0). \tag{3.13}$$

is a linear problem. Therefore for any $E'(1) = E'(1,t)$, there exists a solution ϕ of (3.13) with $\phi(\cdot, 0) = 0$. From ϕ we obtain $e = \epsilon - \phi$ by (3.11), with $e = E$. In the following lemmas we derive for the error e H_1-estimates.

Lemma 2. (Ohm [3]) Let $E > 0$ be fixed and $E'(1) = E'(1,t)$ be measurable and bounded by 1. Then

$$\|\phi\|_{L_\infty(H_1)} \leq c\{\|\dot\epsilon'\|_{L_2(H_{-1})} + \|\epsilon'\|_{L_2(H_{-1})} + \|E'\|_{L_\infty(L_\infty)}\|\epsilon'\|_{L_2(L_2)}\}. \tag{3.14}$$

Here $L_\infty(L_p)$ is the abbreviation for $L_\infty(0, \bar t; L_p)$ and the norm in H_{-1} is defined by

$$\|\psi'\|_{-1} = \sup\{(\psi', \eta')|\ \eta \in \dot H_2 \cap \|\eta''\| \leq 1\}. \tag{3.15}$$

Lemma 3. (Ohm [3]) Assume $r \geq 4$, i.e., the approximation spaces are at least cubic splines. Let the exact solution u of problem $P_u 2$ be sufficiently smooth. Then for any time t fixed

$$\|\epsilon\|_{H_k} + \|\dot\epsilon\|_{H_k} \leq ch^{r-k} \qquad \text{for } k = 0, 1. \tag{3.16}$$

Theorem 2. (Ohm [3]) Let $\bar t > 0$ be chosen properly and assume that the solution of problem $P_u 2$ is sufficiently smooth in $I \times [0, \bar t]$. Then there exists exactly one finite element solution u_h in the neighbourhood of u with optimal order of convergence

$$\|u - u_h\|_{L_\infty(0,\bar t; H_1)} = o(h^{r-1}).$$

Lemma 4. Assume $r \geq 4$, i.e., the approximation spaces are at least cubic splines. Let the exact solution u of problem $P_u 2$ be sufficiently smooth. Then

$$\|\epsilon'\|_{L_\infty} \leq ch^{r-1}. \tag{3.17}$$

Since

$$\|\chi\|_{L_\infty(I)} \leq ch^{-\frac{1}{2}}\|\chi\|_{L_2(I)} \qquad \text{for } \chi \in S_h,$$

by Lemma 2 we get

$$\begin{aligned} \|e'\|_{L_\infty(L_\infty)} &\leq \|\epsilon'\|_{L_\infty(L_\infty)} + \|\phi'\|_{L_\infty(L_\infty)} \\ &\leq \|\epsilon'\|_{L_\infty(L_\infty)} + c\|\phi'\|_{L_\infty(L_2)}h^{-\frac{1}{2}} \\ &\leq \|\epsilon'\|_{L_\infty(L_\infty)} \\ &\quad + ch^{-\frac{1}{2}}\{\|\dot\epsilon'\|_{L_2(H_{-1})} + \|\epsilon'\|_{L_2(H_{-1})} + \|E'\|_{L_\infty(L_\infty)}\|\epsilon'\|_{L_2(L_2)}\}. \end{aligned}$$

By Lemma 3 and Lemma 4 we obtain

$$\|e'\|_{L_\infty(L_\infty)} \leq c(h^{r-1} + h^{r-\frac{3}{2}}\|E'\|_{L_\infty(L_\infty)}).$$

The image e of any E with

$$E' \in B_1 = \{w|\ \|w\|_{L_\infty(L_\infty)} \leq 1\}$$

is contained in B_1 for $h \leq \bar{h}$ with appropriate $\bar{h} < \frac{1}{2c}$. Schauder's fixed point theorem guarantees the existence of an E with $e = E$. Thus we have prove the following theorem.

Theorem 3. Let $\bar{t} > 0$ be chosen properly and assume that the solution of problem P_u2 is sufficiently smooth in $I \times [0, \bar{t}]$. Then there exists exactly one finite element solution u_h in the neighbourhood of u with optimal order of convergence

$$\|u - u_h\|_{L_\infty(0,\bar{t};W^1_\infty)} = o(h^{r-1}).$$

REFERENCES

1. M.R. Ohm, Bull. Kor. Math. Soc., 31, 309–318 (1994).
2. M.R. Ohm, The Journal of Dong-Seo University, 1, 7–15 (1995).
3. M.R. Ohm, JMAA, 199, 231–242 (1996).
4. J. Douglas, T. Dupont and L. Wahlbin, Math. Comp., 29, 475–483 (1975).
5. J. Nitsche, Finite Element Approximations to the One-Dimensional Stefan Problem, c.deBoored., Academic Press, New York (1978) pp. 119–142.
6. J. Nitsche, Computational methods in non-linear Mechanics, 341–360 (1980).
7. W. Walter, Differential- und Integralunglichungen, Springer-Verlag, Berlin (1964) pp. 12-13.
8. M. Wheeler, SIAM J. Numer. Anal., 10, 914-917 (1973).

8th Int. Coll. on Differential Equations, pp. 347–354
D. Bainov (Ed)

Optimal Checking in Reliability Theory

E. OMEY
EHSAL , Stormstraat 2, 1000 Brussels, Belgium,
Tel: (00 32) (0) 2 210 12 11, Fax: (00 32) (0) 2 217 64 64,
e-mail: omedward@prof.ehsal.be

Abstract. When a system fails, it usually fails at a random time T. It may be necessary to check the system from time to time to see whether it has failed or not. It is impossible to check too many times, because with each check, the cost of checking increases. On the other hand, if the system fails at time t_1 and we detect this failure only at time $t_2 > t_1$, then a loss occurs during this time interval. The longer the time between failure and its detection, the greater the loss. This implies that it is useful to check rather frequently.

In order to keep the total expected cost under control it will be necessary to find a balance between the cost of checking and the loss by not detecting a failure. The compromize between this two requirements is called an optimal checking schedule.

In the paper we discuss various checking schedules:

a. checking at fixed times $a, 2a, 3a, \ldots$;

b. checking for the first time at time b and then at fixed times $b + na$, $n = 1, 2, \ldots$;

c. checking at times $t(1)$, $t(2)$, $t(3), \ldots$;

d. checking according to a checking rate function $n(t)$;

e. checking according to a random checking scheme.

Results of this type can be applied in medical health control, reliabilty theory, inventory control, etc.

1. INTRODUCTION

When a system can fail at a random time T, it may be necessary to check the system from time to time to see whether it has failed or not. If the system fails at time t_1 and we detect this failure at time $t_2 > t_1$, then a loss occurs during this time interval. The longer the time between failure and its detection, the greater the loss. This implies that it is useful to check rather frequently. However, there is a cost for each check and so the number of checks must be kept relatively small. In order to keep the total expected costs under control it will be necessary to find the balance between the loss by checking and the loss by not detecting a failure. The compromize between this two requirements is called an optimal checking schedule.

Suppose we have a system such that the failure time T is a continuous random variable with distribution function $F(t)$. From time to time we check the system to see whether it has failed or not. The system is such that we can only see its failure upon checking the system. When the system is working incorrectly, it produces lower quality products.

Given $T = t$, let $N(t) =$ the number of checks in the interval $[0, t]$, let $c(1)$ denote the cost per check and let $c(I)$ denote the cost per time unit due to a failing system. If the system fails at time $T = t$, then the corresponding cost equals $C(t) = c(1)(N(t) + 1) + c(I)k(t)$ where $k(t)$ is the time between failure and its detection. The expected cost is $E = E(C(T))$. Note that we have $N(t)$ checks in which we find the system working and we need an extra check to find the system failing.

An optimum checking procedure is a specification of checking times $0 = t(0) < t(1) < t(2) < \ldots$ for which E is minimized.

2. CHECKS AT FIXED TIMES

2.1. Checks at times $a, 2a, \ldots, na, \ldots$

Suppose that we check the system at fixed times $a, 2a, \ldots, na, \ldots$. If the system fails at time t, we have to check the system $n+1$ times, where n is such that $na < t < (n+1)a$. Hence $n = [t/a]$ and the cost $C(t)$ equals

$$C(t) = (n+1)c(1) + c(I)k(t),$$

where $k(t) = (n+1)a - t$. Hence

$$C(T) = c(1)\frac{T}{a} + (c(1) + ac(I))k(T)/a.$$

It follows that

$$EC(T) = c(1)E\left(\frac{T}{a}\right) + (c(1) + ac(I))E\left(\frac{k(T)}{a}\right).$$

In order to minimize $E = E(C(T))$ w.r.t. the parameter a, we assume that the time $k(T)$ between failure and its detection equals $a/2$. Hence E approximately equals

$$E = c(1)E\left(\frac{T}{a}\right) + \frac{c(1) + ac(I)}{2}.$$

Minimizing w.r. to a we find that a^* satisfies

$$a^{*2} = \frac{2c(1)E(T)}{c(I)} \tag{2.1.1}$$

and consequently $E^* = \dfrac{c(1)}{2} + (2c(1)c(I)E(T))^{1/2}$.

Remark. Using the approximation $k(T) \approx a/2$ we find that $C(T) = c(1)\dfrac{T}{a} + \dfrac{c(1) + ac(I)}{2}$. Hence the variance of $C(T)$ is $\mathrm{Var}\,(C(T)) = \dfrac{c^2(1)\mathrm{Var}\,(T)}{a^2}$.

Examples. 1. Suppose that T is a r.v. with mean $E(T) = 500$.
If $c(1) = 10$ and $c(I) = 1$, we find $a^* = 100$ and $E^* = 105$.
If $c(1) = 10$ and $c(I) = 50$, we find $a^* = 14.14$ and $E^* = 712$.

2. Suppose that T is exponentially distributed with mean value $E(T) = 1/\alpha$. In this case we can calculate $E([T/a])$ exactly. We find that $E([T/a]) = \exp(-\alpha a)/(1 - \exp(-\alpha a))$ and consequently that $E = \{\alpha c(1) + c(I)(\alpha a - 1 + \exp(-\alpha a))\}/\{\alpha(1 - \exp(-\alpha a))\}$. Setting the derivative of E w.r.t. a equal to zero, we find the equation $\exp(\alpha a) - 1 - \alpha a = \alpha c(1)/c(I)$. Using a Taylor expansion we find that a^* satisfies (cf.(2.1.1))

$$a^{*^2} \approx \frac{2c(1)}{\beta c(I)}, \qquad \text{cf. (2.1.1).}$$

Numerical calculations show that this approximation is close to the exact solution.

2.2. Checks at times $b, b + a, b + 2a, \ldots$

Often systems have a small probability of failing in the beginning of their life. Only after some time defects occur with a higher probability. It may be interesting to start checking after a certain period b.

If we check for the first time at time $b > 0$ and from then on we check at times $b + a, b + 2a, \ldots$ we find:

i) if $t < b : C(t) = c(1) + c(I)(b - t)$;

ii) if $t > b : C(t) = c(1) + c(1)(t - b)/a + \{c(1) + ac(I)\}k(t)$ where $k(t) = [(t - b)/a] - (t - b)/a + 1$.

As in the previous case, for $t > b$, we approximate $k(t)$ by 1/2. Some calculations show that $E = E(C(T))$ approximately equals

$$E(C(T)) = \frac{c(1)}{2} + \left(c(1) + \frac{ac(I)}{2}\right)(1 - F(b)) +$$

$$+ c(1)\frac{E(T) - \int_0^b (1 - F(t))dt}{a} + c(I)\int_0^b F(t)dt.$$

Setting the derivatives of E w.r.t. a and b equal to zero we find that

$$a^2 c(I)\frac{1 - F(b)}{2} = c(1)\left(E(T) - \int_0^b (1 - F(t))dt\right) \tag{2.2.1}$$

and

$$\left(c(I) + \frac{c(1)}{a}\right)(1 - F(b)) + (c(1) + c(I)a)\frac{f(b)}{2} = c(I) \tag{2.2.2}$$

where we assumed that T has a density $f(t)$.

From (2.2.1) we find that

$$a^{*^2} = \frac{2c(1)\left(E(T) - \int_0^b (1 - F(t))dt\right)}{c(I)(1 - F(b))}. \tag{2.2.3}$$

Remark. Using (2.2.2) in the expression for E we find that

$$E(a^*, b) = \frac{c(1)}{2} + c(I) \int_0^b F(t)dt + c(1)(1 - F(b)) +$$

$$+ \sqrt{2c(1)c(I)(1 - F(b)) \left(E(T) - \int_0^b (1 - F(t))dt \right)}.$$

As b increases, the first term becomes larger while the last terms decrease. An optimal value of b should exist.

Examples. 1. If T is exponentially distributed with $E(T) = 1/\alpha$, then (2.2.1) and (2.2.2) lead to ${a^*}^2 = 2c(1)/\alpha c(I)$,

$$b^* = \frac{1}{\alpha} \log \left(\frac{c(I) + \alpha c(1)/2 + \sqrt{2\alpha c(1)c(I)}}{c(I)} \right),$$

and

$$E^* = \frac{c(1)}{2} + c(I)b^*.$$

Suppose for example that $E(T) = 500$. If $c(1) = 10$, $c(I) = 1$, we find $a^* = 100$, $b^* = 95.31$ and $E^* = 100.31$. If $c(1) = 10$, $c(I) = 50$, we find $a^* = 14.14$, $b^* = 14.04$ and $E^* = 707$.

2. If it is known that $F(t) = 0$ for $t < K$, one can choose $b = K$ as a first checking time. In this case we find

$$E = 2c(1) + (E(T) - K)\frac{c(1)}{a} + c(I)\frac{a}{2}$$

and an optimal a can be found as before.

3. If T is PARETO-distributed, for $t > 0$ and $\alpha > 1$, we have $\log(1 - F(t)) = -\alpha \log(1 + t)$. From (2.2.1) we obtain

$${a^*}^2 = \frac{2c(1)(1 + b)E(T)}{c(I)},$$

where $E(T) = 1/(\alpha - 1)$.

Using (2.2.2), b^* can be calculated numerically.

2.3. Checks at times $t(1), t(2), \ldots$

Suppose that the failure time distribution $F(t)$ has a density $f(t)$ and that we check at times $t(1), t(2), \ldots$.

In this case we have

$$E = \int_0^\infty \left\{ c(1)(N(t) + 1) + c(I)k(t) \right\} dF(t)$$

or

$$E = \sum_{k=0}^\infty \int_{t_k}^{t_{k+1}} \left\{ c(1)(k + 1) + c(I)(t_{k+1} - t) \right\} f(t)dt.$$

Taking derivatives w.r.t. $t(k)$, we find the equations

$$d(k) = \frac{F(t(k)) - F(t(k-1))}{f(t(k))} - \frac{c(1)}{c(I)} \tag{2.3.1}$$

where $d(k) = t(k+1) - t(k)$. When $f(t(k)) = 0$, then $t(k)$ is the last check which has to be performed. Given $t(1)$, we can calculate $t(k)$ by using (2.3.1).

From Barlow and Proshan [1] we recall:

Lemma. (Barlow and Proshan [1], Theorems 3.2 and 3.3)

a) If $f(x) > 0$ for $x > 0$ and if the failure time distribution satisfies the following condition:

$$\text{(C)} \qquad \text{for all} \quad y > 0, \qquad \frac{F(x+y) - F(x)}{f(x)} \quad \text{is decreasing in } x.$$

Then the optimum checking intervals $d^*(k)$ are nonincreasing.

b) Suppose the conditions of (a) hold and suppose $f(x+y)/f(x)$ is strictly decreasing for $x \geq 0$, $y > 0$.

Let $t^*(k)$ and $t(k)$ denote the optimum checking times and an arbitrary other checking schedule respectively.

If $t(1) > t^*(1)$, then for some integer n, $d(n) > d(n-1)$.

If $t(1) < t^*(1)$, then for some integer n, $d(n) < 0$.

Under the conditions of the previous result Barlow and Proshan use the following algorithm for obtaining the optimum inspection schedule:

1. Choose $t(1)$ so that $c(1) = c(I) \int_0^{t_1} F(t)dt$;
2. Use (2.3.1) to calculate $t(2), t(3), \ldots$ and $d(n)$;
3. If for some n, $d(n) > d(n-1)$, then reduce $t(1)$ and repeat step 2; if for some n, $d(n) < 0$, then increase $t(1)$ and repeat step 2;
4. Continue.

3. CHECKS ACCORDING TO A DENSITY $n(t)$

Let $n(t)$ denote a nonnegative function and let $n(t)dt$ denote the number of checks per time unit.

In this case the number of checks in the interval $[0, t]$ equals $N(t)$ where

$$N(t) = \int_0^t n(t)dt.$$

The time between two checks is equal to $1/n(t)$. In practice, the first check then happens at time $t(1) = 1/n(0)$; the next checks occur at times $t(k+1) = t(k) + 1/n(t(k))$.

If failure occurs at time $T = t$, the cost of checking equals $C(t) = c(1)N(t) + c(I)/2n(t)$ where we assumed that a failure which occurs at time a and is detected at time b induces an average cost $c(I)(b-a)/2$. It follows that

$$E = c(1)E(N(T)) + \frac{c(I)E\left(\frac{1}{n(T)}\right)}{2}.$$

We seek the function $n(t)$ that minimizes E. To this end we rewrite E as

$$E = \int_0^\infty (c(1)N(t) + c(I)/2N'(t))f(t)dt =$$

$$= \int_0^\infty H(N(t), N'(t), t)dt.$$

The usual Euler equation leads to

$$\frac{\partial H}{\partial N} - \frac{d}{dt}\left(\frac{\partial H}{\partial N'}\right) = 0$$

so that $c(1)f(t) + \frac{d}{dt}\left(\frac{c(I)f(t)}{2n^2(t)}\right) = 0.$

Integrating with respect to t leads to

$$\frac{c(I)f(t)}{2n^2(t)} = c(1)(a - F(t)),$$

where a is an integrating constant.

It follows that $n^2(t) = \dfrac{c(I)f(t)}{2c(1)(a - F(t))}$.

Obviously a should satisfy $F(t) < a$.

Using the expression for $n(t)$ in E, we find that

$$E = \sqrt{\frac{c(1)c(I)}{2}} \int_0^\infty \frac{(1 - F(t) + a - F(t))\sqrt{f(t)}}{\sqrt{a - F(t)}} dt.$$

For any value of F, the integrand in the expression for E is minimized when $a = 1$. It follows that

$$E^* = \sqrt{\frac{c(1)c(I)}{2}} \int_0^\infty \sqrt{1 - F(t)}\sqrt{f(t)}dt$$

and the corresponding value of $n(t)$ is given by:

$$n^{*2}(t) = \frac{c(I)f(t)}{2c(1)(1 - F(t))}.$$

Examples. 1. In the case where T is exponentially distributed with $E(T) = 1/\alpha$, we find that $n^2(t) = c(I)\alpha/2c(1)$.

Hence, the time between two checks is $1/n(t)$, a constant. Note that this corresponds to the results of section 2.1.

2. In the case where T follows a $UNIF(0, K)$-distribution, we find that $n^2(t) = c(I)/(2c(1)(K - t))$.

3. For the Pareto-case we find $n^{*2}(t) = c(I)\alpha/2c(1)(1 + t)$.

4. RANDOM CHECKS

Suppose that we check the system at times which are determined by a random scheme. We consider two policies.

4.1. Policy 1. Let $A, A(1), A(2), \ldots$ denote a sequence of i.i.d. random variables independent of T, and define $S(n)$ by $S(n) = A(1) + A(2) + \ldots + A(n)$. In this randomized policy we check the system at times $S(n)$.

If $T = t$ we have to check the system $N(t) + 1$ times where $N(t)$ denotes the value of n for which $S(n) \leq t < S(n+1)$.

The costs in this case are as follows. Given $T = t$ we have

$$C(t|T = t) = c(1)(N(t) + 1) + c(I)(S(N(t) + 1) - t).$$

In probability theory the process $N(t)$ is well-known and it is called the renewal counting process. It is well-known that

$$H(t) = E(N(t)) = \sum_{1}^{\infty} U^{*^n}(t)$$

where $U(t)$ is the distribution function of the r.v. A and where U^{*^n} denotes the n-fold convolution of U with itself.
Using Wald's lemma we find that $E(S(N(t) + 1)) = E(A)(H(t) + 1)$, where $E(A)$ is the mean interchecking time.
We find that

$$E(C(t|T = t)) = c(1)(H(t) + 1) + c(I)\{E(A)(H(t) + 1) - t\} =$$
$$= (c(1) + E(A)c(I))(H(t) + 1) - c(I)t.$$

The next inequalities can be used to estimate $E = E(C(T))$.

Lemma. (Barlow and Proschan [1], Theorem 2.5)
(a) For any renewal process, $H(t) \geq t/E(A) - 1$.
(b) Suppose that for all $t \geq 0$, $\int_t^{\infty}(1 - U(x))dx \leq E(A)(1 - U(t))$,
then $t \leq E(A)(H(t) + 1) \leq t + E(A)$.

Using the Lemma, we obtain

$$c(1)\frac{t}{E(A)} \leq E(C(t|T = t)) \leq c(1)\left(1 + \frac{t}{E(A)}\right) + c(I)E(A)$$

and hence

$$c(1)\frac{E(T)}{E(A)} \leq E(C(T)) \leq c(1) + c(1)\frac{E(T)}{E(A)} + c(I)E(A).$$

Note that using $a = E(A)$, the upperbound is similar to the results of section 2.1.

Example. Suppose that A has an exponential distribution with $E(A) = a$. In this case $H(t) = t/a$ and we have

$$E(C(t|T = t)) = c(1) + c(1)\frac{t}{a} + c(I)a.$$

4.2. Policy 2. A variant of the above setting is based on so-called delayed renewal theory. Let B denote a r.v. and let $A, A(2), A(3), \ldots$ denote i.i.d. r.v. independent of B.

Define $S(1) = B$ and $S(n+1) = S(n) + A(n+1)$. The policy again consists of checking the system at times $S(n)$.

Define $M(t)$ as follows: $M(t) = n$ iff $S(n) \leq t \leq S(n+1)$. This process is known as the delayed renewal process. The mean of $M(t)$ is denoted by $W(t) = E(M(t))$.

As in the previous case we have

$$C(t|T=t) = c(1)(M(t)+1) + c(II)(S(M(t)+1) - t).$$

Since in this case $E(S(M(t)+1)) = E(B) + E(A)W(t)$ it follows that

$$E(C(t|T=t)) = c(1)(W(t)+1) + c(II)(E(B) + E(A)W(t) - t).$$

Now we choose the r.v. B in the following special way. We choose B so that the d.f. $V(x)$ of B equals

$$V(x) = \frac{1}{E(A)} \int_0^x (1 - U(t))dt.$$

In this case (see e.g. Ross [2], Theorem 3.15) it can be shown that $W(t) = t/E(A)$ so that

$$E(C(t|T=t)) = c(1) + c(1)\frac{t}{E(A)} + c(II)E(B).$$

With our special choice of B, it can be calculated that

$$E(B) = \frac{E(A^2)}{2E(A)} = \frac{\sigma^2(A) + E^2(A)}{2E(A)}$$

and it follows (with $E = E(C(T))$) that

$$E = c(1) + c(1)\frac{E(T)}{E(A)} + c(II)\frac{E(A)}{2} + c(II)\frac{\sigma^2(A)}{2E(A)}.$$

In order to minimize E, it seems to be appropriate to choose A so that $\sigma^2(A) = 0$. In this case A is a constant, say $A = a$, and B is a r.v. following a $UNIF(0,a)$-distribution.

ACKNOWLEDGEMENTS

The results presented here were partly supported by the Ministry of Education of the Flemish Government, International Department, Grant 13DA-B/CIA/Bulgarije97/as.

References

1. R.E. Barlow and F. Proschan, Mathematical Theory of Reliability, Wiley, New York, (1965).
2. S.M. Ross, Applied Probability Models with Optimization Applications, Holden-Day, San Francisco, (1970).
3. J.B. Keller, Management Sciences, **21**, 256–260 (1974).

8th Int. Coll. on Differential Equations, pp. 355-362
D. Bainov (Ed)

Integration of Some Constitutive Relations of Plain Strain Elastoplasticity Using Block-Embedded Modified Runge-Kutta Methods

G. PAPAKALIATAKIS
Section of Mechanics, Department of Civil Engineering, School of Engineering, Democritus University of Thrace, GR-67100 Xanthi, Greece

T.E. SIMOS
Section of Mathematics, Department of Civil Engineering, School of Engineering, Democritus University of Thrace, GR-67100 Xanthi, Greece

Abstract. In this paper we solve via Runge-Kutta methods a set of coupled ordinary differential equations in which the constitutive relations of Tresca and Mohr-Coulomb materials in plain strate elastoplasticity has been transformed. In our comparisons we have included a low order block-embedded modified Runge-Kutta method with phase-lag of orders six, eight, ten and twelve. Conclusions on efficiency of the used methods have been made, based on extensive numerical tests.
Keywords: Plain strain elastoplasticity, Tresca constitutive relation, Mohr-Coulomb, constitutive relation, Runge-Kutta methods, phase-lag.

1 Introduction

In the mathematical theory of plasticity two types of theories have been advanced. The one which is called **deformation theory** or **total deformation theory** is based on the hypothesis that the plastic strain is a solely function of the stress. In the other type, which is referred as **flow** or **incremental theory** it is stated that the plastic strain increments are related to the stress state and stress increment. In general these relations are not integrable and the integral depends on the loading path. Using finite element method and incremental techniques, the flow theory can easily be applied to solve plasticity problems.

In incremental theory the constitutive behaviour of an elastoplastic material may be expressed with an incremental stress-stain relation of the form

$$\dot{\sigma} = \mathbf{D_{ep}}\dot{\epsilon} \tag{1}$$

where $\dot{\sigma}$ denotes the stress increment vector, $\dot{\epsilon}$ is the strain increment vector and the superior dot represents a derivative with respect to time. Also $\mathbf{D_{ep}}$ denotes the elastoplastic constitutive matrix, that for the perfectly plastic Tresca and Mohr-Coulomb structural material to be considered in this work depends only on the components of the stress within the element.

Using a finite element computer program based on flow theory of plastisity the nodal displacement increments and the strain increments are computed that corresponds to the nodal forces. Once these are known, the above equation can be used to find the stress increments. This equation constitutes a set of ordinary differential equations and the integration of this set presents difficultes because the matrix $\mathbf{D_{ep}}$ is a highty non-linear function of the stresses.

Assuming that the strain rates are constant, equation (1) can be analysed to a of coupled ordinary differential equations of the form

$$\dot{\sigma} = \mathbf{D_{ep}} \frac{\Delta \epsilon}{\Delta \mathbf{t}} \tag{2}$$

where $\Delta \mathbf{t}$ is the time interval and $\mathbf{t_0} \leq \mathbf{t} \leq \mathbf{t_0} + \Delta \mathbf{t}$. This set of equations defines a standard initial-value problem since $\dot{\sigma}_\mathbf{0} = \dot{\sigma}(\mathbf{t_0})$ and $\Delta \epsilon$ are known and the matrix $\mathbf{D_{ep}}$ is the function of stresses σ. Solving the above system of ordinary differential equations we find an approximation for the stresses.

Sloan and Booker [1] has compared the well known embedded Runge-Kutta method of Fehlberg [2], the embedded Runge-Kutta method of Dormand and Prince [3] and the embedded Runge-Kutta method of England [4]. All these methods are of order four and five. The conclusion of the work of Sloan was that for this class of problems the embedded Runge-Kutta Dormand and Prince method was the most efficient.

The purpose of this paper is the numerical solution, using high-order embedded Runge-Kutta methods and low-order modified Runge-Kutta methods with minimal phase-lag, of the initial-value problems of the form (2). We use for comparison the embedded Runge-Kutta methods of order five and six of Dormand and Prince [5], of Prince and Dormad [6], of Sharp and Verner [7], of Verner [8-11]. We use, also, the embedded Runge-Kutta method of order six and seven of Verner [9,11] and of Sharp and Verner [7] and the embedded Runge-Kutta methods of order seven and eight of Prince and Dormand [6] and of Verner [9]. Finally we use the block-embedded modified Runge-Kutta Fehlberg method with phase-lag of orders 6-12 produced in [12]. This method is based on the well known Runge-Kutta Fehlberg method RKF2(3).

2 Increments stress-strain relations for perfectly plastic materials

During an infinitesimal increment of stress, changes of strain are assumed to be divisible into elastic and plastic parts. Thus

$$\dot{\epsilon} = \dot{\epsilon}^{\mathbf{e}} + \dot{\epsilon}^{\mathbf{P}} \tag{3}$$

where $\dot{\epsilon}^{\mathbf{l}}$ denotes the elastic and $\dot{\epsilon}^{\mathbf{P}}$ the plastic strain increment.

The elastic strain inrements are related to the stress increments by the Hooke's law as usual, by a symmetric matrix of constants $\mathbf{D}$, so that

$$\dot{\epsilon}^{\mathbf{e}} = \mathbf{D}^{-1} \dot{\sigma} \tag{4}$$

Using the normality principle of the Von-Mises flow rule, the plastic strain increment $\dot{\epsilon}^{\mathbf{P}}$ can be written as

$$\dot{\epsilon}^{\mathbf{P}} = \lambda \frac{\partial \mathbf{F}}{\partial \sigma} \tag{5}$$

where **F** denotes the yield surface which for a perfectly plastic material may be written as a function of σ (yielding of material can occur only if the stress σ satisties the yied criterion $\mathbf{F}(\sigma) = \mathbf{0}$). Also *lambda* denotes a proportionality constant.

A reduction of the restriction of the above rule can be obtained by specifying a plastic potential

$$\mathbf{Q} = \mathbf{Q}(\sigma) \tag{6}$$

which defines the plastic strain increment as

$$\dot{\epsilon}^{\mathbf{P}} = \lambda \frac{\partial \mathbf{Q}}{\partial \sigma} \tag{7}$$

The particular case of $\mathbf{Q} = \mathbf{F}$ is known as associated plasticity flow rule and the matrix $\mathbf{D_{ep}}$ is symmetric. Otherwise the material is said to have an non associated flow rule and the matrix $\mathbf{D_{ep}}$ is in general unsymmetric.

Comparing equations (3), (4) and (7) we can write

$$\dot{\epsilon} = \mathbf{D}^{-1}\dot{\sigma} + \lambda \frac{\partial \mathbf{Q}}{\partial \sigma} \tag{8}$$

When plastic yield is occurring the stress are on the yield surface. Differentiating we can write therefore

$$\left\{\frac{\partial \mathbf{F}}{\partial \sigma}\right\}^T \dot{\sigma} = 0 \tag{9}$$

The indeterminate constant *lambda* can be eliminated. This results in an explicit expansion which determines the stress changes in terms of imposed strain changes with

$$\dot{\sigma} = \mathbf{D_{ep}}\dot{\epsilon}$$

$$\mathbf{D_{ep}} = \mathbf{D} - \mathbf{D}\left\{\frac{\partial \mathbf{Q}}{\partial \sigma}\right\}\left\{\frac{\partial \mathbf{F}}{\partial \sigma}\right\}^{\mathbf{T}} \mathbf{D}\left[\left\{\frac{\partial \mathbf{F}}{\partial \sigma}\right\}^{\mathbf{T}} \mathbf{D}\left\{\frac{\partial \mathbf{Q}}{\partial \sigma}\right\}\right]^{-1} \tag{10}$$

The elasto-plastic matrix $\mathbf{D_{ep}}$ takes the place of the elasticity matrix $\mathbf{D_T}$ in incremental analysis.

For the yield surfaces we have the following relations

Tresca relation

$$\mathbf{F} = 2\bar{\sigma} cos\theta - \mathbf{Y}(\mathbf{k}) = \mathbf{0} \tag{11}$$

where $\mathbf{Y}(\mathbf{k})$ is the yield stress from uniaxial tests.

Mohr-Coulomb relation

$$\mathbf{F} = \sigma_m sin\phi + \bar{\sigma} cos\theta - \frac{\bar{\sigma}}{\sqrt{3}} sin\phi sin\theta - c cos\phi = \mathbf{0} \tag{12}$$

where $c(k)$ and $\phi(k)$ are the cohesion and angle of friction, respectively, which could depend on some strain hardening parameter k and

$$\sigma_m = \frac{J_1}{3} = \frac{\sigma_x + \sigma_y + \sigma_z}{3}$$

$$\bar{\sigma} = J_2^{\frac{1}{2}} = \left[\frac{1}{2}\left(\mathbf{s}_x^2 + \mathbf{s}_y^2 + \mathbf{s}_z^2\right) + \tau_{xy}^2 + \tau_{yz}^2 + \tau_{xz}^2\right]^{\frac{1}{2}}$$

$$\theta = \frac{1}{3} sin^{-1}\left[-\frac{3\sqrt{3}J_3}{2\bar{\sigma}^3}\right] \tag{13}$$

with $-\frac{\pi}{6} < \theta < \frac{\pi}{6}$, where

$$\begin{aligned} \mathbf{J}_3 &= \mathbf{s}_x\mathbf{s}_y\mathbf{s}_z + 2\tau_{xy}\tau_{yz}\tau_{xz} - \mathbf{s}_x\tau_{yz}^2 - \mathbf{s}_y\tau_{xz}^2 - \mathbf{s}_z\tau_{xy}^2 \\ \mathbf{s}_x &= \sigma_x - \sigma_m, \ \mathbf{s}_y = \sigma_y - \sigma_m, \ \mathbf{s}_z = \sigma_z - \sigma_m \end{aligned} \tag{14}$$

3 Embedded Runge-Kutta methods

To solve the system of coupled ordinary differential equations (2) via embedded Runge-Kutta methods we must estimate at each step size the local truncation error. To do this we must have one method of lower order q and a method of higher order $q+1$. The lower order method produces an approximation of the stress σ, while the higher order method produces an approximation of the stress $\hat{\sigma}$ given by

$$\begin{aligned} \sigma &= \sigma_0 + \sum_{i=1}^{m} b_i \Delta\sigma_i \\ \hat{\sigma} &= \sigma_0 + \sum_{i=1}^{m} \hat{b}_i \Delta\sigma_i \end{aligned} \tag{15}$$

where

$$\begin{aligned} \Delta\sigma_1 &= \mathbf{D}_{ep}(\sigma_0) h_i \Delta\epsilon \\ \Delta\sigma_j &= \mathbf{D}_{ep}\left(\sigma_0 + \sum_{k=1}^{j-1} a_{j,k}\Delta\sigma_k\right) h_n \Delta\epsilon, \ j = 1(1)m \end{aligned} \tag{16}$$

where b_i, $\hat{b}_i$ and $a_{j,k}$, $j = 2(1)m$, $k = 1(1)j-1$, $i = 1(1)m$ are the coefficients of the appropriate Runge-Kutta methods, h_n is the steplength of the current subinterval.

The estimate of the relative error in **sigma** in the current subinterval is given by

$$E_n = \frac{\|\hat{\sigma} - \sigma\|}{\|\hat{\sigma}\|} \tag{17}$$

The current steplength is accepted if the relative error is less than a tolerance TOL specified by the user. If the relative error is greater or equal to the specified tolerance then the steplength is rejected. In both cases, the steplength of the next subinterval h_{n+1} is given by

$$h_{n+1} = w h_n, \ w > 0 \tag{18}$$

where w is given by

$$w = 0.9\left(\frac{TOL}{E_n}\right)^{1/q} \tag{19}$$

where q is the order of the Runge-Kutta method used.

To start the integration we apply the methods (15) with $h_n = 1$. If $E_n \geq TOL$ then this steplength will be rejected and the procedure is repeated without updating the stresses. It is obvious that in this case only an additional evaluation of the matrix $\mathbf{D}_{ep}$ is required. If the steplength is accepted, i.e. if $E_n < TOL$ then the stresses are calculated via methods (15) and also the new steplength of the next subinterval is computed. In our tests the relative error tolerances, TOL, are equal to 10^{-s}, $s = 3(1)6$. But, in this paper we present the results for 10^{-5}. We have similar conclusion for the other tolerances. For the new low order block-embedded modified Runge-Kutta method with phase-lag of orders six, eight, ten and twelve we use the step control procedures defined by (18) and (19) with the following modifications. To start the integration we apply the block-embedded method with phase-lag of order six, with $h_n = 1$. We calculate E_n using the methods of the Block Embedded method defined with phase-lag of orders six and eight. If $E_n \geq TOL$ then we recalculate E_n using the methods of the Block Embedded method with phase-lag of orders eight and ten. If $E_n \geq TOL$ then we recalculate E_n using the methods of the Block Embedded method with phase-lag of orders ten and twelve. If $E_n \geq TOL$ this steplength will be rejected and the procedure is repeated without updating the stresses. It is obvious that in this case only an additional evaluation of the matrix $\mathbf{D_{ep}}$ is required. If the steplength is accepted, i.e. if $E_n < TOL$ using any of the couples of the methods described above then the stresses are calculated via Block-Embedded Runge-Kutta Fehlberg Method RKBEMF2.

4 Numerical tests

In the above section we have described the general embedded Runge-Kutta method for integrating Tresca and Mohr-Coulomb constitutive relations.

The numerical procedure for examing the efficiency of the proposed methods is fully described in [1].

The accuracy of the Runge-Kutta methods is defined from the errors in the elasto-plastic stresses which are computed from the formula

$$E = \frac{\|\sigma - \sigma_{\mathbf{e}}\|_\infty}{\|\sigma_{\mathbf{e}}\|_\infty} \tag{20}$$

where $\|.\|_\infty$ is the maximum norm and $\sigma_{\mathbf{e}}$ are the stresses calculated from the exact solution which for the Tresca constitutive law is fully defined in [1]. For the Mohr-Coulomb constitutive law (in which there isn't analytical solution) the matrix $\sigma_{\mathbf{e}}$ is calculated using a high order Runge-Kutta method (RKDP87 defined in [6]) with constant stepsizes equal to 0.01.

In our numerical tests we measure the average error and the maximum error given by

$$E_{av} = \frac{1}{N}\sum_{n=1}^{N} E_n \tag{21}$$

$$E_{max} = max_{1 \leq n \leq N}\,(E_n) \tag{22}$$

where E_n is the error calculated from (17) for each of the N subintervals.

4.1 Tresca constitutive relation

The various methods are used to integrate the Tresca constitutive law in order to approximate the elastic stress in the plane $Z = 0$. The initial stresses are chosen to be equal to $\sigma_x = \sigma_y = 0$ and $\tau_{xy} = c$. The material parameters are $\nu = 0.49$, $G = 100$ and $c = 1$. We note here that the material parameters are appeared in the elastoplastic constitutive matrix $\mathbf{D_{ep}}$.

In Table 2 we present the average error, the maximum error and the real time of computation for the integration of the Tresca constitutive law using the Runge-Kutta methods presented in Table 1. The relative error tolerance, TOL, is equal to 10^{-5}.

a/a	Method	a/a	Method
MI	RKDP6(5)8M of [6]	MVIII	RKVE6(5)9a of [10]
MII	RKVE5(6) of [9]	MIX	CIRK65 of [7]
MIII	RKDP6(5)8C of [5]	MX	RKVE6(7) of [9]
MIV	RKDP6(5)8S of [5]	MXI	RKVE7(6) of [11]
MV	RKVE6(5)9c of [8]	MXII	RKVE7(8) of [9]
MVI	RKVE6(5)8M of [8]	MXIII	RKDP8(7) of [6]
MVII	RKVE5(6) of [11]	MXIV	CIRK76 of [7]
		MXV	RKBEMF2 produced in [12]

Table 1: Runge-Kutta methods used for the integration of the system of coupled ordinary differential equations

Method	E_{av}	E_{max}	Real time of computation (secnds)
MI	1.410^{-6}	9.710^{-6}	55
MII	3.210^{-6}	1.510^{-5}	65
MIII	7.010^{-7}	4.110^{-6}	85
MIV	4.110^{-7}	2.810^{-6}	50
MV	1.610^{-7}	8.610^{-7}	135
MVI	2.010^{-7}	9.710^{-7}	78
MVII	3.010^{-6}	9.710^{-6}	96
MVIII	7.010^{-8}	2.810^{-7}	81
MIX	1.810^{-7}	1.210^{-6}	108
MX	1.510^{-6}	9.510^{-6}	93
MXI	3.010^{-8}	1.510^{-7}	113
MXII	4.510^{-8}	4.710^{-7}	83
MXIII	6.610^{-8}	1.010^{-6}	58
MXIV	7.510^{-7}	8.710^{-6}	100
MXV	8.910^{-10}	1.110^{-9}	109

Table 2: Average error, maximum error and real time of computation for the integration of the Tresca constitutive law using the Runge-Kutta methods presented in Table 1

Method	E_{av}	E_{max}	Real time of computation (secnds)
MI	7.710^{-7}	8.110^{-6}	31
MII	3.110^{-6}	1.310^{-5}	73
MIII	4.510^{-7}	3.610^{-6}	26
MIV	6.410^{-7}	8.810^{-6}	26
MV	1.510^{-7}	9.110^{-7}	66
MVI	1.310^{-7}	9.510^{-7}	47
MVII	3.010^{-6}	1.210^{-5}	49
MVIII	6.010^{-8}	7.210^{-7}	71
MIX	9.010^{-8}	1.110^{-6}	84
MX	1.810^{-6}	2.110^{-5}	53
MXI	1.710^{-7}	2.410^{-6}	25
MXII	6.010^{-8}	9.510^{-7}	42
MXIII	7.010^{-8}	1.810^{-6}	25
MXIV	5.710^{-7}	1.110^{-6}	48
MXV	6.310^{-10}	4.210^{-9}	75

Table 3: Average error, maximum error and real time of computation for the integration of the Mohr-Coulomb constitutive law using the Runge-Kutta methods presented in Table 1

4.2 Mohr-Coulomb constitutive relation

The various methods are used, also, to integrate the Mohr-Coulomb constitutive law in order to approximate the elastic stress in the plane $Z = 0$. The initial stresses are chosen to be equal to $\sigma_x = \sigma_y = 0$ and $\tau_{xy} = ccos\phi$. The material parameters are $\nu = 0.3$, $G = 100$, $c = 1$ and $\phi = 30^o$.

In Table 3 we present the average error, the maximum error and the real time of computation for the integration of the Mohr-Coulomb constitutive law using the Runge-Kutta methods presented in Table 1. The relative error tolerance, TOL, is equal to 10^{-5}.

From the results presented we have that the most rapid methods are the method RKDP6(5) 8S of [5] for the Tresca criterion and the methods RKVE7(6) of [11] and RKDP8(7) of [6] for the Mohr-Coulomb criterion. The most accurate methods are the RKVE7(6) of [11] and the RKBEMF2 of [12] for the Tresca criterion. We note that for the Mohr-Coulomb criterion the method RKVE7(6) of [11] is also very accurate but the most accurate methods are the methods RKVE6(5)9a of [10], RKVE7(8) of [9] and RKBEMF2 of [12]. Because of the significant interest for the accuracy of the used methods we propose the methods RKVE6(5)9a of [10], RKVE7(8) of [9], RKBEMF2 of [12] and RKVE7(6) of [11] as the most efficient methods for these problems.

References

1. S.W. Sloan and J.R. Booker, Intern. J. Numer. Meth. Engrg., 33, 163–196 (1992).

2. E. Fehlberg, Computing, 6, 61–71 (1970).

3. J.R. Dormand and P.J. Prince, J. Comput. Appl. Math., 6, 19–26 (1980).

4. R. Enlgand, Comput. J, 12, 166–170 (1969).

5. J.R. Dormand and P.J. Prince, J. Comput. Appl. Math., 15, 203–211 (1986).

6. P.J. Prince and J.R. Dormand, J. Comput. Appl. Math., 7, 67–75 (1981).

7. P.W. Sharp and J.H. Verner, SIAM J. Scien. Comput., 31, 1169–1190 (1994).

8. J.H. Verner, A Comparison of Some Runge-Kutta Formula Pairs Using DETEST, Dept. Mathematics and Statistics, Queen's University, Kingston, Ontario, Canada (1990).

9. J.H. Verner, SIAM J. Numer. Anal., 15, 772–790 (1978).

10. J.H. Verner, SIAM J. Numer. Anal., 28, 496–511 (1991).

11. J.H. Verner, Annals Numer. Math., 1, 225–244 (1994).

12. G. Papakaliatakis and T.E. Simos, Numerical Methods for the Integration of Tresca and Mohr-Coulomb Relations in Plane Strain Elastoplasticity, Section of Mechanics, Department of Civil Engineering, University of Thrace (1996).

13. P.J. van der Houwen and B.P. Sommeijer, SIAM J. Numer. Anal., 24, 595–617 (1987).

14. T.E. Simos, Comput. Math. Appl., 26, 43–49 (1993).

15. E. Fehlberg, Low Order Classical Runge-Kutta Formulas with Step-size Control and Their Application to Some Heat Transfer Problems, NASA TR R-315 (1969).

8th Int. Coll. on Differential Equations, pp. 363–371
D. Bainov (Ed)

Numerical Study of the Homicidal Chauffeur Game

V.S. PATSKO
Institute of Mathematics and Mechanics, Ekaterinburg, Russia

V.L. TUROVA
Technical University of Munich, Munich, Germany

Abstract. The Isaacs' homicidal chauffeur differential game is considered. In this game a chauffeur (the pursuer P) minimizes the capture time of a pedestrian (the evader E). The objective of the pedestrian is to avoid the capture or to maximize the capture time. The capture occurs when the distance between the players becomes less or equal than a given value (the capture radius). The pursuer's control is the rate of turn, the evader governs choosing directions of his velocity. The velocity of the pursuer is greater than that one for the evader but his maneuverability is bounded. Numerical solutions to this problem are obtained using an algorithm proposed by the authors for computing level sets of the value function.

Keywords: *Differential games, time-optimal control, homicidal chauffeur game*

Introduction

The homicidal chauffeur game was formulated by R.Isaacs more than thirty years ago [1]. Since that time, many authors studied this problem in various ways. The most complete qualitative solution was given in the PhD dissertation of A.Merz [2]. In this work, the region of parameters of the problem was divided into certain subregions. In each subregion, the structure of the solution was described.

Very often (see [3–5]) the dynamics of the homicidal chauffeur game was used, but the statement of the problem differed from the one of Isaacs. In [3], for example, a surveillance-evasion differential game of degree with the pursuer's detection zone in the shape of a circle was considered.

Many papers on differential games are devoted to the development of algorithms for solving nonlinear differential games in the plane [6–10]. In some of these papers (see, i.e., [10]), the homicidal chauffeur game is used to demonstrate the efficiency of the algorithms.

In this paper, the homicidal chauffeur game is investigated using the algorithm proposed by the authors for computing level sets of the value function. Our methods are based on the general theory of differential games [11, 12]. The algorithm is a natural extension of the algorithms from [13], and exploits ideas of the algorithms [14–16] for linear time-optimal differential games in the plane. Some experience [13, 14, 17, 18] in solving differential games of kind [1] in the plane helps to found out very complicated types of solutions and to verify the solution's validity. The computation results are consistent with those obtained by A.Merz.

It is the well-known fact due to works of R.Isaacs, J.V.Breakwell, and A.Merz that the solution of the homicidal chauffeur game may contain various types of singularities inherent to differential games: switching, dispersal, equivocal, universal, and focal lines.

The our algorithm's peculiarity consisting in the use of reverse time extremal trajectories makes possible to recognize types of singular lines.

Statement of the problem

The dynamics of the homicidal chauffeur game in reduced coordinates has the form

$$\begin{aligned} \dot{x}_1 &= -\frac{w^{(1)}}{R} x_2 \, \varphi + v_1 \\ \dot{x}_2 &= \frac{w^{(1)}}{R} x_1 \, \varphi + v_2 - w^{(1)}, \qquad | \varphi | \leq 1, \quad v \in Q. \end{aligned} \tag{1}$$

Here $(x_1, \ x_2)'$ is the state vector, $w^{(1)}$ and R are constants which have the sense of the pursuer's velocity and the minimal radius of turn, respectively. The objective of the control φ of the first player is to minimize the time of attaining a given terminal set M. The objective of the control $v = (v_1, v_2)'$ of the second player is to maximize this time. So, the payoff of the game is the time of attaining the terminal set.

Let $\theta \geq 0$. The level set (the Lebesgue set) of the value function is denoted by $W(\theta, M)$. This is the set of all points in the plane such that the first player can guarantee the transition of trajectories of the system (1) to the terminal set M within the time θ. In this paper, the basic idea of the algorithm for computing sets $W(\theta, M)$ is described. The classical formulation of the homicidal chauffeur game assumes that the sets M and Q are circles of the radii l (capture radius) and $w^{(2)}$, respectively, with the centers at the origin. It is assumed that $w^{(1)} > w^{(2)}$. With the algorithm proposed, the sets $W(\theta, M)$ can be computed for arbitrary sets M and Q in the plane. This enable us to obtain new types of solutions and to study some interesting cases.

The algorithm

The algorithm is based on ideas of the algorithms for linear time-optimal game problems [14–16]. The set $W(\theta, M)$ is formed via a step-by-step backward procedure giving a sequence of embedded sets

$$W(\Delta, M) \subset W(2\Delta, M) \subset W(3\Delta, M) \subset \ldots \subset W(i\Delta, M) \subset \ldots \subset W(\theta, M). \tag{2}$$

Here Δ is the step of the backward procedure. Each set $W(i\Delta, M)$ consists of all initial points such that the first player brings system (1) into the set $W((i-1)\Delta, M)$ within the time duration Δ. We put $W(0, M) = M$.

The crucial point of the algorithm is the computation of "fronts". The front F_i (Figure 1) is the set of all points of $\partial W(i\Delta, M)$ for which the minimum guaranteeing time of the achievement of $W((i-1)\Delta, M)$ is equal to Δ. For other points of $\partial W(i\Delta, M)$ the optimal time is less than Δ. The line $\partial W(i\Delta, M) \setminus F_i$ possesses the properties of the barrier [1]. The front F_i is designed using the previous front F_{i-1}. For the first step of the backward procedure, F_0 coincides with the usable part [1] Γ_0 of the boundary of M.

Let us explain how the fronts can be constructed. Denote $p(x) = (-x_2, x_1)' \cdot w^{(1)}/R$, $g = (0, -w^{(1)})'$. Using this notation, the equations (1) can be rewritten as follows: $\dot{x} = p(x)\varphi + v + g$. Suppose the front F_{i-1} is a smooth curve. Let x_* be an arbitrary point of F_{i-1}, and ℓ is the normal vector to the front at x_*. Let $\varphi^\circ = \arg\min_{|\varphi| \leq 1} \ell' p(x_*) \varphi$, $v^\circ = \arg\max_{v \in Q} \ell' v$. We call φ°, v° the extremal controls. The controls φ° and v° are chosen from the conditions of minimizing and, respectively, maximizing the projection of the velocity vector of (1) onto the direction ℓ. If the vector x_* is collinear to ℓ, then any control $\varphi \in [-1,1]$ is extremal. If Q is a polygon in the plane, and ℓ is collinear to some normal vector to an edge $[q_1, q_2]$ of Q, then any control $q \in [q_1, q_2]$ is extremal.

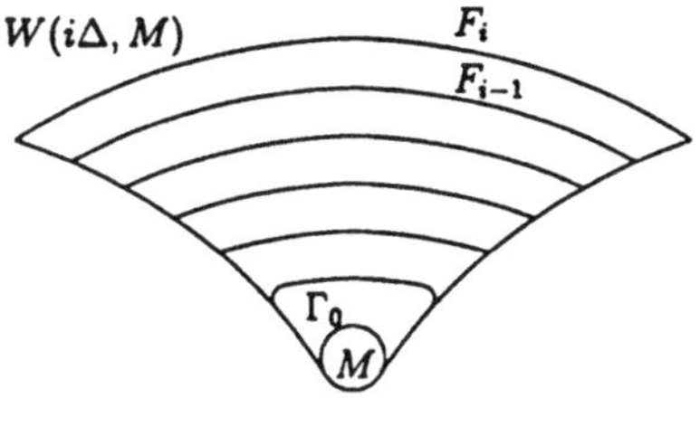

Figure 1: Construction of the sets $W(i\Delta, M)$.

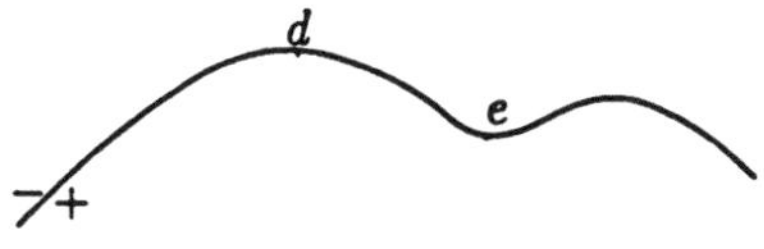

Figure 2: Local convexity and concavity.

After computing the extremal controls, the extremal trajectories issued from the front's points in the reverse time are considered: $x(\tau) = x_* - \tau\ (p(x_*)\varphi^\circ + v^\circ + g)$. The ends of these trajectories at $\tau = \Delta$ are used to form the next front F_i. If the extremal control φ° is not unique at some point $x_* \in F_{i-1}$, then the the segment $\Phi(x_*) = \{ \bigcup_{\varphi^\circ \in [-1,1]} (x_* - \Delta\ (p(x_*)\varphi^\circ + v^\circ + g))\}$ is considered instead of the single point. Similarly, if the extremal control v° is not unique, the segment $\Xi(x_*) = \{ \bigcup_{v^\circ \in [q_1, q_2]} (x_* - \Delta\ (p(x_*)\varphi^\circ + v^\circ + g))\}$ is considered.

For each front, we distinguish points of the local convexity and points of the local concavity. In Figure 2, d is a point of the local convexity, and e is a point of the local concavity. If x_* is a point of the local convexity and the extremal control φ° is not unique, we obtain a local picture like the one shown in Figure 3A after issuing the extremal trajectories from the point x_*. Here, the additional segment $\Phi(x_*)$ is appeared on the new front F_i. If the extremal control v° is not unique, we obtain a local picture similar to the one shown

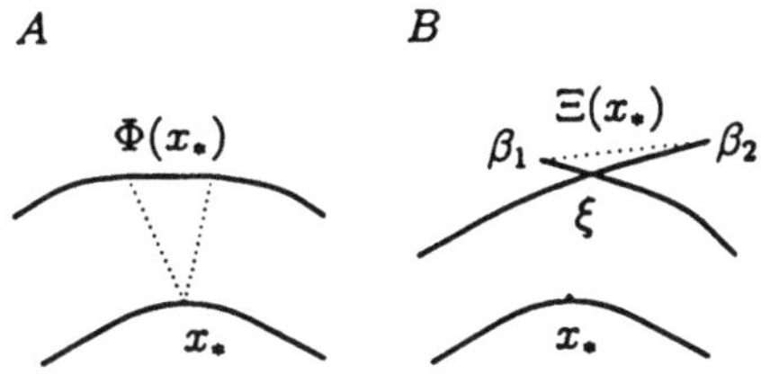

Figure 3: Nonuniqueness of extremal controls in the case of local convexity.

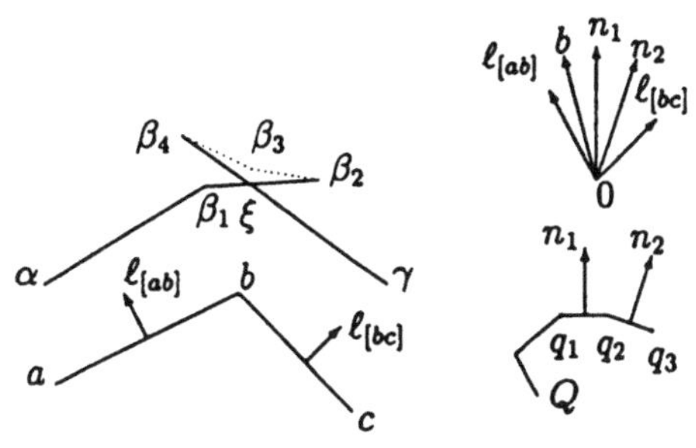

Figure 4: Example of local constructions.

in Figure 3B: the "swallow tail " $\beta_1\xi\beta_2$ does not belong to the new front F_i and it is taken away. For points of the local concavity, there is an inverse situation: if the extremal control φ° is not unique, a swallow tail that should be removed appears; if the extremal control v° is not unique, an additional segment $\Xi(x_*)$ appears on the new front F_i. If both φ° and v° are non-unique, the insertion or the swallow tail arises depending on which of segments $\Phi(x_*)$ or $\Xi(x_*)$ is greater.

In the course of numerical computations, we operate with polygonal lines instead of smooth curves. Two normal vectors to the links $[a, b]$, $[b, c]$ of the polygonal line are considered at each vertex b (Figure 4). The algorithm treats all possible variants of disposition of the vectors $\ell_{[ab]}$, $\ell_{[bc]}$, normals to the edges of Q, and the vector b. In Figure 4, for instance, the case is shown where the vector b is between the vectors $\ell_{[ab]}$, $\ell_{[bc]}$, and the normals n_1, n_2 to the set Q are between the vectors b and $\ell_{[bc]}$. The extremal controls of the players are computed for each of these vectors, and the extremal trajectories are issued from the points a, b, c. The ends of these trajectories computed at $\tau = \Delta$ give a local picture depicted in Figure 4. In the case considered, four extremal trajectories were issued from the point b. Their ends are: $\beta_1, \beta_2, \beta_3, \beta_4$. The segment $[\beta_1, \beta_2]$ appears due to nonuniqueness of the extremal control φ° for the vector b. The segments $[\beta_2, \beta_3]$ and $[\beta_3, \beta_4]$ arise due to nonuniquiness of the extremal control v° for the vectors n_1 and n_2. After removing the swallow tail $\xi\beta_4\beta_3\beta_3\beta_2\xi$, the polygonal line $\alpha\beta_1\xi\gamma$ is obtained to be a fragment of the next front.

More detailed description of such local constructions is given in [17]. The main difference from the case of the linear dynamics [17] is that the extremal control of the first player can change its value not only at front's vertices but also at some interior points of front's links. In the game considered, such a switching may occur only once for each front's link.

Since the algorithm is based on the computation of the extremal trajectories, the singular lines can be obtained (see [17] for the idea).

Examples

In this section, the results of computing the sets $W(\tau, M)$, $\tau = i\Delta$, are discussed.

In all examples presented, the following values of parameters of the problem are used: $w^{(1)} = 0.6$, $w^{(2)} = 0.2$, $R = 2$. The set Q is a 25-polygon inscribed into the circle of radius $w^{(2)}$ with the center at $(0, 0)$. The set M is a 15-polygon inscribed into the circle of radius 0.15. The center of the circle is in $(0, -0.23)$, $(0, 0.2)$, $(0, 0.5)$, $(0, -0.45)$, $(0.35, -0.3)$, $(0.2, 0.3)$ for Figures 5A–F, respectively. The step Δ is 0.0025.

In Figure 5A, the sets $W(\tau, M), \tau = 10\Delta i$, are depicted. The computations are done up to $\tau = 0.95$ when the self-intersection of the front happens. The point of the self-intersection divides the front into two parts: the interior part (adjoining to the gap) and the exterior one (enclosing the whole consruction). For $\tau > 0.95$, the computations should be carried out from the exterior and interior parts of the last front independently.

In Figure 5B (5C) the sets $W(\tau, M), \tau = 11\Delta i$ ($\tau = 12\Delta i$), are depicted. The self-intersection of the front happens at $\tau = 0.7$ ($\tau = 0.6$). The computations are continued from the interior part of the self-intersecting front. The gap between this part of the front and the barrier lines is completly filled out. The computations from the exterior part of the self-intersecting front were not carried out. In Figure 5B, the last interior front corresponds to $\tau = 1.045$. In Figure 5C, the second self-intersection of the front happens at $\tau = 1.045$ in the process of filling out the gap. The closed front arises. The computations from this closed front are done until it shrinks into a point at $\tau = 1.125$.

In Figure 5D, the sets $W(\tau, M), \tau = 8\Delta i$, are shown. The self-intersecting front, which corresponds to $\tau = 0.355$, is drawn with the thick dash line. The computations are continued from both interior and exterior parts of the front. The last interior (exterior) front corresponds to the time $\tau = 0.38$ ($\tau = 0.76$). The sets $W(\tau, M)$ for $0.355 < \tau < 0.38$ are doubly connected.

In Figure 5E, the sets $W(\tau, M), \tau = 13\Delta i$, are shown. The gap between the interior part of the self-intersecting front and the barrier lines is completly filled out as it is in Figure 5B. But interior fronts have much more complicated structure.

The most complicated case is presented in Figure 5F. The sets $W(\tau, M), \tau = 10\Delta i$, are depicted. The self-intersection of the interior front, which corresponds to $\tau = 0.9$ and is drawn with the thick dash line, produces two gaps. The next front consists of three parts: an exterior part (which is not shown), and two interior parts (two loops inside the dash contour). As the result, the sets $W(\tau, M)$ for $0.9 < \tau < 0.95$ are triple connected.

For all Figures 5A–F, barriers are drawn with thick lines.

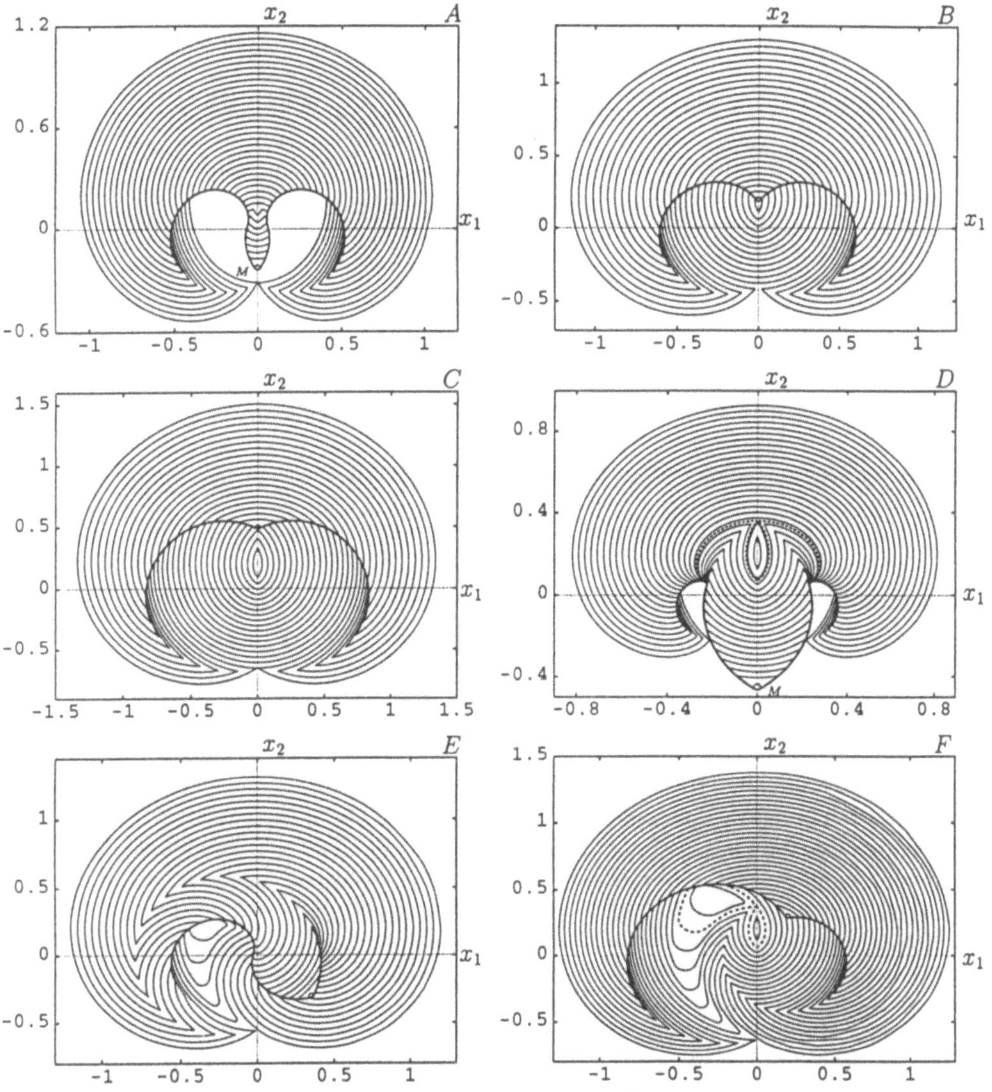

Figure 5: Examples of numerically computed sets $W(i\Delta, M)$.

Semipermeable curves in the homicidal chauffeur game

In [13, 14, 17, 18], algorithms for solving linear and some nonlinear differential games of kind are described. These algorithms are based on the computation of the semipermeable curves [1].

In this section, the results of some analysis of the semipermeable curves in the homicidal chauffeur game will be given. Using these results, one can find the solvability sets of the game of kind for any terminal set M and an arbitrary set Q. Since the set $W(\theta, M)$ converges to the solvability set of the corresponding game of kind as $\theta \to \infty$, solutions to the game of kind can be used for the verification of the computation of the sets $W(\theta, M)$.

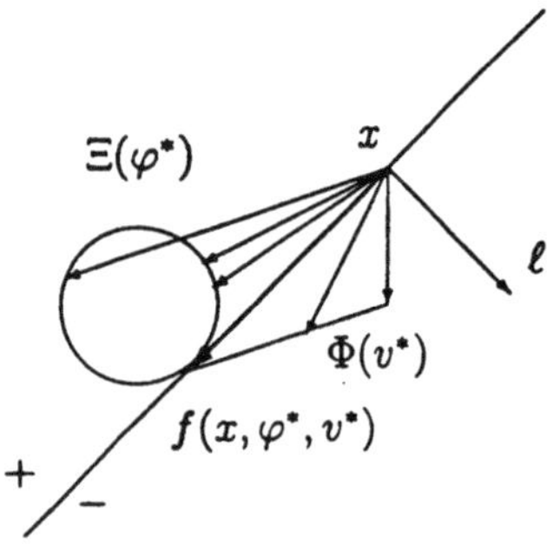

Figure 6: Semipermeable direction.

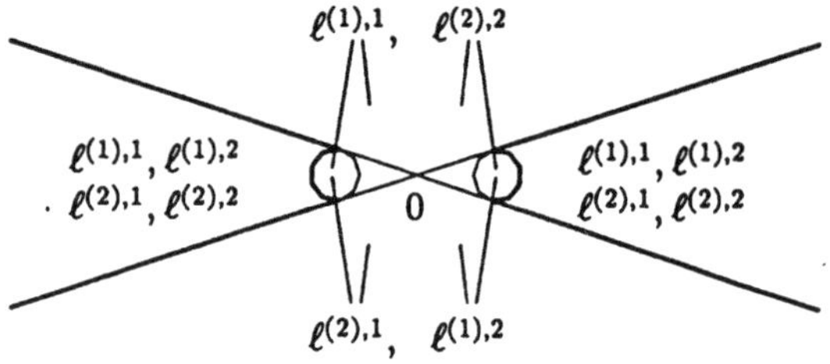

Figure 7: Domains of $\ell^{(j),i}$.

Let us explain now what semipermeable curves mean. Let

$$H(\ell, x) = \min_{\varphi} \max_{v} \ell' f(x, \varphi, v) = \max_{v} \min_{\varphi} \ell' f(x, \varphi, v), \quad x \in R^2, \ell \in R^2,$$

be the Hamiltonian of the game. Here $f(x, \varphi, v) = p(x)\varphi + v + g$. Fix $x \in R^2$ and consider ℓ such that $H(\ell, x) = 0$. Denote $\varphi^* = \arg\min_{\varphi} \ell' f(x, \varphi, v)$, $v^* = \arg\max_{v} \ell' f(x, \varphi, v)$. It holds: $\ell' f(x, \varphi^*, v) \leq 0$ for any $v \in Q$ and $\ell' f(x, \varphi, v^*) \geq 0$ for any $\varphi \in [-1, 1]$. This means that the direction $f(x, \varphi^*, v^*)$ which is orthogonal to ℓ separates the vectograms $\Phi(v^*) = \bigcup_{\varphi \in [-1,1]} f(x, \varphi, v^*)$ and $\Xi(\varphi^*) = \bigcup_{v \in Q} f(x, \varphi^*, v)$ of the first and second players (Figure 6). Such a direction is called semipermeable.

So, the semipermeable directions are defined by the roots of the equation $H(\ell, x) = 0$. We will distinguish the roots "−" to "+" and the roots "+" to "−". When defining these roots, we will suppose that $\ell \in E$ where E is a closed polygonal line around the origin. We say that ℓ_* is a root "−" to "+" if $H(\ell_*, x) = 0$ and $H(\ell, x) < 0$ $(H(\ell, x) > 0)$ for $\ell < \ell_*$ $(\ell > \ell_*)$ that are sufficiently close to ℓ_*. The notation $\ell < \ell_*$ means that the direction of the vector ℓ can be obtained from the direction of the vector ℓ_* using the counterclockwise rotation by the angle not exceeding π. The roots "−" to "+" and the roots "+" to "−" are called the roots of the first and second type, respectively.

One can prove that, in the game considered, the equation $H(\ell, x) = 0$ has at least one root of the first type and one root of the second type. Moreover, it has two roots of the first type and two roots of the second type at most. We denote the roots of the first type by $\ell^{(1),i}(x)$ and the roots of the second type by $\ell^{(2),i}(x)$. One can find the domains of the functions $\ell^{(j),i}(\cdot)$, $j = 1, 2$, $i = 1, 2$. The form of these domains is shown in Figure 7.

It can be proved that the function $\ell^{(j),i}(\cdot)$ satisfies the Lipschitz condition in any closed subset of its domain. So, we can consider the following differential equation

$$dz/dx = \Pi\ell^{(j),i}(x), \tag{3}$$

where Π is the matrix of rotation by the angle $\pi/2$ (the rotation's direction depends on j). Since the tangent at each point of phase trajectories of this equation is a semipermeable direction, the trajectories are semipermeable curves. It means that the first player can keep one side of the curve (say, positive side) and the second player can keep another side (negative side). So, the equation (3) specifies a family $\Lambda^{(j),i}$ of semipermeable curves. The families $\Lambda^{(j),i}$, $j = 1, 2$, $i = 1, 2$, are depicted in Figure 8.

The curves of different families belonging to the same type can be sewed in some cases so that the semipermeability property will be preserved. The procedure for computing the solvability set of the game of kind is based [17, 18] on the issuing two semipermeable curves

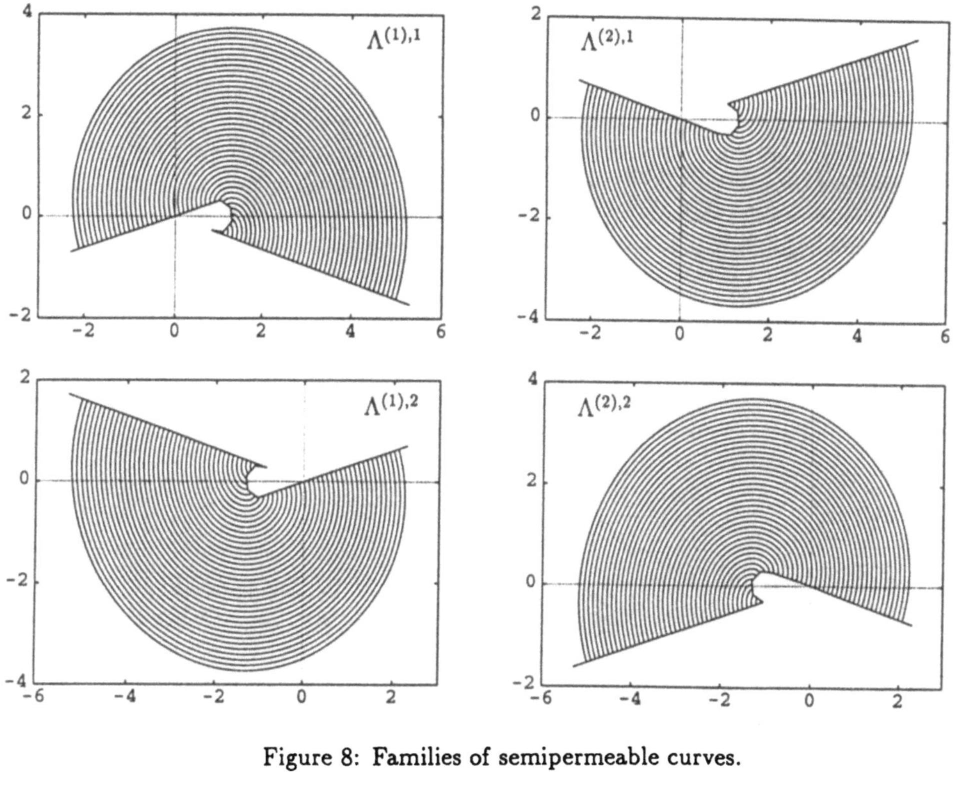

Figure 8: Families of semipermeable curves.

of the first and second type (they are faced each to other with positive sides) from the endpoints of M's usable part, on the sewing semipermeable curves of two different families belonging to the same type, and on the analysis of mutual dispositions of composite curves. Here, it will be shown with some example how this methodology can be used to check the validity of the computation of $W(i\Delta, M)$.

Let us consider the following example. The terminal set M is the regular 25-polygon inscribed into the circle of the radius 0.1 with the center at $(0.2, -0.4)$. The set Q is the triangle with the vertices $(-1.2, 1), (1.2, 1), (0, -6)$. In this example, the solvability set of

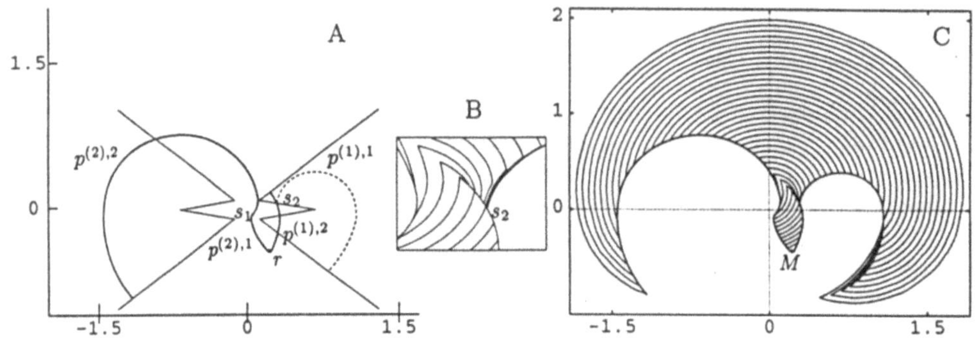

Figure 9: Example with the triangle set Q.

the game of kind is the whole plane. This can be proved using the following consideration of semipermeable curves. The semipermeable curves $p^{(2),1} \in \Lambda^{(2),1}$ and $p^{(1),2} \in \Lambda^{(1),2}$ issued from the endpoints of the usable part of M do not intersect each other before they finish at the boundaries of the corresponding domains (Figure 9A). The conjunction of $p^{(2),1}$ and $p^{(2),2}$ at the point s_1 is smooth. It provides the semipermeability property of $p^{(2),1} \bigcup p^{(2),2}$ at s_1. The composite curve $p^{(2),1} \bigcup p^{(2),2}$ does not intersect $p^{(1),2}$. Though the conjunction of the arc $rs_2 \in p^{(1),2}$ and and the curve $p^{(1),1}$ is non-smooth, the semipermeability property is fulfilled at the conjunction point s_2. The composite semipermeable curves of the first and second types do not intersect each other. Further semipermeable curves are not being produced.

The sets $W(\tau, M)$ computed numerically are shown in Figure 9C. One can see that the curve $p^{(2),1} \bigcup p^{(2),2}$ is one of the two barriers. The curve $p^{(1),2}$ is the other barrier. The structure of the fronts near $p^{(1),1}$ is shown in Figure 9B. The fronts lie above the curve $p^{(1),1}$ even though they come very close to it. More detailed consideration of the behavior of fronts for sufficiently large τ shows that $p^{(1),1}$ is not a barrier.

References

1. R.Isaacs, Differential Games, John Wiley, New York (1965).

2. A.W.Merz, The Homicidal Chauffeur - a Differential Game, PhD dissertation, Stanford University (1971).

3. J.Lewin and J.V.Breakwell, JOTA, 16 (3-4), 339–353 (1975).

4. R.J.Breakwell, in: Differential Games and Applications, P.Hagedorn, H.W.Knobloch, and G.J.Olsder (Eds.), Springer, Berlin (1977) pp. 70–95.

5. J.Lewin and G.J.Olsder, Computers Math. Applic., 18 (1-3), 15–34 (1989).

6. V.N.Ushakov, Engrg. Cybernetics, 18 (4), 16–23 (1981).

7. V.A.Ivanov, A.M.Taras'yev, V.N.Ushakov, and A.P.Khripunov, J. Appl. Math. Mech., 57 (3), 419–425 (1993).

8. A.I.Subbotin, Generalized Solutions of First-Order PDEs: the Dynamical Optimization Perspective, Birkhäuser, Boston (1995).

9. M.Bardi and M.Falcone, SIAM J. Control Optim., 28, 950–965 (1990).

10. P.Cardaliaguet, M.Quincampoix, P.Saint-Pierre, RAIRO-Modelisation-Matematique-et-Analyse-Numerique, 28 (4), 441–461 (1994).

11. N.N.Krasovskii and A.I.Subbotin, Positional Differential Games, Nauka, Moscow (1974).

12. N.N.Krasovskii and A.I.Subbotin, Game-Theoretical Control Problems, Springer-Verlag, New York (1988).

13. A.I.Subbotin and V.S.Patsko (Eds.), Algorithms and Programs for Solving Linear Differential Games, Inst. of Math. and Mech., Sverdlovsk (1984).

14. V.S.Patsko and V.L.Turova, Numerical solution of two-dimensional differential games, Preprint, Inst. of Math. and Mech., Ekaterinburg (1995).

15. V.S.Patsko and V.L.Turova, in: Proc. of UKACC Int. Conf. on Control'96, Exeter, UK, 2–5 September 1996, Conference Publication N 427 (1996) pp. 947–952.

16. V.S.Patsko and V.L.Turova, in: Proc. of the Seventh International Colloquium on Differential Equations, Plovdiv, Bulgaria, 18–23 August 1996, D.Bainov (Ed.), Utrecht, the Netherlands (1997) pp. 327–338.

17. V.S.Patsko, in: Differential Games and Control Problems, A.B.Kurzhanskii (Ed.), Inst. of Math. and Mech., Sverdlovsk (1975) pp. 167–227.

18. V.L.Turova, in: Investigations of Minimax Control Problems, A.I.Subbotin and V.S.Patsko (Eds.), Inst. of Math. and Mech., Sverdlovsk (1985) pp. 91–116.

8th Int. Coll. on Differential Equations, pp. 373-378
D. Bainov (Ed)

Nonlocal Regularization of Protter Problem for the 3-D Tricomi Equation

N.I. POPIVANOV
Department of Mathematics and Informatics, University of Sofia, 1164 Sofia, Bulgaria
nedyu@fmi.uni-sofia.bg

M. SCHNEIDER
Department of Mathematics, University of Karlsruhe, Karlsruhe, Germany
Manfred.Schneider@math.uni-karlsruhe.de

Abstract. In this paper we investigate some boundary value problems in $\mathbb{R}^3$ for the Tricomi equation of mixed type. M.H.Protter initiated the study of these problems, which are three-dimensional analogues of the plane problems, formulated by C. Morawetz, and investigated by C. Morawetz, P. Lax and R. Phillips. It is known that in the hyperbolic part of the domain, for an infinite number of smooth functions in the right-hand side, these problems do not have a classical solutions. Using Friedrichs' theory of symmetric positive operators, non-local problems are concidered, which are regularizers, in some sense, of these ill-posed problems.
Keywords: changing type equations, boundary value problems, generalized solution, Ill-posed problems, regularization methods, non-local problems.
AMS Subject Classification: 35Mxx, 65R30.

1. Introduction and statement of the problem

Denote points of $\mathbb{R}^3$ by $x = (x_1, x_2, t)$ and put $\rho = \sqrt{x_1^2 + x_2^2}$, $\varphi = \arctan(x_2/x_1)$. Let $G = G_+ \cup S_0 \cup G_-$ is a bounded simply connected region, where

$$G_+ = \left\{(x_1, x_2, t) \in \mathbb{R}^3 : 0 < t < (3/4)^{2/3},\ 2/3t^{3/2} < \rho < 1 - 2/3t^{3/2}\right\},$$

S_0 is the disc $S_0 = \{x : t = 0,\ 0 \le \rho \le 1\}$ and $G_- = \{x \in \mathbb{R}^3 : 0 < \rho < 1,\ \ g(\rho) < t < 0\}$. We assume that the rotating surface $\Gamma = \{\ 0 \le \rho \le 1, t = g(\rho)\ \}$ is C^2, lies in the half space $\{t < 0\}$ and meets the plane $\{t = 0\}$ in the curve $\{\rho = 1,\ t = 0\}$, i.e. $g \in C^2[0,1]$, $g(\rho) < 0$, $0 \le \rho < 1$, $g(1) = 0$, $g'(0) = g''(0) = 0$.

We consider in G the Tricomi equation

$$Lu := t\left(u_{x_1x_1} + u_{x_2x_2}\right) - u_{tt} \equiv t\left\{\rho^{-1}\left(\rho u_\rho\right)_\rho + \rho^{-2}u_{\varphi\varphi}\right\} - u_{tt} = f, \tag{1}$$

where f is a given function. We note that the boundary surfaces

$$S_1 = \left\{0 \le t \le (3/4)^{2/3}, \rho = 1 - 2/3t^{3/2}\right\}, S_2 = \left\{0 \le t \le (3/4)^{2/3}, \rho = 2/3t^{3/2}\right\}$$

are characteristics of (1). The Tricomi equation (1) is of changing type in G: it is elliptic in G_-, hyperbolic in G_+ and parabolic on S_0.

Problem P. Find a solution of (1) in G, which satisfies the condition

$$u = 0 \quad \text{on} \quad S_1 \cup \Gamma. \tag{2}$$

Problem P*. Find a solution of (1) in G, with

$$u = 0 \quad \text{on} \quad S_2 \cup \Gamma. \tag{3}$$

Protter [1] formulated these (adjoint) problems as three-dimensional analogues of a plane problem formulated and studied by Morawetz [2] and Lax, Phillips [3]. Didenko [4] proved the existence of solutions, in a function space with a degenerated weight function on $\{t = 0\}$. Some uniqueness results for quasi - regular solutions were obtained by Aziz, Schneider [5]. The Protter problems from [1] are also investigated by Salzman, Schneider [6], Papadakis [7], and others. Sorokina [8] got some interesting results, which we will discuss in §3. But for now there are no general existence results for the Problem P in G. To explain the difficulties in solving Problem P in G, we consider the Tricomi equation (1) in the domain G_+ only, with the boundary equations

$$u = 0 \quad \text{on} \quad S_0 \cup S_1. \tag{4}$$

If we denote this as Problem P_+, then we have for the corresponding adjoint Problem P_+^* the boundary dates $v = 0$ on $S_0 \cup S_2$. Popivanov showed in 1986, that this homogeneous Problem P_+^* has an infinite number of classical solutions

$$v_n(t,\rho,\varphi) = t\rho^{-n}\left(\rho^2 - 4t^3/9\right)^{n-4/3}\left(a_n \cos n\varphi + b_n \sin n\varphi\right), \tag{5}$$

$n \in \mathbb{N}$, $n \geq 4$, where a_n, b_n are arbitrary constants. That means that for the classical solvability of Problem P_+ the function f must at least be orthogonal to all functions v_n.

In [9] the authors gave an appropriate definition for generalized solution of the Problem P_+. They showed that the Problem (1), (4) in case $f = v_n$ has a unique generalized solution, but that this solution has a singularity on the characteristic cone S_2. More precisely, for any $l \in \mathbb{N}$, there exists a function $f_l \in C^l\left(\overline{G}_+\right)$ such that the unique generalized solution $u_l \in C\left(\overline{G}_+ \setminus S_2\right)$ satisfies the estimate

$$\int_{S_{2,\varepsilon}} |u_l|\, ds \geq \varepsilon^{-l}, \quad 0 < \varepsilon < 1, \tag{6}$$

where $S_{2,\varepsilon} = \left\{x \in G_+ : \rho = \varepsilon + 2t^{3/2}/3\right\}$.

This situation can be interpreted in terms of improperly posed (ill-posed) problems. A small perturbation of the right side has a strong effect on the singularity of the solution. For investigations of improperly posed problems numerous regularization methods exist. For our problems we refer especially to Morozov in [10, pp. 154-159] for a nonlocal regularization method for the backward heat equation problem. Instead of considering the ill-posed problem (1), (4), Edmunds, Popivanov [11] studied an appropriate new problem. This nonlocal regularization of the Problem P_+ connects points from G_+ with ones on the characteristic cone S_2. The new problem is such that its solutions are free of singularities on S_2 and in some sense it is a regularizer of the ill-posed Problem P_+.

Having this situation in mind (with regard to the work on ill-posed problems), with the same reason we investigate non-local problems. Let α be a small positive parameter.

Given a point $(t_0, \rho_0, \varphi_0) \in G_+$ we concider the curve $t = C\rho^{-\alpha} - \alpha\rho/(\alpha+1), \varphi = \varphi_0$ through this point, i.e. $C = \{t_0 + \alpha\rho_0/(\alpha+1)\}\rho_0^{\alpha}$. Now let $(p_\alpha(t_0,\rho), q_\alpha(t_0,\rho_0))$ be the point of intersection of two curves

$$\rho = 2t^{3/2}/3, \quad t = C\rho^{-\alpha} - \alpha\rho/(\alpha+1).$$

This construction connect one arbitrary fixed point from G_+ with a unique determined point on the cone S_2.

Instead of studying Problem P we now consider

Problem B. Find a solution $\omega(t,\rho,\varphi)$ of the equation

$$(7) \qquad \begin{cases} (L\omega)(t,\rho,\varphi) - t\rho^{-2}\omega_{\varphi\varphi}(p_\alpha(t,\rho), q_\alpha(t,\rho), \varphi) = f(t,\rho,\varphi) \text{ in } G_+, \\ (L\omega)(t,\rho,\varphi) = f(t,\rho,\varphi) \text{ in } G_-, \end{cases}$$

which satisfies the boundary conditions (2), i.e. $\omega = 0$ on $S_1 \cup \Gamma$.

Note that we have changed the Tricomi equation (1) in the hyperbolic part G_+ only to the equation (7), which has continuous coefficients in G. Equation (7) is non-local in the hyperbolic part G_+ of the domain G, because it involves points with coordinates (t,ρ,φ) and $(p_\alpha(t,\rho), q_\alpha(t,\rho), \varphi)$. We remark here merely that in our non-local Problem B, in the additional term $t\rho^{-2}\omega_{\varphi\varphi}(p_\alpha(t,\rho), q_\alpha(t,\rho), \varphi)$ of (7), the point $(p_\alpha, q_\alpha, \varphi)$ lies just on the characteristic cone S_2, where the singularity, shown by (6), appears in the 'generalized solution' of the original Problem P_+.

2. The nonlocal problem : existence,uniqueness and smothness of the generalized solution

Following the work of Morawetz [2] and Lax and Phillips [3] in the two-dimensional case and Sorokina [8] in the multi-dimensional case, we introduce the weighted Sobolev space

$$(8) \qquad \widetilde{W}_2^1(G) := \left\{\omega : \|\omega\|_{\widetilde{W}_2^1(G)} = \left(\int_G \left(\omega^2 + \omega_t^2 + r\left(\omega_{x_1}^2 + \omega_{x_2}^2\right)\right) dx\right)^{1/2} < \infty\right\},$$

where $r = \sqrt{x_1^2 + x_2^2 + t^2}$.

Definition 1 A function $\omega(t,\rho,\varphi)$ is called a generalized solution of Problem B if: (a) ω, $\dfrac{\partial\omega}{\partial t} \in L_2(G)$, (b) $\omega \in \widetilde{W}_2^1(G_-)$; $\sqrt{r}\dfrac{\partial\omega}{\partial\rho}$, $\sqrt{r}\rho^{-1}\dfrac{\partial\omega^1}{\partial\varphi} \in L_2(G_+)$, where

$$(9) \qquad \omega^1(t,\rho,\varphi) := \omega(t,\rho,\varphi) - \omega(p_\alpha(t,\rho),\, q_\alpha(t,\rho), \varphi);$$

(c) $\omega = 0$ on $S_1 \cup \Gamma$; (d) for any function $v \in C^* := \{v \in C^1(\overline{G}) : v = 0$ on $S_2 \cup \Gamma$ and in some neighbourhood of $(0,0,0)\}$, the equality

$$(10) \qquad \begin{aligned} &\int_G \left(\frac{\partial\omega}{\partial t}\frac{\partial v}{\partial t} - t\frac{\partial\omega}{\partial\rho}\frac{\partial v}{\partial\rho} - fv\right)\rho d\rho d\varphi dt \\ &- \int_{G_-} \frac{t}{\rho^2}\frac{\partial\omega}{\partial\varphi}\frac{\partial v}{\partial\varphi}\rho d\rho d\varphi dt - \int_{G_+} \frac{t}{\rho^2}\frac{\partial\omega^1}{\partial\varphi}\frac{\partial v}{\partial\varphi}\rho d\rho d\varphi dt = 0 \end{aligned}$$

holds.

Theorem 1 *Suppose that the parameter $\alpha > 0$ is so small and the surface Γ is such that the conditions*
(E1) $0 < \alpha < 2/\left(3+\sqrt[3]{6}\right), \min\{g(\rho) : 0 \le \rho \le 1\} < 2/\alpha - 1$,
(E2) $\rho g'(\rho) + \alpha[\rho + |g(\rho)|] > 0,\ \rho \in [0,1]$,
are satisfied. Then for any $f \in L_2(G)$, there exists one and only one generalized solution of problem B. If the function f be such that $\partial^k f/\partial\varphi^k \in L_2(G)$ for $k = 1, \dots, \ell$ and $\ell \in \mathbb{N}$, then there exists a generalized solution ω with

$$(11)\qquad \frac{\partial^k \omega}{\partial \varphi^k} \in \widetilde{W}_2^1(G)\ (k = 0, \dots \ell - 1),\ \frac{\partial^{\ell+1}\omega}{\partial t \partial \varphi^\ell} \in L_2(G),\ \sqrt{r}\frac{\partial^{\ell+1}\omega}{\partial \rho \partial \varphi^\ell} \in L_2(G_+)$$

and

$$(12)\qquad \sum_{k=0}^{\ell-1} \left\| \frac{\partial^k \omega}{\partial \varphi^k} \right\|_{\widetilde{W}_2^1(G)} + \left\| \frac{\partial^{\ell+1}\omega}{\partial t \partial \varphi^\ell} \right\|_{L_2(G)} + \left\| \sqrt{r}\frac{\partial^{\ell+1}\omega}{\partial \rho \partial \varphi^\ell} \right\|_{L_2(G_+)} \le C_\alpha \sum_{k=0}^{\ell} \left\| \frac{\partial^k f}{\partial \varphi^k} \right\|_{L_2(G)}.$$

3. Local and non-local problems, and their connection

We investigate here the connection between solutions of Protter Problem P and those of the nonlocal Problem B.

Definition 2 A function $u \in \tilde{W}_2^1(G)$ is called a generalized solution of Problem P if, and only if, $u = 0$ on $S_1 \cup \Gamma$ and the equality

$$(13)\qquad \int_G \left\{ u_t v_t - t u_\rho v_\rho - t\rho^{-2} u_\varphi v_\varphi - f v \right\} \rho d\rho dt d\varphi = 0$$

holds for all $v \in C^*$.

We shall also consider the following problem:

Problem P_φ. Find a generalized solution of Problem P which satisfies the extra condition

$$(14)\qquad u_\varphi = 0 \text{ on } S_2, \text{i.e.} \int_{S_2} u v_\varphi \, ds = 0 \text{ for all } v \in C_0^\infty(S_2).$$

Remark. For the Tricomi equation (1) Problem P_φ was formulated by Didenko [4]. Sorokina [8] studied the corresponding system of first order PDEs in a multidimensional case for the variables $x \in \mathbb{R}^m$, $m \ge 3$. That work was very helpful for us in the case $m = 3$. Anyway, we would like to say, that in the case $m \ge 4$ we can not follow the arguments there.

We denote by H the set of all $f \in L_2(G)$ for which there exists a generalized solution u_f of Problem P and by H_φ the subspace of functions $f \in H$ for which the solution u_f satisfies (14).

Remark. For Protter's Problem (1), (4) in G_+, we know

$$(15)\qquad \dim(L_2(G_+)/H) = \infty;$$

(see (5)). But in our case here we do not have any information about the structure of spaces H and H_φ.

For the Protter Problem P the uniqueness of the quasi - regular solution were obtained by Aziz, Schneider [5]. But up to now we do not know whether or not the unique solution depends continuously on f. We now investigate the uniqueness and continuous dependence on the data of solutions of Problem P_φ. The non-local Problem B will give some information.

Theorem 2 *We suppose that the surface* Γ *and parameter* α *are such, that the conditions (E1)-(E2) hold. Let* $u \in \tilde{W}_2^1(G)$ *be a generalized solution of Problem* P_φ. *Then it is also a unique generalized solution of Problem B and satisfies the a priori estimate*

$$\|u\|_{\tilde{W}_2^1(G)} \leq C\,\|f\|_{L_2(G)}\,. \tag{16}$$

Corollary 1 *If* $f \in H_\varphi$, *then a generalized solution* u_f *of Problem* P_φ *exists and coincides with the generalized solution* ω_f *of Problem B.*

Some information about the smoothness (with respect to φ) of the generalized solution of Problem P_φ we can obtain, using Theorems 1 and 2.

Corollary 2 *If* $f \in H_\varphi$, *and* $\partial^k f/\partial\varphi^k \in L_2(G)$, *for* $k = 0, ..., l$, *then the conclusions of Theorem 1 hold for a generalized solution* u_f *of Problem* P_φ.

Of course, for $f \in H\backslash H_\varphi$ a generalized solution of Problem P need not coincide with the generalized solution ω_f of Problem B; and if $f \in L_2(G)\backslash H$, there is no generalized solution of Problem P. Thus the non-local Problem B (for which Theorem 1 holds, ensuring uniqueness, existence and differentiability with respect to φ) is, so to speak, a 'regular continuation' of the strongly over-determined Problem P_φ when $f \in L_2(G)\backslash H_\varphi$. In this sense, we may regard Problem B to be a 'non-local regularization' of Problems P and P_φ. All this suggests the following procedure for tackling the ill-posed Problem P. For a given function $f \in L_2(G)$ we first try to solve the nonlocal Problem B. To do that, it is possible first to find the solution $\hat{u} = (u_1, u_2, u_3)$ of the local problem for the corresponding system of PDEs and then to determine a solution ω_f of Problem B by integration of $u_3(t, \rho, \varphi)$.

Theorem 3 *Let* $\hat{u} = (u_1, u_2, u_3)$ *be a strong solution of the problem for the corresponding system of PDEs for* $f \in L_2(G)$. *Let the function* $u_\varphi := x_1u_2 - x_2u_1$ *has a generalized derivative* $\partial u_\varphi/\partial t \in L_{2,loc}(G)$. *Then the solution* ω_f *of the nonlocal Problem B for the function* f *is a solution of the original Problem P too.*

Acknowlegements. This research was partially supported by the Bulgarian NSF under Grant MM-512/95.

References

1. M.H. Protter, J. Rat. Mech. Anal., 3, 435–446 (1954).
2. C.S. Morawetz, Comm. Pure Appl. Math., 11, 315–331 (1958).
3. P.D. Lax and R. Phillips, Comm. Pure Appl. Math., 13, 427–455 (1960).
4. V.P. Didenko, Diff. Equations, 11, 25–28 (1975).
5. A.K. Aziz and M. Schneider, SIAM J. Math. Anal., 10, 913–921 (1979).
6. H. Salzman and M. Schneider, Mh. Math., 84, 237–246 (1977).
7. J.S. Papadakis, Bull. Soc. Math. Grèce, 14, 157-171 (1973).
8. N.G. Sorokina, Diff. Equations, 13, 128–134 (1977).
9. N.I. Popivanov and M. Schneider, J. Math. Anal. Appl., 175, 537–579 (1993).

10. V.A. Morozov, Regularization Methods for Ill-posed Problems, CRC Press, Boca Raton (1993).

11. D.E. Edmunds and N.I. Popivanov, SIAM J. Math. Anal., 29 (1998), (to appear).

8th Int. Coll. on Differential Equations, pp. 379-384
D. Bainov (Ed)

Blow Up and Singularities of the Solutions of Homogeneous Quasilinear Hyperbolic Systems in the Plane

PETER POPIVANOV
Institute of Mathematics, Bulgarian Academy of Sciences, Sofia 1113, Bulgaria

ANGELA SLAVOVA
Department of Mathematics, University of Mining and Geology, Sofia 1100, Bulgaria

Abstract. This paper deals with the singularities of the classical solutions of hyperbolic quasilinear systems in the plane. Continuous solutions with gradient blowing up in a given point are constructed. An application to the semilinear equation of Born-Infeld is proposed.

Keywords: quasilinear hyperbolic system, characteristics, blow up of the solution

1. This paper deals with the singularities of the classical solutions of hyperbolic quasilinear systems in the plane. One of the motivations for developing such a theory is the application in mechanics and specially in fluid gas mechanics. In the famous book [1] very interesting effects arising in supersonic flow and formation of the shock waves are found. We have collected here several results on blow up of the classical solutions and formation of singularities for the solutions of quasilinear hyperbolic systems. To do this we use the well-known hodograph transformation. In the recent books [2,3] the reader can find many theorems concerning the blow up of the solutions of quasilinear and nonlinear hyperbolic systems (See also the well-known paper of John [4] and lectures of Hörmander [5]). We shall follow the so called geometric theory of the blow up. In all our cases it will be assumed that the systems are strictly hyperbolic.

2. Consider now the following quasilinear system:

$$\begin{pmatrix} 1 & 0 \\ 0 & 1 \end{pmatrix} \begin{pmatrix} u \\ v \end{pmatrix}_x + \begin{pmatrix} a(u,v) & b(u,v) \\ c(u,v) & d(u,v) \end{pmatrix} \begin{pmatrix} u \\ v \end{pmatrix}_y = 0, \tag{1}$$

where a, b, c, d are $C^2(C^\infty)$ real-valued functions of their arguments. In some cases it will be supposed that the coefficients are defined in a neighbourhood of the point $(u_0, v_0) \in R^2$ and in other cases it will be assumed that they are defined everywhere.

The system (1) will be called strictly hyperbolic with respect to x iff the algebraic equation in ξ: $\xi^2 + (a+d)\xi\eta + \eta^2(ad-bc) = 0$ has only real distinct roots, i.e. iff the discriminant $D = (a+d)^2 - 4(ad-bc) > 0$ in some neighbourhood of (u_0, v_0).

The characteristic curves of (1) in the Oxy plane are given by:

$$C_+ : \frac{dy_+}{dx} = \xi_+, \quad C_- = \frac{dy_-}{dx} = \xi_-$$

and $\xi_{\pm}(u(x,y),v(x,y))$, $u,v \in C^2$ are the distinct real roots of the equation $\lambda^2-(a+d)\lambda+(ad-bc)=0$. Certainly, the characteristics depend on the unknown solution $(u(x,y),v(x,y)) \in C^2$ in a neighbourhood of the point (x_0,y_0) such that $u(x_0,y_0)=u_0$, $v(x_0,y_0)=v_0$. It is classical that at least in a local domain of (u,v) each strictly hyperbolic system can be rewritten in the form

$$
\begin{aligned}
u_x + a(u,v)u_y &= 0,\\
v_x + b(u,v)v_y &= 0,
\end{aligned}
\tag{2}
$$

$a(u,v) \neq b(u,v)$. To do this a suitable change of the functions (u,v) should be fulfilled in a neighbourhood of (u_0,v_0). Without any loss of generality we can assume that $a(u,v) \neq 0$. So our considerations will be local as the first change of the variables (unknown functions) is purely local. Our second step is to use the so called hodograph transformation. Consider now the mapping

$$
\Omega \ni (x,y) \to (u(x,y),v(x,y)) \in R^2, \tag{3}
$$

where Ω is a small neighbourhood of (x_0,y_0) and the image runs over a tiny neighbourhood of (u_0,v_0). Suppose that the jacobian $j = \begin{vmatrix} u_x & u_y \\ v_x & v_y \end{vmatrix}(x_0,y_0) \neq 0$.

According to the inverse function theorem there exist two smooth functions $x(u,v)$, $y(u,v) \in C^2(\tilde{\Omega})$, where $\tilde{\Omega}$ is a neighbourhood of (u_0,v_0) and such that $J = \begin{vmatrix} x_u & x_v \\ y_u & y_v \end{vmatrix}(u_0,v_0)$ $\neq 0$ and the mapping $\tilde{\Omega} \to (u,v) \to (x(u,v),y(u,v)) \in R^2$ is inverse to (3) in $\tilde{\Omega}$.

Then we have that $u_x = jy_v$, $u_y = -jx_v$, $v_x = -jy_u$, $v_y = jx_u$. So the system (2) takes the form

$$
\begin{aligned}
y_v - a(u,v)x_v &= 0,\\
y_u - b(u,v)x_u &= 0, \quad (u,v) \in \tilde{\Omega}.
\end{aligned}
\tag{4}
$$

Certainly, (4) is a **linear** system with respect to the **arguments** $(u,v) \in \tilde{\Omega}$ and with unknown functions $x(u,v)$, $y(u,v)$.

Eliminating y from (4) we get the next second order linear equation with respect to x:

$$
x_{uv} + \frac{a_u}{a-b}x_v - \frac{b_v}{a-b}x_v = 0, \quad (u,v) \in \tilde{\Omega}. \tag{5}
$$

A new non-degenerate change of the arguments u,v: $\xi = u+v$, $\eta = u-v$ leads us to the second order hyperbolic equation in the plane $O\xi\eta$:

$$
x_{\xi\xi} - x_{\eta\eta} + \frac{a_u - b_v}{a-b}x_\xi - \frac{a_u + b_v}{a-b}x_\eta = 0. \tag{6}
$$

Obviously, (ξ,η) are varying in a small neighbourhood of $\xi_0 = u_0+v_0$, $\eta_0 = u_0 - v_0$.

In the previous equation we have $a(\frac{\xi+\eta}{2},\frac{\xi-\eta}{2})$, $a_u(\frac{\xi+\eta}{2},\frac{\xi-\eta}{2})$ etc. For the sake of simplicity we shall write $a(\xi,\eta)$ instead of $a(\frac{\xi+\eta}{2},\frac{\xi-\eta}{2})$, and $a_u(\xi,\eta)$ instead of $a_u(\frac{\xi+\eta}{2},\frac{\xi-\eta}{2})$.

Evidently, $u = \frac{\xi+\eta}{2}$, $v = \frac{\xi-\eta}{2}$.

We are now able to formulate our first result concerning the strictly hyperbolic system (1). So let $P_0 = (x_0,y_0)$, $Q_0 = (u_0,v_0)$ and $(u,v) \in C^2$ near P_0. In our further investigations (say $a,b,c,d \in C^\infty$ near Q_0) we shall deal with C^∞ smooth solutions.

Theorem 1 *Consider the system (1) with C^2 coefficients a, b, c, d defined in a neighbourhood of Q_0. Then for each neighbourhood U of P_0 we can find a triangle $\overline{P_0MN} \subset U$ and a pair of functions $(u,v) \in C(\overline{P_0MN})$, $(u(P_0), v(P_0)) = Q_0$, $(u,v) \in C^2$ in $\overline{P_0MN} \setminus \{P_0\}$ and such that (u,v) is a solution of (1) in the triangle $\overline{P_0MN} \setminus \{P_0\}$, $u_0|_{MN} \in C^2$, P_0M, P_0N are non-characteristic and non-tangential each to other smooth curves passing through P_0, $MN \parallel O_x$ and $\nabla(u,v)$ blows up at P_0 (see Figure 1).*

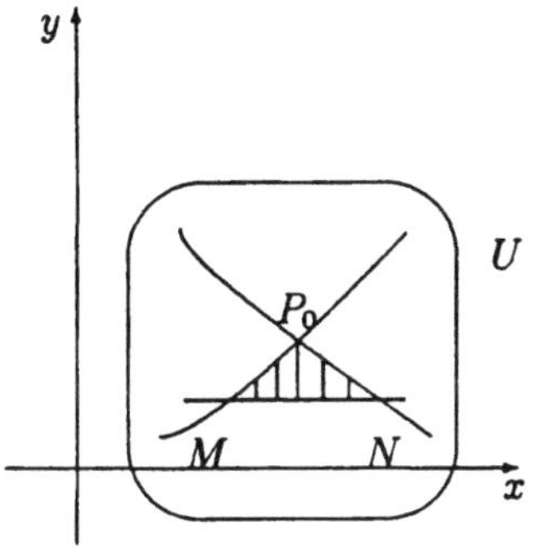

Figure 1

Proof:

Consider the noncharacteristic Cauchy problem for the linear equation (6):

$$x_{\xi\xi} - x_{\eta\eta} + A(\xi,\eta)x_\xi + B(\xi,\eta)x_\eta = 0, \tag{7}$$
$$x|_{\eta=0} = x_0(\xi) \in C^2,$$
$$x_\eta|_{\eta=0} = x_1(\xi) \in C^2,$$

where $A(\xi,\eta) = \frac{a_u - b_v}{a-b}$, $B = -\frac{a_u + b_v}{a-b}$. We point out that both functions $A, B \in C^1$ near (ξ_0, η_0). Without loss of generality let $P_0 = (0,0)$, $Q_0 = (0,0)$.

Assume that we have found somehow two C^2 smooth functions solving the linear system (4) near $(0,0)$. Then $j = \begin{vmatrix} x_u & x_v \\ y_u & y_v \end{vmatrix}_{(u,v)} \neq 0 \Leftrightarrow \begin{vmatrix} x_\xi & x_\eta \\ y_\xi & y_\eta \end{vmatrix}_{(\xi,\eta)} \neq 0$. Having in mind (4) we conclude that $j \neq 0 \Leftrightarrow x_u \neq 0, x_v \neq 0 \Leftrightarrow x_\xi + x_\eta \neq 0$, $x_\xi - x_\eta \neq 0$. Let $x_0(\xi) = \xi^2$, $x_1(\xi) \equiv 0$. There are no difficulties to verify that there exists a unique local solution $x \in C^2$ near $(0,0)$ of the Cauchy problem (7). Moreover, $x(\xi,\eta) = \xi^2 + \eta^2 + o(\xi^2 + \eta^2)$, $(\xi,\eta) \to 0$, $(x_\xi \pm x_\eta)(0,0) = 0$, $x_\xi \pm x_\eta = 2\xi \pm 2\eta + o(|\xi| + |\eta|)$, $(\xi,\eta) \to 0$.

Going back to the (u,v) coordinates we find that in a full neighbourhood of (0,0): $x(u,v) = 2u^2 + 2v^2 + o(u^2 + v^2)$, $(u,v) \to (0,0)$, $x_u = 4u + o(|u| + |v|)$, $(u,v) \to 0$, $x_v = 4v + o(|u| + |v|)$, $(u,v) \to 0$.

Put $v = ku$, $k = const$, $0 < k < 1$. Then $x_u > 0$, $x_v > 0$ in the angle $C : 0 < k_1 \leq \frac{v}{u} \leq k_2 < 1$, $u > 0$. Certainly $x_u(0,0) = x_v(0,0) = 0$, i.e. $j \neq 0$ in the open angle C and j vanishes only at its vertex $(0,0)$. Let us note that $v = k_{1,2}u$ are non-characteristic straight lines for the equation (5). According to (5) $y_{uv} = y_{vu}$, the conditions for existence of a total differential are valid in a neighbourhood of $(0,0)$ and thus there exists a function $y \in C^2$ near $(0,0)$ such that $y(u,v) = 2b(0,0)u^2 + 2a(0,0)v^2 + o(u^2 + v^2)$, $(u,v) \to (0,0)$. The transformation just above obtained namely:

$$\begin{aligned} x(u,v) &= 2u^2 + 2v^2 + o(u^2 + v^2), \ (u,v) \to 0, \\ y(u,v) &= 2b(0,0)u^2 + 2a(0,0)v^2 + o(u^2 + v^2), \ (u,v) \to 0 \end{aligned} \tag{8}$$

is defined and it is C^2 in a small neighbourhood of $(0,0)$. Moreover, there exists an angle $C : \{u > 0, 0 < k_1 \leq \frac{v}{u} \leq k_2 < 1\}$ and such that $x_u > 0$, $x_v > 0$ in $\bar{C} \setminus \{0\}$, $x_u(0,0) = x_v(0,0) = 0$.

The images of the sides $C_{1,2} = \{u \geq 0, v = k_{1,2}u\}$ under (8) are smooth curves $\tilde{C}_{1,2}$ in $R^2_{x,y}$ which are nontangential at $(0,0)$.

As our considerations are local we shall assume that $\frac{b(0,0)}{a(0,0)} \notin [\frac{k_1}{k_2}, \frac{k_2}{k_1}]$ for a suitable choice of the constants $0 < k_1 < k_2 < 1$. We remind to the reader that $a(0,0) \neq 0$ and $\frac{b(0,0)}{a(0,0)} \neq 1$. We claim that the mapping $(u,v) \to (x(u,v), y(u,v))$ defined by (8) is an injection on $\bar{C} \cap \mathcal{O}$ and $\mathcal{O}$ is a small neighbourhood of O. In fact, $(u,v) \neq (0,0)$ implies that $x(u,v) > 0$ while $x(0,0) = 0$. So consider the points $P \neq Q$, $P, Q \in \bar{C} \cap \mathcal{O}$, $P \neq O$, $Q \neq O$ and suppose that $x(P) = x(Q)$, $y(P) = y(Q)$. Having in mind that (8) is C^2 smooth near O we get immediately that there exist two points P^*, P^{**} belonging to the open interval (P,Q) and such that $< \vec{PQ}, \nabla x(P^*) >= 0$, $< \vec{PQ}, \nabla y(P^{**}) >= 0$. Here, $< .,. >$ stands for the scalar product of two vectors, $\vec{PQ} = \vec{OQ} - \vec{OP}$. Put $P^* = (u_1, v_1)$, $P^{**} = (u_2, v_2)$. Then $\frac{v_1}{u_1} = l_1 \in [k_1, k_2]$, $\frac{v_2}{u_2} = l_2 \in [k_1, k_2]$ and $\frac{l_1}{l_2} \in [\frac{k_1}{k_2}, \frac{k_2}{k_1}]$.

According to (8)

$$\nabla x(P^*) = 4u_1(1 + o(1), \frac{v_1}{u_1} + o(1)), \; o(1) \to 0 \;\text{ if }\; u_1 \to 0,$$

$$\nabla y(P^{**}) = 4u_2(b(0,0) + o(1), \frac{v_2}{u_2}a(0,0) + o(1)), \; o(1) \to 0 \;\text{ if }\; u_2 \to 0.$$

On the other hand, $\begin{vmatrix} 1 + o(1) & l_1 + o(1) \\ b(0,0) + o(1) & l_2 a(0,0) + o(1) \end{vmatrix} = a(0,0)l_2 - b(0,0)l_1 + o(1) =$ $= a(0,0)l_1(\frac{l_2}{l_1} - \frac{b(0,0)}{a(0,0)}) + o(1)$, $u_1, v_1 \to 0$. Taking $\mathcal{O}$ sufficiently small we come to the conclusion that $\nabla x(P^*)$ is not colinear to $\nabla y(P^{**})$, i.e. $\vec{PQ} = 0$.

Thus (8) is a homeomorphism from $\bar{C} \cap \bar{\mathcal{O}}$ onto its image $\overline{OMN}$ and it is a diffeomorphism from $(\bar{C} \cap \bar{\mathcal{O}}) \setminus \{0\}$ onto $\overline{OMN} \setminus \{0\}$ (see Figure 2).

To prove the same result for the non-diagonal system (1) it is sufficient to apply the local change of the unknown functions (u,v) transforming locally (1) into (2).

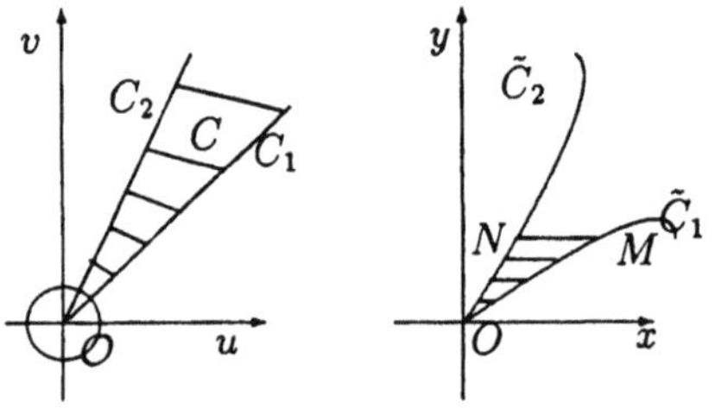

$(x(0), y(0)) = (0,0)$

Figure 2

The observation that $\begin{pmatrix} u_x & u_y \\ v_x & v_y \end{pmatrix} = \begin{pmatrix} x_u & x_v \\ y_u & y_v \end{pmatrix}^{-1}$ in the open triangle OMN shows that $\begin{pmatrix} u_x & u_y \\ v_x & v_y \end{pmatrix} = \frac{1}{4(a(0,0)-b(0,0))} \begin{pmatrix} \frac{a(0,0)}{u} & -\frac{1}{u} \\ -\frac{b(0,0)}{v} & \frac{1}{v} \end{pmatrix} + o(1)$, $(u,v) \to (0,0)$.

Thus $\nabla_{x,y}(u,v)$ blows up at $(0,0)$ as $(x,y) \to (0,0) \Leftrightarrow (u,v) \to (0,0)$.

We shall propose now an example illustrating our main result. This is the so called Born-Infeld equation [6]. So consider the equation:

$$(1-\varphi_t^2)\varphi_{xx}+2\varphi_x\varphi_t\varphi_{xt}-(1+\varphi_x^2)\varphi_{tt}=0. \tag{9}$$

This equation possesses global solutions of the form $\varphi = \Phi(x \pm t)$, $\Phi \in C^2(R^1)$. The equation (9) is hyperbolic if $1+\varphi_x^2-\varphi_t^2>0$. After the changes $\varphi_1 : \xi = x-t$, $\eta = x+t$ and $\varphi_2 : u = \varphi_\xi$, $v = \varphi_\eta$ we see that (9) is equivalent in each simply connected domain in (x,y) to the next system:

$$\begin{aligned} u_\eta - v_\xi &= 0,\\ v^2u_\xi-(1+2uv)u_\eta+u^2v_\eta &= 0. \end{aligned} \tag{10}$$

The hodograph transformation $\varphi_\xi : (\xi,\eta) \to (u,v)$ reduces (10) to:

$$\begin{aligned} \xi_v - \eta_u &= 0,\\ v^2\eta_v+(1+2uv)\xi_v+u^2\xi_u &= 0. \end{aligned} \tag{11}$$

Eliminating η from (11) we obtain:

$$v^2\xi_{vv}+(1+2uv)\xi_{uv}+u^2\xi_{uu}+2v\xi_v+2u\xi_u=0. \tag{12}$$

The linear system (11) and the equation (12) have the same characteristics in the "hyperbolic" domain $1+4uv>0$: $s=C_1$, $r=C_2$, where $s=\frac{\sqrt{1+4uv}\pm1}{2u}$, $r=\frac{\sqrt{1+4uv}\pm1}{2v}$, $u\neq0$, $v\neq0$. Obviously, both systems of characteristics coincide with the straight lines of the type $v=k^2u+k$, $k=const.$, $u=k^2v+k$. Their envelope is the hyperbola $uv=-1/4$.

Below we make the standard change φ_4 of the variables along the characteristics:

$$r=\frac{2u}{1+\sqrt{1+4uv}},\quad s=\frac{2v}{1+\sqrt{1+4uv}},\quad 1+4uv>0. \tag{13}$$

Obviously, the domain $uv \geq -1/4$ is mapped onto the set $1 > rs \geq -1$.

There are no difficulties to verify that (13) is a diffeomorphism between the two open domains and that the hyperbola $rs=1$ is not included into the image of φ_4. In the new coordinates r,s the system (11) takes the form:

$$\begin{aligned} \eta_r + r^2\xi_r &= 0,\\ \xi_s + s^2\eta_s &= 0, \end{aligned}$$

i.e. $\xi_{rs}=0 \Rightarrow \xi = F(r)+G(s)$, F,G being arbitrary and sufficiently smooth functions on R^1. Thus, φ_5:

$$\begin{aligned} \xi &= F(r)-\int s^2G'(s)ds,\\ \eta &= G(s)-\int r^2F'(r)dr \end{aligned} \tag{14}$$

and $F,G \in C^k(R^1)$, $k \geq 3$ are arbitrary real-valued functions.

It is elementary to find the initial function $\varphi \in C^2$ in each simply connected subdomain of $|rs|<1$. More specially, in each rectangle with sides parallel to the coordinate axes we have

$$\varphi(r,s)=\int rF'(r)dr+\int sG'(s)ds. \tag{15}$$

Let us point out that the Jacobian of (14) $J = \frac{D(\xi,\eta)}{D(r,s)} = F'G'(1-r^2s^2)$.

Conclusion: Following the described procedure we can construct infinitely many local solutions of (9), namely: $(u,v) = \varphi_4^{-1}\varphi_5^{-1}(x-t,x+t)$, $\varphi = \varphi(\varphi_5^{-1}(x-t,x+t))$, $F,G \in C^3$, $F'(r_0)G'(s_0) \neq 0$, $|r_0 s_0| < 1$, $\xi_0 = x_0 - t_0$, $\eta_0 = x_0 + t_0$, $(\xi_0,\eta_0) = \varphi_5(r_0,s_0)$ and φ is given by (15).

Analyse now the next case of "global solution" for (9). Thus, $F(r) = \frac{1}{r}$, $r > 0$, $G(s) = \frac{1}{s}$, $s > 0$.

Then

$$\xi = \frac{1+rs}{r}, \quad r > 0, \qquad \eta = \frac{1+rs}{s}, \quad s > 0, \quad sr \leq 1. \tag{16}$$

The mapping (16) is a diffeomorphism between the open domains $0 < r,s$; $rs < 1$ and $\xi\eta > 4$, $\xi > 0$, $\eta > 0$.

In this case the solution φ of (9) is given by the formula

$$\varphi(x,t) = -ln\frac{x^2-t^2-2-\sqrt{(x^2-t^2)^2-4(x^2-t^2)}}{2}$$

in the domain $x > 0$, $|t| < x$, $x^2 - t^2 > 4$.

So we have constructed a solution of (9) with the properties: φ is analytic in the hyperbola $\Pi : x^2 - t^2 > 4$, $x > 2$, $\varphi \in C(\bar{\Pi})$, $\varphi|_{\partial\Pi} = 0$, $\nabla_{x,t}\varphi|_{\partial\Pi} = \infty$.

According to our Theorem 1 we are able to construct a local classical solution φ of (9) having the properties:

$\varphi \in C^2(\overline{P_0MN} \setminus \{P_0\})$, $\varphi \in C^1(\overline{P_0MN})$ and the Hessian of φ blows up at P_0.

References

1. R.Courant, K.Fridrichs, Supersonic Flow and Shock Waves, N.Y. Interscience Publishers Inc. (1948).
2. Li Ta Tsien, Global Classical Solutions for Quasilinear Hyperbolic Systems, Research in Appl. Math. Series, J. Wiley and Sons (1994).
3. S.Alinhac, Blow up for Nonlinear Hyperbolic Equations, Progress in Nonlinear Diff. Equations, v.17. Birkhäuser (1995).
4. F.John, C.P.A.M., 27, 377–405 (1974).
5. L. Hörmander, Lecture Notes in Math., Springer-Verlag, 1256, 214–280 (1986).
6. G. Whitham, Linear and Nonlinear Waves, J. Wiley and Sons (1974).

8th Int. Coll. on Differential Equations, pp. 385-390
D. Bainov (Ed)

Dedicated to the memory of professor Filippo Chiarenza

Dirichlet Problem in Morrey Spaces for Elliptic Equations in Non Divergence Form with VMO Coefficients

MARIA ALESSANDRA RAGUSA
Dipartimento di Matematica, Università di Catania, Viale "A. Doria" 6, 95125 Catania, Italy, e-mail: maragusa@dipmat.unict.it

Keywords: *Uniformly elliptic operator, VMO, Morrey Spaces*

Let Ω be an open bounded set in $\mathbb{R}^n$, $n \geq 2$ with $\partial\Omega$ sufficiently smooth, $0 < \lambda < n$, $1 < p < \infty$.

We say that a function $f \in L^1_{loc}(\Omega)$ belongs to the Morrey Space $L^{p,\lambda}(\Omega)$ if

$$\|f\|^p_{L^{p,\lambda}(\Omega)} \equiv \sup_{x\in\Omega,\ \rho>0} \frac{1}{\rho^\lambda} \int_{B_\rho(x)\cap\Omega} |f(y)|^p dy < \infty$$

where $B_\rho(x)$ is a ball of radius ρ centered at the point x.

We also say that a function f belongs to the the John-Nirenberg space BMO (see [1]), or that f has "bounded mean oscillation", if

$$\|f\|_* \equiv \sup_{B\subset\mathbb{R}^n} \frac{1}{|B|} \int_B |f(x) - f_B| dx < \infty$$

where f_B is the integral average $\frac{1}{|B|}\int_B f(x)dx$ of the function $f(x)$ over the set B, and B belongs to the class of balls of $\mathbb{R}^n$.

Let, for a function $f \in BMO$

$$\eta(r) = \sup_{x\in\mathbb{R}^n,\ \rho\leq r} \frac{1}{|B_\rho|} \int_{B_\rho} |f(x) - f_{B_\rho}| dx.$$

A function $f \in BMO$ belongs to the Sarason class VMO, or f has "vanishing mean oscillation" (see [2]) if $\lim_{r\to 0^+} \eta(r) = 0$.

We recall the following classical Sobolev spaces

$$W^{k,p}(\Omega) = \{f(x):\ D^\alpha f \in L^p(\Omega),\quad |\alpha| \leq k\}, \qquad k \text{ is an integer},$$

equipped with the norm $\|f\|_{W^{k,p}(\Omega)} = \|f\|_{L^p(\Omega)} + \|D^k f\|_{L^p(\Omega)}$. The closure of the space $C_0^\infty(\Omega)$ with respect to the norm in $W^{k,p}(\Omega)$ will be denoted, as usual, by $W_0^{k,p}(\Omega)$. We set $W^{k,p,\lambda}(\Omega)$ the Banach space of functions belonging to $W^{k,p}(\Omega)$ and having k-th order derivative lying in the Morrey space $L^{p,\lambda}(\Omega)$. A natural norm in that space is

$$\|f\|_{W^{k,p,\lambda}(\Omega)} = \|f\|_{L^p(\Omega)} + \|D^k f\|_{L^{p,\lambda}(\Omega)}.$$

In the following we consider the Dirichlet problem

$$(1) \qquad \begin{cases} \mathcal{L}u \equiv \sum_{i,j=1}^n a_{ij}(x) D_{x_i x_j} u = f(x) & \text{a. e. in } \Omega \\ u \in W^{2,p,\lambda}(\Omega) \cap W_0^{1,p}(\Omega), \quad 1 < p < \infty \end{cases}$$

where $\mathcal{L}$ is a uniformly elliptic operator over the bounded domain Ω, $a_{ij} \in VMO \cap L^\infty(\Omega)$ and $f \in L^{p,\lambda}(\Omega)$. Specifically we study regularity results in Morrey spaces of solutions to the above Dirichlet problem.

Local and global regularity properties have been studied by many authors in the last thirty years. If the coefficients a_{ij} are Hölder continuous the regularizing properties of $\mathcal{L}$ in Hölder spaces and unique classical solvability of the Dirichlet problem for $\mathcal{L}u = f$ have been well studied (see [3] chapter 6 and reference therein). If the coefficients a_{ij} are uniformly continuous Agmon, Douglis and Nirenberg in [4] (see also [3]) prove that if $\mathcal{L}u \in L^p(\Omega)$ then the strong solution u of $\mathcal{L}u = f$ belongs to $W^{2,p}(\Omega)$ for all values of p in the range $(1, \infty)$. In the case that the coefficients a_{ij} are not uniformly continuous the problem is less studied because the L^p-theory of $\mathcal{L}$, the strong solvability of the Dirichlet problem for $\mathcal{L}u = f$ do not hold any more.

For two dimensional domains Ω and $p = 2$ Talenti in [5], assuming a_{ij} only measurable and bounded functions, proved that it is possible to obtain isomorphic properties of $\mathcal{L}$ considered as a mapping from $W^{2,2}(\Omega) \cap W_0^{1,2}(\Omega)$ into $L^2(\Omega)$. In the multi-dimensional case ($n \geq 3$) if $a_{ij} \in W^{1,n}(\Omega)$ (see [6]) or a_{ij} satisfy some other kind of assumption (e.g. the "Cordes condition" as in [7]) the local $W^{2,p}$ regularity has been proved for p belonging to a suitable neighborhood of $p = 2$.

A crucial step in the local and global Sobolev regularity of strong solutions to $\mathcal{L}u = f$ is the study of the class of functions VMO made by F. Chiarenza in the paper [8], by Chiarenza, Frasca and Longo in [9] and [10] and by Chiarenza, Franciosi and Frasca in [11]. In these papers the authors suppose $a_{ij} \in VMO \cap L^\infty(\Omega)$, $\mathcal{L}u \in L^p(\Omega)$ and prove that for each value of p in the range $(1, \infty)$ the function u belongs to $W^{2,p}(\Omega)$. It is also studied the well posedness of the Dirichlet problem for $\mathcal{L}u = f$ in $W^{2,p}(\Omega) \cap W_0^{1,p}(\Omega)$. The techniques used in the above mentioned papers allow to extend these results to the case of oblique derivative boundary condition to linear equation (see [12] and [13]) and quasilinear equation ([14] and [15]) with VMO coefficients. Then, if p is sufficiently large, it follows Hölder continuity of the strong solution or of its gradient.Moreover, for small p the Hölder continuity can be obtained if $\mathcal{L}u \in L^{p,\lambda}(\Omega)$ for suitable $\lambda > 0$.

Regularity properties of the operator $\mathcal{L}$ in Morrey spaces if $a_{ij} \in VMO$ are studied by Caffarelli in [16]. He proved that, if $f \in L^{n,n\alpha}_{loc}(\Omega)$ with $\alpha \in (0,1)$, a $W^{2,p}$–viscosity solution of $\mathcal{L}u = f$ belongs to $C^{1+\alpha}_{loc}(\Omega)$. The hypothesis for the function f could not be weakened with $f \in L^{p,\lambda}_{loc}(\Omega)$, $p < n$, $\lambda > 0$ because the method followed in [16] relies on the maximum principle of Aleksandrov-Pucci.

Interior estimates in Morrey spaces for the second derivatives of the $W^{2,p}$–solutions to $\mathcal{L}u = f$, where $a_{ij} \in VMO \cap L^\infty$, are proved in [17]. The authors showed that if the known term $f \in L^{p,\lambda}_{loc}(\Omega)$ the second derivatives of a strong solution u of the above non divergence form equation belong to the same space.

In the following we will deal with talk about the paper [18]. There the authors extend the regularity properties of [16] and [17] from local to global.

Let Ω be an open bounded set in $\mathbb{R}^n$ with $C^{1,1}$ smooth boundary and the coefficients a_{ij} of $\mathcal{L}$ such that

$$\text{(2)} \qquad \begin{cases} a_{ij}(x) \in L^\infty(\Omega) \cap VMO, \quad a_{ij}(x) = a_{ji}(x), \\ \exists\, \kappa > 0: \ \kappa^{-1}|\xi|^2 \le a_{ij}(x)\xi_i\xi_j \le \kappa|\xi|^2 \quad \forall \xi \in \mathbb{R}^n, \text{ a.a. } x \in \Omega. \end{cases}$$

We define $B_r^+ = \{x = (x_1, \dots, x_n) \equiv (x', x_n) \in \mathbb{R}^n: \ x_n > 0\}$.

Let $\Gamma(x,\xi)$ be the normalized fundamental solution of $\mathcal{L}$, i.e.,

$$\Gamma(x,\xi) = \frac{1}{n(2-n)\omega_n\sqrt{\det\{a_{ij}(x)\}}}\left(\sum_{i,j=1}^n A_{ij}(x)\xi_i\xi_j\right)^{(2-n)/2} \quad \text{for a.a. } x \text{ and all } \xi \in \mathbb{R}^n\setminus\{0\},$$

where $A_{ij}(x)$ stand for the entries of the inverse matrix of the matrix $\{a_{ij}(x)\}_{i,j=1,\dots,n}$, and ω_n is the measure of the unit ball in $\mathbb{R}^n$. We set also

$$\Gamma_i(x,\xi) = \frac{\partial \Gamma(x,\xi)}{\partial \xi_i}, \quad \Gamma_{ij}(x,\xi) = \frac{\partial \Gamma(x,\xi)}{\partial \xi_i \partial \xi_j}, \quad M = \max_{i,j=1,\dots,n}\max_{|\alpha|\le 2n}\left\|\frac{\partial^\alpha \Gamma_{ij}(x,\xi)}{\partial \xi^\alpha}\right\|_{L^\infty(\Omega\times\Sigma)}.$$

where Σ is the surfaces of the unit sphere in $\mathbb{R}^n$.

The functions $\Gamma_{ij}(x,\xi)$ are Calderón–Zygmund kernels in the ξ variable.

In the next theorem we prove the boundary regularity in Morrey spaces for second order derivatives of solutions to the Dirichlet problem for the operator $\mathcal{L}$.

Theorem 1. *Let* (2) *be true,* $1 < p < \infty$ *and* $0 < \lambda < n$. *Assume further that* $f \in L^{p,\lambda}(\Omega)$ *and* $u \in W^{2,p} \cap W_0^{1,p}(\Omega)$ *are such that* $\mathcal{L}u = f$ *a.e. in* Ω.

Then $D_{ij}u \in L^{p,\lambda}(\Omega)$ *and moreover, there exists a constant* $c = c(n,\kappa,p,\lambda,M,\partial\Omega)$ *such that*

$$\|D_{ij}u\|_{L^{p,\lambda}(\Omega)} \le c\left(\|u\|_{L^{p,\lambda}(\Omega)} + \|f\|_{L^{p,\lambda}(\Omega)}\right).$$

Let us now give an outline of the proof. The arguments rests on two tools. The first one is related to the following explicit representation formula for the second derivatives of a solution in terms of singular integral operators and commutators with Calderon-Zygmund kernels:

$$\begin{aligned} D_{ij}u(x) = {}& \text{P.V.}\int_{B_r^+} \Gamma_{ij}(x, x-y)\left\{\sum_{h,k=1}^n (a_{hk}(x) - a_{hk}(y))D_{hk}u(y) + \mathcal{L}u(y)\right\}dy \\ & + \mathcal{L}u(x)\int_{|\xi|=1} \Gamma_i(x,\xi)\xi_j\, d\sigma_\xi + I_{ij}(x), \end{aligned}$$

where we have set

$$I_{ij}(x) = \int_{B_r^+} \Gamma_{ij}(x, T(x)-y)\left\{\sum_{h,k=1}^n (a_{hk}(x) - a_{hk}(y))D_{hk}u(y) + \mathcal{L}u(y)\right\}dy$$

for $1 \le i,j < n$;

$$I_{in}(x) = I_{ni}(x) = \int_{B_r^+}\left(\sum_{j=1}^n \Gamma_{ij}(x, T(x)-y)A_j(x)\right)\left\{\sum_{h,k=1}^n (a_{hk}(x) - a_{hk}(y))D_{hk}u(y) + \mathcal{L}u(y)\right\}dy$$

for $1 \le i < n$;

$$I_{nn}(x) = \int_{B_r^+} \left(\sum_{i,j=1}^{n} \Gamma_{ij}(x, T(x)-y) A_i(x) A_j(x) \right) \left\{ \sum_{h,k=1}^{n} (a_{hk}(x) - a_{hk}(y)) D_{hk} u(y) + \mathcal{L}u(y) \right\} dy.$$

Further,

$$T(x,y) = x - \frac{2x_n}{a_{nn}(y)} \boldsymbol{a}_n(y), \qquad T(x) \equiv T(x,x),$$

with $\boldsymbol{a}_n(y) = (a_{in}(y))_{i=1,\dots,n}$ being the last row (column) of the matrix $\boldsymbol{a}(y) = \{a_{ij}(y)\}$, and

$$A(x) = (A_1(x), \dots, A_n(x)) = T(e_n, x) \equiv T((0,\dots,0,1), x).$$

To prove the boundedness in Morrey space of these integral operators we observe that the first integral appearing in the representation formula is sum of singular integral operators and commutators with Calderon-Zygmund kernels.

The second type of operators is not singular, specifically

$$\int_{|\xi|=1} \Gamma_i(x,\xi)\xi_j d\sigma_\xi \in L^\infty(B_r^+)$$

with bound independent on r and the integrals appearing in the definition of I_{ij} are Lebesque integrals. Then, using geometric properties of T and the VMO assumption on the coefficients $a_{ij}(x)$ we will prove the boundedness in $L^{p,\lambda}$–norms of these integral operators. Finally, the second one consists of global a priori estimate in Morrey spaces for the strong solutions of the Dirichlet problem (1), using standard techniques of the theory of PDE.

Combing this results with the $W^{2,p}$–strong solvability of the Dirichlet problem (1) proved in [10] it is possible to prove, in the next theorem, the well-posedness of (1) in the Morrey spaces $W^{2,p,\lambda}(\Omega)$.

Theorem 2. *Let* (2) *be true,* $1 < p < \infty$ *and* $0 < \lambda < n$.

Then for every $f \in L^{p,\lambda}(\Omega)$ *there exists a unique solution of the Dirichlet problem*

$$\text{(3)} \qquad \begin{cases} \mathcal{L}u = f(x) & a.e. \ in \ \Omega, \\ u \in W^{2,p,\lambda}(\Omega) \cap W_0^{1,p}(\Omega). \end{cases}$$

Moreover, there is a constant $c = c(n, \kappa, p, \lambda, M, \eta, \Omega)$ *such that*

$$\|D_{ij}u\|_{L^{p,\lambda}(\Omega)} \le c \|f\|_{L^{p,\lambda}(\Omega)}.$$

Remark 3. We observe that the result obtained in Theorem 1 for the homogeneous Dirichlet problem (3) can be applied in the nonhomogeneous problem. More precisely, if $f \in L^{p,\lambda}(\Omega)$, $\varphi \in W^{2,p,\lambda}(\Omega)$ and

$$\text{(4)} \qquad \begin{cases} \mathcal{L}u = f(x) & \text{a.e. in } \Omega \\ u = \varphi \quad \text{on } \partial\Omega, & u - \varphi \in W_0^{1,p}(\Omega) \end{cases}$$

we have $\mathcal{L}\varphi \in L^{p,\lambda}(\Omega)$, then $u(x) - \varphi(x)$ satisfy the homogeneous Dirichlet problem (3). Then, the strong solutions to (4) will belong to $W^{2,p,\lambda}(\Omega)$.

Remark 4. The results above can applied in the study of Morrey regularity of strong solutions to (4) for general elliptic operators

$$\mathcal{L} \equiv a_{ij}(x)D_{ij} + b_i(x)D_i + c(x)$$

where the coefficients $a_{ij} \in VMO \cap L^\infty$ and the lower order terms $b_i(x)$ and $c(x)$ satisfying suitable integrability properties (see [13], [19] and [20] for the $W^{2,p}$–regularity results of the problem (4).

Finally we observe that it is possible to generalize a result of Caffarelli (see [16]) as follows.

Corollary 5. *Let $u \in W^{2,p}(\Omega)$ be a strong solution to 4 with $f \in L^{p,\lambda}(\Omega)$ and $\varphi \in W^{2,p,\lambda}(\Omega)$. Suppose $n - p < \lambda < n$.*

Then the gradient Du is a Hölder continuous function on $\overline{\Omega}$ with exponent $\alpha = 1 - (n - \lambda)/p$.

Proof. Theorem 1 and well known properties of Morrey spaces (see [21]) for suitable values of $p \in (1, \infty)$ and $\lambda \in (0, n)$ allow us to obtain global Hölder regularity for the gradient Du of the strong solution of (4). ∎

We believe that the estimate the Hölder seminorm of the gradient will be useful in the theory of nonlinear elliptic equations (see [22]).

References

1. F. John and L. Nirenberg, Commun. Pure Appl. Math., 14, 415–426 (1961).
2. D. Sarason, Trans. Amer. Math. Soc., 207, 391–405 (1975).
3. D. Gilbarg and N.S. Trudinger, Elliptic Partial Differential Equations of Second Order, 2nd ed., Springer–Verlag, Berlin (1983).
4. S. Agmon, A. Douglis, L.Nirenberg, Commun. Pure Appl. Math., 12, 623–727 (1959).
5. G. Talenti, Le Matematiche, 21, 339–376 (1966).
6. C. Miranda, Ann. Mat. Pura Appl., 63, 353-386 (1963).
7. S. Campanato, Ann. Scuola Norm. Superiore di Pisa, 21, 701–707 (1967).
8. F. Chiarenza, in: Nonlinear Analysis, Function Spaces and Applications, Vol. 5, Prague (1994).
9. F. Chiarenza, M. Frasca and P. Longo, Ricerche Mat., 40, 149–168 (1991).
10. F. Chiarenza, M. Frasca and P. Longo, Trans. Amer. Math. Soc., 336, 841–853 (1993).
11. F. Chiarenza, M. Franciosi and M. Frasca, Atti Accad. Naz. Lincei, Cl. Sci. Fis. Mat. Nat., 5, 27–32 (1994).
12. G. Di Fazio and D.K. Palagachev, Comment. Math. Univ. Carolinae, 37, 537–556 (1996).
13. A. Maugeri and D.K. Palagachev, Forum Math., 9, (1997), (in print).

14. D.K. Palagachev, Trans. Amer. Math. Soc., 347, 2481–2493 (1995).

15. G. Di Fazio and D.K. Palagachev, Bull. Austral. Math. Soc., 53, 501–513 (1996).

16. L. Caffarelli, Rend. Sem. Mat. Fis. Milano, 58, 253–284 (1988).

17. G. Di Fazio and M.A. Ragusa, J. Funct. Anal., 112, 241–256 (1993).

18. G. Di Fazio, D.K. Palagachev and M. A. Ragusa, Global Morrey regularity of strong solutions to Dirichlet problem for elliptic equations with discontinuous coefficients, preprint.

19. C. Vitanza, Le Matematiche, 47, 177–186 (1992).

20. C. Vitanza, Le Matematiche, 48, 287–296 (1993).

21. S. Campanato, Ricerche Mat., 12, 67–86 (1963).

22. L. Caffarelli, Ann. of Math., 130, 189–213 (1989).

8th Int. Coll. on Differential Equations, pp. 391–398
D. Bainov (Ed)

On the Calculation of Lyapunov and Moment Lyapunov Exponents for Noise Driven Coupled Oscillators

N. SRI NAMACHCHIVAYA
Dept. of Aeronautical and Astronautical Engineering, University of Illinois, Urbana, USA
H. J. VAN ROESSEL
Dept. of Mathematical Sciences, University of Alberta, Edmonton, Alberta, Canada

Abstract. The primary concern in the analysis of dynamical systems is the determination and prediction of steady-state motions and their corresponding stability. The stability analysis of a dynamical system becomes more difficult when the system depends on a random process. Since most practical dynamical systems are subject to some form of noise, the determination of conditions under which the resulting *stochastic* system is stable (in some sense) is of great interest.

In the study of stability of solutions of random dynamical systems, both almost-sure stability and p^{th} moment stability have been widely used. Stability in the almost-sure sense is determined by the sign of the maximal *Lyapunov exponent* whereas p^{th} moment stability is determined by the *moment Lyapunov exponent.*

The dynamical system we consider consists of two coupled oscillators driven by real noise:

$$\ddot{q}_i + \omega_i^2 q_i + 2\varepsilon^2\zeta\omega_i\dot{q}_i + \varepsilon\sum_{j=1}^{2} k_{ij}q_j f(\xi(t)) = 0, \qquad i,j = 1,2, \tag{1}$$

where the q_i's are generalized coordinates, ω_i is the ith natural frequency, $\varepsilon^2\zeta$ represents a small viscous damping coefficient and the stochastic term $\varepsilon\xi(t)$ is a small-intensity, real-noise process.

Two perturbative techniques for the evaluation of the Lyapunov exponent in the presence of a small intensity noise have been developed by the authors [1, 2]: a *forward* method and a *backward* method. However, since the backward method has the advantage that it also allows for the determination of the moment Lyapunov exponent, it is this latter method that will be described in this paper. Before proceeding with the details of the backward method, a brief summary of some of the theoretical results required is given.

1. Theoretical Results

Consider the following random dynamical system:

$$\dot{x}(t) = A(\xi(t))x(t), \qquad x(0) = x_0, \qquad x_0 \in \mathbb{R}^d \setminus \{0\}, \tag{2}$$

$$d\xi = X_0(\xi)\,dt + \sum_{i=1}^{r} X_i(\xi) \circ dW_i, \qquad \xi \in M, \tag{3}$$

where M is a connected Riemannian manifold and $\xi(t)$ is a stationary ergodic diffusion process with associated generator

$$G = X_0 + \frac{1}{2}\sum_{i=1}^{r} X_i^2.$$

By introducing polar coordinates in $\mathbb{R}^d$ (Khas'minskii transformation)

$$\rho = \ln \|x\|, \qquad s = \frac{x}{\|x\|}, \qquad s \in S^{d-1}$$

Eq. (2) becomes

$$\dot{\rho} = q(\xi, s), \qquad \rho(0) = \rho_0, \tag{4}$$

$$\dot{s} = h(\xi, s), \qquad s(0) = s_0, \tag{5}$$

$$d\xi = X_0(\xi)\, dt + \sum_{i=1}^{r} X_i(\xi) \circ dW_i, \tag{6}$$

where $q(\xi, s) = s^T A(\xi) s$ and $h(\xi, s) = (A(\xi) - q(\xi, s)I)s$. Notice that $\{\xi, s\}$ is decoupled from ρ and form a diffusive process with generator $\mathcal{L} = G + h\frac{\partial}{\partial s}$.

The exponential growth rate of $x(t; x_0)$, the *Lyapunov exponent* $\lambda(x_0)$, and the exponential growth rate of the pth-moment, the *moment Lyapunov exponent* $g(p; x_0)$ become

$$\lambda(x_0) = \lim_{t\to\infty} \frac{1}{t} \int_0^t q(\xi(\tau), s(\tau))\, d\tau \quad \text{and} \quad g(p; x_0) = \lim_{t\to\infty} \frac{1}{t} \ln \mathbb{E}[e^{p\int_0^t q(\xi(\tau), s(\tau))\, d\tau}]$$

respectively.

We now make two *non-degeneracy* assumptions on the vector fields associated with the problem:

$$\dim LA(X_1, \cdots, X_r) = \dim M, \tag{7}$$

$$\dim LA(X_0 + h\frac{\partial}{\partial s} + \frac{\partial}{\partial t}, X_1, \cdots, X_r) = \dim M + d, \tag{8}$$

where $LA(Z)$ denotes the Lie algebra generated by the set Z of vector fields. Condition (7) implies that there exists a unique smooth positive invariant density ν that satisfies $G^*\nu = 0$, and condition (8) implies that there exists a unique smooth positive invariant density μ that satisfies $\mathcal{L}^*\mu = 0$. We now state the main result due to Arnold et al. [3].

Theorem 1 *Assuming conditions (7) and (8) we have*

1. $\lambda(x_0) = \lambda_{max} = \langle q, \mu \rangle$;

2. $g(p, x_0) = g(p)$ *is the principal eigenvalue of the operator* $L(p) := \mathcal{L} + p\, g(p)$. *Moreover,* $g : \mathbb{R} \to \mathbb{R}$ *is convex, analytic and increasing with* $g(0) = 0$ *and* $g'(0) = \lambda$.

These are the results on which the perturbative methods presented in the subsequent sections of this paper are based.

2. Application to Coupled Oscillators

Using the transformation $q_i = x_{2i-1}$, $\dot{q}_i = \omega_i x_{2i}$, $i = 1, 2$, Eq. (1) may be represented by the following system of first order differential equations:

$$\dot{x} = Ax + \varepsilon f(\xi(t))Bx, \tag{9}$$

where A and B are 4×4 constant matrices given by

$$A = \begin{bmatrix} 0 & \omega_1 & 0 & 0 \\ -\omega_1 & -2\varepsilon^2\zeta\omega_1 & 0 & 0 \\ 0 & 0 & 0 & \omega_2 \\ 0 & 0 & -\omega_2 & -2\varepsilon^2\zeta\omega_2 \end{bmatrix}, \qquad B = \begin{bmatrix} 0 & 0 & 0 & 0 \\ -k_{11}/\omega_1 & 0 & -k_{12}/\omega_1 & 0 \\ 0 & 0 & 0 & 0 \\ -k_{21}/\omega_2 & 0 & -k_{22}/\omega_2 & 0 \end{bmatrix}. \tag{10}$$

We assume the noise process is governed by

$$d\xi = \alpha(\xi)\,dt + \sigma(\xi)\circ dW. \tag{11}$$

In order to make the problem tractable we make the following assumptions:

(A1): the generator of the noise process, G, has *an isolated, simple zero eigenvalue*; and

(A2): the natural frequencies ω_1 and ω_2 are *noncommensurable*.

Assumption (A1) implies that the only solution of $Gu = 0$ is $u \equiv const$. Applying the Khas'minskii–like transformation

$$\begin{aligned} x_1 &= e^{\rho}\cos\phi_1\cos\theta, & x_2 &= -e^{\rho}\sin\phi_1\cos\theta,\\ x_3 &= e^{\rho}\cos\phi_2\sin\theta, & x_4 &= -e^{\rho}\sin\phi_2\sin\theta, \end{aligned} \tag{12}$$

to Eq. (9) yields the following set of equations for the amplitude ρ, phase variables (ϕ_1, ϕ_2, θ) and noise process ξ:

$$\dot\rho = q^{\varepsilon}(\phi_1, \phi_2, \theta, \xi), \tag{13}$$

$$\dot\phi_i = \omega_i + h_i^{\varepsilon}(\phi_1, \phi_2, \theta, \xi),\quad \dot\theta = s^{\varepsilon}(\phi_1, \phi_2, \theta, \xi),\quad d\xi = \alpha(\xi)dt + \sigma(\xi)\circ dW_t,$$

where

$$\begin{aligned}
q^{\varepsilon}(\phi_1, \phi_2, \theta, \xi) &= \frac{1}{4}\varepsilon f(\xi) q^1(\phi_1, \phi_2, \theta) + \varepsilon^2 q^2(\phi_1, \phi_2, \theta),\\
h_i^{\varepsilon}(\phi_1, \phi_2, \theta, \xi) &= \frac{1}{2}\varepsilon f(\xi) h_i^1(\phi_1, \phi_2, \theta) + \varepsilon^2 h_i^2(\phi_1, \phi_2, \theta), \qquad i = 1, 2,\\
s^{\varepsilon}(\phi_1, \phi_2, \theta, \xi) &= \frac{1}{4}\varepsilon f(\xi) s^1(\phi_1, \phi_2, \theta) + \varepsilon^2 s^2(\phi_1, \phi_2, \theta),\\
q^1(\phi_1, \phi_2, \theta) &= q_0^1(\phi_1, \phi_2) + q_c^1(\phi_1, \phi_2)\cos 2\theta + q_s^1(\phi_1, \phi_2)\sin 2\theta,\\
q^2(\phi_1, \phi_2, \theta) &= q_0^2(\phi_1, \phi_2) + q_c^2(\phi_1, \phi_2)\cos 2\theta,\\
s^1(\phi_1, \phi_2, \theta) &= s_0^1(\phi_1, \phi_2) + s_c^1(\phi_1, \phi_2)\cos 2\theta + s_s^1(\phi_1, \phi_2)\sin 2\theta,\\
s^2(\phi_1, \phi_2, \theta) &= s_s^2(\phi_1, \phi_2)\sin 2\theta,\\
h_1^1(\phi_1, \phi_2, \theta) &= h_{1(0)}^1(\phi_1) + h_{1(\theta)}^1(\phi_1, \phi_2)\tan\theta,\\
h_2^1(\phi_1, \phi_2, \theta) &= h_{2(0)}^1(\phi_2) + h_{2(\theta)}^1(\phi_1, \phi_2)\cot\theta,
\end{aligned}$$

$$\begin{aligned}
q_0^1(\phi_1, \phi_2) &= p_{11}\sin 2\phi_1 + p_{22}\sin 2\phi_2, & s_0^1(\phi_1, \phi_2) &= p_{21}^{-}\sin\phi^{+} - p_{21}^{+}\sin\phi^{-},\\
q_c^1(\phi_1, \phi_2) &= p_{11}\sin 2\phi_1 - p_{22}\sin 2\phi_2, & s_c^1(\phi_1, \phi_2) &= p_{21}^{+}\sin\phi^{+} - p_{21}^{-}\sin\phi^{-},\\
q_s^1(\phi_1, \phi_2) &= p_{21}^{+}\sin\phi^{+} + p_{21}^{-}\sin\phi^{-}, & s_s^1(\phi_1, \phi_2) &= p_{22}\sin 2\phi_2 - p_{11}\sin 2\phi_1,
\end{aligned}$$

$$\begin{aligned}
h_{1(0)}^1(\phi_1) &= p_{11}(1 + \cos 2\phi_1), & h_{1(\theta)}^1(\phi_1, \phi_2) &= p_{12}(\cos\phi^{+} + \cos\phi^{-}),\\
h_{2(0)}^1(\phi_2) &= p_{22}(1 + \cos 2\phi_2), & h_{2(\theta)}^1(\phi_1, \phi_2) &= p_{21}(\cos\phi^{+} + \cos\phi^{-}),
\end{aligned}$$

$$\begin{aligned}
q_0^2(\phi_1, \phi_2) &= -\frac{1}{2}[\delta_1(1 - \cos 2\phi_1) + \delta_2(1 - \cos 2\phi_2)],\\
q_c^2(\phi_1, \phi_2) &= -\frac{1}{2}[\delta_1(1 - \cos 2\phi_1) - \delta_2(1 - \cos 2\phi_2)],\\
s_s^2(\phi_1, \phi_2) &= \frac{1}{2}[\delta_1(1 - \cos 2\phi_1) - \delta_2(1 - \cos 2\phi_2)],\\
h_i^2(\phi_i) &= -\delta_i\sin 2\phi_i.
\end{aligned}$$

In the above expressions, $p_{21}^{\pm} = p_{21} \pm p_{12}$, $\phi^{\pm} = \phi_1 \pm \phi_2$, $\delta_i = \zeta\omega_i$ and $p_{ij} = k_{ij}/\omega_i$. The processes $(\phi_1, \phi_2, \theta, \xi)$ alone form a diffusive Markov process, the associated generator of which is given by

$$\mathcal{L}^\varepsilon = G(\xi) + [\omega_1 + h_1^\varepsilon(\phi_1, \phi_2, \theta, \xi)]\frac{\partial}{\partial\phi_1} + [\omega_2 + h_2^\varepsilon(\phi_1, \phi_2, \theta, \xi)]\frac{\partial}{\partial\phi_2} + s^\varepsilon(\phi_1, \phi_2, \theta, \xi)\frac{\partial}{\partial\theta}.$$

The two solution procedures, the forward and backward methods, are based on the two results of the main theorem in the previous section:

- *Forward Method* — first solve $\mathcal{L}^{\varepsilon *}\mu^\varepsilon = 0$ to get μ^ε, then compute $\lambda^\varepsilon = \langle q^\varepsilon, \mu^\varepsilon \rangle$.
- *Backward Method* — solve $L^\varepsilon(p)\varphi^\varepsilon(p) = g^\varepsilon(p)\varphi^\varepsilon(p)$ to get $g^\varepsilon(p)$, then $\lambda^\varepsilon = g^{\varepsilon\prime}(0)$.

It is clear that the backward method allows for the determination of both the Lyapunov exponent and the moment Lyapunov exponent. The details of the backward method are described in the next section.

3. The Backward Method

In general, explicit expressions for the Lyapunov exponent λ^ε and the moment Lyapunov exponent $g^\varepsilon(p)$ are difficult to obtain. We can make use of the fact that g^ε is analytic in p to differentiate $L^\varepsilon(p)\varphi^\varepsilon(p) = g^\varepsilon(p)\varphi^\varepsilon(p)$ and set $p = 0$ to get:

$$\mathcal{L}^\varepsilon\psi^\varepsilon = \lambda^\varepsilon - q^\varepsilon, \tag{14}$$

where $\phi^\varepsilon(0)$ has been normalised to unity and $\psi^\varepsilon = \partial\varphi^\varepsilon/\partial p|_{p=0}$. Reapeating the procedure leads to:

$$\mathcal{L}^\varepsilon\eta^\varepsilon = g^{\varepsilon\prime\prime}(0) - 2(q^\varepsilon - \lambda^\varepsilon)\psi^\varepsilon, \tag{15}$$

where $\eta^\varepsilon = \partial^2\varphi^\varepsilon/\partial p^2|_{p=0}$. We now proceed to solve Eq. (14) to obtain the Lyapunov exponent λ^ε and Eq. (15) to obtain the moment Lyapunov exponent.

3.1. Lyapunov Exponent

Assuming ε to be small, we consider asymptotic expansions for ψ^ε and λ^ε:

$$\psi^\varepsilon \sim \psi_0 + \psi_1\varepsilon + \psi_2\varepsilon^2, \qquad \lambda^\varepsilon \sim \lambda_0 + \lambda_1\varepsilon + \lambda_2\varepsilon^2.$$

Inserting these expansions into Eq. (14) leads to the following sequence of equations:

$$\mathcal{L}^0\psi_0 = \lambda_0, \qquad \mathcal{L}^0\psi_1 = \lambda_1 - \frac{1}{4}fq^1 - \mathcal{L}^1\psi_0, \qquad \mathcal{L}^0\psi_2 = \lambda_2 - q^2 - \mathcal{L}^1\psi_1 - \mathcal{L}^2\psi_0,$$

where

$$\mathcal{L}^0 = G(\xi) + \sum_{i=1}^{2}\omega_i\frac{\partial}{\partial\phi_i},$$

$$\mathcal{L}^1 = \frac{1}{4}f(\xi)s^1(\phi_1, \phi_2, \theta)\frac{\partial}{\partial\theta} + \frac{1}{2}f(\xi)\sum_{i=1}^{2}h_i^1(\phi_1, \phi_2, \theta)\frac{\partial}{\partial\phi_i},$$

$$\mathcal{L}^2 = s^2(\phi_1, \phi_2, \theta)\frac{\partial}{\partial\theta} + \sum_{i=1}^{2}h_i^2(\phi_1, \phi_2, \theta)\frac{\partial}{\partial\phi_i}.$$

The Solution to $\mathcal{O}(1)$.

The solvability condition for the $\mathcal{O}(1)$ equation is $\langle\lambda_0, u\rangle = 0 \quad \forall u \in \ker(\mathcal{L}^{0*})$. It is easily seen that $\ker(\mathcal{L}^{0*}) \supset \{\nu(\xi)C(\theta); C \text{ is an arbitrary function}\}$ and so the solvability condition reduces to $\lambda_0 = 0$. Using assumption (A1), we get

$$\psi_0 = \psi_0(\theta, \omega_1\phi_2 - \omega_2\phi_1).$$

But, with the noncommensurability assumption (A2), this reduces to

$$\psi_0 = \psi_0(\theta),$$

a function of θ which has yet to be determined.

The Solution to $\mathcal{O}(\varepsilon)$.

Inserting the above expression for ψ_0 into the equation for ψ_1 results in

$$\mathcal{L}^0\psi_1 = \lambda_1 - \frac{1}{4}f(\xi)[q^1(\phi_1,\phi_2,\theta) + s^1(\phi_1,\phi_2,\theta)\psi_0'] . \tag{16}$$

The solvability condition: $\lambda_1 = \frac{1}{4}\langle f(q^1 + s^1\psi_0'), u\rangle$ for all $u \in \ker(\mathcal{L}^{0*})$ becomes

$$\lambda_1 = \frac{1}{16\pi^2}\langle f(\xi)R_1(\phi_1,\phi_2,\theta)\,,\ \nu(\xi)C(\theta)\rangle = 0, \tag{17}$$

where

$$R_1(\phi_1,\phi_2,\theta) = q_0^1(\phi_1,\phi_2) + q_c^1(\phi_1,\phi_2)\cos 2\theta + q_s^1(\phi_1,\phi_2)\sin 2\theta.$$

The last equality in Eq. (17) follows from the fact that $R_1(\phi_1,\phi_2,\theta)$ is 2π–periodic in ϕ_1 and ϕ_2. Hence, the solution of Eq. (16) is

$$\psi_1(\phi_1,\phi_2,\theta,\xi) = \frac{1}{4}\int_0^\infty K(\xi,\tau)R_1(\phi_1+\omega_1\tau, \phi_2+\omega_2\tau,\theta)\,d\tau, \tag{18}$$

where

$$K(\xi,\tau) = \int_M f(\eta)\nu(\eta)g(\xi,\tau;\eta,0)\,d\eta$$

and g is the solution of

$$\frac{\partial g}{\partial t} = G^* g, \qquad g(\xi,0;\eta,0) = \delta(\xi-\eta).$$

The Solution to $\mathcal{O}(\varepsilon^2)$.

Employing the above results, the equation for ψ_2 becomes

$$\mathcal{L}^0\psi_2 = \lambda_2 - q^2(\phi_1,\phi_2,\theta) - \mathcal{L}^1\psi_1(\phi_1,\phi_2,\theta,\xi) - s^2(\phi_1,\phi_2,\theta)\psi_0'. \tag{19}$$

The solvability condition is

$$\left\langle \lambda_2 - q^2 - \mathcal{L}^1\psi_1 - s^2\psi_0', u\right\rangle = 0 \qquad \forall u \in \ker(\mathcal{L}^{0*}),$$

which, after some calculation, reduces to

$$\int_0^{\pi/2}\left\{\lambda_2 - Q(\theta) - \Phi(\theta)\psi_0'(\theta) - \frac{1}{2}\Psi^2(\theta)\psi_0''(\theta)\right\}C(\theta)\,d\theta = 0, \tag{20}$$

where

$$
\begin{aligned}
Q(\theta) &= \Psi^2(\theta) + \frac{1}{2}(\Lambda_1 - \Lambda_2)\cos 2\theta + \frac{1}{2}(\Lambda_1 + \Lambda_2) + \frac{1}{8}p_{21}p_{12}(S(\omega^+) - S(\omega^-)),\\
\Psi^2(\theta) &= A\cos^2 2\theta + B\cos 2\theta + C,\\
A &= \frac{1}{32}\left\{(p_{21}^+)^2 S(\omega^+) + (p_{21}^-)^2 S(\omega^-) - \sum_{i=1}^{2} p_{ii}^2 S(2\omega_i)\right\},\\
B &= \frac{1}{16}(p_{21}^2 - p_{12}^2)(S(\omega^+) + S(\omega^-)),\\
C &= \frac{1}{32}\left\{(p_{21}^-)^2 S(\omega^+) + (p_{21}^+)^2 S(\omega^-) + \sum_{i=1}^{2} p_{ii}^2 S(2\omega_i)\right\},\\
S(\omega) &= 2\int_0^\infty \int_M f(\xi)K(\xi,\tau)\cos\omega\tau\, d\xi\, d\tau, \qquad \Lambda_i = -\delta_i + \frac{1}{8}p_{ii}^2 S(2\omega_i), \quad i = 1, 2.
\end{aligned}
$$

Since Eq. (20) must hold for arbitrary $\mathcal{C}(\theta)$, the bracketed quantity must vanish identically. This leads to the following differential equation for ψ_0:

$$
\frac{1}{2}\Psi^2(\theta)\psi_0'' + \Phi(\theta)\psi_0' = \lambda_2 - Q(\theta). \tag{21}
$$

This equation has a solvability condition of its own, namely

$$
\int_0^{\pi/2} (\lambda_2 - Q(\theta))\,\mathcal{F}(\theta)\,d\theta = 0, \tag{22}
$$

where $\mathcal{F}(\theta)$ is the solution of the adjoint problem

$$
\frac{1}{2}\frac{d^2}{d\theta^2}\left(\Psi^2(\theta)\mathcal{F}\right) - \frac{d}{d\theta}\left(\Phi(\theta)\mathcal{F}\right) = 0, \tag{23}
$$

with appropriate boundary conditions for θ based on the Feller boundary classification. The solution to Eq. (23) is given in the previous section. Inserting $\mathcal{F}(\theta)$ into Eq. (22) yields

$$
\lambda_2 = \frac{1}{2}\left\{-(\Lambda_1 - \Lambda_2)\kappa + \Lambda_1 + \Lambda_2 + \frac{1}{4}p_{21}p_{12}[S(\omega^+) - S(\omega^-)]\right\}. \tag{24}
$$

The solution of Eq. (21) is $\psi_0(\theta) = \frac{1}{2}\ln\left[\tan^\kappa\theta \sin 2\theta\right]$.

3.2. Moment Lyapunov Exponent

The primary advantage of the backward method is that it affords us a method to obtain an asymptotic approximation of the moment Lyapunov exponent as well. We expand η^ε and $V^\varepsilon = g^{\varepsilon\prime\prime}(0)$ in an asymptotic expansion in Eq. (15) for small ε to get the following sequence of equations:

$$
\mathcal{L}^0\eta_0 = V_0 + 2\lambda_0\psi_0, \tag{25}
$$

$$
\mathcal{L}^0\eta_1 = V_1 + 2\left[\lambda_0\psi_1 - (\frac{1}{4}fq^1 - \lambda_1)\psi_0\right] - \mathcal{L}^1\eta_0, \tag{26}
$$

$$
\mathcal{L}^0\eta_2 = V_2 + 2\left[\lambda_0\psi_2 - (\frac{1}{4}fq^1 - \lambda_1)\psi_1 - (q^2 - \lambda_2)\psi_0\right] - \mathcal{L}^1\eta_1 - \mathcal{L}^2\eta_0. \tag{27}
$$

The Solution to $\mathcal{O}(1)$.

Since $\lambda_0 = 0$, it follows that $V_0 = 0$ and the leading equation reduces to

$$\mathcal{L}^0 \eta_0 = 0. \tag{28}$$

Hence, as for the case of ψ_0, $\eta_0 = \eta_0(\theta)$, an unknown function of θ. However, the analysis will stop short of an explicit determination of η_0 since the ultimate goal, the determination the moment Lyapunov exponent to $\mathcal{O}(\varepsilon^2)$, is achievable without an explicit expression for η_0.

The Solution to $\mathcal{O}(\varepsilon)$.

Employing the previously obtained results, the equation to $\mathcal{O}(\varepsilon)$ becomes

$$\mathcal{L}^0 \eta_1 = V_1 - \frac{1}{2} f(\xi)[q^1(\phi_1, \phi_2, \theta)\psi_0(\theta) - s^1(\phi_1, \phi_2, \theta)\eta_0'(\theta)]. \tag{29}$$

Applying the solvability condition to the above equation yields

$$V_1 = \frac{1}{16\pi^2} \left\langle f(\xi)\hat{R}_1(\phi_1, \phi_2, \theta)\, ,\, \nu(\xi)C(\theta) \right\rangle = 0, \tag{30}$$

where

$$\begin{aligned}\hat{R}_1(\phi_1, \phi_2, \theta) = 2 &\left[q_0^1(\phi_1, \phi_2) + q_c^1(\phi_1, \phi_2)\cos 2\theta + q_s^1(\phi_1, \phi_2)\sin 2\theta\right]\psi_0(\theta) \\ &+ [s_0^1(\phi_1, \phi_2) + s_c^1(\phi_1, \phi_2)\cos 2\theta + s_s^1(\phi_1, \phi_2)\sin 2\theta]\eta_0'(\theta).\end{aligned}$$

Again, the last equality in Eq. (30) follows from the fact that $\hat{R}_1(\phi_1, \phi_2, \theta)$ is 2π-periodic in ϕ_1 and ϕ_2. The equation for η_1 becomes

$$\mathcal{L}^0 \eta_1 = -\frac{1}{4} f(\xi)\hat{R}_1(\phi_1, \phi_2, \theta). \tag{31}$$

The solution is

$$\eta_1(\phi_1, \phi_2, \theta) = \frac{1}{4}\int_0^\infty K(\xi, \tau)\hat{R}_1(\phi_1 + \omega_1\tau, \phi_2 + \omega_2\tau, \theta)\, d\tau. \tag{32}$$

The Solution to $\mathcal{O}(\varepsilon^2)$.

The equation for η_2 becomes

$$\begin{aligned}\mathcal{L}^0 \eta_2 = V_2 + 2 &\left[(\lambda_2 - q^2(\phi_1, \phi_2, \theta))\psi_0(\theta) - \frac{1}{4} f(\xi) q^1(\phi_1, \phi_2, \theta)\psi_1(\theta)\right] \\ &- \mathcal{L}^1 \eta_1(\phi_1, \phi_2, \theta, \xi) - s^2(\phi_1, \phi_2, \theta)\eta_0'.\end{aligned} \tag{33}$$

The right hand side of the above equation contains an arbitrary function of θ as well as the unknown V_2. The solvability condition for this equation is

$$\left\langle V_2 + 2\left[(\lambda_2 - q^2)\psi_0 - \frac{1}{4} f q^1 \psi_1\right] - \mathcal{L}^1\eta_1 - s^2\eta_0'\, ,\, u\right\rangle = 0 \qquad \forall u \in \ker(\mathcal{L}^{0*}). \tag{34}$$

Again, after some manipulation, this reduces to

$$\int_0^{\pi/2} \left\{V_2 - \hat{Q}(\theta) - \Phi(\theta)\eta_0'(\theta) - \frac{1}{2}\Psi^2(\theta)\eta_0''(\theta)\right\} C(\theta)\, d\theta = 0, \tag{35}$$

where $\hat{Q}(\theta) = \hat{Q}_0(\theta) + \hat{Q}_1(\theta)\psi_0(\theta) + \hat{Q}_2(\theta)\psi_0'(\theta)$, and

$$\hat{Q}_0(\theta) = F + E\cos 2\theta - A\cos^2 2\theta, \qquad \hat{Q}_1(\theta) = 2\Psi^2(\theta) + (\Lambda_1 - \Lambda_2)(\kappa + \cos 2\theta),$$

$$\hat{Q}_2(\theta) = (B - E + 2A\cos 2\theta)\sin 2\theta, \qquad E = \frac{1}{16}\left\{p_{11}^2 S(2\omega_1) - p_{22}^2 S(2\omega_2)\right\},$$

$$F = \frac{1}{32}\left\{(p_{21}^+)^2 S(\omega^+) + (p_{21}^-)^2 S(\omega^-) + \sum_{i=1}^{2} p_{ii}^2 S(2\omega_i)\right\}.$$

Equation (35) must hold for arbitrary $C(\theta)$; this yields

$$\frac{1}{2}\Psi^2(\theta)\eta_0'' + \Phi(\theta)\eta_0' = V_2 - \hat{Q}(\theta). \tag{36}$$

Once again, this equation has its own solvability condition which guarantees the existence of a solution for η_0. This condition is

$$\int_0^{\pi/2} (V_2 - \hat{Q}(\theta))\mathcal{F}(\theta)\, d\theta = 0,$$

where $\mathcal{F}(\theta)$ is given by Eq. (23). Using the expression for ψ_0' and integrating yields

$$V_2 = \kappa(B - E) + F - C + c\int_0^{\pi/2} \left[\hat{Q}_1(\theta)\psi(\theta) + \Psi^2(\theta) + 2A\kappa\cos 2\theta\right]\mathcal{F}(\theta)\, d\theta. \tag{37}$$

The results may be summarized as follows:

Theorem 2 *Assuming conditions (7), (8), (A1) and (A2) we have an asymptotic expansion of the moment Lyapunov exponent of Eq. (9) given by*

$$g^{\varepsilon}(p) = \varepsilon^2\left\{p\lambda_2 + \frac{p^2}{2}V_2\right\} + o(\varepsilon^2 p^2) \tag{38}$$

where λ_2 and V_2 are given by Eqs. (24) and (37), respectively.

References

1. N. Sri Namachchivaya and H. J. Van Roessel, J. Stat. Phys., 71(3/4), 549–567 (1993).

2. N. Sri Namachchivaya, H. J. Van Roessel, and M. M. Doyle, SIAM J. Appl. Math., 56(5), 1400–1423 (1996).

3. L. Arnold, W. Kliemann, and E. Oeljeklaus, in: Lyapunov Exponents, L. Arnold (Ed.), volume 1186 of Lecture Notes in Mathematics, Springer-Verlag, New York (1986) pp. 85–125.

8th Int. Coll. on Differential Equations, pp. 399–407
D. Bainov (Ed)

A Unified Theory of Some Families of Elliptic Integrals Arising in Radiation Field and Other Problems

H.M. SRIVASTAVA
Department of Mathematics and Statistics, University of Victoria, Victoria, British Columbia V8W 3P4, Canada
E-Mail: hmsri@uvvm.uvic.ca

Abstract. A number of families of generalized elliptic integrals, and indeed also definite integrals of such families with respect to their modulus (or complementary modulus), are known to arise naturally in various seemingly diverse physical contexts including, for example, certain radiation field problems, studies of crystallographic minimal surfaces, theory of scattering of acoustic or electromagnetic waves by means of an elliptic disk, studies of elliptical crack problems in fracture mechanics, and so on. Motivated essentially by these and many other potential avenues of their applications, we aim at presenting here a systematic account of the theory of these families of elliptic-type integrals in a unified and generalized manner. We also point out relevant connections of the general theory presented here with the results obtained by numerous earlier workers on the theory and applications of the aforementioned families of elliptic-type integrals.
Key Words and Phrases: *Elliptic integrals, radiation fields, fracture mechanics, Appell functions, Lauricella functions, hypergeometric functions, Born series, Kampé de Fériet functions, Gauss's summation theorem.*
1991 Mathematics Subject Classification. Primary 33E05, 33C65; Secondary 33C75, 78A40, 78A45.

1. Introduction and Preliminaries

In the usual notations, let $K(k)$ and $E(k)$ denote, respectively, the *complete* elliptic integrals of the first and second kind with *modulus* k (*cf.*, *e.g.*, Byrd and Friedman [1]; see also Abramowitz and Stegun [2, Chapter 17]). Also let $\kappa := \sqrt{1-k^2}$ denote the *complementary modulus.* In his systematic study of some problems involving high energies in quantum mechanics, by evaluating the first term in a certain Born series in two different ways and comparing the resulting expressions, Barton [3] found the integral formula:

$$\int_0^1 k^\mu\, K(\kappa)\, dk = \frac{\pi}{4}\left\{\frac{\Gamma(\frac{1}{2}\mu+\frac{1}{2})}{\Gamma(\frac{1}{2}\mu+1)}\right\}^2 \qquad (\kappa := \sqrt{1-k^2};\ \mu > -1). \tag{1.1}$$

Subsequently, while addressing Barton's problem of finding a *direct* proof of his formula (1.1), Bushell [4] not only proved Barton's formula (1.1) *directly*, but also derived a number of additional results analogous to (1.1), thereby extending several known integral formulas recorded, for example, by Byrd and Friedman [1, p. 274] (see also Müller [5] and Kaplan [6]). Furthermore, Bushel! [4] gave a generalization of Barton's integral formula (1.1) in the form:

$$\int_0^1 k^\mu H(\kappa,\gamma)\,dk = \frac{\pi}{4}\frac{\Gamma(\frac{1}{2}\mu+\frac{1}{2})\,\Gamma(\gamma+\frac{1}{2}\mu+\frac{1}{2})}{\Gamma(\frac{1}{2}\mu+1)\,\Gamma(\gamma+\frac{1}{2}\mu+1)} \qquad (\kappa := \sqrt{1-k^2};\ \mu > -1), \tag{1.2}$$

where [4, p. 2, Equation (2.2)]

$$H(k,\gamma) := \int_0^1 (1-t^2)^{-\frac{1}{2}}\,(1-k^2t^2)^{\gamma-\frac{1}{2}}\,dt \qquad (\gamma \geq 0), \tag{1.3}$$

so that, obviously, $H(k,0) = K(k)$ and $H(k,1) = E(k)$.

By making use of (1.2), Bushell [4, Section 4] proved a general theorem which he applied to deduce numerous further results including, for example, an interesting integral formula (*cf.* [4, p. 5, Corollary]) associated with the generalized hypergeometric function ${}_pF_q$ (*cf.*, *e.g.*, Erdélyi *et al.* [7, Chapter 4]; see also Srivastava and Karlsson [8, p. 19 *et seq.*]). Some further integral formulas analogous to (1.2), also proven by Bushell [4, p. 3, Equations (2.5) and (2.6)], can indeed be shown to follow from each other with the aid of a familiar transformation for ${}_3F_2(1)$ (*cf.*, *e.g.*, Hardy [9, p. 499, Equation (6.1)]), which incidentally is usually proven by appealing to Gauss's summation theorem [7, p. 104, Equation 2.8 (46)].

Motivated by the usefulness of each of the above-mentioned integral formulas, as demonstrated by (among others) Barton [3] and Bushell [4], we begin this presentation by proving a general theorem which unifies as well as extends *all* of these integral formulas. Next we present a systematic and unified account of the theory of various families of generalized elliptic integrals which were studied recently by many authors. We also indicate relevant connections of the results presented here with those given by earlier workers on the theory of the aforementioned families of elliptic-type integrals.

We should like to remark in passing that an excellent source of simpler cases of the integral formulas of the types considered here happens to be the recent work by Prudnikov *et al.* [10] in which most (if not all) of the classical results on the subject can be found. More importantly, various interesting families of generalized elliptic integrals, and indeed also definite integrals of such families with respect to their modulus (or complementary modulus), are known to arise naturally in a number of seemingly diverse physical contexts; for instance, in the analysis of certain radiation field problems (*cf.* Berger and Lamkin [11]; see also Epstein and Hubbell [12]), in the studies of crystallographic minimal surfaces (*cf.*, *e.g.*, Cvijović and Klinowski [13]), in the theory of scattering of acoustic or electromagnetic waves by means of an elliptic disk (*cf.* Björkberg and Kristensson [14]; see also Kristensson [15]), in the studies of elliptical crack problems in fracture mechanics (*cf.*, *e.g.*, Mayrhofer and Fischer [16]), and so on.

2. A General Theorem and Its Consequences

We begin by observing that (since $|k| < 1$) the definition (1.3) would remain valid even when the parameter γ is unrestricted, in general. Thus, when $\gamma \in \mathbb{R}$ (or, more generally, when $\gamma \in \mathbb{C}$), it is easily seen from (1.3) that $H(k,-1) = E(k)/(1-k^2)$.

A unification (and generalization) of each of the aforementioned integral formulas (considered, for example, by Barton [3], Bushell [4], and others) is contained in the following theorem (*cf.* Srivastava [17]; see also [18, pp. 336–337]):

Theorem. *If* $\Phi(z) := \sum_{n=0}^{\infty} a_n z^n$ $(|z| < z_0;\ z_0 > 0)$ *and* $\sum_{n=1}^{\infty} a_n/n^{1+\frac{1}{2}\sigma}| < \infty$ $(\Re(\sigma) > -2)$, *then*

$$\int_0^1 k^\rho \kappa^\sigma \Phi(zk) H(\zeta\kappa, \gamma)\, dk = \frac{\pi}{4}\Gamma\left(\frac{1}{2}\sigma + 1\right) \sum_{n=0}^{\infty} \frac{\Gamma[\frac{1}{2}(\rho + n + 1)]}{\Gamma[\frac{1}{2}(\rho + \sigma + n + 3)]} a_n z^n \cdot {}_3F_2\left[\begin{matrix} \frac{1}{2} - \gamma,\ \frac{1}{2}\sigma + 1,\ \frac{1}{2}; \\ \frac{1}{2}(\rho + \sigma + n + 3),\ 1; \end{matrix}\ \zeta^2\right] \quad \left(\kappa := \sqrt{1 - k^2}\right), \tag{2.1}$$

provided further that $\Re(\rho) > -1$ *and* $|\zeta| < 1$ *(or* $|\zeta| = 1$ *and* $\Re(\rho + 2\gamma) > -1)$.

In the special case when $\sigma = 0$, the Clausenian hypergeometric function ${}_3F_2$ occurring on the right-hand side of (2.1) would reduce at once to the Gauss hypergeometric function ${}_2F_1$ which, for $\zeta = 1$, can be summed by means of the Gauss summation theorem which we referred to in the preceding section, and we thus obtain a general integral formula (*cf.*, *e.g.*, [18, p. 338, Equation (2.13)]) whose *further* special case when $\rho = 0$ and $z = 1$ happens to be the main theorem in Bushell's paper [4, p. 5, Equation (4.1)]. Numerous other consequences of (4.1) are also noteworthy.

3. A Brief Historical Introduction to Various Other Families of Generalized Elliptic-Type Integrals

Motivated by their importance or potential for applications in radiation physics, several recent works were devoted exclusively to the study of various interesting generalizations of the complete elliptic integrals $K(k)$ and $E(k)$. For example, over three decades ago, Epstein and Hubbell [12] (and, in a sequel, Weiss [19]) studied the following family of elliptic integrals:

$$\Omega_j(k) := \int_0^\pi (1 - k^2 \cos\theta)^{-j-\frac{1}{2}}\, d\theta \quad \left(j \in \mathbb{N}_0;\ 0 \leq k < 1\right), \tag{3.1}$$

which were encountered in a Legendre polynomial expansion method (*cf.* Berger and Lamkin [11]) when applied to certain problems involving computation of the radiation field off-axis from a uniform circular disk radiating according to an arbitrary distribution law (see also Hubbell *et al.* [20]). This family of elliptic integrals was further extended by Kalla and Al-Saqabi [21] to allow the parameter j in (3.1) to take on *complex* values. In fact, it is easily observed from the definition (1.3) that

$$H(k, \gamma) = \frac{(2 - k^2)^{\gamma - \frac{1}{2}}}{2^{\gamma + \frac{1}{2}}}\, \Omega_{-\gamma}\left(\frac{k}{\sqrt{2 - k^2}}\right) \qquad (k^2 < 1;\ \gamma \in \mathbb{C}). \tag{3.2}$$

Many subsequent contributions to the study of the Epstein-Hubbell generalized elliptic-type integral $\Omega_j(k)$ $(j \in \mathbb{N}_0)$, defined by (3.1), include the recent works by Evans *et al.* [22] and Todorov and Hubbell [23] (see also Al-Zamel [24], Srivastava [25], and Al-Zamel and Kalla [26]). And, as we just observed above, a rather *mild* extension of the Epstein-Hubbell integral (3.1) to allow the parameter j in $\Omega_j(k)$ to take on *complex* values was considered by Kalla and Al-Saqabi [21]. Kalla [27], on the other hand, studied a further generalization of $\Omega_\mu(k)$ $(\mu \in \mathbb{C})$ in the form:

$$S_\mu(k, \nu) := \int_0^\pi \frac{\sin^{2\nu}\theta}{(1 - k^2\cos\theta)^{\mu + \frac{1}{2}}}\, d\theta \quad \left(0 \leq k < 1;\ \mu \in \mathbb{C};\ \Re(\nu) > -\tfrac{1}{2}\right), \tag{3.3}$$

the *additional* constraint $\Re(\mu) > -\frac{1}{2}$, imposed by many earlier authors (*cf.*, *e.g.*, [24], [26], [28], [29], [22], [30], [21], [27], [31]–[33] and [34]) upon the definitions of $\Omega_\mu(k)$, $S_\mu(k,\nu)$, and some other related elliptic-type integrals mentioned below, being *unnecessary*.

Kalla *et al.* ([32] and [33]) (and, subsequently, Glasser and Kalla [30]) developed a systematic study of the following family of elliptic-type integrals:

$$R_\mu(k,\alpha,\gamma) := \int_0^\pi \frac{\cos^{2\alpha-1}\left(\frac{\theta}{2}\right)\sin^{2\gamma-2\alpha-1}\left(\frac{\theta}{2}\right)}{(1-k^2\cos\theta)^{\mu+\frac{1}{2}}}\,d\theta \tag{3.4}$$
$$(0 \leq k < 1;\ \Re(\gamma) > \Re(\alpha) > 0;\ \mu \in \mathbb{C}).$$

Kalla and Al-Saqabi [31] studied the family of integrals [*cf.* Definition (3.3)]:

$$K_\mu(k,m) := \int_0^\pi \frac{\cos^{2m}\theta}{(1-k^2\cos\theta)^{\mu+\frac{1}{2}}}\,d\theta \quad (0 \leq k < 1;\ m \in \mathbb{N}_0;\ \mu \in \mathbb{C}), \tag{3.5}$$

Al-Saqabi [28], on the other hand, considered an interesting unification of these two families of elliptic-type integrals [$S_\mu(k,\nu)$ and $K_\mu(k,m)$] in the form:

$$B_\mu(k,m,\nu) := \int_0^\pi \frac{\cos^{2m}\theta\sin^{2\nu}\theta}{(1-k^2\cos\theta)^{\mu+\frac{1}{2}}}\,d\theta \tag{3.6}$$
$$\left(0 \leq k < 1;\ m \in \mathbb{N}_0;\ \mu \in \mathbb{C};\ \Re(\nu) > -\tfrac{1}{2}\right),$$

Apparently unaware of the aforementioned developments, especially about the elliptic-type integral $\Omega_\mu(k)$ ($\mu \in \mathbb{C}$), Das [35] (in the same year as Bushell [4]) studied the complete elliptic integrals $K(k)$ and $E(k)$ in a unified form $H_\nu(k)$, where [*cf.* Equation (3.2)]:

$$H_\nu(k) = \frac{(2-k^2)^\nu}{2^{\nu+1}}\,\Omega_{-\nu-\frac{1}{2}}\left(\frac{k}{\sqrt{2-k^2}}\right) = H\left(k,\ \nu+\tfrac{1}{2}\right) \quad (k^2 < 1;\ \nu \in \mathbb{C}). \tag{3.7}$$

The following family of elliptic-type integrals:

$$F_{m,n}(k) := (1-k^2)^{n/2}\int_0^{2\pi} \frac{\cos 2m\theta}{(1-k^2\cos^2\theta)^{n/2}}\,d\theta \tag{3.8}$$

studied by Björkberg and Kristensson [14] (see also Kristensson [15]) occurs in the theory of scattering of acoustic or electromagnetic waves by means of an elliptic disk with semi-axes a and b and eccentricity $k := \sqrt{1-(b/a)^2}$ ($a > b > 0$). Motivated by the definition (3.8), and by the Epstein-Hubbell integral $\Omega_j(k)$ ($j \in \mathbb{N}_0$) defined by (3.1), Mohamed-Murid [36] presented a detailed study of an elliptic-type integral $H(\nu,m,\kappa)$ (see also Bromberg [37]). In fact, both $\Omega_\mu(k)$ and $F_{m,n}(k)$ are contained in $H(\nu,m,\kappa)$ as special cases. More interestingly, we have the relationship (as observed earlier by Srivastava and Bromberg [38, p. 30, Equation (9)]):

$$H(\nu,m,k^2) = \frac{(\nu)_m\,k^{2m}}{2^{m-1}\left(\frac{1}{2}\right)_m}\,S_{\nu+m}(k,m) \quad \left(\nu \in \mathbb{R};\ m \in \mathbb{N}_0;\ 0 \leq k < 1\right). \tag{3.9}$$

Next we recall yet another family of elliptic-type integrals (studied by Siddiqi [34]) in the form:

$$\Lambda_\nu(\alpha,k) := \int_0^\pi \frac{\exp\left[\alpha\sin^2\left(\frac{\theta}{2}\right)\right]}{(1-k^2\cos\theta)^{\nu+\frac{1}{2}}}\,d\theta \quad (0 \leq k < 1;\ \alpha,\nu \in \mathbb{R}), \tag{3.10}$$

which does not seem to reduce, under suitable choices for the parameters α and ν, to any other of the aforementional generalizations of the complete elliptic integrals $K(\kappa)$ and $E(\kappa)$ except possibly $\Omega_\nu(k)$ when $\alpha = 0$.

Mayrhofer and Fischer [16] considered the following family of elliptic-type integrals:

$$\Xi_n^{(\ell,m)}(k) := \int_0^\pi \frac{\sin^\ell \theta \, \cos^m \theta}{\left(1 - k^2 \sin^2 \theta\right)^{n+\frac{1}{2}}} \, d\theta \qquad \left(0 \leq k^2 < 1; \; \ell, m, n \in \mathbb{N}_0\right), \tag{3.11}$$

which may be shown to satisfy the relationship:

$$\Xi_n^{(\ell,m)}(k) = \frac{2^{n-\frac{1}{2}} \left[1 + (-1)^m\right]}{(2-k^2)^{n+\frac{1}{2}}} R_n\left(\frac{ik}{2-k^2}, \frac{m+1}{2}, \frac{\ell+m+2}{2}\right). \tag{3.12}$$

Mayrhofer and Fischer [16] also considered the elliptic-type integral $G_n(k)$, where

$$G_n(k) = \frac{1}{\sqrt{2}\,(2-k^2)^{\frac{3}{2}}} H\left(\frac{3}{2}, n, \frac{k^2}{2-k^2}\right) = \frac{1}{4(1-k^2)^{\frac{3}{2}}} F_{n,3}(k). \tag{3.13}$$

Srivastava and Siddiqi [39] presented a systematic study of the following unification (and generalization) of *all* of the aforementioned families of elliptic-type integrals:

$$\Lambda_{\lambda,\mu}^{(\alpha,\beta)}(\rho; k) := \int_0^\pi \cos^{2\alpha-1}\left(\frac{\theta}{2}\right) \sin^{2\beta-1}\left(\frac{\theta}{2}\right) \frac{\left[1 - \rho \sin^2\left(\frac{\theta}{2}\right)\right]^{-\lambda}}{(1 - k^2 \cos\theta)^{\mu+\frac{1}{2}}} \, d\theta$$

$$\left(k^2 < 1; \; \Re(\alpha) > 0; \; \Re(\beta) > 0; \; \lambda, \mu \in \mathbb{C}; \; |\rho| < 1\right), \tag{3.14}$$

where it is tacitly assumed that $\rho \in \mathbb{C}$ whenever $\lambda = -m \; (m \in \mathbb{N}_0)$. As a matter of fact, yet another interesting special case of $\Lambda_{\lambda,\mu}^{(\alpha,\beta)}(\rho; k)$ is the complete elliptic integral of the *third* kind (*cf.*, *e.g.*, Byrd and Friedman [1, p. 10, Entry 110.08]; Gradshteyn and Ryzhik [40, pp. xxx and 905]), and we have

$$\Pi(\alpha^2, k) = \frac{1}{(1-\alpha^2)\sqrt{2(2-k^2)}} \Lambda_{1,0}^{(\frac{1}{2},\frac{1}{2})}\left(\frac{\alpha^2}{\alpha^2-1}; \frac{k}{\sqrt{2-k^2}}\right)$$
$$\left(k^2 < 1; \; -\infty < \alpha^2 < \infty \quad (\alpha^2 \neq 1)\right). \tag{3.15}$$

We remark in passing that none of the aforementioned earlier generalizations of elliptic-type integrals includes the complete elliptic integral of the *third* kind as a special case. Furthermore, the unification and generalization (3.14), studied rather systematically in [39] and [18], may itself prove to be useful in various radiation field and other problems in many of the areas of study, which we listed and referred to above. These earlier studies make use of the Appell functions ([41] and [42]), Mellin-Barnes contour integrals (*cf.* [43]), generating functions [44], fractional calculus (*cf.* [45], [46], and [47]), and so on (see also [48]).

4. Further Generalizations of the Srivastava-Siddiqi Elliptic-Type Integral $\Lambda_{\lambda,\mu}^{(\alpha,\beta)}(\rho; k)$

The following *further* generalization of the Srivastava-Siddiqi elliptic-type integral $\Lambda_{\lambda,\mu}^{(\alpha,\beta)}(\rho; k)$, defined by (3.14), was considered recently by Galué [49] (and, *independently*, also by Kalla and Tuan [50]):

$$\Lambda_{(\lambda,\gamma,\mu)}^{(\alpha,\beta)}(\rho,\delta;k) := \int_0^\pi \frac{\cos^{2\alpha-1}\left(\frac{\theta}{2}\right)\sin^{2\beta-1}\left(\frac{\theta}{2}\right)}{(1-k^2\cos\theta)^{\mu+\frac{1}{2}}} \cdot\left[1-\rho\sin^2\left(\frac{\theta}{2}\right)\right]^{-\lambda}\left[1+\delta\cos^2\left(\frac{\theta}{2}\right)\right]^{-\gamma} d\theta \tag{4.1}$$

$$(0 \leq k < 1;\ \Re(\alpha) > 0;\ \Re(\beta) > 0;\ \lambda,\mu,\gamma \in \mathbb{C};\ |\rho| < 1;\ |\delta| < 1),$$

where it is tacitly assumed that $\rho \in \mathbb{C}$ whenever $\lambda = -m$ $(m \in \mathbb{N}_0)$ and that $\delta \in \mathbb{C}$ whenever $\delta = -n$ $(n \in \mathbb{N}_0)$. Clearly, we have

$$\Lambda_{\lambda+\gamma,\mu}^{(\alpha,\beta)}(\rho;k) = (1-\rho)^{-\gamma}\,\Lambda_{(\lambda,\gamma,\mu)}^{(\alpha,\beta)}\left(\rho,\ \frac{\rho}{1-\rho};\ k\right), \tag{4.2}$$

and similarly for $\Lambda_{\lambda,\mu+\gamma}^{(\alpha,\beta)}(\rho;k)$, which evidently assert the *redundancy* of one of the two additional parameters γ *and* δ which were introduced by Galué [49], Kalla and Tuan [50], and presumably other authors as well.

In view of the above observations, it would seem to be natural to consider a much deeper generalization of all these elliptic-type integrals in the form:

$$\Omega_{\lambda_1,\cdots,\gamma_r;\mu_1,\cdots,\mu_s}^{(\alpha,\beta;\xi,\eta)}(\rho_1,\cdots,\rho_r;\sigma_1,\cdots,\sigma_s) := \int_0^\pi \cos^{2\alpha-1}\left(\frac{\theta}{2}\right)\sin^{2\beta-1}\left(\frac{\theta}{2}\right) \cdot\prod_{j=1}^r\left\{\left[1-\rho_j\sin^{2\xi}\left(\frac{\theta}{2}\right)\right]^{-\lambda_j}\right\}\prod_{j=1}^s\left\{\left[1-\sigma_j\cos^{2\eta}\left(\frac{\theta}{2}\right)\right]^{-\mu_j}\right\} d\theta \tag{4.3}$$

$$\Big(\Re(\alpha) > 0;\ \Re(\beta) > 0;\ \lambda_j,\ \mu_\ell \in \mathbb{C}\ (j = 1,\cdots,r;\ \ell = 1,\cdots,s);$$

$$\xi > 0;\ \eta > 0;\ |\arg(1-\rho_j)| < \pi\ \ (j = 1,\cdots,r);\ |\arg(1-\sigma_\ell)| < \pi\ \ (\ell = 1,\cdots,s)\Big),$$

where the aforementioned restrictions on $\rho_1,\cdots,\rho_r$ and $\sigma_1,\cdots,\sigma_s$ can be waived when λ_j $(j = 1,\cdots,r)$ and μ_ℓ $(\ell = 1,\cdots,s)$ take on negative integer values, respectively, the role of the modulus k being suitably incorporated in any of the ρ and σ parameters.

Explicit hypergeometric representations of the generalized elliptic-type integral (4.3) would involve Kampé de Fériet and Lauricella functions in several variables.

When $r = s = 1$, the definition (4.3) would obviously correspond to that in (3.14). Furthermore, (4.3) can easily be reduced to the definition (4.1) when $r = s - 1 = 1$. The details are fairly straightforward.

Acknowledgements

The present investigation was supported, in part, by the *Natural Sciences and Engineering Research Council of Canada* under Grant OGP00007353.

References

1. P.F. Byrd and M.D. Friedman, Handbook of Elliptic Integrals for Engineers and Scientists, Second Edition (Revised), Die Grundlehren der Mathematischen Wissenschaften in Einzeldarstellungen, Band 67, Springer-Verlag, New York, Heidelberg, and Berlin (1971).

2. M. Abramowitz and I.A. Stegun (Editors), Handbook of Mathematical Functions With Formulas, Graphs, and Mathematical Tables, Applied Mathematics Series, Vol. **55**, National Bureau of Standards, Washington, D.C. (1964).

3. G. Barton, Proc. Roy. Soc. London Ser. A, 388A, 445-456 (1983).

4. P.J. Bushell, Math. Proc. Cambridge Philos. Soc., 101, 1-5 (1987).

5. K.F. Müller, Arch. Elektrotech., 17, 336-353 (1926).

6. E.L. Kaplan, J. Math. and Phys., 29, 69-75 (1950).

7. A. Erdélyi, W. Magnus, F. Oberhettinger and F.G. Tricomi, Higher Transcendental Functions, Vol. I, McGraw-Hill, New York, Toronto, and London (1953).

8. H.M. Srivastava and P.W. Karlsson, Multiple Gaussian Hypergeometric Series, Halsted Press (Ellis Horwood Limited, Chichester), John Wiley and Sons, New York, Chichester, Brisbane, and Toronto (1985).

9. G.H. Hardy, Proc. Cambridge Philos. Soc., 21, 492-503 (1923).

10. A.P. Prudnikov, Yu. A. Brychkov, and O.I. Marichev, Integrals and Series, Vol. 3: More Special Functions, Gordon and Breach, New York (1990).

11. M.J. Berger and J.C. Lamkin, J. Nat. Bur. Standards, 60, 109-116 (1958).

12. L.F. Epstein and J.H. Hubbell, J. Res. Nat. Bur. Standards B: Math. and Math. Phys., 67B, 1-17 (1963).

13. D. Cvijović and J. Klinowski, Proc. Roy. Soc. London Ser. A, 444A, 525-532 (1994).

14. J. Björkberg and G. Kristensson, Canad. J. Phys., 65, 723-734 (1987).

15. G. Kristensson, J. Sound and Vibration, 103, 487-498 (1985).

16. K. Mayrhofer and F.D. Fischer, Z. Angew. Math. Mech., 74, 265–273 (1994).

17. H.M. Srivastava, J. Phys. A: Math. Gen., 28, 2305–2312 (1995).

18. H.M. Srivastava, in: New Frontiers in Algebras, Groups and Geometries. Part III: New Frontiers in Functional Analysis, (Grigorios T. Tsagas, Editor), Hadronic Press, Palm Harbor, Florida (1996) pp. 329–387.

19. G.H. Weiss, J. Res. Nat. Bur. Standards B: Math. and Math. Phys., 68B, 1-2 (1964).

20. J.H. Hubbell, R.L. Bach, and R.J. Herbold, J. Res. Nat. Bur. Standards C: Engrg. and Instrument., 65C, 249-264 (1961).

21. S. Kalla and B. Al-Saqabi, Resultate Math., 20, 507-516 (1991).

22. J.D. Evans, J.H. Hubbell, and V.D. Evans, Appl. Math. Comput., 53, 173-189 (1993).

23. P.G. Todorov and J.H. Hubbell, Appl. Math. Comput., 61, 157-161 (1994).

24. A. Al-Zamel, in: Transform Methods and Special Functions, (Proceedings of the First International Workshop held at Bankya [Bulgaria] on August 12–17, 1994) (P. Rusev, I. Dimovski, and V. Kiryakova, Editors), SCT Publishing Company, Singapore (1995) pp. 1–7.

25. R. Srivastava, Appl. Math. Comput., 69, 255–262 (1995).

26. A. Al-Zamel and S. Kalla, Appl. Math. Comput., 77, 9–32 (1996).

27. S.L. Kalla, in: Mathematical Structures — Computational Mathematics — Mathematical Modelling, Vol. 2 (B. Sendov, Editor), Bulgarian Academy of Sciences, Sofia (1984) pp. 216-219.

28. B.N. Al-Saqabi, Hadronic J., 10, 331–337 (1987).

29. B. Al-Saqabi and S.L. Kalla, Rev. Brasileira Fís., 18, 135-155 (1988).

30. M.L. Glasser and S.L. Kalla, Rev. Técn. Fac. Ingr. Univ. Zulia, 12, 47–49 (1989).

31. S.L. Kalla and B. Al-Saqabi, Rev. Brasileira Fís., 16, 145-156 (1986).

32. S.L. Kalla, S. Conde, and J.H. Hubbell, Applicable Anal., 22, 273-287 (1986).

33. S.L. Kalla, C. Leubner, and J.H. Hubbell, Applicable Anal., 25, 269-274 (1987).

34. R.N. Siddiqi, Rev. Brasileira Fís., 19, 137-147 (1989).

35. J. Das, Bull. Calcutta Math. Soc., (Festschrift for M. Dutta) 73–82 (1987).

36. A.H. Mohamed-Murid, Numerical Computation of Some Complete Elliptic Integrals of the First and Second Kinds, M.S. Thesis, Iowa State University of Science and Technology, Ames, Iowa (1986).

37. S. Bromberg, Rev. Técn. Fac. Ingr. Univ. Zulia, 15, 29-35 (1992).

38. H.M. Srivastava and S. Bromberg, Math. Comput. Modelling, 21, 29–38 (1995).

39. H.M. Srivastava and R.N. Siddiqi, Radiation Phys. Chem., 46, 303–315 (1995).

40. I.S. Gradshteyn and I.M. Ryzhik, Table of Integrals, Series, and Products, Corrected and Enlarged Edition prepared by A. Jeffrey (and incorporating the Fourth Edition prepared by Yu. V. Geronimus and M. Yu. Tseytlin), Academic Press, New York, London, Toronto, and Tokyo (1980).

41. P. Appell and J. Kampé de Fériet, Fonctions Hypergéométriques et Hypersphériques; Polynômes d'Hermite, Gauthier-Villars, Paris (1926).

42. H.M. Srivastava and B.R.K. Kashyap, Special Functions in Queuing Theory and Related Stochastic Processes, Academic Press, New York and London (1982).

43. E.T. Whittaker and G.N. Watson, A Course of Modern Analysis: An Introduction to the General Theory of Infinite Processes and of Analytic functions; With an Account of the Principal Transcendental Functions, Fourth edition, Cambridge University Press, Cambridge, London, and New York (1927).

44. H.M. Srivastava and H.L. Manocha, A Treatise on Generating Functions, Halsted Press (Ellis Horwood Limited, Chichester), John Wiley and Sons, New York, Chichester, Brisbane, and Toronto (1984).

45. K.S. Miller and B. Ross, An Introduction to the Fractional Calculus and Fractional Differential Equations, John Wiley and Sons, New York, Chichester, Brisbane, Toronto, and Singapore (1993).

46. K.B. Oldham and J. Spanier, The Fractional Calculus: Theory and Applications of Differentiation and Integration to Arbitrary Order, Academic Press, New York and London (1974).

47. S.G. Samko, A.A. Kilbas and O.I. Marichev, Fractional Integrals and Derivatives: Theory and Applications, Gordon and Breach Science Publishers, Reading, Tokyo, Paris, Berlin, and Langhorne (Pennsylvania) (1993).

48. D.M. Timuş and H.M. Srivastava, Radiation Phys. Chem., 49, 319-326 (1997).

49. L. Galué, Algebras Groups Geom., 14, 83–105 (1997).

50. S.L. Kalla and Vu Kim Tuan, Comput. Math. Appl., 32, 49–55 (1996); Also published in: Recent Advances in Applied Mathematics, (Proceedings of an International Workshop held at Kuwait University on May 4-7, 1996), Kuwait Foundation for the Advancement of Sciences, State of Kuwait (1996) pp. 273-280.

51. A. Erdélyi, W. Magnus, F. Oberhettinger and F.G. Tricomi, Higher Transcendental Functions, Vol. II, McGraw-Hill, New York, Toronto, and London (1953).

52. A.P. Prudnikov, Yu. A. Bryčkov and O.I. Maričev, Integrals and Series: Supplementary Chapters (in Russian), "Nauka", Moscow (1986).

53. E.D. Rainville, Special Functions, Macmillan, New York, 1960; Reprinted by Chelsea Publishing Company, Bronx, New York (1971).

54. M. El-Gabali, Rev. Acad. Canaria Cienc., 6, 105–118 (1994).

8th Int. Coll. on Differential Equations, pp. 409–414
D. Bainov (Ed)

Computing Derivatives of Eigenvectors Corresponding to Multiple Eigenvalues of Nonsymmetric Matrix Functions

ROGER C. E. TAN
Department of Mathematics, National University of Singapore, Kent Ridge Crescent, Singapore 119260

ALAN L. ANDREW
School of Mathematics, La Trobe University, Bundoora, Victoria 3083, Australia

Abstract. Consider the eigenvalue problem

$$A(\rho)\mathbf{x}_i(\rho) = \lambda_i(\rho)B(\rho)\mathbf{x}_i(\rho)$$

where A and B are analytic, generally nonsymmetric, $n \times n$ matrix-valued functions depending on the parameter $\rho \in \mathbb{R}$, and λ_i and $\mathbf{x}_i$ are the eigenvalues and corresponding eigenvectors of the generalized eigenvalue problem. This paper concerns the numerical computation of the derivatives $\lambda_i'(\rho_0)$ and $\mathbf{x}_i'(\rho_0)$ at some point ρ_0 where $B(\rho_0)$ is nonsingular and, for some $r > 1$ and all $i > r$, $\lambda_1(\rho_0) = \cdots = \lambda_r(\rho_0) \neq \lambda_i(\rho_0)$. Numerical estimates of these derivatives give important information on eigenvalue and eigenvector sensitivity which is required in many engineering applications such as optimum design of structures, mode tracking and model updating. We present some new algorithms for the calculation of these derivatives and show that the methods presented here can be used for some problems for which existing methods fail. Numerical examples are also given to demonstrate the viability of the proposed algorithms.

Keywords. multiple eigenvalues, derivatives (sensitivities) of eigenvalues and eigenvectors.

1. Introduction

In this paper, we consider the eigenvalue problem

$$A(\rho)\mathbf{x}_i(\rho) = \lambda_i(\rho)B(\rho)\mathbf{x}_i(\rho) \quad i = 1, \ldots, n \tag{1}$$

where A and B are nonsymmetric analytic, $n \times n$ matrix-valued functions depending on the parameter $\rho \in \mathbb{R}$, and λ_i and $\mathbf{x}_i$ are the eigenvalues and corresponding eigenvectors of the generalized eigenvalue problem.

The solution of optimal design problems in engineering often requires the computation of derivatives, λ_i' and $\mathbf{x}_i'$, of the eigenvalues, λ_i, and eigenvectors, $\mathbf{x}_i$ at ρ_0. This computation is the subject of many papers. (See [1], [2], [3], [4] for some references.) Most consider only simple eigenvalues [5], [6], [7]. However, it is well known that eigenvalues often coalesce at an optimum design.(See [8] and [9].)

We consider here the numerical computation of $\lambda_i'(\rho_0)$ and $x_i'(\rho_0)$ at some ρ_0 such that, for some $r > 1$ and all $i > r$,

$$\lambda_1(\rho_0) = \cdots = \lambda_r(\rho_0) \neq \lambda_i(\rho_0),$$

under the assumption that all required derivatives exist throughout some neighbourhood of ρ_0. We assume also that $B^{-1}(\rho)A(\rho)$ exists and is nondefective in some neighbourhood of ρ_0.

First derivatives of repeated eigenvalues are relatively easy to compute. Differentiation of (1) shows that

$$(A(\rho) - \lambda_i(\rho)B(\rho))\mathbf{x}_i'(\rho) = (\lambda_i(\rho)B'(\rho) - A'(\rho))\mathbf{x}_i(\rho) + \lambda_i'(\rho)B(\rho)\mathbf{x}_i(\rho). \tag{2}$$

Suppose $\lambda_1(\rho_0)$ is an eigenvalue of the matrix $A^{-1}(\rho_0)B(\rho_0)$ with multiplicity r, and the $n \times r$ matrices $\hat{X}, \hat{Y}$ satisfy

$$A(\rho_0)\hat{X} = \lambda_1(\rho_0)B(\rho_0)\hat{X} \quad \text{and} \quad \hat{Y}^*A = \lambda_1(\rho_0)\hat{Y}^*B(\rho_0),$$

with their columns spanning the the same subspaces as $\{\mathbf{x}_1(\rho_0), \ldots, \mathbf{x}_r(\rho_0)\}$ and $\{\mathbf{y}_1(\rho_0), \ldots, \mathbf{y}_r(\rho_0)\}$ respectively, and $\hat{Y}^*B(\rho_0)\hat{X}$ is nonsingular. Then the derivatives $\lambda_1'(\rho_0)$ are the eigenvalues of the matrix

$$(\hat{Y}^*B(\rho_0)\hat{X})^{-1}\left[\hat{Y}^*\left((A'(\rho_0) - \lambda_1(\rho_0)B'(\rho_0))\hat{X}\right]\right..$$

The relationship between the differentiable matrix $X = (\mathbf{x}_1, \ldots, \mathbf{x}_r)$ and the matrix $\hat{X}$, whose columns span the same space as the columns of $X(\rho_0)$, is discussed in [10] and, in more detail, in [11].

Early attempts to compute derivatives of the corresponding eigenvectors met with mixed success [10]. [12] gives a good method for computing $X'(\rho_0)$ in the case in which the $\lambda_i'(\rho_0)$ are well separated for symmetric matrices A and B with $B(\rho_0)$ positive definite. In [13] the result of [12] is extended to allow first derivatives of repeated eigenvalues to coincide provided the second derivatives are well separated. For the symmetric case, we have developed in [11] a more general method having no restrictions on eigenvalue derivatives, which is also simpler than the method of [13]. Other contributions of [11] include a method for computation of higher order derivatives of $\mathbf{x}_i$ and detailed proofs of two important theorems which are used in the work here. It also addresses issues concerning close eigenvalues and stability.

As remarked in [11], the algorithms developed there can be extended to handle cases involving nonsymmetric matrices A and B using the subspace spanned by the left eigenvectors corresponding to the repeated eigenvalues but details of how this may be implemented were not elaborated. In this paper we illustrate how the algorithms in [11] may be extended to nonsymmetric problems by giving details of the case in which repeated eigenvalues have repeated first and second order derivatives but well separated third derivatives. For the symmetric problem, this case was considered in [14].

2. The Algorithm

Since the corresponding $\mathbf{x}_i'(\rho_0)$ are indeterminate if $\lambda_1(\rho) = \cdots = \lambda_r(\rho)$ throughout an open neighbourhood of ρ_0, we consider only the (common) case in which, for all $\rho \neq \rho_0$ in some neighbourhood of ρ_0, the $\lambda_i(\rho)$ are simple. The derivatives of eigenvectors depend on the normalizing condition used. In [11] we used a normalizing condition which is appropriate in the symmetric case where $B(\rho_0)$ is positive definite. That condition is not appropriate for the more general problem considered here and hence we use instead the condition $\mathbf{x}_i^*(\rho_0)\mathbf{x}_i(\rho_0) = \hat{\mathbf{x}}_i^*\hat{\mathbf{x}}_i = 1, \quad i = 1, \ldots r$. Algorithm 2.1 can be readily modified to deal with other normalizing conditions by making appropriate changes to the diagonal elements of the matrix C in Step 5. Normalization is considered in more detail in [1].

Algorithm 2.1 below computes X, X' and $\Lambda^{(i)}$ for $i = 1, \ldots 4$ in the special case where an eigenvalue of multiplicity r also has repeated first and second order derivatives of multiplicity r at ρ_0, i.e.

$$\Lambda^{(i)}(\rho_0) = \lambda^{(i)}(\rho_0)I, \qquad i = 0, 1, 2$$

and the third order derivatives of the r repeated eigenvalues are well separated. (Note that even this simple case cannot be dealt with by the methods of [12] or [13].)

In what follows, all functions are evaluated at ρ_0.

Algorithm 2.1

Step 1. Find a solution V_1 of

$$W_0 V_1 = -W_1 \hat{X}$$

where

$$W_0 := A - \lambda B,\ W_1 := Z_0 - \mu_1 B,\ Z_0 := A' - \lambda B',$$

and μ_1 is the mean of the diagonal elements of $(\hat{Y} B \hat{X})^{-1} \hat{Y}^* Z_0 \hat{X}$.

(The coefficient matrix W_0 is singular, but existence of solutions of this equation and the corresponding equation in Step 2 can be established by the same argument as given in [11], and these solutions can be computed by the methods described in [11].)

Step 2. Find a solution V_2 of

$$W_0 V_2 = -T_1 - W_2 \hat{X}$$

where

$$W_2 := Z_1 - \mu_2 B,\ Z_1 := A'' - \lambda B'' - 2\mu_1 B',\ T_1 := 2W_1 V_1,$$

and μ_2 is the mean of the diagonal elements of $(\hat{Y} B \hat{X})^{-1} \hat{Y}^* (Z_1 \hat{X} + T_1)$.

Step 3. Using the *eig* command of MATLAB or otherwise, compute a matrix $U = (\mathbf{u}_1, \ldots, \mathbf{u}_r)$ and a diagonal matrix Λ_3 such that

$$U \Lambda_3 U^* = (\hat{Y} B \hat{X})^{-1} \hat{Y}^* (Z_2 \hat{X} + T_2),$$

where

$$Z_2 := A''' - \lambda B''' - 3\mu_1 B'' - 3\mu_2 B',$$

and

$$T_2 := 3W_2V_1 + 3W_1V_2.$$

Then it can be shown that μ_1 and μ_2 are the (repeated) first and second derivatives respectively of λ at ρ_0,

$$\Lambda_3 = \operatorname{diag}(\lambda_1'''(\rho_0), \ldots, \lambda_r'''(\rho_0))$$

and

$$\hat{X}U = X(\rho_0) = (\mathbf{x}_1(\rho_0), \ldots, \mathbf{x}_r(\rho_0)).$$

Step 4. Find a solution V_3 of

$$W_0V_3 = -Z_2\hat{X} + BX\Lambda_3U^* - T_2.$$

Compute

$$D := (\hat{Y}BX)^{-1}\hat{Y}^*\left\{(A^{(4)} - \lambda B^{(4)} - 4\mu_1B''' - 6\mu_2B'')X\right.$$
$$\left.+4(Z_2V_1U - BV_1U\Lambda_3 - B'X\Lambda_3) + 6W_2V_2U + 4W_1V_3U\right\}.$$

Then, as in [11], D can be shown to have ith diagonal element $\lambda_i^{(4)}(\rho_0)$.

Step 5. Compute the matrix C whose diagonal elements are

$$c_{ii} = -\mathbf{x}_i^*V_1\mathbf{u}_i - \sum_{\substack{p=1\\p\neq i}}^{r} c_{pi}\mathbf{x}_i^*\mathbf{x}_p, \quad i = 1, \ldots, r,$$

and whose other elements are given by

$$c_{ij} := \frac{d_{ij}}{4(\lambda_j''' - \lambda_i''')}, \quad i \neq j,$$

where c_{ij} and d_{ij} are the elements in the ith row and jth column of C and D respectively. An argument similar to that in [12] shows that this gives the required result $X' = V_1U + XC$. If a different normalizing condition is used, the diagonal elements of C must generally be defined differently.

As suggested for the symmetric case in [14], after Step 1, we may proceed to assume that $\Lambda'(\rho_0) = \lambda'(\rho_0)I$ if $\|(\hat{Y}B\hat{X})^{-1}\hat{Y}^*Z_0\hat{X}\| \leq \epsilon\|(\hat{Y}B\hat{X})^{-1}Z_0\|$, where ϵ is say 10 times the machine epsilon. Numerical results in [11] have indicated ϵ could be significantly much larger. (See [11] for details.) Otherwise, the structure of $(\hat{Y}B\hat{X})^{-1}\hat{Y}Z_0\hat{X}$ gives details on which the λ_i' are simple, and the method of [10] can be used for this. Also, information regarding Λ'' may be obtained by checking $(\hat{Y}B\hat{X})^{-1}\hat{Y}^*(Z_1\hat{X}+T_1)-\mu_2I$ after Step 2. Details for the general case including computation of higher order derivatives of eigenvectors corresponding to repeated eigenvalues with repeated derivatives of finite order will be dealt with in a later paper.

3. Numerical Example

We illustrate the procedure for implementing the algorithms by considering a 4×4 eigenvalue problem with closed form solution. Here we choose A to be of the form

$$A = SDS^{-1}$$

where

$$S = \begin{bmatrix} \alpha & 1 & 1 & 1 \\ 1 & 2\alpha & 1 & 0 \\ 2 & 1 & 3\alpha & 1 \\ 0 & 1 & 1 & 4\alpha \end{bmatrix},$$

$$D = \text{diag}(3\alpha, \alpha^3 + \alpha^2 - 1, 4\alpha^2 - 3\alpha, 2\alpha)$$

and $B = I$.
The eigenvalues for this example are the diagonal elements of D with the corresponding eigenvectors given by the columns of S.
For $\alpha = 1$,

$$A = \begin{bmatrix} 7.000000 & -2.000000 & -1.000000 & -1.000000 \\ 5.428571 & -0.714286 & -0.857143 & -1.142857 \\ 11.428571 & -3.714286 & -0.857143 & -2.142857 \\ 2.285714 & -1.142857 & -0.571429 & 1.571429 \end{bmatrix}$$

giving repeated eigenvalues

$$\lambda_1 = 3, \; \lambda_2 = \lambda_3 = 1, \; \lambda_4 = 2,$$

with repeated first and second order eigenvalue derivatives

$$\lambda_1' = 3, \; \lambda_2' = \lambda_3' = 5, \; \lambda_4' = 2,$$

$$\lambda_1'' = 0, \; \lambda_2'' = \lambda_3'' = 8, \; \lambda_4'' = 0,$$

$$\lambda_1''' = 0, \; \lambda_2''' = 6, \; \lambda_3''' = 0, \; \lambda_4''' = 0.$$

The two left and right eigenvectors of A corresponding to the repeated eigenvalues computed by MATLAB are the columns of

$$\hat{X} = \begin{bmatrix} -0.322811 & -0.177708 \\ -0.413008 & -0.750116 \\ -0.788042 & 0.611691 \\ -0.322811 & -0.177708 \end{bmatrix} \quad \hat{Y} = \begin{bmatrix} -0.859929 & 0.168243 \\ 0.434936 & -0.904534 \\ 0.212497 & 0.368145 \\ 0.161858 & -0.134097 \end{bmatrix}.$$

Also, we compute

$$V_1 = \begin{bmatrix} 0 & 0 \\ 0 & 0 \\ -1.058631 & -1.105532 \\ 0 & 0 \end{bmatrix},$$

the matrix

$$U = \begin{bmatrix} -0.979192 & -0.884038 \\ 0.154295 & -0.521005 \end{bmatrix}$$

and

$$C = \begin{bmatrix} -0.750000 & -2.618615 \\ 0.000000 & 1.428571 \end{bmatrix}.$$

Hence,

$$X' = V_1 U + \hat{X} U C = \begin{bmatrix} -0.216506 & -0.215980 \\ -0.216506 & 0.323970 \\ 0.216506 & -0.215980 \\ -0.216506 & -0.215980 \end{bmatrix}.$$

Note that the first and second columns of this matrix are the derivatives of the normalized third and second columns respectively of S.

Acknowledgement

Part of this work was carried out while the first author visited La Trobe University supported by a grant from the Australian Research Council.

References

1. A. L. Andrew, K.-W. E. Chu and P. Lancaster, SIAM J. Matrix Anal. Appl., 14, 903–926 (1993).
2. A. L. Andrew and R. C. E. Tan, AIAA J., 34, 2214–2216 (1996).
3. R. T. Haftka and H. M. Adelman, Structural Optimization, 1, 137–151 (1989).
4. J. E. Mottershead and M. I. Friswell, J. Sound Vibration, 167, 347–375 (1993).
5. R. B. Nelson, AIAA J., 14, 1201–1205 (1976).
6. R. C. E. Tan and A. L. Andrew, IMA J. Numer. Anal., 9, 111–122 (1989).
7. R. C. E. Tan, A. L. Andrew and F. M. L. Hong, Comm. Numer. Meth. Engrg., 10, 1–9 (1994).
8. E. J. Haug, K. K. Choi and V. Komkov, Design Sensitivity Analysis of Structural Systems, Academic Press, New York (1986).
9. A. P. Seyranian, E. Lund and N. Olhoff, Structural Optimization, 8, 207–227 (1994).
10. W. C. Mills-Curran, AIAA J., 26, 867–871 (1988).
11. A. L. Andrew and R. C. E. Tan, Math. Res. Paper, La Trobe University, No. 96-14 (1996).
12. W. C. Mills-Curran, AIAA J., 28, 1846 (1990).
13. J. Shaw and S. Jayasuriya, AIAA J., 30, 850–852 (1992).
14. A. L. Andrew and R. C. E. Tan, Z. Angew. Math. Mech., 76, 467–468 (1996).

8th Int. Coll. on Differential Equations, pp. 415-422
D. Bainov (Ed)

Error Estimate of Elliptic Integral by Chebyshev Collocation Method

KUMIKO TSUJI
Kyushu Teikyo Junior College, 4-3-124 Shin-Katsutachi-Machi Ohmuta 836, Japan

Abstract. The error estimate of the collocation Chebyshev approximation (Cheb-C) is obtained based on the orthogonality of finite sum with respect to the maximum norm of higher derivatives of the functions. The elliptic integral is approximated by a connection of Cheb-C, whose intervals are controlled so as to be decrease the approximation error in extraordinary small in the neighborhood of logarithmic singular point.

Keywords: Chebyshev polynomial, elliptic integral

1 Introduction

1.1 Simple pendulum

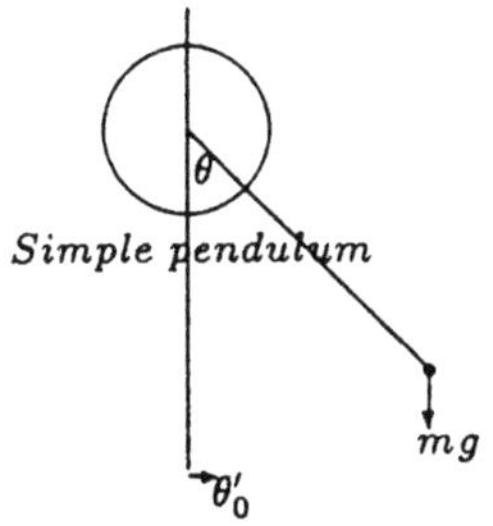

The angle velocity $d\theta/ds$ is given by $(\frac{d\theta}{ds})^2 = (\theta_0')^2 - \frac{4g}{l}sin^2\frac{\theta}{2}$, where θ_0' is the initial angle velocity. The time s from $\theta = 0$ to the stationary point is given as

$$s = \int_0^{2sin^{-1}(\theta_0'\sqrt{l/(4g)}\ )} \frac{d\theta}{\sqrt{(\theta_0')^2 l/(4g) - sin^2(\theta/2)}}\sqrt{\frac{l}{4g}}.$$

Put $\zeta = (\theta_0')^2 l/(4g)$, $\phi = sin^{-1}(\frac{1}{\sqrt{\zeta}}\ sin\frac{\theta}{2})$ and $K(\zeta) = s\sqrt{g/l}$. The integral $K(\zeta)$ is called the complete elliptic integral of first kind :

$$K(\zeta) = \int_0^{\pi/2} \frac{d\phi}{\sqrt{1-\zeta sin^2\phi}}.$$

$K(\zeta)$ is monotone increasing in ζ. $K(\zeta)$ is ∞ at $\zeta = 1$.

The Taylor series is

$$K(\zeta) = \frac{\pi}{2}\sum_{n=0}^{\infty} K_n\zeta^n, \ where\ K_n = [\frac{1\ 3\ 5\ldots(2n-3)(2n-1)}{2\ 4\ldots(2n-2)2n}]^2. \tag{1}$$

The series (1) converges for $0 \leq \zeta < 1$. However convergence rate is slow as $\zeta \to 1$, since $K(1) = \infty$. In the book [1] by Polubarinova-Kochina, P. Ya, (1) is used to evaluate $K(\zeta)$ for ζ in $0 \leq \zeta \leq \frac{1}{2}$. For $1/2 \leq \zeta < 1$, putting $\zeta = 1 - \eta$, $K(1-\eta)$ is expressed using log function as

$$K(1-\eta) = -\frac{1}{\pi}K(\eta)log\ \eta + P(\eta) + T\cdot K(\eta),\ on\ \frac{1}{2} \leq 1-\eta < 1.$$

$P(\eta) = \sum_{n=1}^{\infty} P_n\eta^n$ with $P_n = -2K_n\sum_{m=1}^{2n}\frac{(-1)^{m-1}}{m}$; $T = 0.882542400610611D0$.

1.2 Collocation Chebyshev series

Chebyshev polynomial with degree n is a polynomial with degree n $T_n(t) = cos\ n(cos^{-1}t)$ i.e. $T_n(cos\ \theta) = cos\ n\theta$. Cheb-C is a polynomial of degree N such that $f_N(t) = \sum{}''^{N}_{n=0} p_nT_n(t)$, where the coefficients are obtained by orthogonality of finite sum at collocation points $cos\ \frac{\pi i}{N}$:

$$p_r = \frac{2}{N}\sum{}''^{N}_{i=0} f(cos\frac{\pi i}{N})cos\frac{\pi r i}{N}.$$

Here $\sum''$ denotes the sum weighted 1/2 at $n = 0, N, 2N, \ldots$.

1.3 Truncation error for collocation Chebyshev series restricted on subinterval $[b, a]$

In [2], the truncation error of Cheb-C is obtained as $f(cos\ \theta) - f_N(cos\ \theta) =$

$$\sum_{k=1}^{\infty}\{\sum_{n=(2k-1)N+1}^{2kN-1} q_n(cos\ n\theta - cos(2kN-n)\theta)$$

$$+\sum{}''^{(2k+1)N}_{n=2kN} q_n(cos\ n\theta - cos(n-2kN)\theta)\}.$$

Here $f = \sum'' q_nT_n(t)$. In [3], error bound of finite Chebyshev series is obtained with respect to k th derivative $f^{(k)}$:

$$| f - f_N |_\infty \leq \sqrt{2} \| (I - P_{N-k})f^{(k)} \|_q /\{\sqrt{N}N(N-1)\ldots(N-k+2)\}.$$

Here $f = a_0 + \sqrt{2}\sum_{n=1}^{\infty} a_nT_n$; $P_Nf = a_0 + \sqrt{2}\sum_{n=1}^{N} a_nT_n$.

$$\| f \|_q = \sum_{n=0}^{\infty} a_n^2 = \frac{1}{\pi}\int_0^{\pi} f^2(cos\ \theta)\ d\theta \leq | f |_\infty.$$

$$\| (I - P_{N-k})f^{(k)} \|_q \leq | f^{(k)} |.$$

By combining the theorems in [2] and [3], the following theorem is obtained concerning the truncation error of Cheb-C.

Theorem 1.1 *Let $f(\eta) \in C^\infty$ on [0,1/2]. $\eta \in [b,a]$ be transformed to t by $\epsilon(t)$ on [-1,1]. Here*

$$\epsilon(t) = \frac{(a-b)t + (a+b)}{2}.$$

Let $f(\epsilon(t))$ be approximated by the Cheb-C $f_N(t)$ on [-1,1]. Then:

$$| f(\epsilon(t)) - f_N(t) |_\infty = \frac{2\sqrt{2}}{\sqrt{N}N(N-1)\ldots(N-k+2)} | \frac{d^k f(\epsilon(t))}{d\eta^k} |_\infty (\frac{a-b}{2})^k.$$

1.4 Connection of Cheb-C for an elliptic integral

In the present paper, $K(\varsigma)$ is approximated by connection of Chebyshev series. Let [0,1] be divided into two subintervals [0,1/2] and [1/2,1] . The elliptic integral $K(\varsigma)$ on [0,1/2] is approximated by a Chebyshev series on J=[-1,1] by transformation $\varsigma = (t+1)/4$. [0,1/2] for η in (2) is divided into the subintervals:

(2) $$[0.1/2, 1/2], \ldots, [0.1^{i+1}/2, 0.1^i/2], \ldots.$$

in such way the length of subintervals shrinks as $\frac{1}{2}(0.1^i - 0.1^{i+1})$ and finally converges to zero. $K(1-\eta)$ on [0,1/2] is approximated by a connection of Cheb-C's, where η is transformed to t by $\eta = \epsilon^i(t)$ on J. Here $a = 0.1^i/2$ and $b = 0.1^{i+1}/2$. Applying theorem concerning the truncation error, it is shown that the approximation error decreases as i increases nevertheless of singularity of log function. The singularity of log function for $K(1-\eta)$ diminishes by scale transformations of variable η in the subintervals to t in $[-1,1]$.

1.5 Main result: Truncation error of Cheb-C for $log\ \epsilon^i(t)$

Let $q(t) = log\ \epsilon^i(t)$. $d^kq/dt^k = c_k\epsilon^i(t)^{-k}\delta^k$, where $\delta = (0.1^i - 0.1^{i+1})/2$ and $c_k = (-1)^{k-1}(k-1)!$. Put $e_N = | q(t) - q_N(t) |_\infty$.

(3) $$e_N \leq \frac{2\sqrt{2}\ 4.5^k}{\sqrt{N}N(N-1)\ldots(N-k+2)} | c_k | .$$

For k=12 and N=80,$e_N \leq 0.2081062658D-05$; $| c_k | = 0.3991680000D+08$; $\frac{2\sqrt{2}}{\sqrt{N}N(N-1)\ldots(N-k+2)} = 0.7561000611D-21$.

1. Since $max\ | \frac{d^k log\ \epsilon(t)}{dt^k} | = (k-1)!(\frac{a-b}{2b})^k$ for $\epsilon(t) = (a-b)t/2+(a+b)/2$, $max\ | \frac{d^k log\ \epsilon(t)}{dt^k} |$ increases in the order $1/b^k$ as $b \to 0$, if the length a-b of interval is constant as $b \to 0$. The uniform convergence of e_N to 0 as $N \to \infty$ does not assured for $\eta \to 0$, since $\frac{2\sqrt{2}}{\sqrt{N}N(N-1)\ldots(N-k+2)}$ converges to 0 independently in η.

2. If the length a-b of interval converges to 0 in the order 0.1^{l+1}, the derivative of $d^k log \epsilon^i(t)/dt^k$ is uniformly bounded in k :

   (4) $$| \frac{d^k log \epsilon(t)}{dt^k} |_\infty \leq (k-1)!(\frac{0.1^i - 0.1^{i+1}}{0.1^{i+1}2})^k = 4.5^k(k-1)!$$

   If N is fixed and the length of interval decreases in the order 0.1^{l+1} as $\eta \to 0$, the derivative $\frac{d^k log\ \eta}{d\eta^k}$ does not converge to zero and accuracy does not assured for fixed N as $\eta \to 0$.

3. Accuracy increases uniformly in N and independently in i (i.e. in η).

2 Error bound of Cheb-C with respect to derivative function

Let Chebyshev series of C^∞ function be expressed in two ways with coefficients q_n and a_n and derivative function $\dot{f}$ be expressed as $\dot{f} = b_0 + \sqrt{2}\sum_1^\infty b_n T_n(t)$.

(1) $$q_0/2 = a_0; q_n = \sqrt{2}a_n = \sqrt{2}\frac{b_{n-1} - b_{n+1}}{2n} \ for \ n \not\equiv 0(mod.N);$$

$$q_n/2 = \sqrt{2}a_n = \sqrt{2}\frac{b_{n-1} - b_{n+1}}{2n} \ for \ n \equiv 0 \ (mod.N).$$

Put $e(\theta) = f(cos\ \theta) - f_N(cos\ \theta)$. The error $e(\theta)$ is estimated in the following theorem.

Theorem 2.1 *1.* $| e(\theta) |_\infty \leq \frac{2\sqrt{2}}{\sqrt{N}} \| (I - P_{N-1})\dot{f} \|_q$

2. For $f \in C^2$, $| e(\theta) | \leq \frac{2\sqrt{2}}{\sqrt{N}} \| (I - P_{N-1})\dot{f} \|_q \leq \frac{2\sqrt{2}}{\sqrt{N}N} \| (I - P_{N-2})\ddot{f} \|_q$.

3. $| e(\theta) |_\infty \leq 2\sqrt{2} \| (I - P_{N-k})f^{(k)} \|_q /\{\sqrt{N}N(N-1)\dots(N-k+2)\}$

Proof.$e(\theta) = 2\sum_{k=1}^\infty\{\sum_{n=(2k-1)N+1}^{2kN-1} q_n sin\ kN\theta\ sin(kN-n)\theta$

$$+\sum{}''^{(2k+1)N}_{n=2kN} q_n sin\ (-kN)\theta\ sin(n-kN)\theta\}$$

$e(\theta) = 2\sqrt{2}\sum_{k=1}^\infty\{\sum_{n=(2k-1)N+1}^{2kN-1} \frac{b_{n-1}-b_{n+1}}{2n} sin\ kN\theta\ sin(kN-n)\theta$

$$+\sum{}''^{(2k+1)N}_{n=2kN} \frac{b_{n-1} - b_{n+1}}{2n} sin\ (-kN)\theta\ sin(n-kN)\theta\}$$

$$e(\theta)^2 \leq 8[\sum_{n=N+1}^\infty \frac{| b_{n-1} - b_{n+1} |}{2n}]^2 \leq 8 \sum_{n=N+1}^\infty \frac{1}{n^2} \sum_{n=N}^\infty b_n^2.$$

$$| e(\theta) | \leq 2\sqrt{2}/\sqrt{N} \| (I - P_{N-1})\dot{f} \|_q, \ by \ \sum_{n=N+1}^\infty \frac{1}{n^2} \leq \int_N^\infty \frac{du}{u^2} = \frac{1}{N}.$$

Q.E.D.

3 Truncation error of Taylor series

1. The truncation error of $K(\zeta)$ on $[0, 1/2]$ is bounded by $\pi 2^{-M-1}$:

$$\pi/2 \ | \sum_{n=M+1}^\infty K_n\zeta^n |_\infty \leq \pi 2^{-M-1}.$$

For M=26, $\pi 2^{-M-1} = \pi 2^{-27} = .2340668926827D - 7 \ where \ D - 7 = \times 10^{-7}$.

2. The truncation error of $TK(\eta)$ on $[0, 1/2]$ is bounded by $\pi 2^{-M-1}$:

$$T\pi/2 \ | \sum_{n=M+1}^\infty K_n\eta^n |_\infty \leq \pi 2^{-M-1}.$$

3. The truncation error of $P(\eta)$ on $[0, 1/2]$ is bounded by the value

$$2 \ | \sum_{n=M+1}^\infty K_n \sum_{m=1}^{2n} \frac{(-1)^{m-1}}{m} \eta^n |_\infty \leq 2^{-M+1} \ (2^{-25} = 0.29802322387695D - 7).$$

The logarithmic singularity in truncation error of $K(\eta) log\ \eta$ is possible to be smaller by connection of Cheb-C as M increases. The truncation error of $K(\eta) log\ \eta$ on $[\frac{0.1^{i+1}}{2}, \frac{0.1^i}{2}]$ is $|\ log\ (0.1^{i+1}/2)\ |\ \sum_{n=M+1}^{\infty}(0.1^i/2)^n$ and bounded by the following values.

(1) $$|\ log(0.1/2)\ |\ 2^{-M} = 0.446398895D - 7\ for\ i = 0\ and\ M = 26.$$

(2) $$|\ log\ (0.1^2/2)\ |\ (0.1^2/2)10\frac{(0.1/2)^M}{1-(0.1/2)} = 0.4155320060D - 34\ for\ i \geq 1.$$

4 Truncation error for Cheb-C

Let $g(t) = \sum_{n=0}^{M} K_n\ \epsilon(t)^n$. $|\ g(t) - g_N(t)\ |$ be evaluated separating $p(t)$ and $q(t)$, where $p(t) = \sum_{n=k+1}^{M} K_n\ \epsilon(t)^n$; $q(t) = \sum_{n=0}^{k} K_n\ \epsilon(t)^n$. It holds that $(\epsilon(t)^n)^{(k)} = c_k^{(n)}\epsilon(t)^{n-k}\delta^k$ for $n \geq k+1$, where $c_k^{(n)} = n(n-1), \ldots, (n-k+1)$. Put $e_p =|\ p(t) - p_N(t)\ |_\infty$. Then

(1) $$e_p \leq \frac{2\sqrt{2}\ 2^{-2k}}{\sqrt{N}N(N-1)\ldots(N-k+2)}\sum_{n=k+1}^{M} c_k^n 2^{-(n-k)}.$$

$(\epsilon(t)^n)^{(k)} = 0\ for\ k > n \geq 0$; $c_k^{(k)} = k!$. Put $e_q =|\ q(t) - q_N(t)\ |_\infty$. Then

(2) $$e_q \leq \frac{2\sqrt{2}2^{-2k}c_k^{(k)}}{\sqrt{N}N(N-1)\ldots(N-k+2)}.$$

For $N = 80$, $k = 12$ and $M = 26$, $e_p \leq 0.1148279682D-15$; $e_q \leq 0.2158720129D-19$.

5 Truncation error for Cheb-C

Let $g(t) = \sum_{n=0}^{M} K_n\ \epsilon(t)^n log\ \epsilon(t)$. $|\ g - g_N\ |$ is obtained in this section. Truncation error for $K_0\ log\epsilon(t)$ is obtained already as 0.2D-05.

Let $p(t) = \sum_{n=k+1}^{M} K_n\ \epsilon(t)^n log\ \epsilon(t)$. Put $d_1^{(n)} = n$; $d_{k+1}^{(n)} = d_k^{(n)}(n-k)$; $c_1^{(n)} = 1$; $c_{k+1}^{(n)} = c_k^{(n)}(n-k) + d_k^{(n)}$. Put $e_p =|\ p(t) - p_N(t)\ |_\infty$. Then $e_p \leq$

(1) $$0.45/2\times\ |\ log\ (0.1/2)\ |\ \frac{2\sqrt{2}\ (0.9/4)^{k-1}}{\sqrt{N}N(N-1)\ldots(N-k+2)}\sum_{n=k+1}^{M}|\ d_k^{(n)}\ |\ 2^{-(n-k)}$$

(2) $$+\frac{2\sqrt{2}\ (0.9/4)^k}{\sqrt{N}N(N-1)\ldots(N-k+2)}\sum_{n=k+1}^{M}|\ c_k^{(n)}\ |\ 2^{-(n-k)}.$$

For k=12 and M=26, (1) + (2) = $0.1099856842D - 15$;
(1) = $0.7755487436D - 16$; (2) = $0.324308098383D - 16$

Let $k \geq n \geq 1$. $q^{(k)}/\delta^k = c_k^{(k)} + d_k^{(k)} log\epsilon(t)$; $q^{(k)}/\delta^k = \tilde{d_k^{(n)}}\epsilon(t)^{-(k-n)}\ for\ k \geq n+1$, where $\tilde{d_k^{(n)}} = d_n^{(n)}(-1)^{(k-n)}(k-n-1)!$. Let $q(t) = \sum_{n=1}^{k} K_n\ \epsilon(t)^n log\ \epsilon(t)$. Put $e_q = |q(t) - q_N(t)|_\infty$. Then $e_q \leq$

(3) $$\frac{2\sqrt{2}\ (0.9/4)^k}{\sqrt{N}N(N-1)\ldots(N-k+2)}\ |\ c_k^{(k)}\ |$$

$$(4) \qquad +0.45/2 \mid log\ 0.1/2 \mid \frac{2\sqrt{2}\ (0.9/4)^{k-1}}{\sqrt{N}N(N-1)\ldots(N-k+2)} \mid d_k^{(k)} \mid .$$

$$(5) \qquad + \frac{2\sqrt{2}}{\sqrt{N}N(N-1)\ldots(N-k+2)} \sum_{n=1}^{k-1} \mid \tilde{d}_k^{(n)} \mid (0.9/4)^n (0.2/0.9)^{-(k-n)}.$$

$$(3) + (4) + (5) = 0.9604242234D - 08$$

$(3) = 0.6096863256D - 20$; $(4) = 0.5667879972D - 19$; $(5) = 0.9604242233D - 8$.
Collocation Chebyshev coefficients for $K(\zeta)$ on $0 \leq \zeta \leq 1/2$.

```
 0.3397772427Q+01 0.1400309049Q+00 0.1333844954Q-01 0.1579004633Q-02
 0.2066230450Q-03 0.2864222596Q-04 0.4121843522Q-05 0.6089458449Q-06
 0.9172650852Q-07 0.1402376412Q-07 0.2168950686Q-08 0.3384038841Q-09
 0.5310415216Q-10 0.8348288886Q-11 0.1306922720Q-11 0.2019421979Q-12
 0.3041902688Q-13 0.4396133683Q-14 0.5979735814Q-15 0.7489355714Q-16
 0.8425325644Q-17 0.8273239597Q-18 0.6847832215Q-19 0.4561604281Q-20
 0.2282748444Q-21 0.7602381180Q-23 0.1260408475Q-24 0.1950004069Q-32
 0.1296150853Q-31-0.4285194126Q-32-0.1150743142Q-31 0.5811493607Q-32
-0.5041121629Q-32 0.1138706080Q-31-0.5748900884Q-32 0.1261484114Q-32
 0.6923718150Q-32 0.7010384998Q-32 0.1495003119Q-31 0.4737787663Q-32
 0.1270632281Q-31-0.2249967658Q-31 0.1506077217Q-31-0.1052520715Q-31
 0.6451865314Q-33 0.2243708385Q-32-0.1373669533Q-31 0.1395817727Q-31
 0.1670744227Q-31 0.1688077596Q-31 0.5055566104Q-33 0.4100304852Q-31
-0.8892981518Q-32-0.2116597009Q-31-0.3769526384Q-31-0.9514093925Q-32
 0.3086302736Q-32 0.2760339093Q-31 0.1436262256Q-31 0.9369649179Q-32
-0.3813341290Q-32 0.1588892204Q-33 0.1627410803Q-32-0.1786300023Q-32
 0.3789267166Q-32 0.8228535687Q-32-0.1319262012Q-31-0.1408336272Q-31
-0.1383299183Q-31 0.3140710257Q-31 0.4935195482Q-32 0.2274041782Q-31
-0.2937524648Q-31 0.1006298396Q-31 0.2881191197Q-31 0.2704005642Q-31
 0.3315006917Q-31-0.9562242174Q-32-0.5349270421Q-32-0.8421128682Q-32
 0.1174817266Q-32
```

Collocation Chebyshev coefficients for $K(1-\eta)$ on $0 \leq \eta < 1/2$.

```
i=  0
 0.4457867111Q+01-0.4720080962Q+00 0.1302233212Q+00-0.4579411636Q-01
 0.1795699470Q-01-0.7488419360Q-02 0.3248644141Q-02-0.1448603119Q-02
 0.6591391942Q-03-0.3046030575Q-03 0.1424993094Q-03-0.6732935459Q-04
 0.3207471010Q-04-0.1538577496Q-04 0.7423895494Q-05-0.3600384302Q-05
 0.1753827427Q-05-0.8576570986Q-06 0.4208592422Q-06-0.2071548250Q-06
 0.1022472527Q-06-0.5059294973Q-07 0.2509052224Q-07-0.1246877234Q-07
 0.6208061664Q-08-0.3096278394Q-08 0.1546738113Q-08-0.7738097152Q-09
 0.3876552422Q-09-0.1944507449Q-09 0.9765368336Q-10-0.4909632534Q-10
 0.2470928452Q-10-0.1244785354Q-10 0.6276631886Q-11-0.3167624552Q-11
 0.1599906129Q-11-0.8087042075Q-12 0.4090736333Q-12-0.2070682107Q-12
 0.1048842675Q-12-0.5315916936Q-13 0.2695900742Q-13-0.1367966186Q-13
 0.6945144574Q-14-0.3527858484Q-14 0.1792896102Q-14-0.9115993202Q-15
 0.4637128647Q-15-0.2359839639Q-15 0.1201424370Q-15-0.6119047634Q-16
 0.3117726052Q-16-0.1589104453Q-16 0.8102542578Q-17-0.4132747979Q-17
 0.2108627618Q-17-0.1076215283Q-17 0.5494547216Q-18-0.2806038295Q-18
 0.1433441178Q-18-0.7324646020Q-19 0.3743777013Q-19-0.1914018560Q-19
```

```
0.97879468880Q-20-0.50066008070Q-20 0.25615155530Q-20-0.13108428210Q-20
0.67096660250Q-21-0.34351452180Q-21 0.17590605620Q-21-0.90096193880Q-22
0.46155513280Q-22-0.23650756690Q-22 0.12123451360Q-22-0.62199123210Q-23
0.31999755970Q-23-0.16627298720Q-23 0.89545730990Q-24-0.54260118280Q-24
0.22035567600Q-24
i=  1
0.66343486140Q+01-0.50942434920Q+00 0.13432156980Q+00-0.46625531940Q-01
0.18180744130Q-01-0.75588069360Q-02 0.32730564240Q-02-0.14576508000Q-02
0.66265696030Q-03-0.30602109760Q-03 0.14308728200Q-03-0.67578723040Q-04
0.32182447560Q-04-0.15433046020Q-04 0.74449089700Q-05-0.36098306340Q-05
0.17581153070Q-05-0.85962004380Q-06 0.42176461040Q-06-0.20757519480Q-06
0.10244359860Q-06-0.50685151030Q-07 0.25134028800Q-07-0.12489392080Q-07
0.62178736170Q-08-0.31009646460Q-08 0.15489838840Q-08-0.77488931140Q-09
0.38817573230Q-09-0.19470235540Q-09 0.97775618620Q-10-0.49155555350Q-10
0.24738118240Q-10-0.12461918750Q-10 0.62835061280Q-11-0.31709903420Q-11
0.16015569040Q-11-0.80951514350Q-12 0.40947260740Q-12-0.20726478550Q-12
0.10498125220Q-12-0.53207081090Q-13 0.26982705630Q-13-0.13691397290Q-13
0.69509625040Q-14-0.35307458490Q-14 0.17943305270Q-14-0.91231262940Q-15
0.46406790970Q-15-0.23616084350Q-15 0.12023063220Q-15-0.61234488160Q-16
0.31199241020Q-16-0.15902030460Q-16 0.81080373970Q-17-0.41354982550Q-17
0.21100051250Q-17-0.10769056760Q-17 0.54980095790Q-18-0.28077757520Q-18
0.14343135690Q-18-0.73290288610Q-19 0.37459801390Q-19-0.19151265980Q-19
0.97935225130Q-20-0.50094078530Q-20 0.25629294450Q-20-0.13115553240Q-20
0.67132581770Q-21-0.34369570380Q-21 0.17599748060Q-21-0.90142345960Q-22
0.46178821050Q-22-0.23662532730Q-22 0.12129404330Q-22-0.62229245740Q-23
0.32015040630Q-23-0.16635128750Q-23 0.89587268350Q-24-0.54284913190Q-24
0.22045560990Q-24
i=  2
0.89121259660Q+01-0.51786525850Q+00 0.13487132740Q+00-0.46721687360Q-01
0.18205224930Q-01-0.75663058040Q-02 0.32756182970Q-02-0.14585912690Q-02
0.66302024440Q-03-0.30616685380Q-03 0.14314750450Q-03-0.67604194150Q-04
0.32193427950Q-04-0.15437855160Q-04 0.74470435990Q-05-0.36107890240Q-05
0.17585498720Q-05-0.85981879870Q-06 0.42185620790Q-06-0.20761769400Q-06
0.10246343660Q-06-0.50694461440Q-07 0.25138419840Q-07-0.12491472240Q-07
0.62188630590Q-08-0.31014370330Q-08 0.15492101860Q-08-0.77499806570Q-09
0.38822814910Q-09-0.19472768740Q-09 0.97787891960Q-10-0.49161515740Q-10
0.24741019180Q-10-0.12463333560Q-10 0.62841974720Q-11-0.31713287800Q-11
0.16017228660Q-11-0.80959665880Q-12 0.40951270640Q-12-0.20728453950Q-12
0.10499099700Q-12-0.53211894550Q-13 0.26985086200Q-13-0.13692576020Q-13
0.69515468060Q-14-0.35310358010Q-14 0.17944745600Q-14-0.91238424690Q-15
0.46410355370Q-15-0.23617859940Q-15 0.12023948480Q-15-0.61238905550Q-16
0.31201447000Q-16-0.15903132930Q-16 0.81085887830Q-17-0.41357742170Q-17
0.21101433350Q-17-0.10769749410Q-17 0.54983569280Q-18-0.28079500460Q-18
0.14344010790Q-18-0.73294684800Q-19 0.37462011110Q-19-0.19152377280Q-19
0.97940816910Q-20-0.50096893580Q-20 0.25630712310Q-20-0.13116267710Q-20
0.67136183690Q-21-0.34371387050Q-21 0.17600664720Q-21-0.90146973150Q-22
0.46181157770Q-22-0.23663713300Q-22 0.12130001140Q-22-0.62232265210Q-23
0.32016572990Q-23-0.16635913660Q-23 0.89591431520Q-24-0.54287399350Q-24
0.22046562480Q-24
i=  3
0.11210849860Q+02-0.51926666520Q+00 0.13493029640Q+00-0.46731510370Q-01
0.18207700220Q-01-0.75670611230Q-02 0.32758758440Q-02-0.14586857080Q-02
```

```
 0.6630566967Q-03-0.3061814715Q-03 0.1431535418Q-03-0.6760674686Q-04
 0.3219452814Q-04-0.1543833692Q-04 0.7447257406Q-05-0.3610885004Q-05
 0.1758593388Q-05-0.8598386994Q-06 0.4218653785Q-06-0.2076219487Q-06
 0.1024654225Q-06-0.5069539340Q-07 0.2513885935Q-07-0.1249168044Q-07
 0.6218962087Q-08-0.3101484309Q-08 0.1549232833Q-08-0.7750089492Q-09
 0.3882333946Q-09-0.1947302224Q-09 0.9778912011Q-10-0.4916211216Q-10
 0.2474130946Q-10-0.1246347512Q-10 0.6284266646Q-11-0.3171362643Q-11
 0.1601739472Q-11-0.8096048145Q-12 0.4095167183Q-12-0.2072865159Q-12
 0.1049919720Q-12-0.5321237612Q-13 0.2698532436Q-13-0.1369269394Q-13
 0.6951605261Q-14-0.3531064809Q-14 0.1794488969Q-14-0.9123914116Q-15
 0.4641071195Q-15-0.2361803757Q-15 0.1202403704Q-15-0.6123934745Q-16
 0.3120166768Q-16-0.1590324322Q-16 0.8108643941Q-17-0.4135801823Q-17
 0.2110157161Q-17-0.1076981870Q-17 0.5498391674Q-18-0.2807967480Q-18
 0.1434409833Q-18-0.7329512455Q-19 0.3746223215Q-19-0.1915248844Q-19
 0.9794137625Q-20-0.5009717517Q-20 0.2563085413Q-20-0.1311633918Q-20
 0.6713654398Q-21-0.3437156876Q-21 0.1760075641Q-21-0.9014743597Q-22
 0.4618139148Q-22-0.2366383138Q-22 0.1213006084Q-22-0.6223256711Q-23
 0.3201672649Q-23-0.1663599219Q-23 0.8959184695Q-24-0.5428764831Q-24
 0.2204666312Q-24
i=  4
 0.1351290708Q+02-0.5194646646Q+00 0.1349362672Q+00-0.4673249501Q-01
 0.1820794803Q-01-0.7567136710Q-02 0.3275901613Q-02-0.1458695155Q-02
 0.6630603432Q-03-0.3061829337Q-03 0.1431541457Q-03-0.6760700219Q-04
 0.3219463819Q-04-0.1543838511Q-04 0.7447278791Q-05-0.3610894604Q-05
 0.1758597741Q-05-0.8598406898Q-06 0.4218662957Q-06-0.2076223742Q-06
 0.1024656211Q-06-0.5069548660Q-07 0.2513890331Q-07-0.1249170126Q-07
 0.6218971991Q-08-0.3101489037Q-08 0.1549235098Q-08-0.7750100376Q-09
 0.3882339191Q-09-0.1947304759Q-09 0.9778924293Q-10-0.4916217180Q-10
 0.2474133849Q-10-0.1246348928Q-10 0.6284273564Q-11-0.3171366029Q-11
 0.1601741132Q-11-0.8096056302Q-12 0.4095171195Q-12-0.2072867135Q-12
 0.1049920695Q-12-0.5321242428Q-13 0.2698534818Q-13-0.1369270574Q-13
 0.6951611107Q-14-0.3531067710Q-14 0.1794490410Q-14-0.9123921280Q-15
 0.4641074761Q-15-0.2361805533Q-15 0.1202404590Q-15-0.6123939164Q-16
 0.3120168975Q-16-0.1590325425Q-16 0.8108649457Q-17-0.4135804584Q-17
 0.2110158544Q-17-0.1076982563Q-17 0.5498395149Q-18-0.2807969224Q-18
 0.1434410708Q-18-0.7329516853Q-19 0.3746225426Q-19-0.1915249956Q-19
 0.9794143218Q-20-0.5009720332Q-20 0.2563086831Q-20-0.1311634632Q-20
 0.6713658001Q-21-0.3437158693Q-21 0.1760076558Q-21-0.9014748224Q-22
 0.4618141484Q-22-0.2366384318Q-22 0.1213006683Q-22-0.6223259709Q-23
 0.3201674202Q-23-0.1663600006Q-23 0.8959188841Q-24-0.5428767433Q-24
 0.2204667292Q-24
```

References

1. P. Ya. Polubarinova-Kochina, Theory of Groundwater Movement, Princeton Univ. Press (1962) pp. 286.

2. C.W. Clenshaw and A.R. Curtis, Numerische Math., 2, 197–205 (1960).

3. M. Urabe, Numerische Math., 9, 341–366 (1967).

8th Int. Coll. on Differential Equations, pp. 423–430
D. Bainov (Ed)

Construction of Singularities for Nonlinear Wave Equations

MIKIO TSUJI
Department of Mathematics, Kyoto Sangyo University, Kamigamo, Kita-ku, Kyoto 603, Japan
E-mail: mtsuji@cc.kyoto-su.ac.jp

Key words and phrases. Nonlinear equations, contact structure, weak solutions, singularities of smooth mappings.

1 Introduction

We consider the Cauchy problem for nonlinear hyperbolic partial differential equations of second order. Then the Cauchy problem does not generally admit a classical solution in the large, that is to say, singularities generally appear in finite time. The typical example of singularity is "shock wave". Our problem is to extend the solution beyond the singularities. Though there are many papers concerning this subject, the existence and uniqueness of weak solution is even now unsolved. In this talk we consider the above problem from geometrical point of view.

Let us explain our method briefly. First we lift the solution surface into cotangent space so that the lifted surface does not contain singularities. To do so, we need to solve certain hyperbolic system of first order. Though it is not easy to solve the system, we can sometimes get the smooth solution in the large. As the singularites do not appear in the solution of the system, the extension of the solution is unique in cotangent space. The method of integration of second order partial differential equations will be studied in §2.

Next we project the lifted solution surface to the base space. In a neighbourhood of a point where this projection is regular, we can get a smooth solution. This is just the characteristic method for second order partial differential equations. In a neighbourhood of a point where the mapping is singular, we use the singularity theory of smooth mappings to construct a solution with singularities.

Finally we apply our method to certain nonlinear wave equation in §3 and §4, and also to certain system of conservation law which is called as "p-system". Then we see that one can not construct a weak solution in the sense of Lax by the characteristic method. This may suggest us that the definition of weak solution may not be appropriate.

If the equations are of first order, we could construct reasonable weak solutions by the above method. For example, refer to J. Guckenheimer, C. Jennings, M. Tsuji [1], [2], S. Nakane [3], [4], S. Izumiya [5], S. Izumiya and G. T. Kossioris [6], [7], etc. etc.

2 Integration of Monge-Ampère equations

In this section, we will study the method of integration of second order nonlinear partial differential equations. As the equations of second order are too general, we will treat the equations of Monge-Ampère type as follows:

$$F(x,y,z,p,q,r,s,t) = Ar + Bs + Ct + D(rt - s^2) - E = 0 \tag{2.1}$$

where $p = \partial z/\partial x, q = \partial z/\partial y, r = \partial^2 z/\partial x^2, s = \partial^2 z/\partial x \partial y$, and $t = \partial^2 z/\partial y^2$. Here we assume that A, B, C, D and E are real smooth functions of (x, y, z, p, q). Partial differential equations of second order which appear in physics and geometry are often written in the above form. The aim of this talk is to consider the structure of singularities of solutions of (2.1) in the case where equation (2.1) is hyperbolic. The definition of "hyperbolicity" will be given later. For our aim, we have to represent the solutions explicitely. First we explain the characteristic method developed principally by D. Darboux ([8]) and E. Goursat ([9]) "from our point of view". The principal idea of Darboux and Goursat is to reduce the solvability of (2.1) to the integration of first order partial differential equations, though their theory is local. Now we recall the notion of "strip". Let

$$\Gamma : (x,y,z,p,q) = (x(\alpha), y(\alpha), z(\alpha), p(\alpha), q(\alpha)), \quad \alpha \in \mathbf{R}^1,$$

be a smooth curve in $\mathbf{R}^5 = \{(x,y,z,p,q)\}$. We say that it satisfies "strip condition" if it holds

$$\frac{dz}{d\alpha}(\alpha) = p(\alpha)\frac{dx}{d\alpha}(\alpha) + q(\alpha)\frac{dy}{d\alpha}(\alpha) . \tag{2.2}$$

Let Γ be any strip in $\mathbf{R}^5$, and consider equation (2.1) in an open neighbourhood of Γ. As a "characteristic" strip means that one can not determine the values of the second order derivatives of solution along the strip, we have the following

Definition 2.1 *A curve Γ in $\mathbf{R}^5 = \{(x,y,z,p,q)\}$ is a "characteristic strip" if it satisfies (2.2) and*

$$\det \begin{bmatrix} F_r & F_s & F_t \\ \dot{x} & \dot{y} & 0 \\ 0 & \dot{x} & \dot{y} \end{bmatrix} = F_t \dot{x}^2 - F_s \dot{x}\dot{y} + F_r \dot{y}^2 = 0 \tag{2.3}$$

where $F_t = \partial F/\partial t, F_s = \partial F/\partial s, F_r = \partial F/\partial r, \dot{x} = dx/d\alpha$ and $\dot{y} = dy/d\alpha$.

Denote the discriminant of (2.3) by Δ, then we get $\Delta = F_s^2 - 4F_rF_t = B^2 - 4(AC + DE)$. If $\Delta < 0$, equation (2.1) is called to be elliptic. If $\Delta > 0$, equation (2.1) is hyperbolic. In this note, we will treat the equations of hyperbolic type. More precisely, we assume $\Delta \geq 0$ and also $D \neq 0$. Let λ_1 and λ_2 be the solutions of $\lambda^2 + B\lambda + (AC + DE) = 0$. Then the characteristic strip satisfies the following equations:

$$dz - pdx - qdy = 0, \quad Ddp + Cdx + \lambda_1 dy = 0, \quad Ddq + \lambda_2 dx + Ady = 0, \tag{2.4}$$

or, replacing λ_1 and λ_2 in (2.4), it holds

$$dz - pdx - qdy = 0, \quad Ddp + Cdx + \lambda_2 dy = 0, \quad Ddq + \lambda_1 dx + Ady = 0. \tag{2.5}$$

Let us denote $\omega_0 = dz - pdx - qdy$, $\omega_1 = Ddp + Cdx + \lambda_1 dy$ and $\omega_2 = Ddq + \lambda_2 dx + Ady$. Exchanging λ_1 and λ_2 in ω_1 and ω_2, we define ϖ_1 and ϖ_2 by $\varpi_1 = Ddp + Cdx + \lambda_2 dy$ and $\varpi_2 = Ddq + \lambda_1 dx + Ady$. Take an exterior product of ω_1 and ω_2, and also of ϖ_1 and

ϖ_2. Next substitute into their products the contact relations $\omega_0 = 0$, $dp = rdx + sdy$ and $dq = sdx + rdy$. Then we get

$$\omega_1 \wedge \omega_2 = \varpi_1 \wedge \varpi_2 = D\Big\{Ar + Bs + Ct + D(rt - s^2) - E\Big\} dx \wedge dy. \tag{2.6}$$

In a space whose dimension is greater than two, the decomposition as above is not possible in general. Here we introduce the notion of "first integral".

Definition 2.2 *A function $V = V(x, y, z, p, q)$ is called "first integral" of $\{\omega_0, \omega_1, \omega_2\}$ if $dV \equiv 0 \mod\{\omega_0, \omega_1, \omega_2\}$.*

Remark We can easily see that a function $V = V(x, y, z, p, q)$ is the "first integral" of (2.4) (or (2.5)) if it is constant on any solution of (2.4) (or of (2.5) respectively).

G. Darboux [8] and E. Goursat [9] had considered equation (2.1) under the assumption that (2.4), or (2.5), has at least two independent first integrals. We denote them by u and v. Then we get the following

Proposition 2.3 *Assume that $\lambda_1 \neq \lambda_2$, and that (2.4) has two independent first integrals $\{u, v\}$. Then there exists a function $k = k(x, y, z, p, q) \neq 0$ satisfying*

$$du \wedge dv = k\, \omega_1 \wedge \omega_2 = kD\{Ar + Bs + Ct + D(rt - s^2) - E\} dx \wedge dy\,. \tag{2.7}$$

If equation (2.1) is written as (2.7), it would be obvious that (2.4), or (2.5), has two independent first integrals $\{u, v\}$. Therefore the representation (2.7) gives the characterization of "Monge-Ampère equations of Darboux-Goursat type". Let $\{u, v\}$ be two independent first integrals of (2.4). For any function g of two variables whose gradient does not vanish, $g(u, v) = 0$ is called an "intermediate integral" of (2.1). Let C_0 be an initial strip defined in $\mathbf{R}^5 = \{(x, y, z, p, q)\}$. If the strip C_0 is not characteristic, we can find an "intermediate integral" $g(u, v)$ which vanishes on C_0. Here we put $g(u, v) = f(x, y, z, p, q)$. The Cauchy problem for (2.1) satisfying the initial condition C_0 is to look for a solution $z = z(x, y)$ of (2.1) which contains the strip C_0, i.e., the two dimensional surface $\{(x, y, z(x, y), \partial z/\partial x(x, y), \partial z/\partial y(x, y))\}$ in $\mathbf{R}^5$ contains the strip C_0. The representation (2.7) assures that, as $du \wedge dv = 0$ on a surface $g(u, v) = 0$, a smooth solution of $f(x, y, z, \partial z/\partial x, \partial z/\partial y) = 0$ satisfies equation (2.1). Therefore we get the following

Theorem 2.4 ([8], [9]) *Assume that the initial strip C_0 is not characteristic. Then a function $z = z(x, y)$ is a solution of the Cauchy problem for (2.1) with the initial condition C_0 if and only if it is a solution of $f(x, y, z, \partial z/\partial x, \partial z/\partial y) = 0$ satisfying the same initial condition C_0.*

Next we will study the method of integration of Monge-Ampère equations (2.1) in the case where (2.4), and (2.5) also, has not two independent first integrals. We suppose $D \neq 0$ for simplicity, though it is not indispensable for our study. The essential condition for our following discussion is $\Delta \neq 0$. Then we can write equation (2.1) by the product of one forms as (2.6). Here we take heuristic approach to study the method of integration of (2.1). Let us suppose that a solution is represented by two parameters, i.e.,

$$x = x(\alpha, \beta), y = y(\alpha, \beta), z = z(\alpha, \beta), p = p(\alpha, \beta), q = q(\alpha, \beta). \tag{2.8}$$

Then ω_i and ϖ_i $(i = 1, 2)$ are written as $\omega_i = c_{i1}d\alpha + c_{i2}d\beta$, $\varpi_i = d_{i1}d\alpha + d_{i2}d\beta$ $(i = 1, 2)$. Hence we have $\omega_1 \wedge \omega_2 = (c_{11}c_{22} - c_{12}c_{21})\, d\alpha \wedge d\beta$ and $\varpi_1 \wedge \varpi_2 = (d_{11}d_{22} - d_{12}d_{21})\, d\alpha \wedge d\beta$.

As $\omega_1 \wedge \omega_2 = \varpi_1 \wedge \varpi_2 = 0$ on the surface $\{F = 0\}$, a sufficient condition so that (2.8) is a solution of (2.1) is

$$c_{11} = c_{21} = d_{12} = d_{22} = 0. \tag{2.9}$$

Adding the contact relation $dz = pdx + qdy$ to (2.9), we get a system of first order partial differential equation as follows:

$$\begin{cases} \dfrac{\partial z}{\partial \alpha} - p\dfrac{\partial x}{\partial \alpha} - q\dfrac{\partial y}{\partial \alpha} = 0 \\ D\dfrac{\partial p}{\partial \alpha} + C\dfrac{\partial x}{\partial \alpha} + \lambda_1\dfrac{\partial y}{\partial \alpha} = 0, \quad D\dfrac{\partial q}{\partial \alpha} + \lambda_2\dfrac{\partial x}{\partial \alpha} + A\dfrac{\partial y}{\partial \alpha} = 0 \\ D\dfrac{\partial p}{\partial \beta} + C\dfrac{\partial x}{\partial \beta} + \lambda_2\dfrac{\partial y}{\partial \beta} = 0, \quad D\dfrac{\partial q}{\partial \beta} + \lambda_1\dfrac{\partial x}{\partial \beta} + A\dfrac{\partial y}{\partial \beta} = 0 \end{cases} \tag{2.10}$$

If $(x(\alpha,\beta), y(\alpha,\beta), z(\alpha,\beta), p(\alpha,\beta), q(\alpha,\beta))$ satisfies system (2.10), one can prove $\partial z/\partial\beta - p\partial x/\partial\beta - q\partial y/\partial\beta = 0$. Therefore we do not need to add this to (2.10). This means that (2.10) is the "determined" system. The local solvability of (2.10) is already proved by H. Lewy and J. Hadamard. Let us denote the solution of the Cauchy problem for (2.10) by $(x(\alpha,\beta), y(\alpha,\beta), z(\alpha,\beta), p(\alpha,\beta), q(\alpha,\beta))$. We can prove that, if the initial strip is not characteristic, the Jacobian $D(x,y)/D(\alpha,\beta)$ does not vanish in a neighbourhood of the initial strip. Therefore we can uniquely solve the system of equations $x = x(\alpha,\beta)$, $y = y(\alpha,\beta)$ with respect to (α,β). Then the solution of (2.1) with the initial condition is given by $z(x,y) = z(\alpha(x,y), \beta(x,y))$.

3 Nonlinear wave equations

Let us consider the Cauchy problem for nonlinear wave equations as follows:

$$\partial^2 z/\partial x^2 - (\partial/\partial y) f(\partial z/\partial y) = 0 \qquad \text{in} \quad \{x > 0, y \in \mathbf{R}^1\} \equiv \mathbf{R}^2_+ \tag{3.1}$$

$$z(0,y) = z_0(y), \quad (\partial z/\partial x)(0,y) = z_1(y) \qquad \text{on} \quad \{x = 0, y \in \mathbf{R}^1\} \tag{3.2}$$

where $f(q)$ is in $C^\infty(\mathbf{R}^1)$ and $f'(q)$ is positive. Here $z = z(x,y)$ is an unknown function of $(x,y) \in \mathbf{R}^2$, and we assume that the initial functions $z_i(y)$ $(i = 0,1)$ are sufficiently smooth. Equation (3.1) is also of Monge-Ampère type. For example, if we may put $A = 1, B = D = E = 0$, and $C = -f'(q)$ in (2.1), then we get (3.1).

It is well known that the Cauchy problem (3.1)-(3.2) does not have a classical solution in the large. For example, see N. F. Zabusky [10] for the case where $f'(q) = (1 + \varepsilon q)^{2\alpha}$ and P. D. Lax [11] for 2×2 hyperboric systems of conservation law. After them, many people have considered the life-span of classical solutions. As the number of papers on this subject is too many, we do not give here historical remarks.

Our problem is how to extend the solution of (3.1) after the apperance of sigularities. In this section, as we will consider this problem from the physical point of view, we will look for single-valued solutions of (3.1). The first question is what kinds of singularities may appear. In [10], N. J. Zabusky showed that the first order derivatives remain bounded, and that the second order derivatives tend to infinity in finite time. Therefore, for equations which have similar properties, we are led to the following

Definition 3.1 *Let $z = z(x,y)$, $(\partial z/\partial x)(x,y)$ and $(\partial z/\partial y)(x,y)$ be bounded and measurable. The function $z = z(x,y)$ is a weak solution of* (3.1)-(3.2) *if it satisfies equation* (3.1)

in distribution sense, i.e.,

$$\int_{\mathbf{R}^2_+} \{ \frac{\partial z}{\partial x}\frac{\partial \varphi}{\partial x} - f(\frac{\partial z}{\partial y})\frac{\partial \varphi}{\partial y} \} dxdy + \int_{\mathbf{R}^1} z_1(y)\varphi(0,y)\, dy = 0 \tag{3.3}$$

and $z(0,y) = z_0(y)$ *for all* $\varphi(x,y) \in C_0^\infty(\mathbf{R}^2)$.

Remark In Definition 3.1, $\partial z/\partial x$ and $\partial z/\partial y$ are the derivatives of $z = z(x,y)$ in classical sense, not in distribution sense. Therefore, even if $z = z(x,y)$ may have jump discontinuities, Dirac measure does not appear in $\partial z/\partial x$ and $\partial z/\partial y$.

Let $z = z(x,y)$ be a weak solution of (3.1) in the sense of Definition 3.1. If $(\partial z/\partial x)(x,y) \equiv p(x,y)$ and $(\partial z/\partial y)(x,y) \equiv q(x,y)$ have jump discontinuities along a curve $y = \gamma(x)$, we get by (3.3) the jump condition of Rankine-Hugoniot type as follows:

$$[p]\dot{\gamma} + [f(q)] = 0, \tag{3.4}$$

where [ ] means the quantity of difference, i.e., $[p] = p(x,\gamma(x)+0) - p(x,\gamma(x)-0)$. Moreover suppose that $z = z(x,y)$ is continuous along the curve $y = \gamma(x)$, though $p(x,y)$ and $q(x,y)$ have jump discontinuities along it. Then, differentiating $z(x,\gamma(x)+0) = z(x,\gamma(x)-0)$ with respect to x, we get

$$[q]\dot{\gamma} + [p] = 0. \tag{3.5}$$

This means that the curve $y = \gamma(x)$ must satisfy two different differential equations (3.4) and (3.5). Therefore it seems to us that one can not generally prove the existence of continuous weak solutions of (3.1). But we can construct discontinuous weak solutions of (3.1) in the sense of Definition 3.1 by the characteristic method.

It is well known that equation (3.1) can be transformed into a system of conservation law for which P. D. Lax [11] gave the definition of weak solutions. Then we can show that weak solutions of (3.1) can not be transformed to weak solutions of systems of conservation law. By these trials, we have some question on the definition of weak solution of systems of conservation law. We think that the best method to explain our considerations is to present some concrete example. Therefore, in §4, we will solve an example exactly without stating our results in general form.

4 Example

In this section we will construct an exact solution of the Cauchy problem as follows:

$$\partial^2 z/\partial x^2 - (1+\epsilon q)^{-4}\partial^2 z/\partial y^2 = 0 \qquad \text{in} \quad \{x > 0, y \in \mathbf{R}^1\}, \tag{4.1}$$

$$z(0,y) = \sin y, \quad (\partial z/\partial x)(0,y) = \epsilon^{-1}(1+\epsilon\cos y)^{-1} \quad \text{on} \quad \{x = 0, y \in \mathbf{R}^1\}, \tag{4.2}$$

where $0 < \epsilon = \text{constant} \le 1/2$. We can get the solution of (4.1)-(4.2) by solving the following Cauchy problem:

$$\partial z/\partial x - \epsilon^{-1}\{1 + \epsilon(\partial z/\partial y)\}^{-1} = 0 \qquad \text{in} \quad \{x > 0, y \in \mathbf{R}^1\}, \tag{4.3}$$

$$z(0,y) = \sin y \qquad \text{on} \quad \{x = 0, y \in \mathbf{R}^1\}. \tag{4.4}$$

Concerning the geometric meaning of (4.3), see G. Darboux [8], E. Goursat [9] and M. Tsuji [12], [13]. The characteristic equations for (4.3)-(4.4) are given by

$$dx/d\beta = 1, \quad dy/d\beta = (1+\epsilon q)^{-2}, \quad dz/d\beta = p + q(1+\epsilon q)^{-2}, \quad dp/d\beta = dq/d\beta = 0, \tag{4.5}$$

$$x(0)=0,\ \ y(0)=\alpha,\ \ z(0)=\sin\alpha,\ \ p(0)=\epsilon^{-1}(1+\epsilon\cos\alpha)^{-1},\ \ q(0)=\cos\alpha. \tag{4.6}$$

The solutions of (4.5)-(4.6) are written by

$$y=\alpha+(1+\epsilon\cos\alpha)^{-2}x \tag{4.7}$$

$$z=\sin\alpha+\epsilon^{-1}(1+2\epsilon\cos\alpha)(1+\epsilon\cos\alpha)^{-2}x \tag{4.8}$$

$$p=\epsilon^{-1}(1+\epsilon\cos\alpha)^{-1},\qquad q=\cos\alpha. \tag{4.9}$$

When $(\partial y/\partial\alpha)(x,\alpha)\neq 0$, we can uniquely solve the equation $y=y(x,\alpha)$ with respect to α and denote it by $\alpha=\alpha(x,y)$. Substituting $\alpha=\alpha(x,y)$ into $z=z(x,\alpha), p=p(\alpha)$ and $q=q(\alpha)$, we define $z(x,y)=z(x,\alpha(x,y))$, $p(x,y)=p(\alpha(x,y))$ and $q(x,y)=q(\alpha(x,y))$. Then it holds $(\partial/\partial x)z(x,y)=p(x,y)$ and $(\partial/\partial y)z(x,y)=q(x,y)$. Moreover we see that $z=z(x,y)$ satisfies (4.1))-(4.2) in a neighbourhood of $x=0$. As (4.9) means $p(x,y)=(\partial z/\partial x)(x,y)=\epsilon^{-1}\{1+\epsilon(\partial z/\partial y)\}^{-1}=\epsilon^{-1}(1+\epsilon\ q)^{-1}$, we get

$$(\partial/\partial x)p+3^{-1}\epsilon^2(\partial/\partial y)p^3=0 \qquad \text{in} \quad \{x>0, y\in\mathbf{R}^1\}, \tag{4.10}$$

$$p(0,y)=\epsilon^{-1}(1+\epsilon\cos y)^{-1} \qquad \text{on} \quad \{x=0, y\in\mathbf{R}^1\}. \tag{4.11}$$

As $0<\epsilon\leq 1/2$, (4.10)-(4.11) is a single convex conservation law. The characteristic strip of (4.10)-(4.11) is same to the one of (4.3)-(4.4), i.e., it is written by (4.7) and (4.9). Calculating $(\partial y/\partial\alpha)(x,\alpha)$, we can easily find a point (x^0,α^0) where $(\partial y/\partial\alpha)(x,\alpha)$ vanishes for the first time, i.e., $(\partial y/\partial\alpha)(x,\alpha)\neq 0$ for $x<x^0$ and $\alpha\in\mathbf{R}^1$, and $(\partial y/\partial\alpha)(x^0,\alpha^0)=0$. By elementary calculation, we see that, for $x>x^0$, $\alpha=\alpha(x,y)$ takes three values in a neighbourhood of (x^0,y^0) where $y^0=y(x^0,\alpha^0)$. Denote them by $\alpha=\alpha_1(x,y)<\alpha_2(x,y)<\alpha_3(x,y)$, and define $z_i(x,y)\equiv z(x,\alpha_i(x,y))$, $p_i(x,y)\equiv p(\alpha_i(x,y))$, and $q_i(x,y)\equiv q(\alpha_i(x,y))$ $(i=1,2,3)$. Then the discontinuity curve $y=\gamma(x)$ of $p=p(x,y)$ is obtained as a solution of the following Cauchy problem

$$\frac{d\gamma}{dx}=\frac{1}{3}\epsilon^2\ \frac{p_3(x,\gamma)^3-p_1(x,\gamma)^3}{p_3(x,\gamma)-p_1(x,\gamma)}\quad,\quad x>x^0, \tag{4.12}$$

$$\gamma(x^0)=y^0. \tag{4.13}$$

The Cauchy problem (4.12)-(4.13) has a unique solution $\gamma=\gamma(x)$, though the right hand side is not Lipschitz continuous in a neighbourhood of (x^0,y^0). See M. Tsuji [2]. Now we define the solution $z=z(x,y)$ of (4.1)-(4.2) by

$$z(x,y)=\begin{cases} z_1(x,y) & (y<\gamma(x))\ , \\ z_3(x,y) & (y>\gamma(x))\ . \end{cases}$$

Then we can prove

$$z(x,\gamma(x)-0)-z(x,\gamma(x)+0)=z_1(x,\gamma(x))-z_3(x,\gamma(x))\neq 0,\quad x>x^0. \tag{4.14}$$

Hence we take the derivatives of $z=z(x,y)$ in classical sense, i.e.,

$$\frac{\partial z}{\partial x}(x,y)=\begin{cases} p_1(x,y) & (y<\gamma(x))\quad , \\ p_3(x,y) & (y>\gamma(x))\quad . \end{cases} \tag{4.15}$$

Since $y=\gamma(x)$ is the solution of (4.12)-(4.13), $z=z(x,y)$ satisfies $[p]\dot\gamma-[3^{-1}\epsilon^2p^3]=[p]\dot\gamma+[f(q)]=0$ where $f(q)=-3^{-1}\epsilon^{-1}(1+\epsilon q)^{-3}$. Hence $z=z(x,y)$ is a weak solution of (4.1)-(4.2) in the sense of Definition 3.1. Moreover we can prove that $y=\gamma(x)$ exists in the large, and that another jump discontinuity curve $y=\gamma_n(x)$ starts at a point $(x^0,y^0+2n\pi)$ for any integer n, because all data are the periodic functions of period 2π. Then it holds

$$\gamma_n(x)-\gamma(x)=2n\pi\ ,\quad x>x^0\ .$$

Theorem 4.1 *The Cauchy probrem* (4.1)-(4.2) *admits a weak solution in the large in the sense of* Definition 3.1. *Moreover the weak solution is analytic except on the curves* $y = \gamma_n(x)$ *and it does not satisfy* $[q]\dot{\gamma} + [p] = 0$ *along* $y = \gamma_n(x)$ *for any integer* n.

At today's point, we do not have any criterion on the uniqueness of weak solutions of (3.1). Therefore we do not insist that our solution is reasonable. But it has several nice properties. The first one is that the jump discontinuity of $p = p(x, y)$ along the curve $y = \gamma(x)$ satisfies the entropy condition for the single convex conservation law (4.10)-(4.11). The second one is that $z = z(x, y)$ is analytic except on the curve $y = \gamma_n(x)$. We think that the analytic extension is acceptable in physical sciences.

Finally we will give a relation between nonlinear wave equation and certain system of conservation law. Let us transform the example (4.1) to 2×2 system of conservation law. Put $U(x, y) =^t (p, q), F(U) =^t (f(q), p)$ and $U_0 = (\epsilon^{-1}(1 + \epsilon\cos y)^{-1}, \cos y)$ where $f(q) = -3^{-1}\epsilon^{-1}(1 + \epsilon q)^{-3}$. Then the Cauchy problem (4.1)-(4.2) is expressed by the following form:

$$(\partial/\partial x)U - (\partial/\partial y)F(U) = 0 \quad \text{in} \quad \{x > 0, y \in \mathbf{R}^1\}, \tag{4.16}$$

$$U(0, y) = U_0(y) \quad \text{on} \quad \{x = 0, y \in \mathbf{R}^1\}. \tag{4.17}$$

P. D. Lax [11] gave the definition of weak solutions of (4.16)-(4.17) as follows;

Definition 4.2 *A bounded and measurabule* 2*-vector function* $U = U(x,y)$ *is a weak solution of* (4.16)-(4.17) *if it satisfies* (4.16)-(4.17) *in distribution sense, i.e.,*

$$\int_{\mathbf{R}^2_+} \{U(x,y)\frac{\partial\varphi}{\partial x}(x,y) - F\ (U)\frac{\partial\varphi}{\partial y}(x,y)\}dxdy + \int_{\mathbf{R}^1} U_0(y)\varphi(0,y)dy = 0 \tag{4.18}$$

for any two-vector function $\varphi(x, y) \in C_0^\infty(\mathbf{R}^2)$.

If $U = U(x,y)$ is a weak solution of (4.16) which has jump discontinuity along a curve $y = \gamma(x)$, we get the jump condition of Rankine-Hugoniot as follows:

$$[p]\dot{\gamma} + [f(q)] = 0 \tag{4.19}$$

$$[q]\dot{\gamma} + [p] = 0. \tag{4.20}$$

Equations (4.19) and (4.20) are same to (3.4) and (3.5) respectively. We see by the characteristic method that a solution of (4.16)-(4.17) is written by the same characteristic curves as (4.7) and (4.9), i.e.,

$$y = \alpha + (1 + \epsilon\cos\alpha)^{-2}x \tag{4.21}$$

$$p = \epsilon^{-1}(1 + \epsilon\cos\alpha)^{-1}, \quad q = \cos\alpha. \tag{4.22}$$

Here we repeat the same calculation concerning (4.21) and (4.22) just as done for (4.7), (4.8) and (4.9). Then we can prove that it is impossible to construct a weak solution of (4.16)-(4.17) by a family of characteristic curves (4.21)-(4.22), because (4.14) means that (4.20) does not hold. This might suggest us that we had better to introduce another definition of weak solutions of (4.16)-(4.17), because it seems to us that the solution constructed by the characteristic method would be acceptable.

Acknowledgements

The author's research was partially supported by Grant-in-Aid for Scientific Research (C) (No.07640261), Ministry of Education, Science and Culture.

References

1. M. Tsuji, J. Math. Kyoto Univ., 26, 299–308 (1986).
2. M. Tsuji, Ann. Inst. H. Poincaré - Analyse Nonlnéaire, 7, 505-523 (1990).
3. S. Nakane, SIAM J. Math. Anal., 19, 1391–1408 (1988).
4. S. Nakane, J. Mat. Soc. Japan, 43, 89–100 (1991).
5. S. Izumiya, Adv. Studies in Math., 22, 89–100 (1993).
6. S. Izumiya and G. T. Kossioris, J. Diff. Eq., 118, 166–193 (1995).
7. S. Izumiya and G. T. Kossioris, Banach Center Publications, 33, 127–148 (1996).
8. G. Darboux, Leçon sur la Théorie Générale des Surfaces, tome 3, Gauthier-Villars, Paris (1894).
9. E. Goursat, Leçons sur L'intégration des Équations aux Dérivées Partielles du Second Ordre, tome 1, Hermann, Paris (1896).
10. N. J. Zabusky, J. Math. Physics, 3, 1028–1039 (1962).
11. P. D. Lax, Comm. Pure Appl. Math., 10, 537-566 (1957).
12. M. Tsuji, Bull. Sci. Math., 119, 433–457 (1995).
13. M. Tsuji, Publications of Banach Center, 39, 161–170 (1997).
14. B. Gaveau, Bull. Sci. Math., 113, 407–442 (1989).
15. P. D. Lax, J. Math. Physics, 5, 611-613 (1964).

8th Int. Coll. on Differential Equations, pp. 431–439
D. Bainov (Ed)

Iterative Methods for Total Least Squares Problems

TIANRUO YANG
Department of Computer and Information Science, Linköping University, S-581 83, Linköping, Sweden
E-mail: tiaya@ida.liu.se

Abstract. To accelerate efficiently the algorithms for total least squares problems, we propose the following two Newton iteration algorithms with the solution of least squares problems as the starting vector. The iterative solution of the linear systems can be easily accelerated by preconditioning techniques. A linesearch strategy is combined with Newton's methods to ensure global convergence. Some numerical experiments are presented as well.

1 Introduction

Given an over-determined system of equation $Ax = b$, the Total Least Squares (TLS) problem, in its simplest form, is to find the smallest perturbation to A and b to make the system of equations compatible. Specifically, we find an error matrix E and vector r such that for some vector x

$$\min_{E,r} \|(E,\ r)\|_F, \qquad (A+E)x = b + r.$$

The vector x corresponding to the optimal (E, r) is called the TLS solution.

The most common algorithms to compute the TLS solution are based on Singular Value Decomposition (SVD), a computationally expensive matrix decomposition [1]. A very complete survey of computational aspects and analysis of the TLS problem is given in [2]. Recently Van Huffel [3] presents some iterative methods based on inverse iteration and Chebyshev acceleration which deal with the computation of a basis of a singular subspace associated with the smallest singular values and compares them by analyzing their convergence properties, the convergence rate and the operation counts per iteration step.

In [4] we propose a Krylov subspace method which use a Lanczos bidiagonalization process when the coefficient matrix is large and sparse and also problems with multiple right hand side. A drawback of the Krylov subspace method is that it is not clear how preconditioning can be applied. To accelerate efficiently the algorithm for total least squares problems, we propose the following two Newton's methods with the solution of least squares problems as the starting vector, which can be easily accelerated by preconditioning techniques.

The paper is organized as follows. In section 2, we will describe the Newton's methods fully. The corresponding local and global convergence are presented in section 3 and 4, respectively. Finally some numerical experiments are described as well.

2 Newton's method

It has been shown that the total least squares solution $x = x_{TLS}$ satisfies the linear system [1]

$$\begin{pmatrix} A^TA & A^Tb \\ b^TA & b^Tb \end{pmatrix} \begin{pmatrix} x_{TLS} \\ -1 \end{pmatrix} = \sigma^2 \begin{pmatrix} x_{TLS} \\ -1 \end{pmatrix},$$

where $\sigma = \sigma_{n+1}$ is the smallest singular value of (A, b). Solving the last equation for $\sigma^2 = b^Tr$ gives a system of n nonlinear equations in x

$$f(x) = -A^Tr - x(b^Tr) = 0, \quad r = b - Ax.$$

We will attempt to solve this system by Newton's method, cf. Dongarra [5].

We have $\Delta r = -A\Delta x$, and

$$f(x + \Delta x) = -A^TA\Delta x - \Delta x(b^Tr) + x(b^TA)\Delta x + \Theta((\Delta x)^2).$$

Hence the derivative $J(x) = f'(x) \in \Re^{n\times n}$ is

$$J(x) = A^TA - \sigma^2 I + xc^T, \quad c = A^Tb, \quad \sigma^2 = b^Tr.$$

In Newton's method we compute successive approximations $x_{k+1} = x_k + \delta x_k$, where x_0 is an initial approximation, and

$$J(x_k)\delta x_k = -f_k, \quad f_k = -f(x_k). \tag{1}$$

For TLS problems it is natural to take $x_0 = (A^TA)^{-1}A^Tb$, the standard least squares solution.

We write the Jacobian $f' = J = J_1(x) + J_2(x)$, where

$$J_1(x_k) = A^TA - \sigma^2 I, \quad \sigma^2 = b^T(b - Ax_k), \quad J_2(x_k) = x_kc^T.$$

Hence $J_1(x_k)$ is symmetric and $J_2(x_k)$ is a matrix of rank one. To solve (1) we get from the Sherman-Morrison formula

$$\delta x_k = -(J_1 + x_kc^T)^{-1}f_k = -J_1^{-1}f_k + \alpha J_1^{-1}x_kc^TJ_1^{-1}f_k,$$

where $\alpha = 1/(1 + c^TJ_1^{-1}x_k)$. It follows that the solution to the system (1) can be written in the form

$$\delta x_k = d_k - \frac{b^T(Ad_k)}{1 + b^T(Az_k)}z_k, \quad d_k = -J_1^{-1}f_k, \quad z_k = J_1^{-1}x_k, \tag{2}$$

where $J_1(x)$ is evaluated at x_k. Thus we need to solve two linear systems with the symmetric, but possibly indefinite, matrix $A^TA - (r_k^Tb)I$. The corresponding algorithm can be described as follows:

ALGORITHM Newton Iteration I

$$
\begin{aligned}
&x = x_{LS};\\
&\text{for} \quad k = 1, 2, \ldots \quad \text{do}\\
&\quad r = b - Ax;\\
&\quad \sigma^2 = r^T b;\\
&\quad f = -A^T r - \sigma^2 x;\\
&\quad \eta = (f^T f)^{1/2};\\
&\quad \text{solve} \quad (A^T A - \sigma^2 I)d = -f;\\
&\quad \text{solve} \quad (A^T A - \sigma^2 I)z = x;\\
&\quad \beta = b^T(Ad)/(1 + b^T(Az));\\
&\quad x = x + d - \beta z;\\
&\text{endfor}
\end{aligned}
$$

In a slightly different approach we consider σ as an independent variable and write the equations as

$$f(x, \sigma) = \begin{pmatrix} f_1(x, \sigma) \\ f_2(x, \sigma) \end{pmatrix} = \begin{pmatrix} -A^T r - \sigma^2 x \\ -b^T r + \sigma \end{pmatrix} = 0.$$

We have (neglecting second order quantities)

$$\begin{pmatrix} \Delta f_1 \\ \Delta f_2 \end{pmatrix} = J \begin{pmatrix} \Delta x \\ 2\sigma \Delta\sigma \end{pmatrix}, \quad J = \begin{pmatrix} J_1 & -x \\ b^T A & 1 \end{pmatrix},$$

where $J_1(\sigma) = A^T A - \sigma^2 I$. The Jacobian matrix J can be factored as

$$J = \begin{pmatrix} J_1 & 0 \\ b^T A & \tau \end{pmatrix} \begin{pmatrix} I & -z \\ 0 & 1 \end{pmatrix}, \quad \tau = 1 + b^T A z,$$

where $J_1 z = x$. Hence with

$$\begin{pmatrix} f_{1k} \\ f_{2k} \end{pmatrix} = \begin{pmatrix} f_1(x_k, \sigma^{(k)}) \\ f_2(x_k, \sigma^{(k)}) \end{pmatrix},$$

Newton's method can be written

$$\begin{pmatrix} J_1 & 0 \\ b^T A & \tau \end{pmatrix} \begin{pmatrix} d_1 \\ d_2 \end{pmatrix} = \begin{pmatrix} -f_{1k} \\ -f_{2k} \end{pmatrix}, \quad \begin{pmatrix} I & -z \\ 0 & 1 \end{pmatrix} \begin{pmatrix} \delta x_k \\ \delta(\sigma^{(k)})^2 \end{pmatrix} = \begin{pmatrix} d_1 \\ d_2 \end{pmatrix}.$$

We have $\delta(\sigma^{(k)})^2 = d_2$, $\delta x_k = d_1 - d_2 z$, where

$$J_1 d_1 = -f_{1k}, \quad \tau = 1 + b^T A z, \quad d_2 = -(f_2 + b^T A d_1)/\tau.$$

Again, in each iteration step we have to solve two linear system with system matrix $J_1 = A^T A - \sigma^2 I$. Since we want to converge to the smallest singular value of A it is natural to take $\sigma^{(0)} = 0$.

ALGORITHM Newton Iteration II

$$
\begin{aligned}
& x = x_{LS}; \quad \sigma^2 = 0; \\
& \text{for} \quad k = 1, 2, \ldots \quad \text{do} \\
& \quad r = b - Ax; \\
& \quad f_1 = -A^T r - \sigma^2 x; \\
& \quad f_2 = -b^T r + \sigma^2; \\
& \quad \rho = (f_1^T f_1 + (f_2)^2)^{1/2}; \\
& \quad \text{solve} \quad (A^T A - \sigma^2 I) z = x; \\
& \quad \text{solve} \quad (A^T A - \sigma^2 I) d_1 = -f_1; \\
& \quad d_2 = -(f_2 + b^T A d_1)/(1 + b^T A z); \\
& \quad \sigma^2 = \sigma^2 + d_2; \\
& \quad x = x + d_1 - d_2 z; \\
& \text{endfor}
\end{aligned}
$$

Note that because of the special starting values the first iteration simplifies. Since $x = x_{TLS}$ we get $f_1 = 0$, $f_2 = -r^T r$. Further $\sigma = d_\lambda = r^T r/(x^T x + 1)$, which is the Rayleigh quotient of x.

3 Local Convergence

We make the following assumptions about $f(x)$

1. There exists a $x_{TLS} \in \Re^N$ with $f(x_{TLS}) = 0$.
2. f is continuously differentiable in a neighborhood of x_{TLS}.

The following theorem is shown in [6].

Theorem 1 *Assume that f satisfies the above assumptions. There exists $\epsilon > 0$ such that if $\|x_0 - x_{TLS}\| \le \epsilon$, then the sequence of Newton iterates $\{x_n\}$ converges to x_{TLS}. Moreover, the convergence is linear in the sense that*

$$\|x_{n+1} - x_{TLS}\|_* \le t\|x_n - x_{TLS}\|_*,$$

where $\|y\|_* \equiv \|J(x_{TLS})y\|$.

Note that the Jacobian $J(x) = A^T A - \sigma^2 I + x c^T$ with $\sigma^2 = b^T r$ and $c = A^T b$ is Lipschitz continuous near x_{TLS} with constant ψ where $\psi = 2\hat{\sigma}_1\|b\|_2$, $\hat{\sigma}_1$ is the largest singular value of A. Before presenting the quadratic convergence of the iterates x_n, we need to prove the following lemma where $\hat{\sigma}_n$ and σ_{n+1} are the singular values of A and (A, b) respectively.

Lemma 2

$$\|J^{-1}(x_{TLS})\|_2 \le \frac{2}{\hat{\sigma}_n^2 - \sigma_{n+1}^2}. \tag{3}$$

Proof: The solution of total least squares problems x_{TLS} is

$$x_{TLS} = (A^T A - \sigma_{n+1}^2 I)^{-1} A^T b.$$

As described before $J = J_1 + J_2$, where $J_1 = A^T A - b^T r I = A^T A - \sigma_{n+1}^2 I$. Hence,

$$\|J_1^{-1}(x_{TLS})\|_2 = \frac{1}{\hat{\sigma}_n^2 - \sigma_{n+1}^2}.$$

From the Sherman-Morrison formula, we express $J^{-1}(x_{TLS})$ as

$$J^{-1}(x_{TLS}) = J_1^{-1}(x_{TLS}) - E,$$

where $E = 1/(1 + b^T A J_1^{-1} x_{TLS}) J_1^{-1}(x_{TLS}) x_{TLS} (A^T b)^T J_1^{-1}(x_{TLS})$.
Since $b^T A J_1^{-1} = x_{TLS}^T$, the corresponding norm of E is bounded

$$\begin{aligned} \|E\|_2 &\leq \alpha \|J_1^{-1}(x_{TLS})\|_2 \|x_{TLS}\|_2^2 \\ &= \frac{\|x_{TLS}\|_2^2}{1 + \|x_{TLS}\|_2^2} \|J_1^{-1}(x_{TLS})\|_2, \end{aligned}$$

which leads to (3) directly. ■

Now the quadratic convergence of the iterates x_n can be stated in the following theorem:

Theorem 3 *Assume that f satisfies the above assumptions and J is Lipschitz continuous near x_{TLS} with constant ψ. Then there exists an $\epsilon > 0$ such that if $\|x_0 - x_{TLS}\| \leq \epsilon$, then the sequence of Newton iterates $\{x_n\}$ converges to x_{TLS} quadratically in the sense that*

$$\|x_{k+1} - x_{TLS}\| \leq \beta\psi \|x_k - u_{TLS}\|^2, \qquad k = 0, 1, 2, \dots,$$

where

$$\beta = \frac{2}{\hat{\sigma}_n^2 - \sigma_{n+1}^2}, \qquad \psi = 2\hat{\sigma}_1 \|b\|_2.$$

Furthermore,

$$\lim_{k \to \infty} \sup \frac{\|x_{k+1} - x_{TLS}\|_2}{\|x_k - x_{TLS}\|_2^2} \leq \beta\psi.$$

Proof: The proof follows by a simple modification of the proof of Theorem 5.2.1 of Dennis and Schnabel [7]. ■

4 Global convergence

Newton's method by itself may often fail to converge if the initial guess x_0 is far away from the root. To enhance the robustness of the hybrid Krylov algorithm considered in the previous section, we will consider a global strategy based on a linesearch backtracking procedure described fully in [4].

Dennis and Schnabel [7] originally suggest using a global strategy for finding the root x_{TLS} that is based on a globally convergent method for the problem

$$\min_{x \in \Re^N} F(x) = \frac{1}{2} f(x)^T f(x). \tag{4}$$

A *descent direction* for F at the current approximation x is any vector p such that

$$\nabla F(x)^T p < 0,$$

where $\nabla F(x) = (\partial f/\partial x_1(u), \dots, \partial f/\partial x_N(u))^T$. An easy computation shows that

$$\nabla F(x) = J(x)^T f(x),$$

and p is a descent direction for F at x if

$$f(x)^T J(x) p < 0.$$

For such a direction, one can show that there exists certain $\lambda_0 > 0$ such that $F(x+\lambda p) < F(x)$ for all $0 < \lambda \le \lambda_0$.

If $\bar{\delta}$ is an approximation solution of (1), then

$$f^T J\bar{\delta} = -f^T f - f^T \bar{r}, \tag{5}$$

where $\bar{r} = -f - J\bar{\delta}$ is the residual associated with $\bar{\delta}$. Thus, $\bar{\delta}$ will be a descent direction for F at x whenever $|f^T\bar{r}| < f^T f$. In particular, if $\|\bar{r}\|_2 < \|f\|_2$, then $\bar{\delta}$ is a descent direction. This result was also given in [8] and is restated in the following proposition.

Proposition 4 *A sufficient condition for p to be a descent direction for F at x is that*

$$\|f(x) + J(x)p\|_2 \le \|f(x)\|_2. \tag{6}$$

An important condition to guarantee global convergence is proposed by Dennis and Schnabel [7]. Given the current Newton iterate $x = x_n$ and a descent direction p, we want to take a step in the direction of p that yields an acceptable x_{n+1}. We will define a step $\delta = \lambda p$ to be acceptable if both of the Goldstein-Armijo [7] conditions are met, namely

$$F(x+\lambda p) \le F(x) + \alpha\lambda\nabla F(x)^T p, \tag{7}$$

and

$$F(x+\lambda p) \ge F(x) + \beta\lambda\nabla F(x)^T p, \tag{8}$$

for given scalars α, β satisfying $0 < \alpha < \beta < 1$. In the following, we will refer to these two conditions as the $\alpha-$ and $\beta-$conditions, respectively. For a given descent direction p, the next result shows that there exist points $x + \lambda p$ satisfying (7) and (8).

Theorem 5 *Let F be continuously differentiable. Let x_n be Newton iterates and δx_n be the direction provided by (2) such that $\|f + J\delta x_n\|_2 < \|f\|_2$. Then given $0 < \alpha < \beta < 1$, there exist $\lambda_u > \lambda_l > 0$ such that $x_n + \lambda\delta x_n$ satisfy $\alpha-$ and $\beta-$condition for any $\lambda \in (\lambda_l, \lambda_u)$.*

Proof: Since $\alpha < 1$, $F(x_n + \lambda\delta x_n) < F(x_n) + \alpha\lambda\nabla F(x_n)^T\delta x_n$ for all $\lambda > 0$ sufficiently small. Since $F(x_n + \lambda\delta x_n)$ is bounded below, there exists some smallest positive $\hat{\lambda}$ such that

$$F(\hat{x}) = F(x_n + \hat{\lambda}\delta x_n) = F(x_n) + \alpha\hat{\lambda}\nabla F(x_n)^T\delta x_n. \tag{9}$$

Thus any $x \in (x_n, \hat{x})$ satisfies $\alpha-$condition. By the mean value theorem, there exists $\bar{\lambda} \in (0, \hat{\lambda})$ such that

$$\begin{aligned} F(\hat{x}) - F(x_n) &= \nabla F(x_n + \bar{\lambda}\delta x_n)^T(\hat{x} - x_n) \\ &= \nabla F(x_n + \bar{\lambda}\delta x_n)^T\hat{\lambda}\delta x_n, \end{aligned} \tag{10}$$

and so from (9) and (10),

$$\nabla F(x_n + \bar{\lambda}\delta x_n)^T \delta x_n = \alpha \nabla F(x_n)^T \delta x_n > \beta \nabla F(x_n)^T \delta x_n \tag{11}$$

since $\alpha < \beta$ and $\nabla F(x_n)^T \delta x_n < 0$. By the continuity of ∇F, (11) still holds for λ in some interval $(\lambda_l. \lambda_u)$ about $\bar{\lambda}$. Therefore, if we restrict (λ_l, λ_u) to be in $(0, \hat{\lambda})$, $x_n + \lambda \delta x_n$ satisfies $\alpha-$ and $\beta-$condition for any $\lambda \in (\lambda_l, \lambda_u)$. ■

Next, the following theorem originally proposed by Wolfe [9, 10] gives a sufficient condition for global convergence in a weak sense.

Theorem 6 *Let F as we defined be continuously differentiable function. Then, given any starting vector, there exists sequence x_n, $k = 1, 2, \dots$, obeying $\alpha-$ and $\beta-$condition. In addition this sequence is such that either*

$$\nabla F(x_n)^T \delta x_n < 0 \tag{12}$$

or

$$\nabla F(x_n) = 0 \quad and \quad \delta x_n = 0, \tag{13}$$

for each $n > 0$, where $\delta x_n = x_{n+1} - x_n$. Furthermore, for any such sequence, either

$$\begin{array}{ll} (a) & \nabla F(x_n) = 0 \text{ for some } n \geq 0, \text{ or} \\ (b) & \lim_{n\to\infty} \frac{\nabla F(x_n)^T \delta x_n}{\|\delta x_n\|_2} = 0. \end{array}$$

This is a quite important results although the conditions are too weak to imply that the sequence generated from our algorithm x_n converges.

5 Numerical tests

The following tests were performed in MATLAB on a SUN workstation using double precision. The set of test problems denoted $PS(m, n, d, p)$ was generated, as described in Paige and Saunders [11], as

$$A = Y \begin{pmatrix} D \\ 0 \end{pmatrix} Z^T \in \mathbf{R}^{m \times n}, \qquad Y = I - 2yy^T, \quad Z = I - 2zz^T.$$

Here y and z are Householder vectors of appropriate dimension, generated by

$$y_i = \sin(4\pi i/m), \qquad z_i = \cos(4\pi i/n),$$

followed by normalization so that $\|y\| = \|z\| = 1$. The singular values are taken to be

$$D = diag\,(n, n/2, \dots, n/2^{n-1}).$$

The multiple right hand side is taken to be $m \times p$ matrix of ones. In particular, the one-dimensional problems corresponds to $p = 1$. The error perturbation of $(A\ b)$ is $(E\ r)$ with the form of sR where s is between 10^{-4} and 10^{-3} and R is a randomly generated matrix whose elements are uniformly distributed in the interval $(0, 1)$.

As the exact solution of the total least squares problems we have taken is computed by the singular value decomposition. In the tests, we use the conjugate gradient method CGTLS to solve the linear equation of the following general form

$$(A^T A - \sigma^2 I)y = g,$$

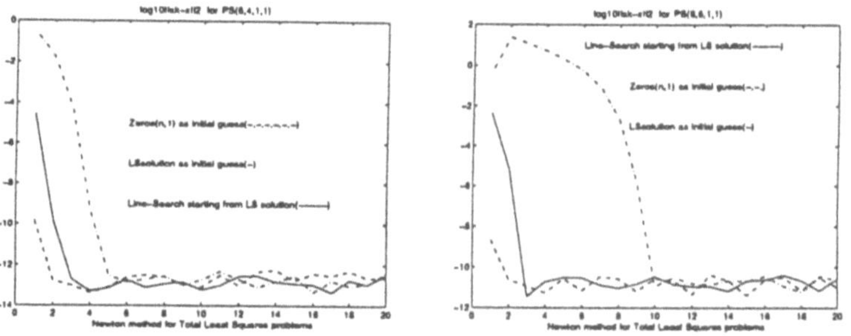

Figure 1: Newton's method I for PS(6,4,1,1) and PS(8,6,1,1)

which appears in algorithms where σ^2 is known. This method can be derived similarly to CGLS for the linear least square problems [4]. In [4], we have referred the work done by Elfving [12] which demonstrated that CGLS1 (denoted as CGLS throughout the thesis) is the most accurate and stable version compared with several other implementations of the conjugate gradient method. Similarly the method CGTLS can be derived which avoids forming $A^T A$ and recurs the residual $r_k = b - Ax_k$. Here the least squares solution used as starting vector was computed by CGLS method.

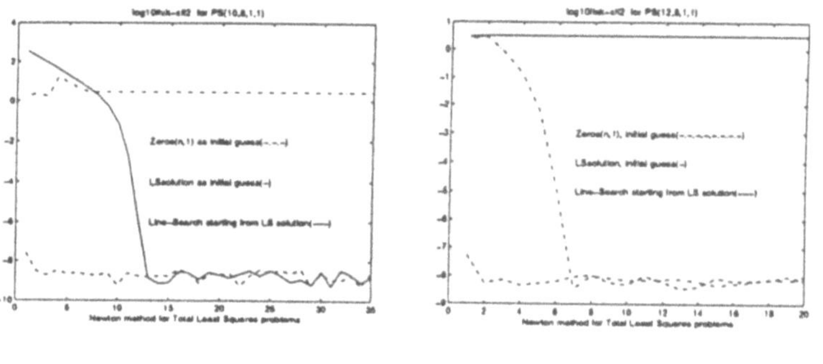

Figure 2: PS(10,8,1,1) and PS(12,8,1,1) with bigger error perturbation

The first set of numerical experiments plotted in Figures 1 shows the performance of Newton's method I with $d = 1/2 \times 10^{-3}$ using two different initial vectors, zero(n,1) and the LS solution. We also show the performance of Newton's method I with the linesearch technique with $\alpha = 1/4$ and $\beta = 3/4$, starting from the LS solution. From these experiments, it is easy to see that the closer the starting vector is to the exact solution, the faster it converges. Further Newton's method I with linesearch starting from the LS solution is an even better choice with rather few iterations and rapid convergence. For perturbations of this size, the method with the LS solution as initial guess is also a good method achieving almost the same performance as with linesearch strategy. We will prefer to use the LS solution as starting vector in further algorithms.

The second set of numerical experiments plotted in Figures 2 shows results of Newton's method I with zero(n,1) and the LS solution as starting vector will not converge, using a bigger error perturbation with $d = 10^{-1}$. Meanwhile, we also can see that the performance with linesearch strategy starting from the LS solution is rather good. The convergence is almost as fast as for the smaller value of d in Figures 1.

For the Newton's method II, we can get the similar conclusions with the Newton's method I.

References

1. G. H. Golub and C. F. Van Loan, SIAM Journal on Numerical Analysis, 17, 883–893 (1980).
2. S. Van Huffel and J. Vandewalle, in: Frontiers in Applied Mathematics, Vol. 9, SIAM, Philadelphia (1991).
3. S. Van Huffel, Linear Algebra and its Applications, 154/156, 675–709 (1991).
4. T. Yang, Licentiate Thesis LiU-TEK-LIC-1996:25, Linköping University, 581 83, Linköping, Sweden (1996).
5. J. J. Dongarra, SIAM Journal on Scientific and Statistical Computing, 4, 712–719 (1983).
6. R. S. Dembo, S. C. Eisenstat and T. Steihaug, SIAM Journal on Numerical Analysis, 19, 400–408 (1982).
7. J. E. Dennis and R. B. Schnabel, Numerical Methods for Unconstrained Optimization and Nonlinear Equations, Computational Mathematics. Prentice-Hall, Englewood Cliffs, NJ (1983).
8. P. Brown and Y. Saad, SIAM Journal on Scientific and Statistical Computing, 11, 450–481 (1990).
9. P. Wolfe, SIAM Review, 11, 226–235 (1969).
10. P. Wolfe, SIAM Review, 13, 185–188 (1971).
11. C. C. Paige and M. A. Saunders, ACM Transactions on Mathematical Software, 8, 43–71 (1982).
12. T. Elfving, Technical Report LiTH-MAT-R-78-3, Department of Mathematics, Linköping University (1978).

8th Int. Coll. on Differential Equations, pp. 441–449
D. Bainov (Ed)

Theoretical Performance Analysis of the IQMR Method on Distributed Memory Computers with Different Network Topologies

TIANRUO YANG
Dept. of Computer Science, Linköping University, S-581 83, Linköping, Sweden
tiaya@ida.liu.se

HAI-XIANG LIN
Dept. of Technical Mathematics and Informatics, Delft University of Technology, P.O. Box 356, 2600 GA Delft, The Netherlands
lin@dutind3.twi.tudelft.nl

Abstract. For the solutions of linear systems of equations with unsymmetric coefficient matrices, we have proposed an improved version of the quasi-minimal residual (IQMR) method [1] by using the Lanczos process as a major component combining elements of numerical stability and parallel algorithm design. The algorithm is derived such that all inner products and matrix-vector multiplications of a single iteration step are independent and communication time required for inner product can be overlapped efficiently with computation time. In this paper, we mainly present the qualitative analysis of the parallel performance with *Store-and-Forward routing* and *Cut-Through routing* schemes and topologies such as *ring, mesh, hypercube and balanced binary tree*. Theoretically it is shown that the hypercube topology can give us the best parallel performance with regards to parallel efficiency, speed-up, and runtime, respectively.

1 Introduction

One of the fundamental task of numerical computing is the ability to solve linear systems with nonsymmetric coefficients. These systems arise very frequently in scientific computing, for example from finite difference or finite element approximations to partial differential equations, as intermediate steps in computing the solution of nonlinear problems or as subproblems in linear and nonlinear programming.

One of them, the quasi-minimal residual (QMR) algorithm [2], uses the Lanczos process [3] with look-ahead, a technique developed to prevent the process from breaking down in case of numerical instabilities, and in addition imposes a quasi-minimization principle. This combination leads to a quite efficient algorithm among the most frequently and successfully used iterative methods. This method is widely used for very large and sparse problems, which in turn are often solved on massively parallel computers.

On massively parallel computers, the basic time-consuming computational kernels of QMR are usually: inner products, vector updates, matrix-vector multiplications. In many situations, especially when matrix operations are well-structured, these operations are suitable for implementation on vector and share memory parallel computers [4]. But for parallel distributed memory machines, the matrices and vectors are distributed over the processors, so that even when the matrix operations can be implemented efficiently by parallel operations, we still can not avoid the global communication, i.e. communication of all processors, required for inner product computations. Vector updates are perfectly parallelizable and, for large sparse matrices, matrix-vector multiplications can be implemented with communication between only nearby processors. The bottleneck is usually due to inner products enforcing global communication. The detailed discussions on the communication problem on distributed memory systems can be found in [5, 6]. These global communication costs become relatively more and more important when the number of the parallel processors is increased and thus they have the potential to affect the scalability of the algorithm in a negative way [5, 6].

Recently, we have proposed a new improved two-term recurrences Lanczos process [1] without look-ahead as the underlying process of QMR. The algorithm is reorganized without changing the numerical stability so that all inner products and matrix-vector multiplications of a single iteration step are independent and communication time required for inner product can be overlapped efficiently with computation time. Therefore, the cost of global communication on parallel distributed memory computers can be significantly reduced. The resulting IQMR algorithm maintains the favorable properties of the Lanczos process while not increasing computational costs. In [7], we also propose a theoretical model of computation and communications phases to allow us to give a qualitative analysis of the parallel performance on a massively distributed memory computer with two-dimensional grid topology. The efficiency, speed-up, and runtime are expressed as functions of the number of processors scaled by the number of processors that gives the minimal runtime for the given problem size. This provides a natural way to analyze the performance characteristics for the range of the number of processors that can be used effectively. The model not only shows clearly the dramatic influence of global communication on the performance, but also evaluates effectively the improvements in the performance due to the communication reductions by overlapping. The model also provides useful insight in the scalability of IQMR method. But it is still limited by assumptions on the communication model. In this paper, we mainly investigate the effect of *Store-and-Forward routing* and *Cut-Through routing* and topologies such as *ring, mesh, hypercube and balanced binary tree.* Theoretically it is shown that the hypercube topology can give us the best parallel performance with regards to parallel efficiency, speed-up, and runtime, respectively.

The paper is organized as follows. We will describe briefly the improved Lanczos process and the resulting improved quasi-minimal residual (IQMR) method in section 2 and 3, respectively. In section 4, the parallel performance model is presented including the communication model and assumptions for computation time and communication costs. Finally the theoretical complexity analysis of parallel performance with *Store-and-Forward routing* and *Cut-Through routing* and different topologies such as *ring, mesh, hypercube and balanced binary tree* is described fully.

2 The Improved Lanczos Process

The classical unsymmetric Lanczos process [8] based on three-term recurrence reduces a matrix A to tridiagonal form T using a similarity transformation which leads to the following three relations that serve to derive the unsymmetric Lanczos process:

$$W^TV = I, \quad AV = VT, \quad A^TW = WT^T, \tag{1}$$

where I is the identity matrix and T is a tridiagonal matrix. This process leads to two inner products per iteration which require global communication on massively parallel distributed memory computers. Some improvements on the reduction of global communication required by inner products have been investigated in [9].

Although Lanczos used the similar technique built on coupled two-term recurrence in the early of 1950's, the majority of papers are dealing with the three-term recurrences process, until, recently, Freund et al. [10] reused this idea to improve numerical stability. They claimed that, the latter variant of the Lanczos process, may be numerically more stable. That is why we pursue further on this unsymmetric Lanczos process with two-recurrences as underlying process of QMR method.

Recently, Bucker et al. [11, 12] propose a new parallel version of the quasi-minimal residual (QMR) method based on the coupled two-term recurrences Lanczos process without look-ahead strategy. The algorithm is derived that both generated sequences of Lanczos vectors are scalable and there is only a single global synchronization point per iteration. Based on the similar idea, we have proposed a new improved two-term recurrences Lanczos process without look-ahead technique [1, 13] as follows:

Algorithm 1 Improved Lanczos Process

$p_0 = q_0 = u_0 = 0,\ \gamma_1 = (\tilde{v}_1, \tilde{v}_1), \mu_1 = 0, \xi_1 = (\tilde{w}_1, \tilde{w}_1),$
$s_1 = A^T\tilde{w}_1, \rho_1 = (\tilde{w}_1, \tilde{v}_1), \varepsilon_1 = (s_1, \tilde{v}_1), \tau_1 = \frac{\varepsilon_1}{\rho_1};$
for $n = 1, 2, \ldots$ **do**
  $q_n = \frac{1}{\xi_n}s_n - \frac{\gamma_n\mu_n}{\xi_n}q_{n-1};$
  $\tilde{w}_{n+1} = q_n - \frac{\tau_n}{\xi_n}\tilde{w}_n;$
  $s_{n+1} = A^T\tilde{w}_{n+1};$
  $t_n = A\tilde{v}_n;$
  $u_n = \frac{1}{\gamma_n}t_n - \mu_n u_{n-1};$
  $\tilde{v}_{n+1} = u_n - \frac{\tau_n}{\gamma_n}\tilde{v}_n;$
  $p_n = \frac{1}{\gamma_n}\tilde{v}_n - \mu_n p_{n-1};$
  $\gamma_{n+1} = (\tilde{v}_{n+1}, \tilde{v}_{n+1});$
  $\xi_{n+1} = (\tilde{w}_{n+1}, \tilde{w}_{n+1});$
  $\rho_{n+1} = (\tilde{w}_{n+1}, \tilde{v}_{n+1});$
  $\varepsilon_{n+1} = (s_{n+1}, \tilde{v}_{n+1});$
  $\mu_{n+1} = \frac{\gamma_n\xi_n\rho_{n+1}}{\gamma_{n+1}\tau_n\rho_n};$
  $\tau_{n+1} = \frac{\varepsilon_{n+1}}{\rho_{n+1}} - \gamma_{n+1}\mu_{n+1};$
end for

Now we can schedule the process algorithm, without changing the numerical stability so that all inner products and matrix-vector multiplications of a single iteration step are independent and communication time required for inner product can be overlapped efficiently with computation time. The framework of this improved Lanczos process based on two-term recurrences is described. Under the assumptions, the improved Lanczos process can be efficiently parallelized as follows:

- The inner products of a single iteration step (11), (12), (13) and (14) are independent.
- The matrix-vector multiplications of a single iteration step (6) and (7) are independent.
- The communications required for the inner products (11), (12), (13) and (14) can be overlapped with the update for p_n in (10).

Therefore, the cost of communication time on parallel distributed memory computers can be significantly reduced.

Since the biorthogonality relationship are used to derive the algorithm. We can conversely show that, in exact arithmetic, the vectors $\tilde{v}_i$ and $\tilde{w}_i$ generated by the above algorithm are biorthogonal.

Theorem 1 *Assume no breakdown occurs, the vectors $\tilde{v}_i$ and $\tilde{w}_i$ generated by Improved Lanczos Process satisfy*

$$\tilde{w}_i^T \tilde{v}_j = \begin{cases} 0 & i \neq j, \\ \rho_i \neq 0 & i = j. \end{cases}$$

3 The Improved Quasi-Minimal Residual Method

The improved two-term recurrences Lanczos process without look-ahead technique then is used as a major component to a Krylov subspace method for solving a system of linear equations

(2) $$Ax = b, \quad \text{where} \quad A \in \Re^{n \times n} \quad x, b \in \Re^n.$$

In each step, it produces approximation x_n to the exact solution of the form

(3) $$x_n = x_0 + \mathcal{K}_n(r_0, A), \quad n = 1, 2, \ldots$$

Here x_0 is any initial guess for the solution of linear systems, $r_0 = b - Ax_0$ is the initial residual, and $\mathcal{K}_n(r_0, A) = span\{r_0, Ar_0, \ldots, A^{n-1}r_0\}$, is the n-th Krylov subspace with respect to r_0 and A.

Given any initial guess x_0, the n-th Improved QMR iterate is of the form

(4) $$x_n = x_0 + V_n z_n,$$

where V_n is generated by the improved unsymmetric two-term recurrences Lanczos process without look-ahead technique [1, 13] and z_n is determined by a quasi-minimal residual property.

The result is an improved QMR method (IQMR) based on coupled two-term recurrences with scaling of both sequence of Lanczos vectors where numerical stability can be maintained so that all inner products and matrix-vector multiplications of a single iteration step are independent and communication time required for inner product can be overlapped efficiently with computation time. The framework of this improved QMR method is depicted in Algorithm 2.

Algorithm 2 The IQMR Method

1: $\tilde{v}_1 = \tilde{w}_1 = r_0 = b - Ax_0, \lambda_1 = 1, \kappa_0 = 1,$

2: $p_0 = q_0 = u_0 = d_0 = f_0 = 0, \gamma_1 = (\tilde{v}_1, \tilde{v}_1), s_1 = A^T\tilde{w}_1, \xi_1 = (\tilde{w}_1, \tilde{w}_1),$

3: $\rho_1 = (\tilde{w}_1, \tilde{v}_1), \varepsilon_1 = (s_1, \tilde{v}_1), \mu_1 = 0, \tau_1 = \frac{\varepsilon_1}{\rho_1};$

4: **for** $n = 1, 2, \ldots$ **do**

5: $\quad q_n = \frac{1}{\xi_n} s_n - \frac{\gamma_n \mu_n}{\xi_n} q_{n-1};$

6: $\quad \tilde{w}_{n+1} = q_n - \frac{\tau_n}{\xi_n} \tilde{w}_n;$

7: $\quad s_{n+1} = A^T \tilde{w}_{n+1};$

8: $\quad t_n = A\tilde{v}_n;$

9: $\quad u_n = \frac{1}{\gamma_n} t_n - \mu_n u_{n-1};$

10: $\quad \tilde{v}_{n+1} = u_n - \frac{\tau_n}{\gamma_n} \tilde{v}_n;$

11: $\quad p_n = \frac{1}{\gamma_n} \tilde{v}_n - \mu_n p_{n-1};$

12: **if** $(r_{n-1}, r_{n-1}) < \mathrm{tol}$ **then**

13: quit

14: **else**

15: $\quad \gamma_{n+1} = (\tilde{v}_{n+1}, \tilde{v}_{n+1});$

16: $\quad \xi_{n+1} = (\tilde{w}_{n+1}, \tilde{w}_{n+1});$

17: $\quad \rho_{n+1} = (\tilde{w}_{n+1}, \tilde{v}_{n+1});$

18: $\quad \varepsilon_{n+1} = (s_{n+1}, \tilde{v}_{n+1});$

19: $\quad \mu_{n+1} = \frac{\gamma_n \xi_n \rho_{n+1}}{\gamma_{n+1} \tau_n \rho_n};$

20: $\quad \tau_{n+1} = \frac{\varepsilon_{n+1}}{\rho_{n+1}} - \gamma_{n+1} \mu_{n+1};$

21: $\quad \theta_n = \frac{\tau_n^2 (1 - \lambda_n)}{\lambda_n \tau_n^2 + \gamma_{n+1}^2};$

22: $\quad \kappa_n = \frac{-\gamma_n \tau_n \kappa_{n-1}}{\lambda_n \tau_n^2 + \gamma_{n+1}^2};$

23: $\quad \lambda_n = \frac{\lambda_{n-1} \tau_{n-1}^2}{\lambda_{n-1} \tau_{n-1}^2 + \gamma_n^2};$

24: $\quad d_n = \theta_n d_{n-1} + \kappa_n p_n;$

25: $\quad f_n = \theta_n f_{n-1} + \kappa_n u_n;$

26: $\quad x_n = x_{n-1} + d_n;$

27: $\quad r_n = r_{n-1} - f_n;$

28: **end if**

29: **end for**

Under the assumptions, the improved QMR method using Lanczos algorithm as underlying process can be efficiently parallelized as follows:

- The inner products of a single iteration step (12), (15), (16), (17) and (18) are independent.
- The matrix-vector multiplications of a single iteration step (7) and (8) are independent.
- The communications required for the inner products (12), (15), (16), (17) and (18) can be overlapped with the update for p_n in (11).

Therefore, the cost of communication time on parallel distributed memory computers can be significantly reduced.

4 The Performance Model

Based on these mathematical background described above, we will make the following assumptions suggested in [14, 6, 7] for our performance model. First, the model assumes perfect load balance and the processors are configured as *ring, d-dimensional mesh with wrap-around connection, hypercube and balanced binary tree*, respectively. For *d-dimensional mesh* without wrap-around connection, the basic communication operation increase at most by a factor of four. In this model, each processor holds a sufficiently large number of successive rows of the matrix, and the corresponding sections of the vectors involved. That is, our problems have a strong data locality. Secondly, we can compute the inner products (reduction) in two steps because the vectors are distributed over the processor topology. The computation of an inner product is executed by all processors simultaneously without communication or locally and these partial results are combined by a global sum operation and accumulated at a single destination processor, called *single node accumulation* (SNA). The second phase consists of reversing the directions and sequence of messages sending the final results from this single processor to all other processors, called *single-node broadcast* (SNB).

With regards to the communication model, the *Store-and-Forward routing* (SF) and *Cut-Through routing* (CT) scheme [15] are discussed in our qualitative performance analysis. In SF routing, when a message is traversing a path with multiple links, each intermediate processor on the path forwards the message to the next processor after it has received and stores the entire message. In CT routing, each intermediate processor on the path does not wait for the entire message to arrive before forwarding the message to the next processor. As soon as a smaller unit called a *flit* is received at an intermediate processor, the flit is passed on to the next one. The corresponding communication cost for SNB and inner products with SF and CT routing can be summarized as follows: where the

Table 1: Communication Operation Cost with SF and CT Routings

Topologies	Operations	SF Routing	CT Routing
Ring	SNB	$(t_s + t_w m)\lceil \frac{p}{2} \rceil$	$(t_s + t_w m)\log_2 p + t_h(p-1)$
	Inner	$(t_s + 5t_w)p$	$2(t_s + 5t_w)\log_2 p + 2t_h(p-1)$
Mesh	SNB	$d(t_s + t_w m)\lceil \frac{p^{1/d}}{2} \rceil$	$(t_s + t_w m)\log_2 p + dt_h(p^{1/d}-1)$
	Inner	$d(t_s + 5t_w)p^{1/d}$	$2(t_s + 5t_w)\log_2 p + 2dt_h(p^{1/d}-1)$
Hypercube	SNB	$(t_s + t_w m)\log_2 p$	$(t_s + t_w m)\log_2 p$
	Inner	$2(t_s + 5t_w)\log_2 p$	$2(t_s + 5t_w)\log_2 p$
Tree	SNB		$(t_s + t_w m + t_h(\log_2 p + 1))\log_2 p$
	Inner		$2(t_s + 5t_w) + t_h(\log_2 p + 1))\log_2 p$

t_s is the start-up time that required to initiate a message transfer at the sending processor. The per-word time t_w is the time each word takes to traverse two directly-connected processors. The per-hop time t_h is the time needed by the header of a message to travel between two directly-connected processors. The communication cost for SNB with CT routing on the hypercube we list is not optimal. The optimal one is $2(t_s \log_2 p + t_w m)$.

In the following part we will describe a simple performance model including the computation time and communication cost for the main kernels as we presented before based on our assumptions. These two important terms are used in our paper suggested in [6]:

- Communication Cost: The term to indicate all the wall-clock time spent in communication, that is not overlapped with useful computation.

- Communication Time: The term to refer to the wall-clock time of the whole communication.

In the non-overlapped communication, the communication time and the communication cost are the same term.

The IQMR algorithm contains three distinct computational tasks per iteration

- Two simultaneous matrix-vector products, $A\tilde{v}_n$ and $A^T\tilde{w}_{n+1}$ whose computation time are given $2t_{fl}N/P$.

- Five simultaneous inner products, $(\tilde{v}_{n+1}, \tilde{v}_{n+1})$, $(\tilde{w}_{n+1}, \tilde{w}_{n+1})$, $(\tilde{w}_{n+1}, \tilde{v}_{n+1})$, $(s_{n+1}, \tilde{v}_{n+1})$ and (r_{n-1}, r_{n-1}) whose computation time are given by $(2n_z - 1)t_{fl}N/P$.

- Nine vector updates, q_n, $\tilde{w}_{n+1}$, u_n, $\tilde{v}_{n+1}$, p_n, d_n, f_n, x_n and r_n which are given $2t_{fl}N/P$,

where N/P is the local number of unknown of a processor, t_{fl} is the average time for a double precision floating point operation and n_z is the average number of non-zero elements per row of the matrix.

The complete (local) computation time for the IQMR method is given approximately by the following equation:

$$T_{comp}^{IQMR} = (19 + 2n_z)\frac{N}{P}t_{fl} = f(m)\frac{N}{p}. \tag{5}$$

The corresponding global accumulation and broadcast time for 5 simultaneous inner product of IQMR method, $g(m)$, is given as follows:

Table 2: Communication Cost with SF and CT Routings

	Topologies	SF Routing	CT Routing
$g(m)$	Ring	$(t_s + 5t_w)p$	$2(t_s + 5t_w)\log_2 p + 2t_h(p-1)$
	Mesh	$d(t_s + 5t_w)p^{1/d}$	$2(t_s + 5t_w)\log_2 p + 2dt_h(p^{1/d} - 1)$
	Hypercube	$2(t_s + t_w 5)\log_2 p$	$2(t_s + 5t_w)\log_2 p$
	Tree	$2(t_s + 5t_w)\log_2 p$	$2(t_s + 5t_w) + t_h(\log_2 p + 1))\log_2 p$

5 Theoretical Performance Analysis

In this section, we will focus on the theoretical analysis of the parallel performance of IQMR method where the efficiency, speed-up and runtime are expressed as functions of the number of processors scaled by the number of processors that gives the minimal runtime for the given problem size.

The total runtime for IQMR is given by the following equation:

$$T_p^{IQMR} = T_{comp}^{IQMR} + T_{comm}^{IQMR} = f(m)\frac{N}{p} + g(m) \tag{6}$$

This equation shows that for sufficiently large p the communication time will dominate the total runtime.

Let p_{max} denote as the number of processors that minimizes the total runtime T_p for any p processors of IQMR method. The percentage of the computation time in the whole runtime $E_p = T_1/pT_p$ is defined as parallel efficiency. The speed-up S_p for any p processors is $S_p = T_1/T_p$ and α be the fraction $\alpha = p/p_{max}$ and $\omega = d\alpha^{\frac{d+1}{d}}$. Then the theoretical performance for SF routing can be summarized as follows:

Table 3: Theoretical Parallel Performance with SF Routing

Topologies	p_{max}	E_p	S_p	T_p
Ring	$(\frac{f(m)N}{t_s+5t_w})^{1/2}$	$\frac{1}{1+\alpha^2}$	$\frac{p}{1+\alpha^2}$	$\frac{1+\alpha^2}{\alpha}(t_s + 5t_w)f(m)N)^{1/2}$
Mesh	$(\frac{f(m)N}{t_s+5t_w})^{\frac{d}{d+1}}$	$\frac{1}{1+\omega}$	$\frac{p}{1+\omega}$	$\frac{1+\omega}{\alpha}((t_s + 5t_w)^d f(m)N)^{\frac{1}{d+1}}$
Hypercube	$\frac{\ln 2}{2}\frac{f(m)N}{t_s+5t_w}$	$\frac{1}{1+\alpha \ln p}$	$\frac{p}{1+\alpha \ln p}$	$\frac{1+\alpha \ln p}{\alpha}\frac{2(t_s+5t_w)}{\ln 2}$

For CT routing, we can conclude the similar results which can be described in the following theorem:

Theorem 2 *Let p_{max}, runtime T_p, speed-up S_p, and efficiency E_p be defined as before for any p processors with SF and CT routings of the parallel IQMR method. Then we have similar relations described as follows for both SF and CT routing with different network topologies. The runtime T_p:*

$$T_p^{ring} > (T_p^{tree}) > T_p^{mesh} > T_p^{hypercube}, \tag{7}$$

the parallel efficiency E_p:

$$E_p^{ring} < (E_p^{tree}) < E_p^{mesh} < E_p^{hypercube}, \tag{8}$$

the parallel speed-up S_p:

$$S_p^{ring} < (S_p^{tree}) < S_p^{mesh} < S_p^{hypercube}, \tag{9}$$

and the corresponding p_{max}:

$$p_{max}^{ring} < (p_{max}^{tree}) < p_{max}^{mesh} < p_{max}^{hypercube}. \tag{10}$$

References

1. T. Yang and H. X. Lin, in: Proceedings of The International Conference on High Performance Computing and Networking (HPCN-97), April (1997).
2. R. W. Freund and N. M. Nachtigal, Numerische Mathematik, 60, 315–339 (1991).

3. R. W. Freund, M. H. Gutknecht, and N. M. Nachtigal, SIAM Journal on Scientific and Statistical Computing, 14, 137–158 (1993).

4. J. J. Dongarra, I. S. Duff, D. C. Sorensen and H. A. van der Vorst, Solving Linear Systems on Vector and Shared Memory Computers, SIAM, Philadelphia, PA (1991).

5. E. de Sturler, in: Proceedings of the 13th IMACS World Congress on Computational and Applied Mathematics, IMACS, Criterion Press (1991).

6. E. de Sturler and H. A. van der Vorst, Technical Report 832, Mathematical Institute, University of Utrecht, Utrecht, The Netherlands (1994).

7. T. Yang, in: Proceedings of International Conference on Advances in Parallel and Distributed Computing (APDC-97), Shanghai, March (1997).

8. C. Lanczos, Journal of Research of National Bureau of Standards, 45, 255–282 (1950).

9. S. K. Kim and A. T. Chronopoulos, Journal of Computational and Applied Mathematics, 357–374 (1992).

10. R. W. Freund and N. M. Nachtigal, SIAM Journal on Scientific and Statistical Computing, 15, 313–337 (1994).

11. H. M. Bucker and M. Sauren, in: Proceedings of Workshop on Applied Parallel Computing in Industrial Problems and Optimization (Para96), Technical University of Denmark, Lyngby, Denmark, Springer-Verlag, August (1996).

12. H. M. Bucker and M. Sauren, Technical Report KFA-ZAM-IB-9606, Central Institute for Applied Mathematics, Research Centre Julich, Germany, March (1996).

13. T. Yang and H. X. Lin, in: Proceedings of The International Workshop on Computational Science and Engineering (IWCSE'97), May (1997).

14. E. de Sturler, Technical Report CSCS-TR-94-05, Swiss Scientific Computing Center, La Galleria, CH-6928 Manno, Switzerland, May (1994).

15. V. Kumar, A. Grama, A. Gupta and G. Karypis, Introduction to Parallel Computing: Design and Analysis of Algorithms, Benjamin/Cummings, Redwood City (1994).

8th Int. Coll. on Differential Equations, pp. 451-455
D. Bainov (Ed)

Optimal Control of Certain M-D Systems

S. A. BELBAS
Mathematics Department, University of Alabama, Tuscaloosa, AL 35487-0350, USA

Abstract. We consider a continuous, nonlinear, multi-dimensional extension of the Roesser model. We obtain optimality conditions for an associated optimal control problem.
Keywords: Multi-dimensional systems, Roesser model, optimal control.
AMS Subject Classification: 49K20, 35F30.

1 INTRODUCTION

The theory of controlled M-D systems, i.e., systems in which the role of time is played by an M-dimensional parameter (which is traditionally called "multi-dimensional time", although it does not have the physical meaning of time), has received much attention in recent years. This theory has applications to processing of multidimensional signals (in particular, image processing), multi-dimensional version of neural networks, control of systems governed by partial differential equations, etc.

The motivation for the problem considered in this paper comes from two directions:

The Roesser model for linear discrete-time 2-D systems [1, 2], and Cesari's work [3] on optimal control of Dieudonné-Rashevsky systems. (We note that one of the differences between the present work and [3] is that here we do not have a complete set of Dieudonné-Rashevsky equations, but rather a proper subset of a Dieudonné-Rashevsky ensemble.)

The standard Roesser model, in discrete 2-D time, has the form

$$\begin{aligned} x_1(i+1,j) &= A_{11}x_1(i,j) + A_{12}x_2(i,j) + B_1u(i,j); \\ x_2(i,j+1) &= A_{21}x_1(i,j) + A_{22}x_2(i,j) + B_2u(i,j); \\ & \quad 0 \le i \le M, \qquad 0 \le j \le N \end{aligned} \tag{1}$$

The system considered in this paper extends the Roesser model in 4 directions:

(i) we consider a system in continuous time,

(ii) we have a nonlinear model;

(iii) we have an M-D time parameter, with arbitrary $M \ge 2$;

(iv) we have a general M-D time domain (not necessarily a rectangular box).

We obtain necessary conditions for an optimal control problem associated with the extension of the Roesser model.

2 FORMULATION OF THE PROBLEM

We consider a controlled system of Roesser type in M-dimensional time $t = (t_1, \ldots, t_M)$ in the domain $\Omega \subset \mathbb{R}^M$. The domain Ω is open, bounded, with oriented piecewise C^1 boundary $\Gamma = \partial\Omega$. For every two points s, t in $\mathbb{R}^M$, we shall write $s \leq t$ iff $s_i \leq t_i$ for all $i = 1, 2, \ldots, M$. For each $t \in \overline{\Omega}$, we set

$$E(t) \equiv \{s \in \overline{\Omega} : s \leq t\}, \epsilon(t) \equiv \{s \in \overline{\Omega} : s \geq t\} \tag{2}$$

We set

$$\Gamma_0 \equiv \Gamma \bigcap \left(\bigcup_{t \in \overline{\Omega}} E(t) \right), \Gamma_1 \equiv \Gamma \backslash \Gamma_0 \tag{3}$$

The controlled system, with state $y = (y_1, \ldots, y_M)$, each y_i taking values in $\mathbb{R}^{n_i}$, and control u taking values in a compact set $K \subset \mathbb{R}^m$, is described by the system of partial differential equations

$$\frac{\partial y_i(t)}{\partial t_i} = f_i(t, y(t), u(t)) \tag{4}$$

with boundary conditions

$$y_i(s) = y_{i,0}(s) \qquad \text{for } s \in \Gamma_0 \tag{5}$$

The functions f_i are continuous and bounded in all their arguments, continuously differentiable with respect to y and u with bounded derivatives, and u is a measurable bounded function. The system (4-5) can be written in the form of a system of integral equations

$$y_i(t) = y_{i,0}(\gamma_i(t)) + \int_{(\gamma_i(t))_i}^{t_i} f_i(s, y(s), u(s))\, ds_i \tag{6}$$

where $\gamma_i(t) \in \Gamma_0$ is such that

$$E(t) \bigcap \{s : s_j = t_j \forall j \neq i\} = \{s : (\gamma_i(t)) \leq s_i \leq t_i\} \tag{7}$$

and $(\gamma_i(t))$ is the i-th coordinate of $\gamma_i(t)$.

The method of Picard iterations may be applied to the integral equations (6) to show existence of solutions.

An optimal control problem for the system (6) concerns the minimization of a functional J, given by

$$J \equiv \int_\Omega F(t, y(t), u(t))\, dt + \int_{\Gamma_1} \Phi(t, y(t))\, d\sigma(t) \tag{8}$$

where $dt \equiv dt_1 dt_2 \ldots dt_M$ and $d\sigma(t)$ is the element of hypersurface area on Γ. The functions F, Φ are continuous and bounded in all their arguments, with continuous and bounded derivatives with respect to y and (for F) also u.

A formal treatment via a co-state $p(t) \equiv (p_1(t), p_2(t), \ldots, p_M(t))$ and the Lagrangian functional corresponding to (8), i.e.,

$$L \equiv \int_\Omega \left\{ F(t, y(t), u(t)) + \sum_j p_j(t) \left[f_j(t, y(t), y(t)) - \frac{\partial y_i(t)}{\partial t_j} \right] \right\} dt + \int_{\Gamma_1} \Phi(t, y(t)) d\sigma(t) \tag{9}$$

leads to the following Hamiltonian equations:

$$\frac{\partial p_j(t)}{\partial t_j} = -\frac{\partial H}{\partial y_j}; \quad p_j(s) = n_j(s) \frac{\partial \Phi(s, y(s))}{\partial y_j} \text{ for } s \in \tilde{\Gamma}_1 \tag{10}$$

where $\tilde{\Gamma}_1$ is the smooth part of Γ_1, and

$$H = H(t,y,p,u) \equiv F(t,y,u) + \sum_j p_j f_j(t,y,u) \tag{11}$$

The coefficients $n_j(s)$ are the coordinates of the unit outward normal vector to point s on Γ. For an optimal trajectory y^* with corresponding control u^* and cotrajectory p^*, we also have the Pontryagin-type minimum condition

$$H(t,y^*(t),p^*(t),u^*(t)) = \min_{u\in K} H(t,y^*(t),p^*(t),u) \tag{12}$$

Thus the Hamiltonian equations constitute a "reverse" Roesser system, as was to be expected.

3 REPRESENTATION OF THE CO-STATE

We examine the variational equations that result from (4-5) under a variation $v(t)$ of the control $u(t)$. For admissible variations $v(t)$, we denote by $\eta(t) = (\eta_1(t),\ldots,\eta_M(t))$ the corresponding variation of the state y. It is convenient to set

$$\begin{aligned} f_{ij}(t,y(t),u(t)) &\equiv \frac{\partial f_i(t,y(t),u(t))}{\partial y_i}; \\ V_i(t,y(t),u(t),v(t)) &\equiv \frac{\partial f_i(t,y(t),u(t))}{\partial u} v(t) \end{aligned} \tag{13}$$

Then, the equations for η are

$$\frac{\partial \eta_i}{\partial t_i} = \sum_j f_{ij}\eta_j + V_i \eta_i(s) = 0 \text{ for } s \in \Gamma_0 \tag{14}$$

(where, for simplicity, we have omitted the variables in f_{ij},η_i,η_j,V_i).

Since the problem (14) is linear jointly in (η, V), we expect the solution η to be representable as a linear operation on V. We are led to seek a Green's function (matrix-valued) $g = [g_{ij}](1 \le i \le n, 1 \le j \le n)$ such that

$$\eta_i(t) = \sum_j \int_{E(t)} g_{ij}(t,\tau)V_j(\tau)d\tau \tag{15}$$

A little reflection shows that the g_{ij}'s must satisfy

$$\frac{\partial g_{ij}(t,\tau)}{\partial t_i} = \sum_k f_{ik}(t,\ldots)g_{kj}(t,\tau) \tag{16}$$

and certain boundary conditions. In order to describe the boundary conditions, we need the following notation:

$$\begin{aligned} \bar{t}_i(\tau) &\equiv (\tau_1,\tau_2,\ldots,\tau_{i-1},t_i,\tau_{i+1},\ldots,\tau_M); \\ t^{(i)} &\equiv (t_1,t_2,\ldots,t_{i-1},t_{i+1},\ldots,t_M) \in \mathbb{R}^{M-1}; \\ \tau^{(i)} &\equiv (\tau_1,\tau_2,\ldots,\tau_{i-1},\tau_{i+1},\ldots,\tau_M) \in \mathbb{R}^{M-1} \end{aligned} \tag{17}$$

Also, δ_{ij} will be Kronecker's delta, and δ_{M-1} will be Dirac's delta in $M-1$ variables, i.e., if δ_1 stands for the standard Dirac's delta distribution in one dimension, then

$$\begin{aligned} \delta_{M-1}(t^{(i)} - \tau^{(i)}) &= \delta_1(t_1-\tau_1)\delta_1(t_2-\tau_2)\ldots\delta_1(t_{i-1}-\tau_{i-1}) \cdot \\ &\cdot\ \delta_1(t_{i+1}-\tau_{i+1})\ldots\delta_1(t_M-\tau_M) \end{aligned} \tag{18}$$

Then the boundary conditions for the g_{ij}'s are

$$g_{ij}(t,\bar{t}_i(\tau)) = \delta_{ij}\delta_{M-1}(t^{(i)} - \tau^{(i)}) \tag{19}$$

Now, we consider the variation of J which can be expressed as

$$\begin{aligned}\delta J &= \int_\Omega \int_{E(t)} \sum_{i,j} F_i(t,y(t),u(t))g_{ij}(t,\tau)V_j(\tau)d\tau dt + \\ &+ \int_\Omega F_u(t,y(t),u(t))v(t)dt + \\ &+ \int_{\Gamma_1}\int_{E(t)} \sum_{i,j} \Phi_i(t,y(t))g_{ij}(t,\tau)V_j(\tau)d\tau d\sigma(t)\end{aligned} \tag{20}$$

where

$$\begin{aligned}F_i(t,y(t),u(t)) &\equiv \frac{\partial F(t,y(t),u(t))}{\partial y_i}, \\ \Phi_i(t,y(t)) &\equiv \frac{\partial \Phi(t,y(t))}{\partial y_i}\end{aligned} \tag{21}$$

With a change in the order of integration, and substituting $V_j = f_{ju}v$, we find

$$\begin{aligned}\delta J &= \int_\Omega \Big\{ F_u(t,\ldots) + \int_{\epsilon(t)} \sum_{i,j} g_{ij}(\tau,t)F_i(\tau,y(\tau),u(\tau))d\tau \cdot \\ &\cdot\ f_{ju}(t,y(t),u(t)) \Big\} v(t)dt + \\ &+ \int_\Omega \int_{c(t)} \sum_{i,j} g_{ij}(\tau,t)\Phi_i(\tau,y(\tau))d\sigma(\tau)\cdot f_{ju}(t,y(t),u(t))v(t)dt\end{aligned} \tag{22}$$

where

$$c(t) = \Gamma_1 \bigcap \epsilon(t) \tag{23}$$

Consequently, we may take

$$\begin{aligned}p_j(t) &\equiv \int_{\epsilon(t)} \sum_i g_{ij}(\tau,t)F_i(\tau,y(\tau),u(\tau))d\tau + \\ &+ \int_{c(t)} \sum_i g_{ij}(\tau,t)\Phi_i(\tau,y(\tau))d\sigma(\tau)\end{aligned} \tag{24}$$

With the notation $H(t,y(t),p(t),u(t)) \equiv F(t,y(t),u(t)) + \sum_j p_j(t)f_j(t,y(t),u(t))$, we obtain

$$\delta J = \int_\Omega H_u v dt \tag{25}$$

We omit the proof of the validity of the Hamiltonian equations as a consequence of the representations of the costate and δJ.

REFERENCES

1. T. Kaczorek, Two-Dimensional Linear Systems, Springer-Verlag, Berlin (1985).

2. R.P. Roesser, IEEE Trans. Autom. Control, 20, 1-10 (1975).

3. L. Cesari, Arch. Rat. Mech. Analysis, 33, 339-357 (1969).

Addendum: After this paper was completed, I was informed of the work of D. Idczak; in his papers in Comp. in Math. Simul., 41, 87-94 (1996), Second Int. Symp. on Methods and Models in Automation and Robotics, 115-119 (1995), (with S. Walczak) IFAC Workshop MIM'97, 1-5 (1997), Bull. Polish Acad. Sci. - Technical Sci., 43, 227-234 (1995), he has studied various aspects of continuous Roesser systems in rectangular domains, and he has proved, among other things, an extremum principle of Pontryayin's type.

8th Int. Coll. on Differential Equations, pp. 457–459
D. Bainov (Ed)

Optimal Control of Discrete-Time Approximations to Systems with Hysteresis

S. A. BELBAS
Mathematics Department, University of Alabama, Tuscaloosa, AL 35487-0350, USA
I.D. MAYERGOYZ
Electrical Engineering Department, University of Maryland, College Park, MD 20-742, USA

Abstract. We study optimal control problems for discrete-time dynamical systems that include hysteretic nonlinearities of Madelung type. We reduce the problem to the form of a controlled system with control aftereffect; in this form of the problem, a variant of the dynamic programming method is applied.
Keywords: Dynamical system, hysteresis, optimal control, dynamic programming.
AMS Subject Classification: 49L20, 39A12, 47H30.

1 INTRODUCTION

In recent years, there has been increasing interest in modelling and analysis of systems with hysteresis. Mathematical models of hysteresis may be found in [1,2,3], and control issues related to hysteresis have been studied in [4,5,6].

Many systems exhibit various forms of hysteresis, for example, mechanical systems [7], components used in scanning tunneling microscopy [8], actuators used in the control of flexible structures [9], etc. etc. Many of the materials used for actuators and sensors exhibit some type of hysteresis; when such actuators and sensors are used as parts of a control system, then the system as a whole inherits the hysteretic behaviour. It is therefore important to have accurate control techniques for systems with hysteresis nonlinearities.

This paper, and the associated work [10], deal with the application of the method of dynamic programming to discrete-time systems with one type of hysteresis (the so-called Madelung hysteresis model; cf. [1]). The applicability of dynamic programming requires an appropriate modification of the standard dynamic programming method, and it leads to important new questions associated with the fact that the admissible values of one of the ingredients of the dynamic programming equations constitute an open set. The systems that are used to model the hysteresis phenomena are hybrid control systems, i.e., part of the state takes values in a discrete state-space, and they contain the phenomenon that we have called control aftereffect, i.e., at each time instant, the evolution of the system is affected by the values of the control at two time-instants. All these ingredients make the problems considered in this paper substantially different from the classical problems of dynamic programming.

2 FORMULATION OF THE PROBLEM

We consider a controlled system other form

$$y(i+1) = f(i, u(i), u(i), \mathcal{H}u(i)) \tag{1}$$

where $\mathcal{H}$ is a hysteresis operator of Madelung type. For this type of hysteresis operator acting on a space of discrete-time signals, the output $\mathcal{H}u(i)$ can be expressed in terms of 3 parameters $\alpha(i), c(i), u(i)$. Here, $\alpha(i)$ is a Boolean variable that determines whether the output is determined from an increasing or a decreasing branch of a family of hysteresis loops, $c(i)$ is a parameter that selects one loop from a one-parameter family of curves, and $u(i)$ is the value of the control at time i; the detailed description of the evolution of these parameters is given in [10] and is not repeated here. We simply mention here that $\mathcal{H}(i)$ can be expressed as

$$\mathcal{H}u(i) = H(\alpha(i), c(i), u(i)) \tag{2}$$

for an appropriate function H, and that the evolution of α and c is given by a set of equations

$$\begin{aligned} \alpha(i+1) &= g_1(\alpha(i), c(i), u(i), u(i+1)); \\ c(i+1) &= g_2(\alpha(i), c(i), u(i), u(i+1)) \end{aligned} \tag{3}$$

and, again, the precise description of g_1, g_2 is given in [10]; for the purposes of the present paper, the exact nature of g_1, g_2 is not important, but the fact that the evolution of α and c can be expressed in the form (3) is important.

The optimal control problem associated with (1-3) is the minimization of a cost functional

$$J \equiv \sum_{i=0}^{N-1} F(i, y(i), u(i), u(i+1)) + F_0(N, y(N), u(N)) \tag{4}$$

3 THE DYNAMIC PROGRAMMING EQUATIONS

If z stands for the triple (y, α, c), then the evolution of z (taking into account the information of the previous section) has the form

$$z(i+1) = g(i, z(i), u(i), u(i+1)) \tag{5}$$

Thus, we have a type of control system which we may call "a system with control aftereffects" (with a shift in the time-parameter of the control, i.e., by setting $v(i) \equiv u(i+1)$, the system (5) becomes $z(i+1) = g(i, z(i), v(i-1), v(i))$, i.e., a form of delay equation relative to the shifted control v). For this type of system, we apply the ideas of dynamic programming to a functional parametrized by current value of time, state z, and current value of the control; thus, we set, for the parametrized cost functional,

$$\begin{aligned} J(k, y, \alpha, c, a) \equiv & \sum_{i=k}^{N-1} F(i, y(i), u(i), u(i+1)) + \\ & + F_0(N, y(N), u(N)), \\ & \text{subject to } y(k) = y, u(k) = a, \alpha(k) = \alpha, c(k) = c \end{aligned} \tag{6}$$

The associated value function V is defined as

$$V(k, y, \alpha, c, a) \equiv \inf_{u(\cdot)} J(k, y, \alpha, c, a) \tag{7}$$

Then the dynamic programming equation has the form

$$\begin{aligned} V(k,y,\alpha,c,a) &= \inf_{\alpha'}\inf_{c'}\inf_{b\epsilon K(\alpha',c')}\{V(k+1,f(k,y,a,H(\alpha,c,a)),\alpha',c',b)+ \\ &+ F(k,y,a,b)\}; V(N,y,\alpha,c,a) = F_0(N,y,a) \end{aligned} \quad (8)$$

where $K(\alpha',c')$ is the set of admissible values of $u(k+1)$ as a function of $\alpha(k+1)\equiv\alpha'$, $c(k+1)\equiv c'$. The detailed description of K is given in [10], but here we outline the method of obtaining K: the equation (3) give the values of α', c' if the values of α, c, a, b are known; conversely, for given $\alpha, c, a, \alpha', c'$, the inverse image of the variable b from equations (3) gives the set $K(\alpha',c')$.

4 VERIFICATION RESULTS

Under the condition of continuity of all functions appearing in the dynamic programming equations, we have the following:

Theorem 4.1. The function $V(k,y,\alpha,c,a)$, solution of (8), equals the infimum, over all admissible control polices, of the parametrized cost functional $J(k,y,\alpha,c,a)$.

Theorem 4.2. If there is a control policy $u^*(i)$ with associated state trajectory $y^*(i)$ and augmented trajectory $\alpha^*(i), c^*(i)$, so that the infimum is achieved in (4), then the set of quadruples $(y^*(i), \alpha^*(i), c^*(i), u^*(i))$, $0\le i\le N$, is optimal, conditioned on $y^*(0)$, $\alpha^*(0)$, $c^*(0)$, $u^*(0)$.

REFERENCES

1. M.A. Krasnosel'skii, A.V. Pokrovskii, Systems with Hysteresis, Springer-Verlag, Berlin (1989).

2. I. Mayergoyz, Mathematical Models of Hysteresis, Springer-Verlag, New York (1991).

3. A. Visintin, Differential Models of Hysteresis, Springer-Verlag, New York (1994).

4. M.'Brokate, Optimale Steuerung von gewöhnlichen Differentialgleichungen mit Nichtlinearitäten vom Hysteresis-Typ, Verlag Peter Lang, Frankfurt a.M. (1987).

5. G. Tao, P. Kokotović, Adaptive Control of Systems with Actuator and Sensor Nonlinearities, John Wiley and Son, New York (1996).

6. L. White, J.T. Oden, Nonlinear Analysis T.M.A., 13, 459-474 (1989).

7. P.-A. Bliman, M. Sorine, in Proc. Third European Control Conference, Rome (1995) pp. 3788-3794.

8. P. Ge, IEEE Trans. Control Syst. Techn., 4, 209-215 (1996).

9. R.C. Smith, Modeling techniques for magnetostrictive actuators, Report CRSC-TR97-6, North Carolina State University, Raleigh, NC (1997).

10. S.A. Belbas, I.D. Mayergoyz, paper in preparation, 1997.

Bei Fragen zur Produktsicherheit wenden Sie sich bitte an:
If you have any questions regarding product safety,
please contact:

Walter de Gruyter GmbH
Genthiner Straße 13
10785 Berlin
productsafety@degruyterbrill.com